Applied Quantum Mechanics

Written specifically for electrical and mechanical engineers, material scientists and applied physicists, this book takes quantum mechanics out of the theory books and into the real world, using practical engineering examples throughout. Levi's unique, practical approach engages the reader and keeps them motivated with numerous illustrations, exercises, and worked solutions. Starting with some scene setting revision material on classical mechanics and electromagnetics, Levi takes the reader from first principles and Schrödinger's equation on to more advanced topics including scattering, eigenstates, the harmonic oscillator and time-dependent perturbation theory. A CD-ROM is included which contains MATLAB source code to support the text. Quantum mechanics is usually thought of as being a difficult subject to master – this book sets out to prove it doesn't need to be.

Professor Levi joined the USC faculty in mid-1993 after working for 10 years at AT&T Bell Laboratories, Murray Hill, New Jersey. During his 20-year research career he invented hot electron spectroscopy, the microdisk laser, and carried out ground-breaking work in parallel fiber-optic interconnect components in computer and switching systems. At USC he and his team continue to work on cutting edge research projects. He has published numerous scientific papers, several book chapters, and holds 12 US patents.

Applied Quantum Mechanics

A. F. J. Levi

University of Southern California

PUBLISHED BY THE PRESS SYNDICATE OF THE UNIVERSITY OF CAMBRIDGE
The Pitt Building, Trumpington Street, Cambridge, United Kingdom

CAMBRIDGE UNIVERSITY PRESS
The Edinburgh Building, Cambridge CB2 2RU, UK
40 West 20th Street, New York, NY 10011-4211, USA
477 Williamstown Road, Port Melbourne, VIC 3207, Australia
Ruiz de Alarcón 13, 28014 Madrid, Spain
Dock House, The Waterfront, Cape Town 8001, South Africa

http://www.cambridge.org

First published 2003

Printed in the United Kingdom at the University Press, Cambridge

Typefaces Times 10.5/14 pt and Helvetica Neue   *System* LATEX 2$_\varepsilon$ [TB]

A catalog record for this book is available from the British Library

Library of Congress Cataloging in Publication data

Levi, A. F. J. (Anthony Frederic John), 1959–
Applied quantum mechanics / A. F. J. Levi.
       p.   cm.
Includes bibliographical references and index.
ISBN 0-521-81765-X – ISBN 0-521-52086-X (pbk.)
1. Quantum theory.   I. Title.
QC174.12 .L44   2002
530.12 – dc21   2002073608

ISBN 0 521 81765 X hardback
ISBN 0 521 52086 X paperback

. . . Dass ich erkenne, was die Welt
Im Innersten zusammenhält . . .

GOETHE

(Faust. ll 382–384)

Contents

3   Using the Schrödinger wave equation   111

4   The propagation matrix   167

5   Eigenstates and operators

6   The harmonic oscillator

9   The semiconductor laser   415

10   Time-independent perturbation   462

Preface

The theory of quantum mechanics forms the basis for our present understanding of physical phenomena on an atomic and sometimes macroscopic scale. Today, quantum mechanics can be applied to most fields of science. Within engineering, important subjects of practical significance include semiconductor transistors, lasers, quantum optics, and molecular devices. As technology advances, an increasing number of new electronic and opto-electronic devices will operate in ways which can only be understood using quantum mechanics. Over the next 30 years, fundamentally quantum devices such as single-electron memory cells and photonic signal processing systems may well become commonplace. Applications will emerge in any discipline that has a need to understand, control, and modify entities on an atomic scale. As nano- and atomic-scale structures become easier to manufacture, increasing numbers of individuals will need to understand quantum mechanics in order to be able to exploit these new fabrication capabilities. Hence, one intent of this book is to provide the reader with a level of understanding and insight that will enable him or her to make contributions to such future applications, whatever they may be.

The book is intended for use in a one-semester introductory course in applied quantum mechanics for engineers, material scientists, and others interested in understanding the critical role of quantum mechanics in determining the behavior of practical devices. To help maintain interest in this subject, I felt it was important to encourage the reader to solve problems and to explore the possibilities of the Schrödinger equation. To ease the way, solutions to example exercises are provided in the text, and the enclosed CD-ROM contains computer programs written in the MATLAB language that illustrate these solutions. The computer programs may be usefully exploited to explore the effects of changing parameters such as temperature, particle mass, and potential within a given problem. In addition, they may be used as a starting point in the development of designs for quantum mechanical devices.

The structure and content of this book are influenced by experience teaching the subject. Surprisingly, existing texts do not seem to address the interests or build on the computing skills of today's students. This book is designed to better match such student needs.

Some material in the book is of a review nature, and some material is merely an introduction to subjects that will undoubtedly be explored in depth by those interested in pursuing more advanced topics. The majority of the text, however, is an essentially self-contained study of quantum mechanics for electronic and opto-electronic applications.

There are many important connections between quantum mechanics and classical mechanics and electromagnetism. For this and other reasons, Chapter 1 is devoted to a review of classical concepts. This establishes a point of view with which the predictions of quantum mechanics can be compared. In a classroom situation it is also a convenient way in which to establish a uniform minimum knowledge base. In Chapter 2 the Schrödinger wave equation is introduced and used to motivate qualitative descriptions of atoms, semiconductor crystals, and a heterostructure diode. Chapter 3 develops the more systematic use of the one-dimensional Schrödinger equation to describe a particle in simple potentials. It is in this chapter that the quantum mechanical phenomenon of tunneling is introduced. Chapter 4 is devoted to developing and using the propagation matrix method to calculate electron scattering from a one-dimensional potential of arbitrary shape. Applications include resonant electron tunneling and the Kronig–Penney model of a periodic crystal potential. The generality of the method is emphasized by applying it to light scattering from a dielectric discontinuity. Chapter 5 introduces some related mathematics, the generalized uncertainty relation, and the concept of density of states. Following this, the quantization of conductance is introduced. The harmonic oscillator is discussed in Chapter 6 using the creation and annihilation operators. Chapter 7 deals with fermion and boson distribution functions. This chapter shows how to numerically calculate the chemical potential for a multi-electron system. Chapter 8 introduces and then applies time-dependent perturbation theory to ionized impurity scattering in a semiconductor and spontaneous light emission from an atom. The semiconductor laser diode is described in Chapter 9. Finally, Chapter 10 discusses the (still useful) time-independent perturbation theory.

Throughout this book, I have made applications to systems of practical importance the main focus and motivation for the reader. Applications have been chosen because of their dominant roles in today's technologies. Understanding is, after all, only useful if it can be applied.

California   A. F. J. L.
2003

MATLAB® programs

The computer requirements for the MATLAB[1] language are an IBM or 100% compatible system equipped with Intel 486, Pentium, Pentium Pro, Pentium4 processor or equivalent. A CD-ROM drive is required for software installation. There needs to be an 8-bit or better graphics adapter and display, a minimum of 32 MB RAM, and at least 50 MB disk space. The operating system is Windows95, NT4, Windows2000, or WindowsXP.

If you have not already installed MATLAB, you will need to purchase a copy and install it on your computer.

After verifying correct installation of the MATLAB application program, copy the directory AppliedQMmatlab on the CD-ROM to a convenient location in your computer user directory.

Launch the MATLAB application program using the icon on the desktop or from the start menu. The MATLAB command window will appear in your computer screen.

From the MATLAB command window use the path browser to set the path to the location of the AppliedQMmatlab directory. Type the name of the file you wish to execute in the MATLAB command window (do not include the '.m' extension). Press the enter key on the keyboard to run the program.

You will find that some programs prompt for input from the keyboard. Most programs display results graphically with intermediate results displayed in the MATLAB command window.

To edit values in a program or to edit the program itself double click on the file name to open the file editor.

You should note that the computer programs in the AppliedQMmatlab directory are not optimized. They are written in a very simple way to minimize any possible confusion or sources of error. The intent is that these programs be used as an aid to the study of applied quantum mechanics. When required, integration is performed explicitly, and in the simplest way possible. However, for exercises involving matrix diagonalization use is made of special MATLAB functions.

Some programs make use of the functions, chempot.m, fermi.m, mu.m, runge4.m, solve_schM.m, and Chapt9Exercise5.m reads data from the datainLI.txt data input file.

---

[1] MATLAB is a registered trademark of the MathWorks, Inc.

1 Introduction

1.1 Motivation

You may ask why one needs to know about quantum mechanics. Possibly the simplest answer is that we live in a quantum world! Engineers would like to make and control electronic, opto-electronic, and optical devices on an atomic scale. In biology there are molecules and cells we wish to understand and modify on an atomic scale. The same is true in chemistry, where an important goal is the synthesis of both organic and inorganic compounds with precise atomic composition and structure. Quantum mechanics gives the engineer, the biologist, and the chemist the tools with which to study and control objects on an atomic scale.

As an example, consider the deoxyribonucleic acid (DNA) molecule shown in Fig. 1.1. The number of atoms in DNA can be so great that it is impossible to track the position and activity of every atom. However, suppose we wish to know the effect a particular site (or neighborhood of an atom) in a single molecule has on a chemical reaction. Making use of quantum mechanics, engineers, biologists, and chemists can work together to solve this problem. In one approach, laser-induced fluorescence of a fluorophore attached to a specific site of a large molecule can be used to study the dynamics of that individual molecule. The light emitted from the fluorophore acts as a small beacon that provides information about the state of the molecule. This technique, which relies on quantum mechanical photon stimulation and photon emission from atomic states, has been used to track the behavior of single DNA molecules.[1]

Interdisciplinary research that uses quantum mechanics to study and control the behavior of atoms is, in itself, a very interesting subject. However, even within a given discipline such as electrical engineering, there are important reasons to study quantum mechanics. In the case of electrical engineering, one simple motivation is the fact that transistor dimensions will soon approach a size where single-electron and quantum effects determine device performance. Over the last few decades advances in the complexity and performance of complementary metal-oxide–semiconductor (CMOS)

---

[1] S. Weiss, *Science* **283**, 1676 (1999).

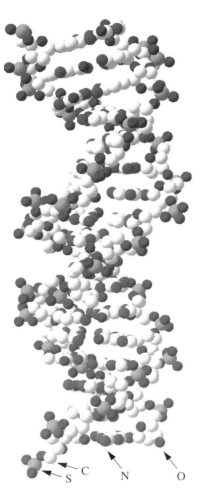

S   C   N   O

Fig. 1.1. Ball and stick model of a DNA molecule. Atom types are indicated.

circuits have been carefully managed by the microelectronics industry to follow what has become known as "Moore's law".[2] This rule-of-thumb states that the number of transistors in silicon integrated circuits increases by a factor of 2 every 18 months. Associated with this law is an increase in the performance of computers. The Semiconductor Industry Association (SIA) has institutionalized Moore's Law via the "SIA Roadmap", which tracks and identifies advances needed in most of the electronics industry's technologies.[3] Remarkably, reductions in the size of transistors and related technology have allowed Moore's law to be sustained for over 35 years (see Fig. 1.2). Nevertheless, the impossibility of continued reduction in transistor device dimensions is well illustrated by the fact that Moore's law predicts that dynamic random access memory (DRAM)

---

[2] G. E. Moore, *Electronics* **38**, 114 (1965). Also reprinted in *Proc. IEEE* **86**, 82 (1998).

[3] http://www.sematech.org.

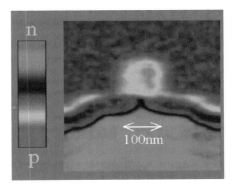

Fig. 1.2. Photograph (left) of the first transistor. Brattain and Bardeen's p–n–p point-contact germanium transistor operated as a speech amplifier with a power gain of 18 on December 23, 1947. The device is a few millimeters in size. On the right is a scanning capacitance microscope cross-section image of a silicon p-type metal-oxide–semiconductor field-effect transistor (p-MOSFET) with an effective channel length of about 20 nm, or about 60 atoms.[4] This image of a small transistor was published in 1998, 50 years after Brattain and Bardeen's device. Image courtesy of G. Timp, University of Illinois.

cell size will be *less* than that of an atom by the year 2030. Well before this end-point is reached, quantum effects will dominate device performance, and conventional electronic circuits will fail to function.

We need to learn to use quantum mechanics to make sure that we can create the smallest, highest-performance devices possible.

Quantum mechanics is the basis for our present understanding of physical phenomena on an atomic scale. Today, quantum mechanics has numerous applications in engineering, including semiconductor transistors, lasers, and quantum optics. As technology advances, an increasing number of new electronic and opto-electronic devices will operate in ways that can only be understood using quantum mechanics. Over the next 20 years, fundamentally quantum devices such as single-electron memory cells and photonic signal processing systems may well become available. It is also likely that entirely new devices, with functionality based on the principles of quantum mechanics, will be invented. The purpose of this book is to provide the reader with a level of understanding and insight that will enable him or her to appreciate and to make contributions to the development of these future, as yet unknown, applications of quantum phenomena.

The small glimpse of our quantum world that this book provides reveals significant differences from our everyday experience. Often we will discover that the motion of objects does not behave according to our (classical) expectations. A simple, but hopefully motivating, example is what happens when you throw a ball against a wall.

---

[4] Also see G. Timp et al. *IEEE International Electron Devices Meeting (IEDM) Technical Digest* p. 615, Dec. 6–9, San Francisco, California, 1998 (ISBN 0780 3477 9).

Of course, we expect the ball to bounce right back. Quantum mechanics has something different to say. There is, under certain special circumstances, a finite chance that the ball will appear on the other side of the wall! This effect, known as tunneling, is fundamentally quantum mechanical and arises due to the fact that on appropriate time and length scales particles can be described as waves. Situations in which *elementary* particles such as electrons and photons tunnel are, in fact, relatively common. However, quantum mechanical tunneling is not always limited to atomic-scale and elementary particles. Tunneling of *large* (macroscopic) objects can also occur! Large objects, such as a ball, are made up of many atomic-scale particles. The possibility that such large objects can tunnel is one of the more amazing facts that emerges as we explore our quantum world.

However, before diving in and learning about quantum mechanics it is worth spending a little time and effort reviewing some of the basics of classical mechanics and classical electromagnetics. We do this in the next two sections. The first deals with classical mechanics, which was first placed on a solid theoretical basis by the work of Newton and Leibniz published at the end of the seventeenth century. The survey includes reminders about the concepts of potential and kinetic energy and the conservation of energy in a closed system. The important example of the one-dimensional harmonic oscillator is then considered. The simple harmonic oscillator is extended to the case of the diatomic linear chain, and the concept of dispersion is introduced. Going beyond mechanics, in the following section classical electromagnetism is explored. We start by stating the coulomb potential for charged particles, and then we use the equations that describe electrostatics to solve practical problems. The classical concepts of capacitance and the coulomb blockade are used as examples. Continuing our review, Maxwell's equations are used to study electrodynamics. The first example discussed is electromagnetic wave propagation at the speed of light in free space, c. The key result – that power and momentum are carried by an electromagnetic wave – is also introduced.

Following our survey of classical concepts, in Chapter 2 we touch on the experimental basis for quantum mechanics. This includes observation of interference phenomenon with light, which is described in terms of the linear superposition of waves. We then discuss the important early work aimed at understanding the measured power spectrum of black-body radiation as a function of wavelength, λ, or frequency, $\omega = 2\pi c/\lambda$. Next, we treat the photoelectric effect, which is best explained by requiring that light be quantized into particles (called photons) of energy $E = \hbar\omega$. Planck's constant $\hbar = 1.0545 \times 10^{-34}$ J s, which appears in the expression $E = \hbar\omega$, is a small number that sets the absolute scale for which quantum effects usually dominate behavior.[5] Since the typical length scale for which electron energy quantization is important usually turns out to be the size of an atom, the observation of discrete spectra for light emitted from excited atoms is an effect that can only be explained using quantum mechanics.

---

[5] Sometimes $\hbar$ is called Planck's *reduced* constant to distinguish it from $h = 2\pi\hbar$.

The energy of photons emitted from excited hydrogen atoms is discussed in terms of the solutions of the Schrödinger equation. Because historically the experimental facts suggested a wave nature for electrons, the relationships among the wavelength, energy, and momentum of an electron are introduced. This section concludes with some examples of the behavior of electrons, including the description of an electron in free space, the concept of a wave packet and dispersion of a wave packet, and electronic configurations for atoms in the ground state.

Since we will later apply our knowledge of quantum mechanics to semiconductors and semiconductor devices, there is also a brief introduction to crystal structure, the concept of a semiconductor energy band gap, and the device physics of a unipolar heterostructure semiconductor diode.

1.2 Classical mechanics

1.2.1 Introduction

The problem classical mechanics sets out to solve is predicting the motion of large (macroscopic) objects. On the face of it, this could be a very difficult subject simply because large objects tend to have a large number of degrees of freedom[6] and so, in principle, should be described by a large number of parameters. In fact, the number of parameters could be so enormous as to be unmanageable. The remarkable success of classical mechanics is due to the fact that powerful concepts can be exploited to simplify the problem. Constants of the motion and constraints may be used to reduce the description of motion to a simple set of differential equations. Examples of constants of the motion include conservation of energy and momentum. Describing an object as rigid is an example of a constraint being placed on the object.

Consider a rock dropped from a tower. Classical mechanics initially ignores the internal degrees of freedom of the rock (it is assumed to be rigid), but instead defines a center of mass so that the rock can be described as a point particle of mass, m. Angular momentum is decoupled from the center of mass motion. Why is this all possible? The answer is neither simple nor obvious.

It is known from experiments that atomic-scale particle motion can be very different from the predictions of classical mechanics. Because large objects are made up of many atoms, one approach is to suggest that quantum effects are somehow averaged out in large objects. In fact, classical mechanics is often assumed to be the macroscopic (large-scale) limit of quantum mechanics. The underlying notion of finding a means to link quantum mechanics to classical mechanics is so important it is called the *correspondence principle*. Formally, one requires that the results of classical mechanics be obtained in the limit $\hbar \rightarrow 0$. While a simple and convenient test, this approach misses

---

[6] For example, an object may be able to vibrate in many different ways.

the point. The results of classical mechanics are obtained because the quantum mechanical wave nature of objects is averaged out by a mechanism called *decoherence*. In this picture, quantum mechanical effects are *usually* averaged out in large objects to give the classical result. However, this is not always the case. We should remember that sometimes even large (macroscopic) objects can show quantum effects. A well-known example of a macroscopic quantum effect is superconductivity and the tunneling of flux quanta in a device called a SQUID.[7] The tunneling of flux quanta is the quantum mechanical equivalent of throwing a ball against a wall and having it sometimes tunnel through to the other side! Quantum mechanics allows large objects to tunnel through a thin potential barrier if the constituents of the object are prepared in a special quantum mechanical state. The wave nature of the entire object must be maintained if it is to tunnel through a potential barrier. One way to achieve this is to have a coherent superposition of constituent particle wave functions.

Returning to classical mechanics, we can now say that the motion of macroscopic material bodies is *usually* described by classical mechanics. In this approach, the linear momentum of a rigid object with mass m is $\mathbf{p} = m \cdot d\mathbf{x}/dt$, where $\mathbf{v} = d\mathbf{x}/dt$ is the velocity of the object moving in the direction of the unit vector $\hat{\mathbf{x}}$. Time is measured in units of seconds (s), and distance is measured in units of meters (m). The magnitude of momentum is measured in units of kilogram meters per second (kg m s^{-1}), and the magnitude of velocity (speed) is measured in units of meters per second (m s^{-1}). Classical mechanics assumes that there exists an inertial frame of reference for which the motion of the object is described by the differential equation

$$\mathbf{F} = d\mathbf{p}/dt = m \cdot d^2\mathbf{x}/dt^2 \tag{1.1}$$

where the vector $\mathbf{F}$ is the force. The magnitude of force is measured in units of newtons (N). Force is a vector field. What this means is that the particle can be subject to a force the magnitude and direction of which are different in different parts of space.

We need a new concept to obtain a measure of the forces experienced by the particle moving from position $\mathbf{r}_1$ to position $\mathbf{r}_2$ in space. The approach taken is to introduce the idea of *work*. The work done moving the object from point 1 to point 2 in space along a path is *defined* as

$$W_{12} = \int_{\mathbf{r}=\mathbf{r}_1}^{\mathbf{r}=\mathbf{r}_2} \mathbf{F} \cdot d\mathbf{r} \tag{1.2}$$

where $\mathbf{r}$ is a spatial vector coordinate. Figure 1.3 illustrates one possible trajectory for a particle moving from position $\mathbf{r}_1$ to $\mathbf{r}_2$. The definition of work is simply the integral of the force applied multiplied by the infinitesimal distance moved in the direction of the force for the complete path from point 1 to point 2. For a *conservative* force field, the work W_{12} is the same for any path between points 1 and 2. Hence, making use of

---

[7] For an introduction to this see A. J. Leggett, *Physics World* **12**, 73 (1999).

Fig. 1.3. Illustration of a classical particle trajectory from position $\mathbf{r}_1$ to position $\mathbf{r}_2$.

Fig. 1.4. Illustration of a closed-path classical particle trajectory.

the fact $\mathbf{F} = d\mathbf{p}/dt = m \cdot d\mathbf{v}/dt$, one may write

$$W_{12} = \int_{\mathbf{r}=\mathbf{r}_1}^{\mathbf{r}=\mathbf{r}_2} \mathbf{F} \cdot d\mathbf{r} = m \int d\mathbf{v}/dt \cdot \mathbf{v} dt = \frac{m}{2} \int \frac{d}{dt}(v^2) dt \qquad (1.3)$$

so that $W_{12} = m(v_2^2 - v_1^2)/2 = T_2 - T_1$, where the scalar $T = mv^2/2$ is called the kinetic energy of the object.

For conservative forces, because the work done is the same for any path between points 1 and 2, the work done around any *closed path*, such as the one illustrated in Fig. 1.4, is always zero, or

$$\oint \mathbf{F} \cdot d\mathbf{r} = 0 \qquad (1.4)$$

This is always true if force is the gradient of a *single-valued* spatial scalar field where $\mathbf{F} = -\nabla V(\mathbf{r})$, since $\oint \mathbf{F} \cdot d\mathbf{r} = -\oint \nabla V \cdot d\mathbf{r} = -\oint dV = 0$. In our expression, $V(\mathbf{r})$ is called the potential. Potential is measured in volts (V), and potential energy is measured in joules (J) or electron volts (eV). If the forces acting on the object are conservative, then total energy, which is the sum of kinetic and potential energy, is a constant of the motion. In other words, total energy $T + V$ is conserved.

Since kinetic and potential energy can be expressed as functions of the variable's position and time, it is possible to define a *Hamiltonian* function for the system, which is $H = T + V$. The Hamiltonian function may then be used to describe the dynamics of particles in the system.

For a nonconservative force, such as a particle subject to frictional forces, the work done around any closed path is not zero, and $\oint \mathbf{F} \cdot d\mathbf{r} \neq 0$.

Let's pause here for a moment and consider some of what has just been introduced. We think of objects moving due to something. Forces cause objects to move. We have introduced the concept of force to help ensure that the motion of objects can be described as a simple process of *cause and effect*. We imagine a force field in three-dimensional space that is represented mathematically as a continuous, integrable vector field, $\mathbf{F}(\mathbf{r})$. Assuming that time is also continuous and integrable, we quickly discover that in a conservative force-field energy is conveniently partitioned between a kinetic and a potential term and total energy is conserved. By simply representing the total energy as a function or Hamiltonian, $H = T + V$, we can find a differential equation that describes the dynamics of the object. Integration of the differential equation of motion gives the trajectory of the object as it moves through space.

In practice, these ideas are very powerful and may be applied to many problems involving the motion of macroscopic objects. As an example, let's consider the problem of finding the motion of a particle mass, m, attached to a spring. Of course, we know from experience that the solution will be oscillatory and so characterized by a frequency and amplitude of oscillation. However, the power of the theory is that we can use the formalism to obtain relationships among all the parameters that govern the behavior of the system.

In the next section, the motion of a classical particle mass m attached to a spring and constrained to move in one dimension is considered. The type of model we will be considering is called the simple harmonic oscillator.

1.2.2   The one-dimensional simple harmonic oscillator

Figure 1.5 illustrates a classical particle mass m attached to a lightweight spring that obeys Hooke's law. Hooke's law states that the displacement, x, from the equilibrium position, $x = 0$, is proportional to the force on the particle. The proportionality constant is κ and is called the spring constant. In this example, we ignore any effect due to the finite mass of the spring by assuming its mass is small relative to the particle mass, m.

To calculate the frequency and amplitude of vibration, we start by noting that the total energy function or Hamiltonian for the system is

$$H = T + V \tag{1.5}$$

where potential energy is $V = \frac{1}{2}\kappa x^2 = \int_0^x \kappa x' dx'$ and kinetic energy is $T = m(dx/dt)^2/2$, so that

$$H = \frac{1}{2}m\left(\frac{dx}{dt}\right)^2 + \frac{1}{2}\kappa x^2 \tag{1.6}$$

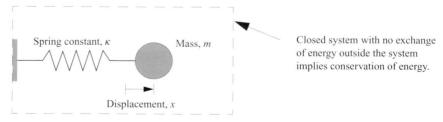

Fig. 1.5. Illustration showing a classical particle mass m attached to a spring and constrained to move in one dimension. The displacement of the particle from its equilibrium position is x. The box drawn with a broken line indicates a closed system.

The system is *closed*, so there is no exchange of energy outside the system. There is no dissipation, energy in the system is a constant, and

$$\frac{dH}{dt} = 0 = m\frac{dx}{dt}\frac{d^2x}{dt^2} + \kappa x\frac{dx}{dt} \tag{1.7}$$

so that the *equation of motion* can be written as

$$\boxed{\kappa x + m\frac{d^2x}{dt^2} = 0} \tag{1.8}$$

The solutions for this second-order linear differential equation are

$$x(t) = A\cos(\omega_0 t + \phi) \tag{1.9}$$

$$\frac{dx(t)}{dt} = -\omega_0 A\sin(\omega_0 t + \phi) \tag{1.10}$$

$$\frac{d^2x(t)}{dt^2} = -\omega_0^2 A\cos(\omega_0 t + \phi) \tag{1.11}$$

where A is the amplitude of oscillation, ω_0 is the frequency of oscillation measured in radians per second (rad s^{-1}), and ϕ is a fixed phase. We may now write the potential energy and kinetic energy as

$$V = \frac{1}{2}\kappa A^2\cos^2(\omega_0 t + \phi) \tag{1.12}$$

and

$$T = \frac{1}{2}m\omega_0^2 A^2\sin^2(\omega_0 t + \phi) \tag{1.13}$$

respectively. Total energy $E = T + V = m\omega_0^2 A^2/2 = \kappa A^2/2$ since $\sin^2(\theta) + \cos^2(\theta) = 1$ and $\kappa = m\omega_0^2$. Clearly, an increase in total energy increases amplitude $A = \sqrt{2E/\kappa} = \sqrt{2E/m\omega_0^2}$, and an increase in κ, corresponding to an increase in the stiffness of the spring, decreases A. The theory gives us the relationships among all the parameters of the classical harmonic oscillator: κ, m, A, and total energy.

We have shown that the classical simple harmonic oscillator vibrates in a single *mode* with frequency ω_0. The vibrational energy stored in the mode can be changed continuously by varying the amplitude of vibration, A.

Suppose we have a particle mass $m = 0.1$ kg attached to a lightweight spring with spring constant $\kappa = 360$ N m^{-1}. Particle motion is constrained to one dimension, and the amplitude of oscillation is observed to be $A = 0.01$ m. In this case, the frequency of oscillation is just $\omega_0 = \sqrt{\kappa/m} = 60$ rad s^{-1}, which is about 9.5 oscillations per second, and the total energy in the system is $E = \kappa A^2/2 = 18$ mJ. We can solve the equation of motion and obtain position, $x(t)$, velocity, $dx(t)/dt$, and acceleration, $d^2x(t)/dt^2$, as a function of time. Velocity is zero when $x = \pm A$ and the particle changes its direction of motion and starts moving back towards the equilibrium position $x = 0$. The position

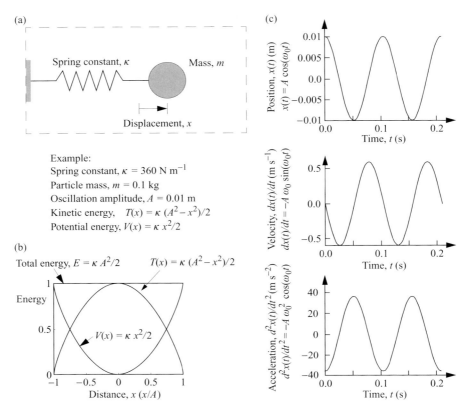

Fig. 1.6. Example predictions for the classical one-dimensional harmonic oscillator involving motion of a particle mass m attached to a lightweight spring with spring constant κ. In this case, the spring constant is $\kappa = 360$ N m^{-1}, particle mass, $m = 0.1$ kg, and the oscillation amplitude is $A = 0.01$ m. (a) Illustration of the closed system showing displacement of the particle from its equilibrium position at $x = 0$. (b) Kinetic energy T and potential energy V functions of position, x. (c) Position, velocity, and acceleration functions of time, t.

$x = \pm A$, where velocity is zero, is called the *classical turning point* of the motion. Peak velocity, $v_{max} = \pm A\omega_0$, occurs as the particle crosses its equilibrium position, $x = 0$. In this case $v_{max} = \pm A\omega_0 = \pm 0.6$ m s^{-1}. Maximum acceleration, $a_{max} = \pm A\omega_0^2$, occurs when $x = \pm A$. In this case $a_{max} = \pm A\omega_0^2 = \pm 36$ m s^{-2}. Figure 1.6 illustrates these results.

Now let's use what we have learned and move on to a more complex system. In the next example we want to solve the equations of motion of an isolated linear chain of particles, each with mass m, connected by identical springs. This particular problem is a common starting-point for the study of lattice vibrations in crystals. The methods used, and the results obtained, are applicable to other problems such as solving for the vibrational motion of atoms in molecules.

We should be clear why we are going to this effort now. We want to introduce the concept of a *dispersion relation*. It turns out that this is an important way of simplifying the description of an otherwise complex system.

The motion of coupled oscillators can be described using the idea that a given frequency of oscillation corresponds to definite wavelengths of vibration. In practice, one plots frequency of oscillation, ω, with inverse wavelength or wave vector of magnitude $q = 2\pi/\lambda$. Hence, the dispersion relationship is $\omega = \omega(\mathbf{q})$. With this relationship, one can determine how vibration waves and pulses propagate through the system. For example, the phase velocity of a wave is $v_q = \omega/q$ and a pulse made up of wave components near a value q_0 often propagates at the group velocity,

$$v_{\mathrm{g}} = \left.\frac{\partial \omega}{\partial q}\right|_{q=q_0} \tag{1.14}$$

If the dispersion relation is known then we can determine quantities of practical importance such as v_q and v_{g}.

1.2.3 The monatomic linear chain

Figure 1.7 shows part of an isolated linear chain of particles, each of mass m, connected by springs. The site of each particle is labeled with an integer, j. Each particle occupies a lattice site with equilibrium position jL, where L is the lattice constant. Displacement from equilibrium of the j-th particle is u_j, and there are a large number of particles in the chain.

Assuming small deviations u_j from equilibrium, the Hamiltonian of the linear chain is

$$H = \sum_j \frac{m}{2}\left(\frac{du_j}{dt}\right)^2 + V_0(0) + \frac{1}{2!}\sum_{jk}\frac{\partial^2 V_0}{\partial u_j u_k}u_j u_k + \frac{1}{3!}\sum_{jkl}\frac{\partial^3 V_0}{\partial u_j u_k u_l}u_j u_k u_l + \cdots \tag{1.15}$$

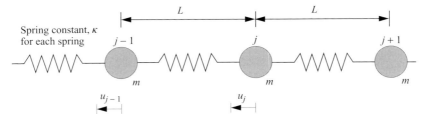

Fig. 1.7. Illustration of part of an isolated linear chain of particles, each of mass m, connected by identical springs with spring constant κ. The site of each particle is labeled relative to the site, j. Each particle occupies a lattice site with equilibrium position jL, where L is the lattice constant. Displacement from equilibrium of the j-th particle is u_j. It is assumed that there are a large number of particles in the linear chain.

The first term on the right-hand side is a sum over kinetic energy of each particle, and $V_0(0)$ is the potential energy when all particles are stationary in the equilibrium position. The remaining terms come from a Taylor expansion of the potential about the equilibrium positions. Each particle oscillates about its equilibrium position and is coupled to other oscillators via the potential.

In the harmonic approximation, the force constant $\kappa_{jk} = (\partial^2 E_0/\partial u_j \partial u_k)|_0$ is real and symmetric so that $\kappa_{jk} = \kappa_{kj}$, and if all springs are identical then $\kappa = \kappa_{jk}$. Restricting the sum in Eqn (1.15) to nearest neighbors and setting $V_0(0) = 0$, the Hamiltonian becomes

$$H = \sum_j \frac{m}{2}\left(\frac{du_j}{dt}\right)^2 + \frac{\kappa}{2}\sum_j \left(2u_j^2 - u_j u_{j+1} - u_j u_{j-1}\right) \qquad (1.16)$$

This equation assumes that motion of one particle is from forces due to the relative position of its nearest neighbors.

The displacement from equilibrium at site j is u_j and is related to that of its nearest neighbor by

$$u_{j\pm1} = u_j e^{\pm iqL} \qquad (1.17)$$

where $q = 2\pi/\lambda$ is the wave vector of a vibration of wavelength, λ. Using Eqn (1.16) and assuming no dissipation in the system, so that $dH/dt = 0$, the equation of motion is

$$m\frac{d^2 u_j}{dt^2} = \kappa(u_{j+1} + u_{j-1} - 2u_j) \qquad (1.18)$$

Second-order differential equations of this type have time dependence of the form $e^{-i\omega t}$, which, on substitution into Eqn (1.18), gives

$$-m\omega^2 u_j = \kappa(e^{iqL} + e^{-iqL} - 2)u_j = -4\kappa \sin^2\left(\frac{qL}{2}\right)u_j \qquad (1.19)$$

From Eqn (1.19) it follows that

$$\omega(q) = \sqrt{\frac{4\kappa}{m}}\sin\left(\frac{qL}{2}\right) \qquad (1.20)$$

This equation tells us that there is a unique nonlinear relationship between the frequency of vibration, ω, and the magnitude of the wave vector, q. This is an example of a *dispersion relation*, $\omega = \omega(\mathbf{q})$.

The dispersion relation for the monatomic linear chain is plotted in Fig. 1.8(a). It consists of a single *acoustic branch*, $\omega = \omega_{\text{acoustic}}(q)$, with maximum frequency $\omega_{\text{max}} = (4\kappa/m)^{1/2}$. Notice that vibration frequency approaches $\omega \to 0$ *linearly* as $q \to 0$. In the long wavelength limit ($q \to 0$), the acoustic branch dispersion relation describing lattice dynamics of a monatomic linear chain predicts that vibrational waves propagate

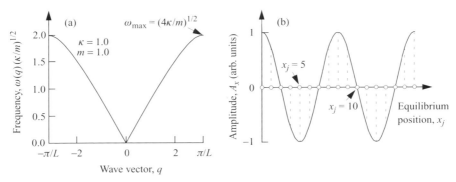

Fig. 1.8. (a) Dispersion relation for lattice vibrations of a one-dimensional monatomic linear chain. The dispersion relation is linear at low values of q. The maximum frequency of oscillation is $\omega_{max} = (4\kappa/m)^{1/2}$. Particles have mass $m = 1.0$, and the spring constant is $\kappa = 1.0$. (b) Amplitude of vibrational motion in the x direction on a portion of the linear chain for a particular mode of frequency ω. Equilibrium position x_j is indicated.

at constant group velocity $v_g = \partial\omega/\partial q$. This is the velocity of sound waves in the system.

Each normal mode of the linear chain is a harmonic oscillator characterized by frequency ω and wave vector q. In general, each mode of frequency ω in the linear chain involves harmonic motion of all the particles in the chain that are also at frequency ω. As illustrated in Fig. 1.8(b), not all particles have the same amplitude. Total energy in a mode is proportional to the sum of the amplitudes squared of all particles in the chain.

The existence of a dispersion relation is significant, and so it is worth considering some of the underlying physics. We start by recalling that in our model there are a large number of atoms in the linear monatomic chain. At first sight, one might expect that the large number of atoms involved gives rise to all types of oscillatory motion. One might anticipate solutions to the equations of motion allowing all frequencies and wavelengths, so that no dispersion relation could exist. However, this situation does not arise in practice because of some important simplifications imposed by *symmetry*. The motion of a given atom is determined by forces due to the relative position of its nearest neighbors. Forces due to displacements of more distant neighbors are not included. The facts that there is only one spring constant and there is only one type of atom are additional constraints on the system. One may think of such constraints as imposing a type of symmetry that has the effect of eliminating all but a few of the possible solutions to the equations of motion. These solutions are conveniently summarized by the dispersion relation, $\omega = \omega(\mathbf{q})$.

1.2.4   The diatomic linear chain

Figure 1.9 illustrates a diatomic linear chain. In this example we assume a periodic array of atoms characterized by a lattice constant, L. There are two atoms per *unit cell*

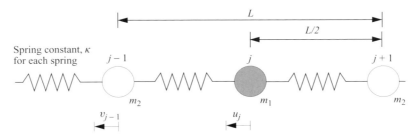

Fig. 1.9. Illustration of an isolated linear chain of particles of alternating mass m_1 and m_2 connected by identical springs with spring constant κ. There are two particles per *unit cell* spaced by $L/2$. One particle in the unit cell has mass m_1, and the other particle has mass m_2. The site of each particle is labeled relative to the site, j. The displacement from equilibrium of particles mass m_1 is u, and for particles mass m_2 it is v.

spaced by $L/2$. One atom in the unit cell has mass m_1 and the other atom has mass m_2. The site of each atom is labeled relative to the site, j. The displacement from equilibrium of particles mass m_1 is u, and for particles mass m_2 it is v. The motion of one atom is related to that of its nearest similar (equal-mass) neighbor by

$$u_{j\pm2} = u_j e^{\pm iqL} \tag{1.21}$$

where $q = 2\pi/\lambda$ is the wave vector of a vibration of wavelength, λ. If we assume that the motion of a given atom is from forces due to the relative position of its nearest neighbors, the equations of motion for the two types of atoms are

$$m_1 \frac{d^2 u_j}{dt^2} = \kappa(v_{j+1} + v_{j-1} - 2u_j) \tag{1.22}$$

$$m_2 \frac{d^2 v_{j-1}}{dt^2} = \kappa(u_j + u_{j-2} - 2v_{j-1}) \tag{1.23}$$

or

$$m_1 \frac{d^2 u_j}{dt^2} = \kappa(1 + e^{iqL})v_{j-1} + 2\kappa u_j \tag{1.24}$$

$$m_2 \frac{d^2 v_{j-1}}{dt^2} = \kappa(1 + e^{-iqL})u_j - 2\kappa v_{j-1} \tag{1.25}$$

Solutions for u and v have time dependence of the form $e^{-i\omega t}$, giving

$$-m_1\omega^2 u_j = \kappa(1 + e^{iqL})v_{j-1} - 2\kappa u_j \tag{1.26}$$

$$-m_2\omega^2 v_{j-1} = \kappa(1 + e^{-iqL})u_j - 2\kappa v_{j-1} \tag{1.27}$$

or

$$(2\kappa - m_1\omega^2)u_j - \kappa(1 + e^{iqL})v_{j-1} = 0 \tag{1.28}$$

$$-\kappa(1 + e^{-iqL})u_j + (2\kappa - m_2\omega^2)v_{j-1} = 0 \tag{1.29}$$

This is a linear set of equations with an intrinsic or (from the German word) *eigen* solution given by the characteristic equation

$$\begin{vmatrix} 2\kappa - m_1\omega^2 & -\kappa(1 + e^{iqL}) \\ -\kappa(1 + e^{-iqL}) & 2\kappa - m_2\omega^2 \end{vmatrix} = 0 \tag{1.30}$$

so that the characteristic polynomial is

$$\omega^4 - 2\kappa\left(\frac{m_1 + m_2}{m_1 m_2}\right)\omega^2 + \frac{2\kappa^2}{m_1 m_2}(1 - \cos(qL)) = 0 \tag{1.31}$$

The roots of this polynomial give the characteristic values, or *eigenvalues*, ω_q.

To understand further details of the dispersion relation for our particular linear chain of particles, it is convenient to look for solutions that are extreme limiting cases. The extremes we look for are $q \to 0$, which is the long wavelength limit ($\lambda \to \infty$), and $q \to \pi/L$, which is the short wavelength limit ($\lambda \to 2L$).

In the *long wavelength limit $q \to 0$*

$$\omega^2\left(\omega^2 - 2\kappa\left(\frac{m_1 + m_2}{m_1 m_2}\right)\right) = 0 \tag{1.32}$$

and solutions are

$$\omega = 0 \quad \text{and} \quad \omega = \left(2\kappa\left(\frac{m_1 + m_2}{m_1 m_2}\right)\right)^{1/2}$$

with the latter corresponding to both atom types beating against each other.

In the *short wavelength limit $q \to \pi/L$*

$$\omega^4 - 2\kappa\left(\frac{m_1 + m_2}{m_1 m_2}\right)\omega^2 + \frac{4\kappa^2}{m_1 m_2} = 0 \tag{1.33}$$

and solutions are

$$\omega_1 = (2\kappa/m_1)^{1/2}$$

corresponding to only atoms of mass m_1 vibrating, and

$$\omega_2 = (2\kappa/m_2)^{1/2}$$

corresponding to only atoms mass m_2 vibrating.

With these limits, it is now possible to sketch a *dispersion relation* for the lattice vibrations. In Fig. 1.10 the dispersion relation $\omega = \omega(q)$ is given for the case $m_1 < m_2$, with $m_1 = 0.5, m_2 = 1.0$, and $\kappa = 1.0$. There is an *acoustic branch* $\omega = \omega_{\text{acoustic}}(q)$ for which vibration frequency linearly approaches $\omega \to 0$ as $q \to 0$, and there is an *optic branch* $\omega = \omega_{\text{optic}}(q)$ for which $\omega \neq 0$ as $q \to 0$.

As one can see from Fig. 1.10, the acoustic branch is capable of propagating low-frequency sound waves the group velocity of which, $v_g = \partial\omega/\partial q$, is a constant for long wavelengths. Typical values for the velocity of sound waves in a semiconductor at room

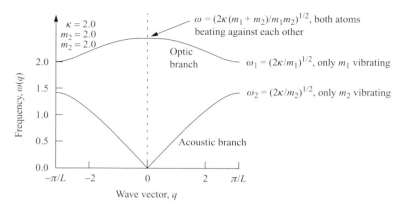

Fig. 1.10. Dispersion relation for lattice vibrations of a one-dimensional diatomic linear chain. Particles have masses $m_1 = 0.5$ and $m_2 = 1.0$. The spring constant is $\kappa = 1.0$.

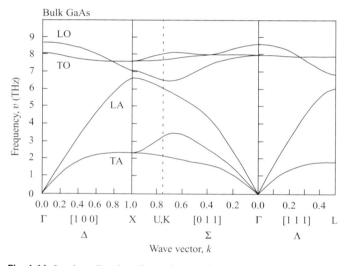

Fig. 1.11. Lattice vibration dispersion relation along principal crystal symmetry directions of bulk GaAs.[8] The longitudinal acoustic (LA), transverse acoustic (TA), longitudinal optic (LO), and transverse optic (TO) branches are indicated.

temperature are $v_g = 8.4 \times 10^3$ m s^{-1} in (100)-oriented Si and $v_g = 4.7 \times 10^3$ m s^{-1} in (100)-oriented GaAs.[9]

For the one-dimensional case, one branch of the dispersion relation occurs for each atom per unit cell of the lattice. The example we considered had two atoms per unit cell, so we had an optic and an acoustic branch. In three dimensions we add extra degrees of freedom, resulting in a total of three acoustic and three optic branches. In our example,

---

[8] Lattice vibration dispersion relations for additional semiconductor crystals may be found in H. Bilz and W. Kress, *Phonon Dispersion Relations in Insulators*, Springer Series in Solid-State Sciences **10**, Springer-Verlag, Berlin, 1979 (ISBN 3 540 09399 0).

[9] For comparison, the speed of sound in air at temperature 0 °C at sea level is 331.3 m s^{-1} or 741 mph.

for a wave propagating in a given direction there is one *longitudinal acoustic* and one *longitudinal optic* branch with atom motion parallel to the wave propagation direction. There are also two *transverse acoustic* and two *transverse optic* branches with atom motion normal to the direction of wave propagation.

To get an idea of the complexity of a real lattice vibration dispersion relation, consider the example of GaAs. Device engineers are interested in GaAs because it is an example of a III-V compound semiconductor that is used to make laser diodes and high-speed transistors. GaAs has the zinc blende crystal structure with a lattice constant of $L = 0.565$ nm. Ga and As atoms have different atomic masses, and so we expect the dispersion relation to have three optic and three acoustic branches. Because the positions of the atoms in the crystal and the values of the spring constants are more complex than in the simple linear chain model we considered, it should come as no surprise that the dispersion relation is also more complex. Figure 1.11 shows the dispersion relation along the principal crystal symmetry directions of bulk GaAs.

1.3 Classical electromagnetism

We now take our ideas of fields and the tools we have developed to solve differential equations from classical mechanics and apply them to electromagnetism. In the following, we divide our discussion between electrostatics and electrodynamics.

1.3.1   Electrostatics

We will only consider stationary distributions of charges and fields. However, to obtain results it is necessary to introduce charge and to build up electric fields slowly over time to obtain the stationary charge distributions we are interested in. This type of adiabatic integration is a standard approach that is used to find solutions to many practical problems.

A basic starting point is the experimental observation that the electrostatic force due to a point charge Q in vacuum (free space) separated by a distance r from charge $-Q$ is

$$\mathbf{F}(\mathbf{r}) = \frac{-Q^2}{4\pi \varepsilon_0 r^2}\hat{\mathbf{r}} \tag{1.34}$$

where $\varepsilon_0 = 8.8541878 \times 10^{-12}$ F m^{-1} is the permittivity of free space measured in units of farads per meter. Force is an example of a vector field the direction of which, in this case, is given by the unit vector $\hat{\mathbf{r}}$. It is a central force because it has no angular dependence. Electrostatic force is measured in units of newtons (N), and charge is measured in coulombs (C). We will be interested in the force experienced by an electron with charge $Q = -e = -1.6021765 \times 10^{-19}$ C.

The force experienced by a charge e in an electric field is $\mathbf{F} = e\mathbf{E}$, where $\mathbf{E}$ is the electric field. Electric field is an example of a vector field, and its magnitude is measured in units of volts per meter ($V\ m^{-1}$). The (negative) potential energy is just the force times the distance moved

$$-eV = \int e\mathbf{E} \cdot d\hat{\mathbf{x}} \tag{1.35}$$

Electrostatic force can be related to potential via $\mathbf{F} = -e\nabla V$, and hence the coulomb potential energy due to a point charge e in vacuum (free space) separated by a distance r from charge $-e$ is

$$eV(r) = \frac{-e^2}{4\pi\varepsilon_0 r} \tag{1.36}$$

The coulomb potential is an example of a scalar field. Because it is derived from a central force, the coulomb potential has no angular dependence and is classified as a central-force potential. The coulomb potential is measured in volts (V) and the coulomb potential energy is measured in joules (J) or electron volts (eV).

When there are no currents or time-varying magnetic fields, the Maxwell equations we will use for electric field $\mathbf{E}$ and magnetic flux density $\mathbf{B}$ are

$$\nabla \cdot \mathbf{E} = \rho/\varepsilon_0\varepsilon_r \tag{1.37}$$

and

$$\nabla \cdot \mathbf{B} = 0 \tag{1.38}$$

In the first equation, ε_r is the relative permittivity of the medium, and ρ is charge density. We chose to define the electric field as $\mathbf{E} = -\nabla V$. Notice that because electric field is given by the negative gradient of the potential, only *differences* in the potential are important. The direction of electric field is positive from positive electric charge to negative electric charge. Sometimes it is useful to visualize electric field as field lines originating on positive charge and terminating on negative charge. The divergence of the electric field is the local charge density. It follows that the flux of electric field lines flowing out of a closed surface is equal to the charge enclosed. This is Gauss's law, which may be expressed as

$$\int_V \nabla \cdot \mathbf{E}\, dV = \oint_S \mathbf{E} \cdot d\mathbf{S} = \int_V (\rho/\varepsilon_0\varepsilon_r) dV \tag{1.39}$$

where the two equations on the left-hand side are expressions for the net electric flux out of the region (Stokes's theorem) and the right-hand side is enclosed charge.

Maxwell's expression for the divergence of the magnetic flux density given in Eqn (1.38) is interpreted physically as there being no magnetic monopoles (leaving the possibility of dipole and higher-order magnetic fields). Magnetic flux density $\mathbf{B}$ is an example of a vector field, and its magnitude is measured in units of tesla (T).

Sometimes it is useful to define another type of electric field called the displacement vector field, $\mathbf{D} = \varepsilon_0\varepsilon_r\mathbf{E}$. In this expression, ε_r is the average value of the relative permittivity of the material in which the electric field exists. It is also useful to define the quantity $\mathbf{H} = \mathbf{B}/\mu_0\mu_r$, which is the magnetic field vector where μ_0 is called the permeability of free space and μ_r is called the relative permeability.

1.3.1.1 The parallel-plate capacitor

Electric charge and energy can be stored by doing work to spatially separate charges Q and $-Q$ in a capacitor. Capacitance is the proportionality constant relating the potential applied to the amount of charge stored. Capacitance is defined as

$$C = \frac{Q}{V}$$

(1.40)

and is measured in units of farads (F).

A capacitor is a very useful device for storing electric charge. A capacitor is also an essential part of the field-effect transistor used in silicon integrated circuits and thus is of great interest to electrical engineers.

We can use Maxwell's equations to figure out how much charge can be stored for every volt of potential applied to a capacitor. We start by considering a very simple geometry consisting of two parallel metal plates that form the basis of a parallel-plate capacitor.

Figure 1.12 is an illustration of a parallel-plate capacitor. Two thin, square, metal plates each of area A are placed facing each other a distance d apart in such a way that $d \ll \sqrt{A}$. One plate is attached to the positive terminal of a battery, and the other plate is attached to the negative terminal of the same battery, which supplies a voltage V. We may calculate the capacitance of the device by noting that the charge per unit area on a plate is ρ and the voltage is just the integral of the electric field between the plates, so

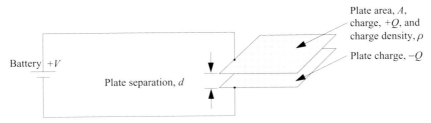

Fig. 1.12. Illustration of a parallel-plate capacitor attached to a battery supplying voltage V. The capacitor consists of two thin, square, metal plates each of area A facing each other a distance d apart.

that $V = |\mathbf{E}| \times d = \rho d / \varepsilon_0 \varepsilon_r$. Hence,

$$C = \frac{Q}{V} = \frac{\rho A}{\rho d / \varepsilon_0 \varepsilon_r} = \frac{\varepsilon_0 \varepsilon_r A}{d} \tag{1.41}$$

where ε_r is the relative permittivity or dielectric constant of the material between the plates. This is an accurate measure of the capacitance, and errors due to fringing fields at the edges of the plates are necessarily small, since $d \ll \sqrt{A}$.

We now consider the values of numbers used in a physical device. A typical parallel-plate capacitor has $d = 100$ nm and $\varepsilon_r = 10$, so the amount of extra charge per unit area per unit volt of potential difference applied is $Q = CV = \varepsilon_0 \varepsilon_r V / d = 8.8 \times 10^{-4}$ C m^{-2}V^{-1} or, in terms of number of electrons per square centimeter per volt, $Q = 5.5 \times 10^{11}$ electrons cm^{-2} V^{-1}. On average this corresponds to one electron per $(13.5$ nm$)^2$ V^{-1}. In a metal, this electron charge might sit in the first 0.5 nm from the surface, giving a density of $\sim 10^{19}$ cm^{-3} or 10^{-4} of the typical bulk charge density in a metal of 10^{23} cm^{-3}. A device of area 1 mm^2 with $d = 100$ nm and $\varepsilon_r = 10$ has capacitance $C = 88$ nF.

The extra charge sitting on the metal plates creates an electric field between the plates. We may think of this electric field as storing energy in the capacitor. To figure out how much energy is stored in the electric field of the capacitor, we need to calculate the current that flows when we hook up a battery that supplies voltage V. The current flow, I, measured in amperes, is simply $dQ/dt = C \cdot dV/dt$, so the instantaneous power supplied at time t to the capacitor is IV, which is just $dQ/dt = C \cdot dV/dt$ times the voltage. Hence, the instantaneous power is $CV \cdot dV/dt$. The energy stored in the capacitor is the integral of the instantaneous power from a time when there is no extra charge on the plates, say time $t' = -\infty$ to time $t' = t$. At $t = -\infty$ the voltage is zero and so the stored energy is

$$\Delta E = \int_{t'=-\infty}^{t'=t} CV \frac{dV}{dt'} dt' = \int_{V'=0}^{V'=V} CV' dV' = \frac{1}{2} CV^2 \tag{1.42}$$

$$\boxed{\Delta E = \frac{1}{2} CV^2} \tag{1.43}$$

Since capacitance of the parallel-plate capacitor is $C = Q/V = \varepsilon_0 \varepsilon_r A/d$ and the magnitude of the electric field is $|\mathbf{E}| = V/d$, we can rewrite the stored energy *per unit volume* in terms of the electric field to give a stored *energy density* $\Delta U = \varepsilon_0 \varepsilon_r |\mathbf{E}|^2 / 2$. Finally, substituting in the expression $\mathbf{D} = \varepsilon_0 \varepsilon_r \mathbf{E}$, one obtains the result

$$\boxed{\Delta U = \frac{1}{2} \mathbf{E} \cdot \mathbf{D}} \tag{1.44}$$

for the energy stored per unit volume in the electric field. A similar result

$$\boxed{\Delta U = \frac{1}{2}\mathbf{B} \cdot \mathbf{H}}$$
(1.45)

holds for the energy stored per unit volume in a magnetic field. These are important results, which we will make use of later in this chapter when we calculate the energy flux density in an electromagnetic wave.

Putting in numbers for a typical parallel-plate capacitor with plate separation $d = 100$ nm, $\varepsilon_r = 10$, and applied voltage V, gives a stored energy density per unit volume of $\Delta U = V^2 \times 4.427 \times 10^3$ J m^{-3}, or an energy density per unit area of $V^2 \times 4.427 \times 10^{-4}$ J m^{-2}.

Magnetic flux density can be stored in an inductor. Inductance, defined in terms of magnetic flux linkage, is

$$L = \frac{1}{I}\int_S \mathbf{B} \cdot d\mathbf{S}$$
(1.46)

where I is the current and S is a specified surface through which magnetic flux passes. Inductance is measured in units of henrys (H). Putting in some numbers, one finds that the external inductance per unit length between two parallel-plate conductors, each of width w and separated by a distance d, is $L = (\mu_0 d/w)$ H m^{-1}. Thus if $d = 1$ μm and $w = 25$ μm, then $L = 5 \times 10^{-8}$ H m^{-1}.

1.3.1.2 The coulomb blockade

Consider the case of a small metal sphere of radius r_1. The capacitance associated with the sphere can give rise to an effect called the coulomb blockade, which may be important in determining the operation of very small electronic devices in the future. To analyze this situation, we will consider what happens if we try to place an electron charge onto a small metal sphere. Figure 1.13 shows how we might visualize an initially distant electron being moved through space and placed onto the metal sphere.

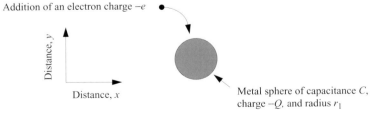

Fig. 1.13. Illustration indicating an electron of charge $-e$ being placed onto a small metal sphere of radius r_1 and capacitance C.

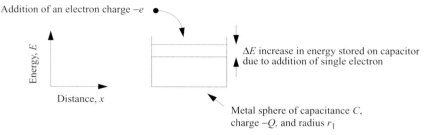

Fig. 1.14. Energy–distance diagram for an electron of charge $-e$ being placed onto a small metal sphere of radius r_1 and capacitance C. In this case, the vertical axis represents the energy stored by the capacitor, and the horizontal axis indicates distance moved by the electron and the size of the metal sphere. ΔE is the increase in energy stored on capacitor due to addition of single electron.

The same idea can be expressed in the energy–distance diagram shown in Fig. 1.14. In this case the vertical axis represents the energy stored by the capacitor, and the horizontal axis can be used to indicate distance moved by the electron and size of the metal sphere.

The coulomb blockade is the discrete value of *charging energy* ΔE needed to place an extra electron onto a capacitor. The charging energy is discrete because the electron has a single value for its charge. The charging energy becomes large and measurable at room temperature in devices that have a very small capacitance. Such devices are usually of nanometer size. As an example, consider a very small sphere of metal. The capacitance of a small sphere can be found by considering two spherical conducting metal shells of radius r_1 and r_2, where $r_1 < r_2$. Assume that there is a charge $+Q$ on the inner surface of the shell radius r_2 and a charge $-Q$ on the outer surface of the shell radius r_1. One may apply Gauss's law

$$\int_V \nabla \cdot \mathbf{E} \, dV = \oint_S \mathbf{E} \cdot d\mathbf{S} = \int_V (\rho/\varepsilon_0 \varepsilon_r) dV \tag{1.47}$$

and show that at radius $r_1 < r < r_2$ the electric flux flowing out of a closed surface is equal to the charge enclosed, so that

$$E_r = \frac{Q}{4\pi \varepsilon_0 \varepsilon_r r^2} \tag{1.48}$$

Since $\mathbf{E} = -\nabla V$, the potential is found by integrating

$$V = -\int_{r_2}^{r_1} E_r dr = -\int_{r_2}^{r_1} \frac{Q}{4\pi \varepsilon_0 \varepsilon_r r^2} dr = \frac{Q}{4\pi \varepsilon_0 \varepsilon_r} \left(\frac{1}{r_1} - \frac{1}{r_2} \right) \tag{1.49}$$

and hence the capacitance is

$$C = \frac{Q}{V} = \frac{4\pi \varepsilon_0 \varepsilon_r}{\left(\dfrac{1}{r_1} - \dfrac{1}{r_2} \right)} \tag{1.50}$$

For an isolated conducting sphere of radius r_1 we let $r_2 \rightarrow \infty$, so that the capacitance is

$$C = 4\pi\varepsilon_0\varepsilon_r r_1 \tag{1.51}$$

Since instantaneous power supplied at time t to a capacitor is just the current $dQ/dt = C \cdot dV/dt$ times the voltage, it follows that the instantaneous power is $CV \cdot dV/dt$. The energy stored in the capacitor is the integral of the instantaneous power from time $t' = -\infty$ to time $t' = t$. At $t = -\infty$ the voltage is zero, and so the charging energy is

$$\Delta E = \int\limits_{t'=-\infty}^{t'=t} CV\frac{dV}{dt'}dt' = \int\limits_{V'=0}^{V'=V} CV'dV' = \frac{1}{2}CV^2 = \frac{Q^2}{2C} \tag{1.52}$$

For a single electron $Q = -e$, so that

$$\Delta E = \frac{e^2}{2C} = \frac{e^2}{8\pi\varepsilon_0\varepsilon_r r_1} \tag{1.53}$$

is the charging energy for a single electron placed onto a metal sphere of radius r_1 embedded in a dielectric with relative permittivity ε_r. For a metal sphere with $r_1 = 10\,\text{nm}$ (20 nm diameter $\sim$70 atoms diameter or $\sim 1.8 \times 10^5$ atoms total) in a dielectric with $\varepsilon_r = 10$ one obtains a capacitance $C = 1.1 \times 10^{-17}$ F and a charging energy $\Delta E = 7.2$ meV. Because electron charge has a single value, one may only add an electron to the metal sphere if the electron has enough energy to overcome the coulomb charging energy ΔE. The effect is called the *coulomb blockade*. Very small single-electron devices such as single-electron transistors[10] and single-electron memory cells[11] exploit this effect. Obviously, in the example we considered, such a device is unlikely to work well in practice since ΔE is less than room-temperature thermal energy $k_B T = 25$ meV, where k_B is the Boltzmann constant and T is the absolute temperature. This thermal background will tend to wash out any discrete coulomb charging energy effects. However, by decreasing the size of the metal particle and (or) reducing the relative permittivity, the value of ΔE can be increased so that $\Delta E > k_B T$ and room-temperature operation of a coulomb blockade device becomes possible. If a metal sphere embedded in a dielectric with $\varepsilon_r = 10$ had a radius $r_1 = 1$ nm (2 nm diameter $\sim$7 atoms diameter $\sim$180 atoms total), then $\Delta E = 72$ meV. If $r_1 = 1$ nm and $\varepsilon_r = 1$, then $\Delta E = 720$ meV.

Notice that we explicitly made use of the fact that electron charge has a single value but that we used a continuously variable charge to derive our expression for the charging energy, ΔE. One might expect a more rigorous theory to avoid this inconsistency.

So far, we have only considered static electric fields. In the world we experience very little can be considered static. Electrons move around, current flows, and electric fields

[10] M. Kenyon et al. *Appl. Phys. Lett.* **72**, 2268 (1998).
[11] K. Yano et al. *Proc. IEEE* **87**, 633 (1999).

change with time. The aim of classical electrodynamics is to provide a framework with which to describe electric and magnetic fields that change over time.

1.3.2   Electrodynamics

Classical electrodynamics describes the spatial and temporal behavior of electric and magnetic fields. Although Maxwell published his paper on electrodynamics in 1864, it was quite some time before the predictions of the theory were confirmed. Maxwell's achievement was to show that time-varying electric fields are intimately coupled with time-varying magnetic fields. In fact, it is not possible to separate magnetic and electric fields – there are only *electromagnetic* fields.

There is no doubt that the classical theory of electrodynamics is one of the great scientific achievements of the nineteenth century. Applications of electrodynamics are so important that it is worth spending some time reviewing a few key results of the theory.

The idea that **E** and **B** are *ordinary vector fields* is an essential element of the classical theory of electromagnetism. Just as with the description of large macroscopic bodies in classical mechanics, quantum effects are assumed to average out. We may think of the classical electromagnetic field as the *classical macroscopic limit* of a quantum description in terms of photons. In many situations, there are large numbers of uncorrelated photons that contribute to the electromagnetic field over the time scales of interest, and so we do not need to concern ourselves with the discrete (quantum) nature of photons.

Often, analysis of a complicated field is simplified by decomposing the field into plane wave components. Plane waves can be represented *spatially* as

$$\sin(kx) = \frac{1}{2i}(e^{ikx} - e^{-ikx}) \tag{1.54}$$

$$\cos(kx) = \frac{1}{2}(e^{ikx} + e^{-ikx}) \tag{1.55}$$

$$e^{ikx} = \cos(kx) + i\sin(kx) \tag{1.56}$$

where $k = 2\pi/\lambda$ is called the wave number, which is measured in inverse meters (m^{-1}), and λ is the wavelength measured in units of meters (m).

Plane waves can be represented *temporally* by

$$e^{-i\omega t} = \cos(\omega t) - i\sin(\omega t) \tag{1.57}$$

where $\omega = 2\pi f$ is the angular frequency measured in radians per second (rad s^{-1}), $f = 1/\tau$ is the frequency measured in cycles per second or hertz (Hz), and τ is the period measured in seconds (s).

Combining the spatial and temporal dependencies, we obtain a function for the field at position **r**, which is a plane wave of the form

$$Ae^{i(\mathbf{k}\cdot\mathbf{r}-\omega t)} \tag{1.58}$$

Table 1.1. *Maxwell equations*

$\nabla \cdot \mathbf{D} = \rho$	Coulomb's law
$\nabla \cdot \mathbf{B} = 0$	No magnetic monopoles
$\nabla \times \mathbf{E} = -\dfrac{\partial \mathbf{B}}{\partial t}$	Faraday's law
$\nabla \times \mathbf{H} = \mathbf{J} + \dfrac{\partial \mathbf{D}}{\partial t}$	Modified Ampère's law

Here, A is the amplitude of the wave, which may be a complex number, and $\mathbf{k}$ is the wave vector of magnitude $|\mathbf{k}| = k$ propagating in the $\hat{\mathbf{k}}$ direction. The fact that Eqn (1.58) is a complex quantity is only a mathematical convenience. Only real values are measured. Because $\mathbf{E}$ and $\mathbf{B}$ are represented by ordinary vectors, one finds the physical field by taking the real part of Eqn (1.58).

Maxwell's equations completely describe classical electric and magnetic fields. The equations in SI-MKS units are given in Table 1.1. In these equations, $\mathbf{D}$ is called the displacement vector field and is related to the electric field $\mathbf{E}$ by $\mathbf{D} = \varepsilon\mathbf{E} = \varepsilon_0\varepsilon_r\mathbf{E} = \varepsilon_0(1 + \chi_e)\mathbf{E} = \varepsilon_0\mathbf{E} + \mathbf{P}$. The displacement vector field $\mathbf{D}$ may also be thought of as the electric flux density, which is measured in coulombs per square meter (C m^{-2}). χ_e is the electric susceptibility, and $\mathbf{P}$ is the electric polarization field. $\mathbf{H}$ is the magnetic field vector, and $\mathbf{B}$ is the magnetic flux density. The convention is that $\mathbf{B} = \mu\mathbf{H} = \mu_0\mu_r\mathbf{H} = \mu_0(1 + \chi_m)\mathbf{H} = \mu_0(\mathbf{H} + \mathbf{M})$, where μ is the permeability, μ_r is the relative permeability, χ_m is the magnetic susceptibility, and $\mathbf{M}$ is the magnetization. $\mathbf{J}$ is the current density measured in amperes per square meter (A m^{-2}), and $\partial\mathbf{D}/\partial t$ is the displacement current density measured in amperes per square meter (A m^{-2}). The permittivity of free space is $\varepsilon_0 = 8.8541878 \times 10^{-12}$ F m^{-1}, and the permeability of free space, measured in henrys per meter is $\mu_0 = 4\pi \times 10^{-7}$ H m^{-1}. The speed of light in free space is $c = 1/\sqrt{\varepsilon_0\mu_0} = 2.99792458 \times 10^8$ m s^{-1}, and the impedance in free space is $Z_0 = \sqrt{\mu_0/\varepsilon_0} = 376.73$ Ω.

The use of vector calculus to describe Maxwell's equations enables a very compact and efficient description and derivation of relationships between fields. Because of this, it is worth reminding ourselves of some results from vector calculus. When considering vector fields $\mathbf{a}$, $\mathbf{b}$, $\mathbf{c}$ and scalar field ϕ, we recall that

$$\nabla \times \nabla\phi = 0 \tag{1.59}$$

$$\nabla \cdot (\nabla \times \mathbf{a}) = 0 \tag{1.60}$$

$$\nabla \times \nabla \times \mathbf{a} = \nabla(\nabla \cdot \mathbf{a}) - \nabla^2\mathbf{a} \tag{1.61}$$

$$\nabla \cdot (\mathbf{a} \times \mathbf{b}) = \mathbf{b} \cdot (\nabla \times \mathbf{a}) - \mathbf{a} \cdot (\nabla \times \mathbf{b}) \tag{1.62}$$

$$\mathbf{a} \times (\mathbf{b} \times \mathbf{c}) = (\mathbf{a} \cdot \mathbf{c})\mathbf{b} - (\mathbf{a} \cdot \mathbf{b})\mathbf{c} \tag{1.63}$$

Another useful relation in vector calculus is the divergence theorem relating volume and surface integrals

$$\int_V \nabla \cdot \mathbf{a} \, d^3 r = \int_S \mathbf{a} \cdot \mathbf{n} \, ds \tag{1.64}$$

where V is the volume and $\mathbf{n}$ is the unit-normal vector to the surface S. Stokes's theorem relates surface and line integrals

$$\int_S (\nabla \times \mathbf{a}) \cdot \mathbf{n} \, ds = \oint_C \mathbf{a} \cdot d\mathbf{l} \tag{1.65}$$

where $d\mathbf{l}$ is the vector line element on the closed loop C.

In vector calculus, the divergence of a vector field is a source, so one interprets $\nabla \cdot \mathbf{B} = 0$ in Maxwell's equations as the absence of sources of magnetic flux density, or, equivalently, the absence of magnetic monopoles. By the same interpretation, we conclude that electric charge density is the source of the displacement vector field, $\mathbf{D}$.

Coulomb's law $\nabla \cdot \mathbf{D} = \rho$ and the modified Ampère's law $\nabla \times \mathbf{H} = \mathbf{J} + \partial \mathbf{D} / \partial t$ given in Table 1.1 imply current continuity. To see this, one takes the divergence of the modified Ampère's law to give

$$\nabla \cdot (\nabla \times \mathbf{H}) = \nabla \cdot \mathbf{J} + \nabla \cdot \frac{\partial \mathbf{D}}{\partial t} \tag{1.66}$$

From vector calculus we know that the term on the left-hand side of Eqn (1.66) is zero, so that

$$0 = \nabla \cdot \mathbf{J} + \nabla \cdot \frac{\partial \mathbf{D}}{\partial t} \tag{1.67}$$

Finally, using Coulomb's law to obtain an expression in terms of current density, $\mathbf{J}$, and charge density, ρ, results in

$$\boxed{0 = \nabla \cdot \mathbf{J} + \frac{\partial \rho}{\partial t}} \tag{1.68}$$

This is an expression of *current continuity*. Physically this means that an increase in charge density in some volume of space is caused by current flowing through the surface enclosing the volume. The current continuity equation expresses the idea that electric charge is a conserved quantity. Charge does not spontaneously appear or disappear, but rather it is transported from one region of space to another by current.

To further illustrate the power of vector calculus, consider Ampère's circuital law. Ampère's circuital law states that a line integral of a static magnetic field taken about any given closed path must equal the current, I, enclosed by that path. The sign convention is that current is positive if advancing in a right-hand screw sense, where the screw rotation is in the direction of circulation for line integration. Vector calculus allows

Ampère's law to be derived almost trivially using Maxwell's equations and Stokes's theorem, as follows:

$$\oint \mathbf{H} \cdot d\mathbf{l} = \int_S (\nabla \times \mathbf{H}) \cdot \mathbf{n} \, ds = \int_S \mathbf{J} \cdot \mathbf{n} \, ds = I \tag{1.69}$$

Having reminded ourselves of Maxwell's equations and the importance of vector calculus, we now apply them to the description of light propagating through a medium. A particularly simple case to consider is propagation of a plane wave through a dielectric medium.

1.3.2.1 Light propagation in a dielectric medium

In a dielectric current density $\mathbf{J} = 0$ because the dielectric has no mobile charge and if $\mu_r = 1$ at optical frequencies then $\mathbf{H} = \mathbf{B}/\mu_0$. Hence, using the equations in Table 1.1, taking the curl of our expression of Faraday's law and using the modified Ampère's law, one may write

$$\nabla \times (\nabla \times \mathbf{E}) = -\frac{\partial}{\partial t}(\nabla \times \mathbf{B}) = -\mu_0 \frac{\partial}{\partial t}(\nabla \times \mathbf{H}) = -\mu_0 \frac{\partial^2}{\partial t^2}\mathbf{D} \tag{1.70}$$

The left-hand term may be rewritten by making use of the relationship $\nabla \times \nabla \times \mathbf{a} = \nabla(\nabla \cdot \mathbf{a}) - \nabla^2 \mathbf{a}$, so that

$$\nabla(\nabla \cdot \mathbf{E}) - \nabla^2 \mathbf{E} = -\mu_0 \frac{\partial^2}{\partial t^2}\mathbf{D} \tag{1.71}$$

For a source-free dielectric $\nabla \cdot \mathbf{D} = 0$ (and assuming ε in the medium is not a function of space – it is isotropic – so that $\nabla \cdot \mathbf{E} = 0$), this becomes

$$\nabla^2 \mathbf{E} = \mu_0 \frac{\partial^2}{\partial t^2}\mathbf{D} \tag{1.72}$$

$$\nabla^2 \mathbf{E}(\mathbf{r}, t) = \frac{\partial^2}{\partial t^2}\mu_0 \varepsilon(t)\mathbf{E}(\mathbf{r}, t) \tag{1.73}$$

where $\varepsilon(t) = \varepsilon_0 \varepsilon_r(t)$ is the complex permittivity function that results in $\mathbf{D}(\mathbf{r}, t) = \varepsilon_0 \varepsilon_r(t)\mathbf{E}(\mathbf{r}, t)$. We can now take the Fourier transform with respect to time to give a *wave equation* for electric field:

$$\nabla^2 \mathbf{E}(\mathbf{r}, \omega) = -\omega^2 \mu_0 \varepsilon_0 \varepsilon_r(\omega)\mathbf{E}(\mathbf{r}, \omega) \tag{1.74}$$

Since the speed of light is $c = 1/\sqrt{\varepsilon_0 \mu_0}$, the wave equation can be written as

$$\boxed{\nabla^2 \mathbf{E}(\mathbf{r}, \omega) = \frac{-\omega^2}{c^2}\varepsilon_r(\omega)\mathbf{E}(\mathbf{r}, \omega)} \tag{1.75}$$

When $\varepsilon_r(\omega)$ is real and positive, the solutions to this wave equation for an electric field propagating in an isotropic medium are just plane waves. The speed of wave propagation is $c/n_r(\omega)$, where $n_r(\omega) = \sqrt{\varepsilon_r(\omega)}$ is the *refractive index* of the material. In the more general case, when relative permeability $\mu_r \neq 1$, the refractive index is

$$n_r(\omega) = \sqrt{\varepsilon_r(\omega)\mu_r(\omega)} = \sqrt{\frac{\varepsilon(\omega)\mu(\omega)}{\varepsilon_0\mu_0}} \qquad (1.76)$$

If ε and μ are both real and positive, the refractive index is real-positive and electromagnetic waves propagate. In nature, the refractive index in a transparent material usually takes a positive value. If one of either ε or μ is negative, the refractive index is imaginary and electromagnetic waves cannot propagate.

It is common for metals to have negative values of ε. Free electrons of mass m in a metal can collectively oscillate at a long-wavelength natural frequency called the plasma frequency, ω_p. In a three-dimensional gas of electrons of density n the plasma frequency is $\omega_p = (ne^2/\varepsilon_0 m)^{1/2}$ which for $m = m_0$ and 10^{21} cm^{-3} $< n < 10^{22}$ cm^{-3} gives 1 eV $< \hbar\omega_p < 4$ eV. At long wavelengths a good approximation for relative permittivity of a metal is $\varepsilon_r(\omega) = 1 - \omega_p^2/\omega^2$. At frequencies above the plasma frequency ε is positive and electromagnetic waves can propagate through the metal. For frequencies below ω_p permittivity is negative, the refractive index is imaginary, electromagnetic waves cannot propagate in the metal and are reflected. This is the reason why bulk metals are usually not transparent to electromagnetic radiation of frequency less than ω_p.

There is, of course, another possibility for which both ε and μ are real and negative. While not usually found in naturally occurring materials, *meta-materials*, artificial structures with behavior not normally occurring in nature, may have negative relative permittivity and negative relative permeability simultaneously over some frequency range. However, although it is interesting to consider the electromagnetic properties of such material, we choose not to do so in this book.

Returning to the simple situation where relative permeability $\mu_r = 1$ and the relative permittivity function $\varepsilon_r(\omega) = 1$, an electric-field plane wave propagating in the $\hat{\mathbf{k}}$-direction with real wave-vector $\mathbf{k}$ and constant complex vector $\mathbf{E}_0$ has a spatial dependence of the form $\mathbf{E}(\mathbf{r}) = \mathbf{E}_0 e^{i\mathbf{k}\cdot\mathbf{r}}$. The amplitude of the wave is $|\mathbf{E}_0|$. The electromagnetic field wave has a *dispersion relation* obtained from the wave equation which, in free space, is $\omega = ck$. This dispersion relation is illustrated in Fig. 1.15. When dispersion $\omega = \omega(\mathbf{k})$ is nonlinear, *phase velocity* ω/k, *group velocity* $\partial\omega/\partial k$, and energy velocity of waves can all be different.

If the electromagnetic wave propagates in a homogeneous dielectric medium characterized by $\mu_r = 1$ and a complex relative permittivity function, then $\varepsilon(\omega) = \varepsilon_0\varepsilon_r(\omega) = \varepsilon_0(\varepsilon_r'(\omega) + \varepsilon_r''(\omega))$, where $\varepsilon_r'(\omega)$ and $\varepsilon_r''(\omega)$ are the real and imaginary parts, respectively, of the frequency-dependent relative permittivity function. In this situation, the dispersion relation shown in Fig. 1.15 is modified, and wave vector $\mathbf{k}(\omega)$ may become complex, giving an electric field

$$\mathbf{E}(\mathbf{r}, \omega) = \mathbf{E}_0(\omega)e^{i\mathbf{k}(\omega)\cdot\mathbf{r}} = \mathbf{E}_0(\omega)e^{i(k'(\omega) + ik''(\omega))\hat{\mathbf{k}}\cdot\mathbf{r}} \qquad (1.77)$$

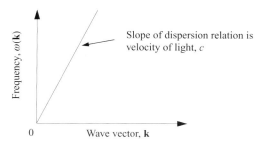

Fig. 1.15. Dispersion relation for an electromagnetic wave in free space. The slope of the line is the velocity of light.

where $k'(\omega)$ and $k''(\omega)$ are the real and imaginary parts, respectively, of the frequency-dependent wave number. The ratio of $k'(\omega)$ in the medium and $k = \omega/c$ in free space is the refractive index. Because in the dielectric we are considering $\mu_r = 1$, the refractive index is

$$n_r(\omega) = \sqrt{\frac{1}{2}\left(\varepsilon_r'(\omega) + \sqrt{\varepsilon_r'^2(\omega) + \varepsilon_r''^2(\omega)}\right)} \tag{1.78}$$

Equation (1.78) is obtained by substituting Eqn (1.77) into Eqn (1.75) and separating the real and imaginary parts of the resulting expression.

The imaginary part of $\mathbf{k}(\omega)$ physically corresponds to an exponential spatial decay in field amplitude $e^{-k''r}$ due to absorption processes.

Returning to the case in which $k''(\omega) = 0$ and $\mu_r = 1$, the refractive index is just $n_r(\omega) = \sqrt{\varepsilon_r'(\omega)}$, and we have a simple oscillatory solution with no spatial decay in the electric field:

$$\mathbf{E}(\mathbf{r}, \omega) = \mathbf{E}_0 e^{-i\omega t} e^{i\mathbf{k}(\omega)\cdot\mathbf{r}} \tag{1.79}$$

and

$$\mathbf{H}(\mathbf{r}, \omega) = \mathbf{H}_0 e^{-i\omega t} e^{i\mathbf{k}(\omega)\cdot\mathbf{r}} \tag{1.80}$$

for the magnetic field vector.

We proceed by recalling Maxwell's equations for electromagnetic waves in free space:

$$\nabla \cdot \mathbf{D} = 0 \tag{1.81}$$

$$\nabla \cdot \mathbf{B} = 0 \tag{1.82}$$

$$\nabla \times \mathbf{E} = -\frac{\partial \mathbf{B}}{\partial t} \tag{1.83}$$

$$\nabla \times \mathbf{H} = \frac{\partial \mathbf{D}}{\partial t} \tag{1.84}$$

The first two equations are divergence equations that require that $\mathbf{k} \cdot \mathbf{E} = 0$ and $\mathbf{k} \cdot \mathbf{B} = 0$. This means that $\mathbf{E}$ and $\mathbf{B}$ are perpendicular (transverse) to the direction of propagation $\hat{\mathbf{k}}$. In 1888 Hertz performed experiments that showed the existence of transverse electromagnetic waves, thereby providing an experimental basis for Maxwell's theory.

There is a relationship between the electric and magnetic field vectors of transverse electromagnetic waves that we can find by considering the two curl equations (Eqn (1.83) and Eqn (1.84)). To find this relationship for electric and magnetic field vectors in free space, we start with the plane-wave expressions for $\mathbf{E}(\mathbf{r}, \omega)$ and $\mathbf{H}(\mathbf{r}, \omega)$ given by Eqn (1.79) and Eqn (1.80). Substituting them into the first curl equation (Eqn (1.83)) and recalling that in free space $\mathbf{H} = \mathbf{B}/\mu_0$ gives

$$\nabla \times \mathbf{E}_0 e^{-i\omega t} e^{i\mathbf{k}(\omega)\cdot\mathbf{r}} = -\mu_0 \frac{\partial}{\partial t} \mathbf{H}_0 e^{-i\omega t} e^{i\mathbf{k}(\omega)\cdot\mathbf{r}} \tag{1.85}$$

$$i\mathbf{k} \times \mathbf{E}_0 e^{-i\omega t} e^{i\mathbf{k}(\omega)\cdot\mathbf{r}} = i\omega\,\mu_0 \mathbf{H}_0 e^{-i\omega t} e^{i\mathbf{k}(\omega)\cdot\mathbf{r}} \tag{1.86}$$

$$i\mathbf{k} \times \mathbf{E} = i\omega\,\mu_0 \mathbf{H} \tag{1.87}$$

Using the fact that the dispersion relation for plane waves in free space is $\omega = ck$ and the speed of light is $c = 1/\sqrt{\varepsilon_0\mu_0}$ leads us directly to

$$\boxed{\mathbf{H} = \sqrt{\frac{\varepsilon_0}{\mu_0}}\,\hat{\mathbf{k}} \times \mathbf{E}} \tag{1.88}$$

or $\mathbf{B}c = \hat{\mathbf{k}} \times \mathbf{E}$, where $\hat{\mathbf{k}} = \mathbf{k}/|\mathbf{k}|$ is the unit vector for $\mathbf{k}$.

The importance of this result is that it is not possible to separate out an oscillating electric or magnetic field. Electric and magnetic fields are related to each other in such a way that there are only *electromagnetic* fields.

One may easily visualize an oscillating transverse electromagnetic wave by considering a plane wave. Figure 1.16 illustrates the magnetic field and the electric field for a plane wave propagating in free space in the x-direction. The shading is to help guide the eye.

Oscillating transverse electromagnetic waves can decay in time and in space. In Fig. 1.17, the temporal decay of an oscillatory electric field and the spatial decay of an oscillatory electric field are shown schematically.

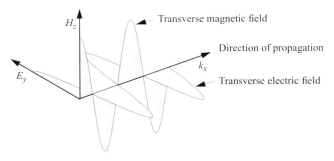

Fig. 1.16. Illustration of transverse magnetic field H_z and electric field E_y of a plane wave propagating in free space in the x direction. The shading is to guide the eye.

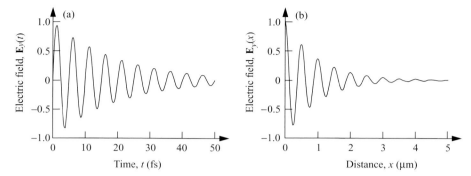

Fig. 1.17. Illustration of: (a) temporal decay of an oscillating electric field; (b) spatial decay of an oscillating electric field.

Figure 1.17(a) illustrates temporal decay of an oscillating electric field propagating in the x direction. In this case an electric field oscillates with period $\tau = 5$ fs, which corresponds to frequency $\omega = 2\pi/\tau = 4\pi \times 10^{14}$ rad s^{-1} and a wavelength $\lambda = 1500$ nm when observed in free space (wave number $k_x = 2\pi/\lambda = 4.2 \times 10^6$ m^{-1}). In the example, the inverse decay time constant is taken to be $\gamma^{-1} = 20$ fs. The function plotted in Fig. 1.17(a) is $\mathbf{E}(t) = \hat{\mathbf{y}}|\mathbf{E}_0|\sin(\omega t)e^{-\gamma t}$, where $|\mathbf{E}_0| = 1$ V m^{-1}.

Figure 1.17(b) illustrates spatial decay of an oscillating electric field. In this case, an electric field oscillates with period $\tau = 5$ fs, which corresponds to frequency $\omega = 2\pi/\tau = 4\pi \times 10^{14}$ rad s^{-1}. The electric field propagates in a dielectric medium characterized by refractive index $n = 3$ and inverse spatial decay length $\gamma_x^{-1} = 10^{-4}$ cm. The function plotted in Fig. 1.17(b) is $\mathbf{E}(x) = \hat{\mathbf{y}}|\mathbf{E}_0|\cos(kx)e^{-\gamma_x x}$, where $|\mathbf{E}_0| = 1$ V m^{-1}. Spatial decay can be independent of temporal decay as, for example, can occur in a high-Q optical resonator.

The facts that only electromagnetic fields exist and that they can grow and decay both spatially and temporally lead us to question how electrodynamics relates to our previous results using electrostatics. In particular, we may ask where the magnetic field is when there is a static electric field in, for example, a parallel-plate capacitor. The answer is that a static electric field can only be formed by movement of charge, and hence by a current. When a transient current flows, a magnetic field is produced. One cannot form a static electric field without a transient current and an associated magnetic field.

1.3.2.2 Power and momentum in an electromagnetic wave

Let's extend what we know so far to obtain the power flux in an electromagnetic wave. This is of practical importance for wireless communication where, for example, we might be designing a receiver for a cellular telephone. In this case, the radio frequency electromagnetic power flux received by the cell phone antenna will help determine the type of amplifier to be used.

The power in an electromagnetic wave can be obtained by considering the response of a test charge e moving at velocity $\mathbf{v}$ in an external electric field $\mathbf{E}$. The rate of work or power is just $e\mathbf{v} \cdot \mathbf{E}$, where $e\mathbf{v}$ is a current. This may be generalized to a continuous distribution of current density $\mathbf{J}$ so that the total power in a given volume is

$$\int_{\text{Volume}} d^3r \mathbf{J} \cdot \mathbf{E} = \int_{\text{Volume}} \left(\mathbf{E} \cdot (\nabla \times \mathbf{H}) - \mathbf{E} \cdot \frac{\partial \mathbf{D}}{\partial t} \right) d^3 r \tag{1.89}$$

where we have used Maxwell's equation $\nabla \times \mathbf{H} = \mathbf{J} + \partial \mathbf{D}/\partial t$. Because this is the power that is extracted from the electromagnetic field, energy conservation requires that there must be a corresponding reduction in electromagnetic field energy in the same volume. Making use of the result from vector calculus $\mathbf{E} \cdot (\nabla \times \mathbf{H}) = \mathbf{H} \cdot (\nabla \times \mathbf{E}) - \nabla \cdot (\mathbf{E} \times \mathbf{H})$ and Maxwell's equation $\nabla \times \mathbf{E} = -\partial \mathbf{B}/\partial t$, we may write

$$\int_{\text{Volume}} d^3r \mathbf{J} \cdot \mathbf{E} = - \int_{\text{Volume}} \left(\nabla \cdot (\mathbf{E} \times \mathbf{H}) + \mathbf{E} \cdot \frac{\partial \mathbf{D}}{\partial t} + \mathbf{H} \cdot \frac{\partial \mathbf{B}}{\partial t} \right) d^3 r \tag{1.90}$$

or in differential form

$$\mathbf{E} \cdot \frac{\partial \mathbf{D}}{\partial t} + \mathbf{H} \cdot \frac{\partial \mathbf{B}}{\partial t} = -\mathbf{J} \cdot \mathbf{E} - \nabla \cdot (\mathbf{E} \times \mathbf{H}) \tag{1.91}$$

Generalizing our previous result for energy density stored in electric fields (Eqn (1.44) and Eqn (1.45)) to electromagnetic waves in a medium with linear response and no dispersion, the total energy density at position $\mathbf{r}$ and time t is just

$$U = \frac{1}{2}(\mathbf{E} \cdot \mathbf{D} + \mathbf{B} \cdot \mathbf{H}) \tag{1.92}$$

(The time averaged energy density is half this value.) After substitution into the differential expression, this gives

$$\boxed{\frac{\partial U}{\partial t} = -\mathbf{J} \cdot \mathbf{E} - \nabla \cdot \mathbf{S}} \tag{1.93}$$

where

$$\boxed{\mathbf{S} = \mathbf{E} \times \mathbf{H}} \tag{1.94}$$

is called the *Poynting vector*. The Poynting vector is the energy flux density in the electromagnetic field. The magnitude of $\mathbf{S}$ is measured in units of $\mathbf{J}$ m^{-2} s^{-1}, and the negative divergence of $\mathbf{S}$ is the flow of electromagnetic energy out of the system. The associated differential equation is just an expression of energy conservation. The rate of change of electromagnetic energy density is given by the rate of loss due to work done by the electromagnetic field density on sources given by $-\mathbf{J} \cdot \mathbf{E}$ and the rate of loss due to electromagnetic energy flow given by $-\nabla \cdot \mathbf{S}$.

Energy flux density is the energy per unit area per unit time flowing in the electromagnetic wave. The energy flux density multiplied by the area the flux is passing through is the power (measured in W or J s^{-1}) delivered by the electromagnetic wave. Suppose we wish to calculate the energy per unit volume of an electromagnetic wave in free space. In this case, there are no sources of current, and the energy per unit volume is just the energy flux density divided by the speed of light. Thus, we may write

$$U = \frac{|\mathbf{S}|}{c} \tag{1.95}$$

Another expression for the Poynting vector $\mathbf{S}$ can be found by eliminating $\mathbf{H}$ from the equation $\mathbf{S} = \mathbf{E} \times \mathbf{H}$. The energy density U is measured in units of J m^{-3} or, equivalently, in units of kg m^{-1} s^{-2}.

Let's consider an electromagnetic plane wave propagating in *free space* for which $\mathbf{E}(\mathbf{r}, \omega)$ and $\mathbf{H}(\mathbf{r}, \omega)$ are given by Eqn (1.79) and Eqn (1.80) respectively. $\mathbf{H}(\mathbf{r}, \omega)$ is related to $\mathbf{E}(\mathbf{r}, \omega)$ via the relationship $\mathbf{H} = \sqrt{\varepsilon_0/\mu_0}\ \hat{\mathbf{k}} \times \mathbf{E}$ given by Eqn (1.88). Substituting this into our expression for the Poynting vector gives

$$\mathbf{S} = \mathbf{E} \times \mathbf{H} = \sqrt{\frac{\varepsilon_0}{\mu_0}}\ \mathbf{E} \times \hat{\mathbf{k}} \times \mathbf{E} \tag{1.96}$$

We apply the result from vector calculus $\mathbf{a} \times (\mathbf{b} \times \mathbf{c}) = (\mathbf{a} \cdot \mathbf{c})\mathbf{b} - (\mathbf{a} \cdot \mathbf{b})\mathbf{c}$, so that

$$\mathbf{S} = \sqrt{\frac{\varepsilon_0}{\mu_0}}((\mathbf{E} \cdot \mathbf{E})\hat{\mathbf{k}} - (\mathbf{E} \cdot \hat{\mathbf{k}})\mathbf{E}) \tag{1.97}$$

For transverse electromagnetic waves, the second term on the right-hand side $(\mathbf{E} \cdot \hat{\mathbf{k}}) = 0$, and so we may write

$$\mathbf{S} = \sqrt{\frac{\varepsilon_0}{\mu_0}}(\mathbf{E} \cdot \mathbf{E})\hat{\mathbf{k}} \tag{1.98}$$

Defining the *impedance of free space*[12] as

$$\boxed{Z_0 = \sqrt{\frac{\mu_0}{\varepsilon_0}} = 376.73\ \Omega} \tag{1.99}$$

our expression for energy flux associated with an electromagnetic field in free space becomes

$$\mathbf{S} = \frac{(\mathbf{E} \cdot \mathbf{E})}{Z_0}\hat{\mathbf{k}} \tag{1.100}$$

This expression is written in the familiar form of power in standard electrical circuit theory, V^2/R. The difference, of course, is that this is an oscillating electromagnetic field. For monochromatic plane waves propagating in the x direction, the Poynting vector may be written

$$\mathbf{S} = \frac{|\mathbf{E}_0|^2}{Z_0}(\cos^2(k_x x - \omega t + \Delta_{\text{phase}}))\hat{\mathbf{k}} \tag{1.101}$$

where Δ_{phase} is a fixed phase. Because the time average of the $\cos^2$ term in Eqn (1.101)

---

[12] It is often convenient to use the approximation $Z_0 = 120 \times \pi\ \Omega$.

is 1/2, the average power per unit area transported by this sinusoidally oscillating electromagnetic field is just

$$\langle \mathbf{S} \rangle = \frac{|\mathbf{E}_0|^2}{2Z_0} \hat{\mathbf{k}} \tag{1.102}$$

Electromagnetic waves carry not only energy, but also momentum. Because electromagnetic waves carry momentum, they can exert a force on a charged particle. The classical force on a test charge e moving at velocity $\mathbf{v}$ is just

$$\mathbf{F} = e(\mathbf{E} + \mathbf{v} \times \mathbf{B}) \tag{1.103}$$

Using Newton's second law for mechanical motion, which relates rate of change of momentum to force (Eqn (1.1)), it is possible to show that the momentum density in an electromagnetic wave is proportional to the energy flux density. The proportionality constant is the inverse of the speed of light squared, $1/c^2$, so that

$$\mathbf{p} = \frac{\mathbf{E} \times \mathbf{H}}{c^2} = \frac{\mathbf{S}}{c^2} \tag{1.104}$$

We could have guessed this result using dimensional analysis. The energy-flux density $\mathbf{S}$ is measured in units of $\mathrm{J\,m^{-2}\,s^{-1}} = \mathrm{kg\,m^2\,s^{-2} \cdot m^{-2}\,s^{-1}}$, so that $\mathbf{S}/c^2$ is measured in units of $\mathrm{J\,m^{-4}\,s} = \mathrm{kg\,m^2\,s^{-2} \cdot m^{-4}\,s} = \mathrm{kg\,m\,s^{-1} \cdot m^{-3}}$, which has the units of momentum per unit volume. For a plane wave, the momentum can be expressed in terms of the energy density as

$$\boxed{\mathbf{p} = \frac{U}{c} \hat{\mathbf{k}}} \tag{1.105}$$

In this equation, $\hat{\mathbf{k}}$ is the unit vector in the direction of propagation of the wave and E is the energy density. The magnitude of the momentum is just

$$|\mathbf{p}| = \frac{1}{c} \cdot \frac{|\mathbf{S}|}{c} = \frac{U}{c} \tag{1.106}$$

$$\boxed{p = \frac{U}{c}} \tag{1.107}$$

When we introduce quantum mechanics, we will make use of this expression to suggest that *if* light energy is quantized then so is the momentum carried by light.

1.3.2.3 Choosing a potential

Electric and magnetic fields are related to a potential in a more complex way than we have discussed so far. In general, Maxwell's equations allow electric and magnetic fields to be described in terms of a scalar potential $V(\mathbf{r}, t)$ and a vector potential $\mathbf{A}(\mathbf{r}, t)$.

To see why this is so, we recall from vector calculus that any vector field $\mathbf{a}$ satisfies $\nabla \cdot (\nabla \times \mathbf{a}) = 0$. Because Maxwell's equations state $\nabla \cdot \mathbf{B} = 0$, it must be possible to choose a vector field $\mathbf{A}$ for which $\mathbf{B} = \nabla \times \mathbf{A}$. Using this expression for $\mathbf{B}$, Faraday's law can now be rewritten

$$\nabla \times \mathbf{E} = -\frac{\partial \mathbf{B}}{\partial t} = -\frac{\partial}{\partial t} \nabla \times \mathbf{A} \tag{1.108}$$

or

$$\nabla \times \left(\mathbf{E} + \frac{\partial \mathbf{A}}{\partial t} \right) = 0 \tag{1.109}$$

Since we know from vector calculus that the curl of the gradient of any scalar field is zero, we may equate the last equation with the gradient of a scalar field, V, where

$$\mathbf{E} + \frac{\partial \mathbf{A}}{\partial t} = -\nabla V \tag{1.110}$$

In general, the scalar and vector fields are functions of space and time, so we are free to choose functions $V(\mathbf{r}, t)$ and $\mathbf{A}(\mathbf{r}, t)$, giving

$$\mathbf{E}(\mathbf{r}, t) = -\nabla V(\mathbf{r}, t) - \frac{\partial}{\partial t} \mathbf{A}(\mathbf{r}, t) \tag{1.111}$$

$$\mathbf{B}(\mathbf{r}, t) = \nabla \times \mathbf{A}(\mathbf{r}, t) \tag{1.112}$$

The exact forms used for $V(\mathbf{r}, t)$ and $\mathbf{A}(\mathbf{r}, t)$ are usually chosen to simplify a specific calculation. To describe this choice one talks of using a particular *gauge*.

Let's see what this means in practice. Suppose we wish to consider a *static* electric field, $\mathbf{E}(\mathbf{r})$. Previously, we chose a gauge where $\mathbf{E}(\mathbf{r}) = -\nabla V(\mathbf{r})$. An interesting consequence of this choice is that, because the static electric field is expressed as a gradient of a potential, the absolute value of the potential need not be known to within a constant. Only differences in potential have physical consequences and therefore meaning. Another choice of gauge is where $\mathbf{A} = -\mathbf{E}t$. In this case, we have a remarkable degree of latitude in our choice of $\mathbf{A}$, since we can add any time independent function to $\mathbf{A}$ without changing the electric field. Other possibilities for the gauge involve combinations of the scalar and vector potential. If we wish to describe magnetic fields, the possible choices of gauge are even greater. In this book we choose a gauge for its simplicity, and we avoid the complications introduced by inclusion of magnetic fields.

1.3.2.4 Dipole radiation

When a current flows through a conductor, a magnetic field exists in space around the conductor. This fact is described by the modified Ampère's law in Maxwell's equations in Table 1.1, and it is the subject of Exercise 1.12. If the magnitude of the current varies over time, then so does the magnetic field. Faraday's law indicates that a changing

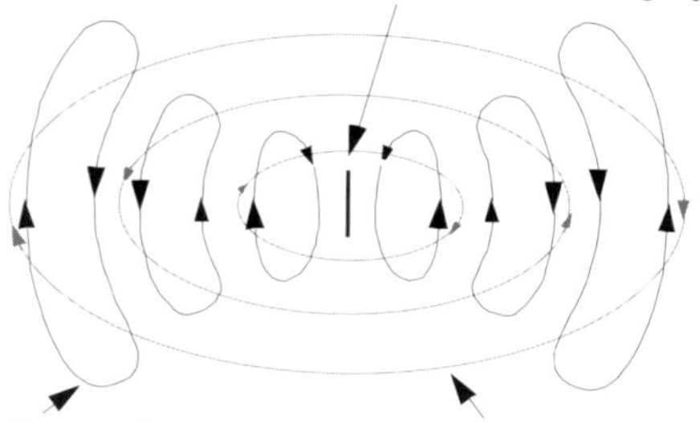

Fig. 1.18. Illustration of a short conductor carrying an oscillating harmonic time-dependent current surrounded by interdependent oscillating **E** and **H** fields. The oscillating current flowing up and down the conducting wire produces electromagnetic waves which propagate in free space.

magnetic field coexists with a changing electric field. Hence, a conductor carrying an oscillating current is always surrounded by interdependent oscillating **E** and **H** fields. As the oscillating current changes its direction in the conductor, the magnetic field also tries to change direction. To do so, the magnetic field that exists in space must first try to disappear by collapsing into the conductor before growing again in the opposite direction. Typically, not all of the magnetic field disappears before current flows in the opposite direction. The portions of the magnetic field and its related electric field that are unable to return to the conductor before the current starts to increase in the opposite direction are propagated away as electromagnetic radiation. In this way, an oscillating current (associated with acceleration and deceleration of charge in a conductor) produces electromagnetic waves. This is illustrated in Fig. 1.18 for a short conducting wire carrying a harmonic time-dependent current.

One may describe radiation due to changes in current in a conductor by considering an element of length r_0 carrying oscillating current $I(t)$. An appropriate arrangement is shown schematically in Fig. 1.19. A small length of conducting wire connects two conducting spheres separated by distance r_0 and oriented in the $\hat{\mathbf{z}}$ direction. Oscillatory current flows in the wire so that $I(t) = I_0 e^{i\omega t}$, where measurable current is the real part of this function. The harmonic time-dependent current is related to the charge on the spheres by $I(t) = \pm dQ(t)/dt$, where $Q(t) = Q_0 e^{i\omega t}$. The plus sign is for the upper sphere, and the minus sign is for the lower sphere. It follows that $I(t) = \pm d(Q_0 e^{i\omega t})/dt = \pm i\omega Q(t)$, so that $Q(t) = \pm I(t)/i\omega$. For equal and opposite charges separated by a small distance, one may define a dipole moment for the harmonic time-dependent source as

$$\mathbf{d} = Q\hat{\mathbf{z}}r_0 = \frac{I r_0 \hat{\mathbf{z}}}{i\omega} \qquad (1.113)$$

If either the current I or the current density $\mathbf{J}$ is known, then we can find the other quantities of interest, such as the total radiated electromagnetic power, P_r. To calculate

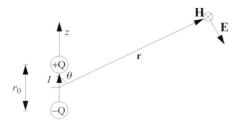

Fig. 1.19. A small length of conducting wire connects two conducting spheres oriented in the z direction that have center-to-center spacing of r_0. Oscillatory current I flows in the wire, charging and discharging the spheres. The magnetic and electric field at position r is indicated.

the quantity P_r, we must solve for the field **H** or **E**. This is done by finding the vector potential **A**.

Considering electromagnetic radiation in free space and substituting

$$\mathbf{H}(\mathbf{r}, t) = \frac{1}{\mu_0} \nabla \times \mathbf{A}(\mathbf{r}, t) \tag{1.114}$$

into the modified Ampère's law gives

$$\nabla \times \mathbf{H} = \frac{1}{\mu_0} \nabla \times \nabla \times \mathbf{A} = \mathbf{J} + \frac{\partial \mathbf{D}}{\partial t} \tag{1.115}$$

Since $\mathbf{D} = \varepsilon_0 \mathbf{E}$ and $\mathbf{E} = -\nabla V - \partial \mathbf{A}/\partial t$, we may write

$$\nabla \times \nabla \times \mathbf{A} = \nabla(\nabla \cdot \mathbf{A}) - \nabla^2 \mathbf{A} = \mu_0 \mathbf{J} + \mu_0 \varepsilon_0 \frac{\partial}{\partial t}\left(-\nabla V - \frac{\partial \mathbf{A}}{\partial t}\right) \tag{1.116}$$

$$\nabla(\nabla \cdot \mathbf{A}) - \nabla^2 \mathbf{A} = \mu_0 \mathbf{J} - \nabla\left(\mu_0 \varepsilon_0 \frac{\partial V}{\partial t}\right) - \mu_0 \varepsilon_0 \frac{\partial^2 \mathbf{A}}{\partial t^2} \tag{1.117}$$

$$\nabla^2 \mathbf{A} - \mu_0 \varepsilon_0 \frac{\partial^2 \mathbf{A}}{\partial t^2} = -\mu_0 \mathbf{J} - \nabla\left(\nabla \cdot \mathbf{A} + \mu_0 \varepsilon_0 \frac{\partial V}{\partial t}\right) \tag{1.118}$$

The definition of vector **A** requires that curl and divergence be defined. While Maxwell's equations force adoption of the curl relationship $\mathbf{H} = (1/\mu_0)\nabla \times \mathbf{A}$, we are free to choose the divergence. In the Lorentz gauge one lets

$$\left(\nabla \cdot \mathbf{A} + \mu_0 \varepsilon_0 \frac{\partial V}{\partial t}\right) = 0 \tag{1.119}$$

The resulting nonhomogeneous wave equation for the vector potential is

$$\nabla^2 \mathbf{A} - \mu_0 \varepsilon_0 \frac{\partial^2 \mathbf{A}}{\partial t^2} = -\mu_0 \mathbf{J} \tag{1.120}$$

The solution for a harmonic time-dependent source is

$$\mathbf{A} = \frac{\mu_0}{4\pi} \int\limits_{\text{volume}} \mathbf{J} \frac{e^{-ikr}}{r} d^3 r \tag{1.121}$$

where the integral over the volume includes the oscillating current density $\mathbf{J}$ of the source. In the equation, $k = 2\pi/\lambda$, where λ is the wavelength of the electromagnetic wave. In our case, we know that the integral over $\mathbf{J}$ is just Ir_0, so that

$$\mathbf{A} = \hat{\mathbf{z}}\frac{\mu_0}{4\pi}Ir_0\frac{e^{-ikr}}{r} \tag{1.122}$$

The vector potential in the $\hat{\mathbf{z}}$ direction is related to the radial distance r and the angle θ in spherical coordinates by

$$A_z = A_r\cos(\theta) - A_\theta\sin(\theta) \tag{1.123}$$

Hence, the radial and angular components of the vector potential are

$$A_r = \frac{\mu_0}{4\pi}Ir_0\frac{e^{-ikr}}{r}\cos(\theta) \tag{1.124}$$

$$A_\theta = \frac{\mu_0}{4\pi}Ir_0\frac{e^{-ikr}}{r}\sin(\theta) \tag{1.125}$$

$$A_\phi = 0 \tag{1.126}$$

Having found expressions for components of the vector potential in spherical coordinates, one proceeds to calculate magnetic and electric fields. The magnetic field is just $\mathbf{H} = (1/\mu_0)\nabla \times \mathbf{A}$, which in spherical coordinates is

$$\mathbf{H} = \frac{1}{\mu_0}\frac{1}{r^2\sin(\theta)}\begin{vmatrix} \hat{\mathbf{r}} & r\theta & r\sin(\theta)\hat{\phi} \\ \dfrac{\partial}{\partial r} & \dfrac{\partial}{\partial\theta} & \dfrac{\partial}{\partial\phi} \\ A_r & rA_\theta & r\sin(\theta)A_\phi \end{vmatrix} \tag{1.127}$$

$$\mathbf{H} = \hat{\mathbf{r}}(0-0) + \theta(0-0) - \hat{\phi}\frac{Ir_0k^2e^{-ikr}}{4\pi}\sin(\theta)\left(\frac{1}{ikr} + \frac{1}{(ikr)^2}\right) \tag{1.128}$$

From Eqn (1.128) one may conclude that H_ϕ is the only component of the magnetic field. The electric field is found from

$$\mathbf{E} = \frac{1}{i\omega\varepsilon_0}\nabla \times \mathbf{H} = \frac{1}{i\omega\varepsilon_0}\left(\hat{\mathbf{r}}\frac{1}{r\sin(\theta)}\frac{\partial}{\partial\theta}(H_\phi\sin(\theta)) - \theta\frac{1}{r}\frac{\partial}{\partial r}(rH_\phi)\right) \tag{1.129}$$

At a large distance, one is many wavelengths from the source so that $kr = 2\pi r/\lambda \gg 1$. This is called the *far-field limit*. In the far-field limit, Eqn (1.128) becomes

$$H_\phi = \frac{iIr_0e^{-ikr}}{4\pi r}k\sin(\theta) = \frac{E_\theta}{Z_0} \tag{1.130}$$

and Eqn (1.129) may be written as

$$E_\theta = \frac{iIr_0e^{-ikr}}{4\pi r}Z_0k\sin(\theta) = Z_0H_\phi \tag{1.131}$$

Notice the consistency with the relation $\mathbf{H} = \sqrt{\varepsilon_0/\mu_0}\,\hat{\mathbf{k}} \times \mathbf{E}$ given by Eqn (1.88), and recall that the impedance of free space is $Z_0 = \sqrt{\mu_0/\varepsilon_0} = 376.73\ \Omega$ (from Eqn (1.99)).

The total time-averaged radiated power P_r from a sinusoidally oscillating electromagnetic field is just $|\mathbf{E}_0|^2/2Z_0$. Hence,

$$P_r = \frac{1}{2}\frac{|E_\theta|^2}{Z_0} = \frac{Z_0}{2}|H_\phi|^2 = \frac{Z_0}{2}\int_{\phi=0}^{\phi=2\pi} d\phi \int_{\theta=0}^{\theta=\pi}\left(\frac{Ir_0e^{-ikr}}{4\pi r}k\sin(\theta)\right)^2 r^2\sin(\theta)d\theta$$

(1.132)

After performing the integral one finds

$$P_r = \frac{Z_0}{12\pi}Ir_0k^2$$

(1.133)

Since the dispersion relation for electromagnetic radiation is $\omega = ck$ and current I is related to the dipole moment $\mathbf{d}$ by $Ir_0\hat{\mathbf{z}} = i\omega\mathbf{d}$, the expression for P_r may be rewritten as

$$P_r = \frac{Z_0}{12\pi}\frac{\omega^4|\mathbf{d}|^2}{c^2}$$

(1.134)

This is the classical expression for total time-averaged electromagnetic radiation from a dipole source. Sometimes one makes the approximation $Z_0 = 120 \times \pi\ \Omega$, so that $P_r = 10\,\omega^4|\mathbf{d}|^2/c^2$.

1.4 Example exercises

Exercise 1.1
Two intelligent players seated at a round table alternately place round beer mats on the table. The beer mats are not allowed to overlap, and the last player to place a mat on the table wins. Who wins and what is the strategy?

Exercise 1.2
Visitors from another planet wish to measure the circumference of the Earth. To do this, they run a tape measure around the equator. How many extra meters are needed if the tape measure is raised one meter above the ground?[13]

Exercise 1.3
A plug can be carved to fit exactly *into* a square hole of side 2 cm, a circular hole of radius $r = 1$ cm, and an isosceles triangular hole with a base 2 cm wide and a height of $h = 2$ cm. What is the smallest and largest *convex* solid volume of the plug?

---

[13] The visitors were not aware that the meter was first defined by the French Academy of Sciences in 1791 as $1/10^7$ of the quadrant of the Earth's circumference running from the North Pole through Paris to the equator.

Exercise 1.4

A one-mile-long straight steel rod lies on flat ground and is attached at each end to a fixed point. During a particularly hot day the rod expands by one foot. Assuming that the expansion causes the rod to describe an arc of a circle between the fixed points, what is the height above the ground at the center of the rod?

Note that one mile is 5280 feet.

Exercise 1.5

The 1925 Cole survey reported an estimate of the original dimensions of the Great Pyramid. The square base had a perimeter of 921.46 m and the vertical apex height was 146.73 m. What was the minimum work done in building the Great Pyramid? How long does it take a 500-MW electric generator to deliver the same amount of work?

In your calculations you may assume that the acceleration due to gravity is $g = 9.8 \ \mathrm{m\,s^{-2}}$ and the density of the stone used is $\rho = 2000 \ \mathrm{kg\,m^{-3}}$.

Exercise 1.6

(a) The rotating blades of a helicopter push air to create the force needed to allow the machine to fly. How much power must be generated by the motor of a 1000-kg helicopter with blades 4 m long?

You may assume that the density of air is $\rho = 1.3 \ \mathrm{kg\,m^{-3}}$.

(b) Estimate the force exerted by a 60-mph wind impinging normally to the side of a suspension bridge with an effective cross-section 500 ft long and 10 ft high.

($1 \ \mathrm{ft^2} \approx 9.3 \times 10^{-2} \ \mathrm{m^2}$ and $1 \ \mathrm{mph} \approx 0.45 \times 10^{-2} \ \mathrm{m\,s^{-1}}$.)

Exercise 1.7

A particle of mass m exhibits classical one-dimensional simple harmonic oscillation of frequency ω_0. What is the maximum kinetic energy of the particle, and how does it depend upon the amplitude of oscillation?

Exercise 1.8

Consider a particle mass m attached to a lightweight spring that obeys Hooke's law. The displacement from equilibrium $x = x_0$ is proportional to the force on the particle $F = -\kappa(x - x_0)$, where κ is the spring constant. The particle is subject to a small external oscillatory force in the $\hat{\mathbf{x}}$ direction so that $F(t) = F_1 \sin(\omega t)$.

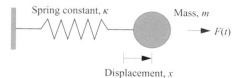

Spring constant, κ Mass, m

$F(t)$

Displacement, x

(a) Adding $F(t)$ to the right-hand side of the equation of motion for a harmonic oscillator

$$\kappa(x - x_0) + m\frac{d^2x}{dt^2} = 0$$

and assuming $x(t) = x_0 + x_1 \sin(\omega t)$, show that

$$x_1(\omega) = \frac{F_1/m}{\omega_0^2 - \omega^2}$$

where $\omega_0^2 = \kappa/m$. Plot $x_1(\omega)$, and interpret what happens as $\omega \rightarrow \omega_0$.

(b) Show that adding a damping term $D \cdot (dx/dt)$ to the left-hand side of the equation of motion used in (a) changes the solution to

$$x_1(\omega) = \frac{F_1/m}{\omega_0^2 - \omega^2 - i(\gamma/2)}$$

where $\gamma(\omega) = 2\omega D/m$.

Assuming that one may replace $\gamma(\omega)$ with a frequency-independent constant $\Gamma = \gamma(\omega_0)/\omega_0 = 2D/m$, plot $|x_1(\omega)|$, phase, $\text{Re}(x_1(\omega))$, and $\text{Im}(x_1(\omega))$ for $\omega_0 = 10$, $\Gamma = 0.5$ and $\Gamma = 10$. Explain the relationship between Γ and the line shape.

Exercise 1.9

The relative atomic mass of Ga is 69.72, the relative atomic mass of As is 74.92, and the frequency of the longitudinal polar-optic lattice vibration in GaAs is $\nu = 8.78\,\text{THz}$ in the long-wavelength limit. Use the solutions for lattice dynamics of a linear chain developed in Section 1.2.4 to estimate the spring constant κ. Use this value to estimate $\nu_1(= \omega_1/2\pi)$ and $\nu_2(= \omega_2/2\pi)$, and then compare this with the experimentally measured values for [100] oriented GaAs.

Exercise 1.10

Microelectromechanical systems (MEMS) feature micron-sized mechanical structures fabricated out of semiconductor material. One such MEMS structure is a cantilever beam shown sketched below.

The lowest-frequency vibrational mode of a long, thin cantilever beam attached at one end is[14]

$$\omega = 3.52\frac{d}{l^2}\sqrt{\frac{E_{\text{Young}}}{12\rho}}$$

where l is the length, d is the thickness, ρ is the density of the beam, and E_{Young}, defined as uniaxial tensile stress divided by strain in bulk material, is Young's modulus.

---

[14] L. D. Landau and E. M. Lifshitz, *Theory of Elasticity*, Butterworth-Heinemann, Oxford, 1986 (ISBN 0 7506 2633 X).

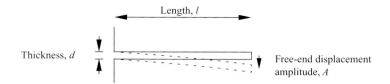

Length, l

Thickness, d

Free-end displacement amplitude, A

(a) A cantilever made of silicon using a MEMS process has dimension $l = 100$ µm, $d = 0.1$ µm, $\rho = 2.328 \times 10^3$ kg m^{-3}, and Young's modulus $E_{Young} = 1.96 \times 10^{11}$ N m^{-2}. Calculate the natural frequency of vibration for the cantilever.

(b) The vibrational energy of a cantilever with width w and free-end displacement amplitude A is

$$\frac{wd^3 A^2 E_{Young}}{6l^3}$$

How much vibrational energy is there in the cantilever of (a) when width $w = 5$ µm and amplitude $A = 1$ µm?

Exercise 1.11

A chemist has developed a spherical dendrimer structure with a core that is a redox active molecule. The charge state of an iron atom at the core can be neutral or $+e$.

Highly fluorescent rhodamine dye molecules, which are incorporated into the dendrimer, can be used to sense the charged state. This works because the fluorescence emission spectrum is sensitive to the local electric field and hence to the charged state of the redox core.

This particular dendrimer, which is 4 nm in diameter and has a relative permittivity of $\varepsilon_r = 2$, is placed on a flat, perfectly conducting metal sheet, which is maintained at zero electrical potential. Calculate the force acting upon the ion core when the fluorescence emission spectrum indicates that the iron atom is in the charged state.

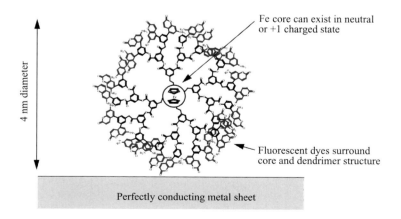

4 nm diameter

Fe core can exist in neutral or +1 charged state

Fluorescent dyes surround core and dendrimer structure

Perfectly conducting metal sheet

Exercise 1.12

The circuit shown below is energized by an input current I from a current source at resonant frequency $\omega = 1/\sqrt{LC}$. Here L is the inductance of the series inductor and C is the capacitance of the series capacitor, respectively. On resonance, the Q of the circuit is large and is given by $Q = \omega L/r$, where r is the resistance of a series resistor. Show that $|V_2| \sim Q|V_1|$ when resistor R is nonzero.

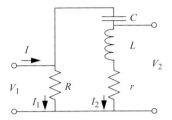

Exercise 1.13

The amplitude of magnetic flux density B in a monochromatic plane-polarized electromagnetic wave traveling in a vacuum is 10^{-6} T. Calculate the value of the total energy density. How is the total energy density divided between the electric and magnetic components?

SOLUTIONS

Solution 1.1

Two people take alternate turns placing beer mats on a table. The beer mats are not allowed to overlap, and the last person to place a mat on the table wins. It should be clear that this is a question about symmetry.

 The first player to place a beer mat on the table is guaranteed to win the game if he or she uses the symmetry of the problem to advantage. Only the first player need be intelligent to guarantee winning. The second player must merely abide by the rules of the game. The first player places the first beer mat in the center of the table and then always places a beer mat symmetrically opposite the position chosen by the other player.

Solution 1.2

Visitors from another planet run a tape measure around the Earth's equator and, having done so, want to know how much extra tape they would need if they raised the tape 1 m above ground level. Since the circumference of a sphere radius R is just $2\pi R$, we conclude that the extra length needed is only $2\pi = 6.28$ m. Because the circumference of a sphere is linear in the radius, the same amount of extra tape is needed *independent* of the Earth's radius.

Solution 1.3

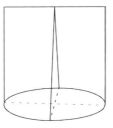

The plug has a circular base of radius $r = 1$ cm, a height of $h = 2$ cm, and a straight top edge of length 2 cm. The smallest convex solid is made if cross-sections perpendicular to the circular base and straight edge are isosceles triangles. If the cork were a cylinder of the same height, cross-sections would be rectangles. Each triangular cross-section is one half the area of the corresponding rectangular cross-section. Since all the triangular cross-sections combine to make up the volume of the plug, the volume must be half that of the cylinder. The cylinder's volume is $\pi r^2 h = 2\pi$ cm^3, so our answer is π cm^3.

The largest convex volume is found by simply slicing the cylinder with two plane cuts to obtain the needed isosceles triangle cross-section. The volume in this case is $(2\pi - 8/3)$ cm^3.

Solution 1.4

A one-mile-long straight steel rod is on flat ground and is attached at each end to a fixed point (one might imagine the rod is part of a rail line). The rod then expands by one foot, causing the rod to describe an arc of a circle between the fixed points. It is our task to find the height above the ground at the center of the rod.

This question involves large numbers (the number of feet in a mile), small numbers (the one-foot expansion), and geometry (the arc of a circle). The best way to proceed is to start by drawing the geometry.

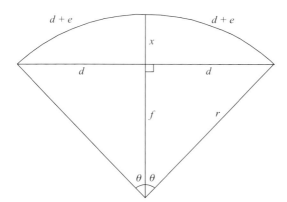

In the above figure the length of the mile-long straight steel rod is $2d$. The length of the rod after expanding by one foot is $2(d + e)$, and the height above the ground at the center of the rod is x. The arc of the circle has radius r. We start by writing down the relationships among the parameters in the figure. This gives four equations, which are

$$r = f + x \tag{1}$$

$$r^2 = d^2 + f^2 \tag{2}$$

$$r\theta = d + e \tag{3}$$

$$\sin(\theta) = \frac{d}{r} \tag{4}$$

Substituting Eqn (1) into Eqn (2) to eliminate f gives

$$r^2 = d^2 + (r - x)^2$$
$$0 = x^2 - 2rx + r^2 - r^2 + d^2$$
$$0 = x^2 - 2rx + d^2 \tag{5}$$

which is quadratic in x. We now need to find the value of θ. To do this, we substitute Eqn (3) into Eqn (4) to eliminate r:

$$\sin(\theta) = \frac{d}{r} = \frac{d}{(d + e)/\theta} = \frac{\theta d}{d + e}$$

$$\theta = \frac{d + e}{d} \sin(\theta)$$

Expanding the $\sin(\theta)$ function gives

$$\theta \sim \frac{d + e}{d}\left(\theta - \frac{\theta^3}{3!} + \frac{\theta^5}{5!} - \cdots\right)$$

Dividing both sides of the equation by θ and retaining terms in θ to second order gives

$$1 \sim \frac{d + e}{d}\left(1 - \frac{\theta^2}{6}\right)$$

so that

$$\theta^2 = 6 - \frac{6d}{d + e} = \frac{6e}{d + e}$$

$$\theta = \sqrt{\frac{6e}{d + e}}$$

The natural unit of length in this problem is one half mile. In this case, we set $d = 1$, so that the expansion in units of one half mile is $e = 1/5280$. Hence,

$$\theta = \sqrt{\frac{6e}{1 + e}} \sim \sqrt{\frac{6}{5280}} = 0.0337 \text{ rad}$$

and from Eqn (3)

$$r = \frac{d + e}{\theta} \sim \frac{1}{0.0337} = 29.7$$

Now we can solve the quadratic Eqn (5), which is in the form

$$0 = ax^2 + bx + c$$

with solution

$$x = \frac{-b \pm \sqrt{b^2 - 4ac}}{2a}$$

In our case, $a = 1$, $b = -2r$ and $c = 1$, so that

$$x = \frac{59.4 \pm \sqrt{59.4^2 - 4}}{2}$$

The roots are

$$x_1 = 59.4$$
$$x_2 = 1/59.4$$

The solution we want is x_2. Converting from units of one half mile to feet gives

$$x = 2640/59.4 = 44.4 \text{ ft}$$

It is quite amazing that a one-foot expansion in an initially straight, mile-long rod results in such a large deflection at the center of the rod. However, our estimate is less than that obtained using the Pythagoras theorem $x^2 + d^2 = (d + e)^2$, for which $x \sim \sqrt{2de} = \sqrt{2640} = 51$ ft.

The reason for the relatively large displacement can be traced back to the nonlinear equations (in this case quadratic) used to describe the physical effect. In this exercise we evaluate the difference in the *square* of two almost identical *large* numbers.

Solution 1.5

We wish to estimate the minimum work done in building the Great Pyramid which has a square base with a perimeter of $l = 921.46$ m and a vertical apex height of $h = 146.73$ m. The square base has side $l/4 = 921.46/4 \sim 230$ m and area $A = 53\,000$ m^2. The density of the stone used is $\rho = 2000$ kg m^{-3}.

Consider distance x measured from the top of the pyramid sketched in the following figure so that the area of a section thickness dx at position x is

$$A(x) = A\left(\frac{x}{h}\right)^2$$

where the area of the base is A. The work done is

$$\int_{x=0}^{x=h} \rho g A \left(\frac{x}{h}\right)^2 (h - x)dx = \left[\frac{\rho g A}{h^2}\left(\frac{hx^3}{3} - \frac{x^4}{4}\right)\right]_{x=0}^{x=h} = \frac{\rho g A h^2}{12}$$

where $g = 9.8$ m s^{-2} is the acceleration due to gravity.

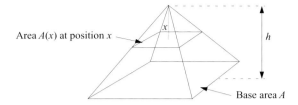

Area $A(x)$ at position x

h

Base area A

Putting in numerical values, area $A = 53\,000$ m^2, height squared $h^2 = 21\,000$ m^2, and

$$\frac{\rho g A h^2}{12} = \frac{2000 \times 9.8 \times 5.3 \times 10^4 \times 2.1 \times 10^4}{12} \sim 1.8 \times 10^{12} \text{ J}$$

A 500-MW electric generator supplies 5×10^8 J s^{-1}, so the time taken to deliver 1.8×10^{12} J is

$$t = \frac{1.8 \times 10^{12}}{5 \times 10^8} = 3.6 \times 10^3 \text{ s}$$

There are 3600 s in an hour, so the time taken is one hour!

As a side note, for those with a general interest in ancient civilizations and their scientific knowledge, if one divides the perimeter of the Great Pyramid by twice the height one gets

$$\frac{l}{2h} = \frac{921.46}{2 \times 146.73} = 3.140$$

which is remarkably close to the value of $\pi = 3.14159$. It seems unlikely that this is mere coincidence.

Solution 1.6

(a) We are going to estimate the power required to keep a helicopter of loaded mass 1000 kg flying. The helicopter blades are 4 m long. We are going to assume that all the air beneath the circle of the blades is moved uniformly downward. For the helicopter to keep flying, the force exerted by air displaced must *at least* counteract the force of gravity on the 1000-kg mass of the helicopter. In classical mechanics, force is the rate of change of momentum $F = dp/dt$, so we have $1000 \times g = \pi r^2 v \rho v$, where $g = 9.8$ m s^{-2} is the acceleration due to gravity, $\rho = 1.3$ kg m^{-3} is the density of air, v is the velocity of the air, and r is the length of the helicopter blades. Hence, we have

$$v^2 \geq \frac{1000g}{\pi r^2 \rho}$$

Power is the rate of change of energy, which is just

$$\frac{1}{2}mv^2 = \frac{1}{2}1000m \cdot \frac{1000g}{\pi r^2 \rho} \sim \frac{1000 \times 1000 \times 9.8}{2 \times \pi \times 16 \times 1.3} \sim 75 \text{ kW}$$

This value of power can be delivered by a small engine capable of delivering about 100 horsepower (1 hp = 745.7 W). Of course, one should remember that this is a minimum requirement and that the engine, gearbox, rotor blades, fuel, controls, airframe, and passengers must have a combined mass of less than 1000 kg.

(b) We use the laws of classical mechanics to estimate the force exerted by a 60-mph wind impinging normal to the side of a suspension bridge with an effective cross-section 500 ft long and 10 ft high. Because the density of air is given in MKS units as $\rho = 1.3$ kg m^{-3}, it makes sense to convert wind speed and the bridge cross-section into MKS units as well.

We use 1 ft$^2 \sim 9.3 \times 10^{-2}$ m^2 and 1 mph ~ 0.45 m s^{-1}. The wind strikes an area $A = 500 \times 10 = 5000$ ft$^2 \sim 465$ m^2. The velocity of the wind is $v = 60$ mph ~ 27 m s^{-1}. The volume of air striking the bridge per second is $V = A \times v = 12555$ m^3 s^{-1}. The mass of air striking the bridge per second is $M = A \times v \times \rho = V \times \rho = 16321$ kg s^{-1}.

The force exerted on the bridge is the rate of change of momentum $F = dp/dt$. We assume that the effective cross-section is calculated in such a way that the final momentum of the air may be taken to be zero. In this case, the force is $F = M \times v = 16321 \times 27 \sim 440\,000$ N, which is about 44 tons of weight. Since $F = A \times \rho \times v^2$, force is proportional to the square of the wind speed. This means that an 85-mph gust of wind would exert twice the force, or about 88 tons of weight.

Solution 1.7

Classical mechanics tells us how the maximum kinetic energy of the particle mass m exhibiting one-dimensional simple harmonic oscillation of frequency ω_0, depends upon the amplitude of oscillation. The maximum velocity of the particle is $v_{max} = \pm A\omega_0$, so the maximum kinetic energy is

$$T_{max} = \frac{1}{2}mv_{max}^2 = \frac{m}{2}A^2\omega_0^2 = A^2\frac{\kappa}{2}$$

It is apparent from this result that the maximum kinetic energy depends upon the oscillation amplitude squared.

Solution 1.8

Consider a particle mass m attached to a lightweight spring that obeys Hooke's law. The displacement from equilibrium $x = x_0$ is proportional to the force on the particle $F = -\kappa(x - x_0)$, where κ is the spring constant. The particle is subject to a small external oscillatory force in the $\hat{x}$ direction such that $F(t) = F_1 \sin(\omega t)$.

(a) Adding $F(t)$ to the right-hand side of the equation of motion for a harmonic oscillator gives

$$\kappa(x - x_0) + m\frac{d^2x}{dt^2} = F_1 e^{-i\omega t}$$

Assuming a solution of the form $x(t) = x_0 + x_1 e^{-i\omega t}$ and substituting into the equation of motion gives

$$\kappa x_1 e^{-i\omega t} - m\omega^2 x_1 e^{-i\omega t} = F_1 e^{-i\omega t}$$

so that

$$x_1(\omega) = \frac{F_1/m}{\omega_0^2 - \omega^2}, \text{ where } \omega_0^2 = \kappa/m$$

As $\omega \to \omega_0$ the system goes into resonance, the oscillation amplitude increases, and $x_1 \to \infty$. To understand this, consider the energy applied to the system by the external force, F. If ω is not equal to ω_0, then within one time period, during portions of the time, energy is imparted to the system (spring and mass) by the force, and for the rest of the time energy is spent by the system to move against the applied force. The applied force and the resulting displacement are not in phase, and the amplitude (energy stored in the system) does not build up. The overall effect is such that energy is not acquired by the system. However, at the resonant frequency, the applied force and the resulting displacement are in phase, and energy is continuously acquired by the system from the external force during the complete cycle. This situation leads to a continuous build up (to infinity) in amplitude. In practice, this does not happen, because all physical systems have a damping term or nonlinearity that limits the oscillation amplitude at resonance (see part (b)).

(b) Adding the damping term $D \cdot (dx/dt)$ to the equation of motion used in (a) gives

$$\kappa(x - x_0) + D\frac{dx}{dt} + m\frac{d^2x}{dt^2} = F(t)$$

Assuming a solution of the form $x(t) = x_0 + x_1 e^{-i\omega t}$ and substituting into the equation of motion gives

$$\kappa x_1 e^{-i\omega t} + -i\omega D x_1 e^{-i\omega t} - \omega^2 m x_1 e^{-i\omega t} = F_1 e^{-i\omega t}$$

so that

$$x_1(\omega) = \frac{F_1/m}{\left(\omega_0^2 - \omega^2\right) - i(\gamma/2)}$$

where $\gamma/2 = \omega D/m$ and $\omega_0^2 = \kappa/m$.

Assuming one may replace $\gamma(\omega)$ with a frequency-independent constant $\Gamma = \gamma(\omega_0)/\omega_0 = 2D/m$, we now plot $|x_1(\omega)|$, phase, $\mathrm{Re}(x_1(\omega))$, and $\mathrm{Im}(x_1(\omega))$, as follows:

$$|x_1(\omega)| = \frac{F_1/m}{\sqrt{\left(\omega_0^2 - \omega^2\right)^2 + (\gamma/2)^2}}$$

$$\mathrm{Re}(x_1(\omega)) = \frac{(F_1/m) \cdot \left(\omega_0^2 - \omega^2\right)}{\left(\omega_0^2 - \omega^2\right)^2 + (\gamma/2)^2}$$

$$\mathrm{Im}(x_1(\omega)) = \frac{(F_1/m) \cdot (\gamma/2)}{\left(\omega_0^2 - \omega^2\right)^2 + (\gamma/2)^2}$$

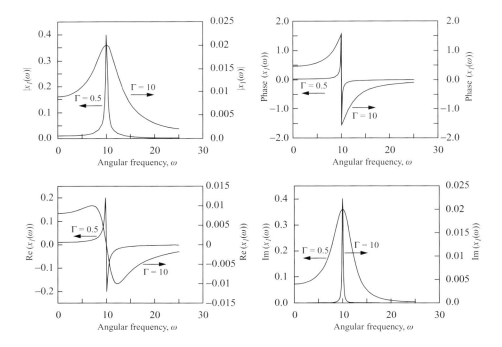

In the figures above we show the results for the two cases: $\Gamma = 0.5$ and $\Gamma = 10$ when $\omega_0 = 10$. The separation in frequencies $(\omega_2 - \omega_1)$ at which the absolute value of x_1 is one half its maximum value of $|x_1(\omega_0)| = 2F_1/m\gamma$ is called the full-width half-maximum (FWHM) or linewidth. If $\Gamma \ll \omega_0$, the line shape is Lorentzian and $1.732 \times \Gamma$ is the value of the FWHM. This is no longer true for an asymmetric line shape that occurs for large values of Γ. An important general feature to notice is that the real part of $x_1(\omega)$ is zero at the resonance frequency ω_0. This resonance is damped by the peak in the imaginary part of $x_1(\omega)$ at ω_0.

If the forcing term $F(t)$ were to suddenly cease, the particle would continue to oscillate until all of the energy was removed by the damping term. In this time-dependent picture, we think about the oscillating state exponentially decreasing in amplitude over a characteristic time, τ. If $\Gamma \ll \omega_0$, the line shape is Lorentzian, and the lifetime, τ, is proportional to the inverse of the FWHM.

An example of a simple harmonic oscillator is a pendulum oscillating at its resonant frequency. This can be used to keep track of time and act as a clock. Another example of a harmonic oscillator is an electrical circuit consisting of a capacitor (spring), an inductor (mass), and a resistor (damping term).

It is worthwhile thinking about what would happen if the damping were negative. For example, in a small-signal analysis, it is possible to have negative differential resistance. If a circuit consists of a capacitor, an inductor, and a device with negative differential resistance, it is possible to create bias conditions under which the circuit will oscillate without any external time-dependent input. The amplitude over which the differential resistance is negative determines the amplitude of the oscillator's output.

Solution 1.9

Given that the relative atomic mass of Ga is 69.72, the relative atomic mass of As is 74.92, and the frequency of the longitudinal polar-optic lattice vibration in GaAs is $\nu = 8.78$ THz in the long-wavelength limit, we will use the solutions for lattice dynamics of a linear chain developed in Section 1.2.4 to estimate the spring constant κ for the system:

$$\kappa = \frac{\omega^2}{2} \frac{m_1 m_2}{m_1 + m_2} = \frac{(2\pi \times 8.78 \times 10^{12})^2}{2} \frac{69.72 \times 74.92 \times m_p}{69.72 + 74.92} = \frac{1.10 \times 10^{29} \times m_p}{2}.$$

We now use this value to estimate $\nu_1 (= \omega_1 / 2\pi)$ and $\nu_2 (= \omega_2 / 2\pi)$ and to compare the results with the experimentally measured values for [100]-oriented GaAs. The results are

$$\nu_1 = \frac{\omega_1}{2\pi} = \frac{1}{2\pi} \left(\frac{2\kappa}{m_1} \right)^{1/2} = \frac{1}{2\pi} \left(\frac{1.10 \times 10^{29} \times m_p}{69.72 \times m_p} \right)^{1/2} = 6.32 \text{ THz}$$

and

$$\nu_2 = \frac{\omega_2}{2\pi} = \frac{1}{2\pi} \left(\frac{2\kappa}{m_2} \right)^{1/2} = \frac{1}{2\pi} \left(\frac{1.10 \times 10^{29} \times m_p}{74.92 \times m_p} \right)^{1/2} = 6.10 \text{ THz}$$

giving a ratio $\nu_1 / \nu_2 = 1.036$. This compares quite well with the experimental values of 7.0 THz and 6.7 THz and the ratio $\nu_1 / \nu_2 = 1.045$.

It is quite surprising how well a simple classical diatomic linear chain model of lattice dynamics compares with the three-dimensional vibration dispersion relation for actual semiconductor crystals such as GaAs.

Solution 1.10

(a) The frequency of vibration of the cantilever beam shown sketched in the following figure with dimensions $l = 100$ μm, $d = 0.1$ μm, density $\rho = 2.328 \times 10^3$ kg m^{-3}, and Young's modulus $E_{Young} = 1.96 \times 10^{11}$ N m^{-2}, is found using the equation

$$\omega = 3.52 \frac{d}{l^2} \sqrt{\frac{E_{Young}}{12\rho}} = \frac{3.52 \times 0.1 \times 10^{-6}}{(100 \times 10^{-6})^2} \sqrt{\frac{1.96 \times 10^{11}}{12 \times 2.328 \times 10^3}} = 9.3 \times 10^4 \text{ rad s}^{-1}$$

The frequency in hertz is $f = \omega / 2\pi = 15$ kHz.

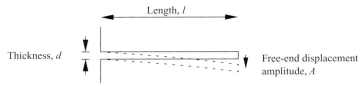

(b) The vibrational energy of the cantilever in (a), with width $w = 5\ \mu$m and free-end displacement amplitude $A = 1\ \mu$m is very small:

$$\frac{wd^3 A^2 E_{\text{Young}}}{6l^3} = \frac{5 \times 10^{-6} \times (0.1 \times 10^{-6})^3 \times (1 \times 10^{-6})^2 \times 1.96 \times 10^{11}}{6 \times (100 \times 10^{-6})^3}$$

$$= 1.63 \times 10^{-16}\ \text{J}$$

Solution 1.11

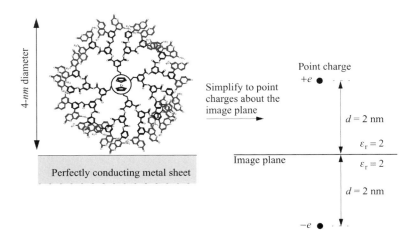

The solution to this exercise is found by considering a point charge $+e$ placed at a distance d from a large perfectly conducting metal sheet. No current can flow in the conductor, so electric field lines from the charge intersect normally to the surface of the conductor. This boundary condition and the symmetry of the problem suggest using the method of images, in which an image charge is placed at a distance d from the position of the sheet and opposite the original charge. The force is then calculated for two oppositely charged point-particles separated by distance $2d$ and embedded in a relative dielectric ε_r. The attractive force is

$$F = \frac{-e^2}{4\pi \varepsilon_0 \varepsilon_r 4d^2}$$

Putting in the numbers for this exercise, we have $\varepsilon_r = 2$ and $d = 2$ nm, giving

$$F = \frac{-(1.602 \times 10^{-19})^2}{4\pi \times 8.854 \times 10^{-12} \times 2 \times 4 \times (2 \times 10^{-9})^2} = -7.2 \times 10^{-12}\ \text{N}$$

where the negative sign indicates an attractive force. A force of a few piconewtons is very small and will have little effect on the relative position of the Fe core atom with respect to the metal sheet.

Solution 1.12

The circuit in the following figure is driven by an input current I from a current source at resonant frequency $\omega = 1/\sqrt{LC}$. We assume that on resonance the Q of the circuit is large and is given by $Q = \omega L/r$, where r is the value of a series resistor.

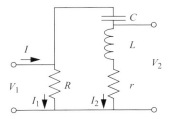

The solution for the current is of the form $I = I_1 e^{i\omega t}$, where $|I_1|$ is the amplitude of the current and $\omega = 1/\sqrt{LC}$ when at resonance. The impedance of the LCr part of the circuit is

$$\frac{-i}{\omega C} + i\omega L + r$$

which on resonance has the value

$$i\omega L - \frac{i}{\omega C} + r = \frac{i(\omega^2 LC - 1) + \omega C r}{\omega C} = r$$

so the equivalent circuit consists of two resistors in parallel with an impedance seen by the input of $rR/(R+r)$. Hence the voltage V_1 is given by

$$V_1 = I\left(\frac{rR}{R+r}\right)$$

The modulus of this voltage is

$$|V_1| = |I_1|\left(\frac{rR}{R+r}\right)$$

The impedance at V_2 is $Z_2 = i\omega L + r$ and the voltage V_2 is given by the current I_2 multiplied by Z_2. The current I_2 flowing through the LCr part of the circuit is given by

$$I_2 = I\left(\frac{R}{R+r}\right)$$

so the voltage V_2 is

$$V_2 = I\left(\frac{R}{R+r}\right)(i\omega L + r)$$

The modulus of this voltage is

$$|V_2| = |I_1| \left(\frac{R}{R+r} \right) \sqrt{\omega^2 L^2 + r^2} = \frac{|V_1|}{r} \sqrt{\omega^2 L^2 + r^2} = |V_1| \sqrt{\frac{\omega^2 L^2}{r^2} + 1}$$

$$|V_2| = |V_1| \sqrt{Q^2 + 1} \sim |V_1| Q$$

This shows that the Q of the circuit can be used to amplify an oscillating voltage.

Solution 1.13

Given that the amplitude of magnetic flux density B in a monochromatic plane-polarized electromagnetic wave traveling in a vacuum is 10^{-6} T, we are asked to calculate the value of the total energy density. We know that energy density is given by

$$U = \frac{1}{2} \varepsilon_0 E^2 + \frac{1}{2\mu_0} B^2$$

and that $E/B = c$ for plane waves in a vacuum. Thus,

$$U = \frac{1}{2} \varepsilon_0 c^2 B^2 + \frac{1}{2\mu_0} B^2$$

and because $c^2 = 1/\varepsilon_0 \mu_0$, we have

$$U = \frac{1}{2\mu_0} B^2 + \frac{1}{2\mu_0} B^2 = \frac{1}{\mu_0} B^2$$

For the magnetic field given,

$$U = \frac{10^{-12}}{4\pi \times 10^{-7}} = 8 \times 10^{-7} \text{ J m}^{-3}$$

The total energy density is divided equally between the electric and magnetic components. The time averaged energy density is half this value.

2 Toward quantum mechanics

2.1 Introduction

It is believed that the basic physical building blocks forming the world we live in may be categorized into particles of matter and carriers of force between matter. All known elementary constituents of matter and transmitters of force are quantized. For example, energy, momentum, and angular momentum take on discrete quantized values. The electron is an example of an elementary particle of matter, and the photon is an example of a transmitter of force. Neutrons, protons, and atoms are composite particles made up of elementary particles of matter and transmitters of force. These composite particles are also quantized. Because classical mechanics is unable to explain quantization, we must learn quantum mechanics in order to understand the microscopic properties of atoms – which, for example, make up solids such as crystalline semiconductors.

Historically, the laws of quantum mechanics have been established by experiment. The most important early experiments involved light. Long before it was realized that light waves are quantized into particles called photons, key experiments on the wave properties of light were performed. For example, it was established that the color of visible light is associated with different wavelengths of light. Table 2.1 shows the range of wavelengths corresponding to different colors.

The connection between optical and electrical phenomena was established by Maxwell in 1864. This extended the concept of light to include the complete electromagnetic spectrum. A great deal of effort was, and continues to be, spent gathering information on the behavior of light. Table 2.2 shows the frequencies and wavelengths corresponding to different regions of the electromagnetic spectrum.

2.1.1 Diffraction and interference of light

Among the important properties of light waves are the abilities of light to exhibit diffraction, linear superposition, and interference. The empirical observation of these phenomena was neatly summarized by the work of Young in 1803. Today, the famous Young's slits experiment can be performed using a single-wavelength, visible laser light

Table 2.1. *Wavelengths of visible light*

Wavelength (nm)	Color
760–622	red
622–597	orange
597–577	yellow
577–492	green
492–455	blue
455–390	violet

Table 2.2. *Spectrum of electromagnetic radiation*

Name	Wavelength (m)	Frequency (Hz)
radio	$> 10^{-1}$	$< 3 \times 10^9$
microwave	$10^{-1} – 10^{-4}$	$3 \times 10^9 – 3 \times 10^{12}$
infrared	$10^{-4} – 7 \times 10^{-7}$	$3 \times 10^{12} – 4.3 \times 10^{14}$
visible	$7 \times 10^{-7} – 4 \times 10^{-7}$	$4.3 \times 10^{14} – 7.5 \times 10^{14}$
ultraviolet	$4 \times 10^{-7} – 10^{-9}$	$7.5 \times 10^{14} – 3 \times 10^{17}$
x-rays	$10^{-9} – 10^{-11}$	$3 \times 10^{17} – 3 \times 10^{19}$
gamma rays	$< 10^{-11}$	$> 3 \times 10^{19}$

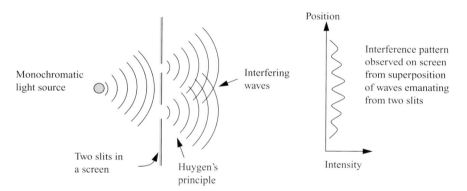

Fig. 2.1. Illustration of the Young's slits experiment. Light from a monochromatic source passing through the slits interferes with itself, and an intensity interference pattern is observed on the viewing screen. The interference pattern is due to the superposition of waves, for which each slit is an effective source. Intensity maxima correspond to electric fields adding coherently (or in phase), and intensity minima correspond to electric fields subtracting coherently.

source, a screen with two slits cut in it, and a viewing screen. Light passing through the slits interferes with itself, and an intensity interference pattern is observed on the viewing screen. See Fig. 2.1. The interference pattern is due to the superposition of waves, for which each slit is an effective coherent source. Hence, the Young's slits interference experiment can be understood using the principle of linear superposition. The wave source at each diffracting slit (Huygen's principle) interferes to create an interference pattern, which can be observed as intensity variations on a screen. Intensity maxima correspond to electric fields adding coherently (or in phase), and intensity minima correspond to electric fields subtracting coherently.

Let's explore this a little more. The linear superposition of two waves at exactly the same frequency can give rise to interference because of a relative phase delay between the waves. For convenience, consider two plane waves labeled $n = 1$ and $n = 2$, with wavelength $\lambda = 2\pi/k$, amplitude E_n, phase ϕ_n, and frequency ω. We do not lose generality by only considering plane waves, because in a linear system we can make any wave from a linear superposition of plane waves. Mathematically, the two waves can be represented as

$$\mathbf{E}_1 = \mathbf{e}_1 |\mathbf{E}_1| e^{i(\mathbf{k}\cdot\mathbf{r}-\omega t)} e^{i\phi_1} \tag{2.1}$$

and

$$\mathbf{E}_2 = \mathbf{e}_2 |\mathbf{E}_2| e^{i(\mathbf{k}\cdot\mathbf{r}-\omega t)} e^{i\phi_2} \tag{2.2}$$

respectively, where $\mathbf{e}_j$ is the unit vector in the direction of the electric field $\mathbf{E}_j$.

The intensity due to the linear superposition of $\mathbf{E}_1$ and $\mathbf{E}_2$ is just

$$\boxed{|\mathbf{E}|^2 = |\mathbf{E}_1 + \mathbf{E}_2|^2 = |\mathbf{E}_1|^2 + |\mathbf{E}_2|^2 + 2|\mathbf{E}_1||\mathbf{E}_2|\cos(\phi)} \tag{2.3}$$

where $\phi = \phi_2 - \phi_1$ is the relative phase between the waves. Our expression for $|\mathbf{E}|^2$ is called the *interference equation*. The linear superposition of the two waves gives a sinusoidal interference pattern in the intensity as a function of phase delay, ϕ.

Plotting the interference equation for the case in which $|\mathbf{E}_1| = |\mathbf{E}_2| = |\mathbf{E}_0|$ in Eqn (2.3), we see in Fig. 2.2 that for $\phi = 0$ there is an intensity maximum

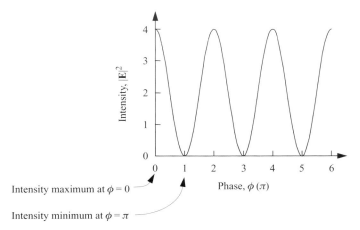

Fig. 2.2. The linear superposition of two waves at exactly the same frequency can give rise to interference if there is a relative phase delay between the waves. The figure illustrates the sinusoidal interference pattern in intensity as a function of phase delay, ϕ, between two equal amplitude waves.

$|\mathbf{E}|^2_{\max} = 4|\mathbf{E}_0|^2$ that is four times the intensity of the individual wave. An intensity minimum $|\mathbf{E}|^2_{\min} = 0$ occurs when $\phi = \pi$. As it stands, this interference pattern is periodic in ϕ and exists over all space. In the more general case, when $|\mathbf{E}_1| \neq |\mathbf{E}_2|$, the interference pattern is still periodic in ϕ, but the intensity maximum has a value $(|\mathbf{E}_1| + |\mathbf{E}_2|)^2$ and the intensity minimum has a value $(|\mathbf{E}_1| - |\mathbf{E}_2|)^2$.

As was just noted, our mathematical model predicts that the interference exists over all space. This is a bit problematic because it seems to be at variance with our everyday experience. Usually we do not see large variations in light intensity due to interference. The reason is that the frequencies of the light waves are not exactly ω (i.e., are not precisely monochromatic). There is a continuous range or spectrum of frequencies about some average value of ω. Because the light is not exactly monochromatic, even if the spectrum is sharply peaked at some value of frequency, there is a linewidth associated with the spectral line typically centered at frequency ω_0. The underlying reason for a finite spectral linewidth is found by considering the temporal behavior of the light wave. By taking the Fourier transform of the continuous spectral line, we can obtain the temporal behavior of the wave.

As an example, suppose we have a laser with light emission at 1500 nm wavelength. The electromagnetic field oscillates at $f = 200\,\text{THz}$ or $\omega_0 = 2\pi f = 1.26 \times 10^{15}$ rad s^{-1}. If the laser is designed to put out a very fast optical pulse at time $t = t_0$, the frequency components in the electromagnetic field will reflect this fact. To be specific, let's assume that the pulse has a Gaussian shape, so that the electromagnetic field can be written as

$$\mathbf{E}_j(t) = \mathbf{e}_j \cos(\omega_0 t) \cdot e^{-(t-t_0)^2/\tau_0^2} \tag{2.4}$$

where τ_0 is proportional to the temporal width of the pulse. The Fourier transform is a Gaussian envelope centered at the frequency ω_0:

$$\mathbf{E}_j(\omega) = \mathbf{e}_j \frac{\tau_0}{\sqrt{2}} \cdot e^{-(\omega-\omega_0)^2 \tau_0^2/4} \tag{2.5}$$

with a spectral power density

$$\mathbf{E}^*(\omega)\mathbf{E}(\omega) = \frac{\tau_0^2}{2} \cdot e^{-(\omega-\omega_0)^2 \tau_0^2/2} \tag{2.6}$$

We can now figure out how the value of τ_0 is related to the width of the spectral power density. It is easy to show that the spectral power density full-width at half-maximum (FWHM) linewidth is related to τ_0 via

$$\Delta\omega_{\text{FWHM}} = \frac{2}{\tau_0} \cdot \sqrt{2\ln(2)} \tag{2.7}$$

In our example, the laser just happens to be designed to put out a very fast optical pulse of Gaussain shape with $\tau_0 = 14.14$ fs. The 200-THz optical field is modulated by a Gaussian envelope to give the electric field as a function of time shown in Fig. 2.3(a).

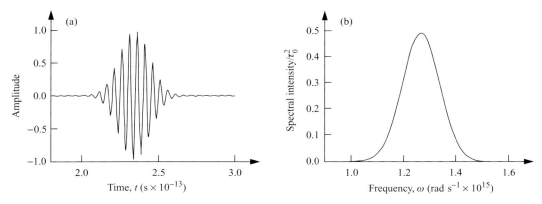

Fig. 2.3. (a) Illustration of a 200-THz electric field modulated by a Gaussian envelope function with $\tau_0 = 14.14$ fs. (b) Spectral line shape centered at $\omega_0 = 1.26 \times 10^{15}$ rad s^{-1} corresponding to the 200-THz oscillating electric field in (a).

The spectral linewidth centered about the frequency $\omega_0 = 2\pi f = 1.26 \times 10^{15}$ rad s^{-1} is determined by the temporal modulation of the pulse. For a Gaussian pulse shape, the FWHM of the time and frequency functions are as follows:

time domain field:

$$\mathbf{E}_j(t) = \mathbf{e}_j \cos(\omega_0 t) \cdot e^{-(t-t_0)^2/\tau_0^2} \quad \Delta t_{\text{FWHM}} = 2\tau_0 \cdot \sqrt{\ln(2)} \tag{2.8}$$

frequency domain field:

$$\mathbf{E}_j(\omega) = \mathbf{e}_j \frac{\tau_0}{\sqrt{2}} \cdot e^{-(\omega-\omega_0)^2\tau_0^2/4} \quad \Delta\omega_{\text{FWHM}} = \frac{4}{\tau_0} \cdot \sqrt{\ln(2)} \tag{2.9}$$

frequency domain intensity:

$$\mathbf{E}^*(\omega)\mathbf{E}(\omega) = \frac{\tau_0^2}{2} \cdot e^{-(\omega-\omega_0)^2\tau_0^2/2} \quad \Delta\omega_{\text{FWHM}} = \frac{2}{\tau_0} \cdot \sqrt{2\ln(2)} \tag{2.10}$$

All oscillators have a finite spectral intensity linewidth, because they must contain transient components from when they were originally turned on. We use the Fourier transform to relate the time domain behavior to the frequency domain behavior.

The fact that the oscillator can only oscillate for a finite time means that the frequency spectrum always has a finite width. This has a direct impact on the observation of interference effects.

At a minimum, one would expect that interference effects can only be observed when the wave and the delayed wave overlap in space. In our example, $\tau_0 = 14.14$ fs, and our pulse has $\Delta t_{\text{FWHM}} = 2\tau_0 \cdot \sqrt{\ln(2)} = 23.5$ fs. We expect that interference between the pulse and the delayed pulse will only be observed for delays approximately equal to or less than Δt_{FWHM}. For a wave moving at the speed of light, this gives a characteristic length $l = \Delta t_{\text{FWHM}} \times c$, which in our case is 7 μm.

Table 2.3. *Relationship between spectral linewidth and coherence time*

Spectral intensity line shape	Spectral width $\Delta\omega_{\text{FWHM}}$
Gaussian	$2(2\pi \ln(2))^{1/2}/\tau_{\text{c}}$
Lorentzian	$2/\tau_{\text{c}}$
rectangular	$2\pi/\tau_{\text{c}}$

In general, the idea of a characteristic time or length over which interference effects can be observed may be formalized using a correlation function. The normalized autocorrelation function is defined as

$$g(\tau) = \frac{\langle f^*(t) f(t+\tau) \rangle}{\langle f^*(t) f(t) \rangle} \tag{2.11}$$

where $f(t)$ is a complex function of time (in our case it is a wave). The value of $|g(\tau)|$ is a measure of the correlation between $f(t)$ and $f(t+\tau)$, where τ is a time delay. For classical monochromatic light, $f(t)$ is of the form $e^{-i\omega t}$, which gives $g(\tau) = e^{-i\omega t}$, so that $|g(\tau)| = 1$. The coherence time is defined as

$$\tau_{\text{c}} = \int\limits_{\tau=-\infty}^{\tau=\infty} |g(\tau)|^2 d\tau \tag{2.12}$$

So if $|g(\tau)| = 1$, the coherence time τ_{c} is infinite and the corresponding coherence length, which is defined as $l_{\text{c}} = \tau_{\text{c}} \times c$, is also infinite. In practice, because the wave source is not purely monochromatic, there is a *coherence length* associated with the nonmonochromaticity. The coherence length gives the spatial scale over which interference from the linear superposition of fields can be observed. For lengths much greater than the coherence length, the phases of different wavelength components can no longer add to create either a maximum or minimum, and all interference effects are effectively washed out.

The relationship between the spectral intensity linewidth $\Delta\omega_{\text{FWHM}}$ and the coherence time τ_{c} for the indicated line shapes is given in Table 2.3.

In the previous few pages we have described some of the properties of light. We were particularly interested in contributions that allow observation of interference effects, and to this end the concept of coherence time was discussed. The reason for our interest is that observation of interference effects between waves is a key attribute of "waviness". We will make use of this when we try to link the behavior of particles and waves using a single unified approach given by quantum mechanics.

So far we have discussed light as a wave. However, there is also experimental evidence that light can behave as a particle. Historically, experimental evidence for this came in two stages. First, measurement of the emission spectrum of thermal light suggested that light is quantized in energy. Second, the photoelectric effect showed that light behaves as a particle that can eject an electron from a metal. In the next few pages we discuss this evidence and its interpretation in terms of quantization of light.

2.1.2 Black-body radiation and evidence for quantization of light

Experimental evidence for the quantization of light into particles called photons initially came from measurement of the emission spectrum of thermal light (called black-body radiation). It was the absolute failure of classical physics to describe the emission spectrum that led to a new interpretation involving the quantization of light.

Application of classical statistical thermodynamics and electromagnetics gives the Rayleigh–Jeans formula (1900) for electromagnetic field radiative energy density emitted from a black body at absolute temperature T as

$$S(\omega) = \frac{k_B T}{\pi^2 c^3} \omega^2 \tag{2.13}$$

Radiative energy density is the energy per unit volume per unit angular frequency, and it is measured in $J\,s\,m^{-3}$. Equation (2.13) predicts a physically impossible infinite radiative energy density as $\omega \to \infty$. This divergence in radiative energy density with decreasing wavelength is called the classical ultraviolet catastrophe. The impossibility of infinite energy as $\omega \to \infty$ was indeed a disaster (or catastrophe) for classical physics. The only way out was to invent a new type of physics, quantum physics.

The black-body radiation spectrum was explained by Planck in 1900. He implicitly (and later explicitly) assumed emission and absorption of discrete energy quanta of electromagnetic radiation, so that $E = \hbar\omega$, where ω is the frequency of the electromagnetic wave and $\hbar$ is a constant. This gives a radiative energy density measured in units of $J\,s\,m^{-3}$:

$$S(\omega) = \frac{\hbar\omega^3}{\pi^2 c^3} \frac{1}{e^{\hbar\omega/k_B T} - 1} \tag{2.14}$$

Equation (2.14) solves the problem posed by the ultraviolet catastrophe and agrees with the classical Rayleigh–Jeans result in the limit of long wavelength electromagnetic radiation ($\omega \to 0$). In both approaches, one assumes thermal *equilibrium* between the radiation and the material bodies (made of atoms). Hence, the radiation has a radiative energy density distribution that is characteristic of *thermal light*.

Figure 2.4 shows the energy density of radiation emitted by unit surface area into a fixed direction from a black body as a function of frequency for three different temperatures, $T = 4800\,K$, $T = 5800\,K$, and $T = 6800\,K$. The Sun has a surface temperature of about $T = 5800\,K$, and a peak in radiative energy density at visible frequency. The figure also shows the results of plotting the classical Rayleigh–Jeans formula to show the ultraviolet catastrophe at high frequencies.

Additional experimental evidence for the quantization of electromagnetic energy in such a way that $E = \hbar\omega$ comes from the photoelectric effect. The photoelectric effect

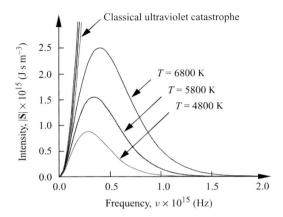

Fig. 2.4. Radiative energy density of black-body radiation emitted by unit surface area into a fixed direction from a black body as a function of frequency ($\nu = \omega/2\pi$) for three different absolute temperatures, $T = 4800\,\mathrm{K}$, $T = 5800\,\mathrm{K}$, and $T = 6800\,\mathrm{K}$. The predictions of the classical Rayleigh–Jeans formula are also plotted to show the ultraviolet catastrophe.

also suggests that light can behave as a particle and have particle–particle collisions with electrons. Because of its importance, we consider this next.

2.1.3 Photoelectric effect and the photon particle

When light of frequency ω is incident on a metal, electrons can be emitted from the metal surface if $\hbar\omega > e\phi$, where ϕ is the work function of the metal. $+e\phi$ is the minimum energy for an electron to escape the metal into vacuum. In addition, such photoelectric-effect experiments show that the *number* of electrons leaving the surface depends upon the light intensity.

This evidence suggests that light can behave as a particle. As illustrated in Fig. 2.5(a), when a particle of light is incident on a metal, the collision can cause an electron particle to be ejected from the surface. The maximum excess kinetic energy of the electron leaving the surface is observed in experiments to be $T_{\mathrm{max}} = \hbar\omega - e\phi$, where $\hbar$ is the

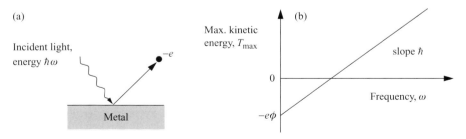

Fig. 2.5. (a) Light of energy $\hbar\omega$ can cause electrons to be emitted from the surface of a metal. (b) The maximum kinetic energy of emitted electrons is proportional to the frequency of light, ω. The proportionality constant is $\hbar$.

slope of the curve in Fig. 2.5(b). The maximum kinetic energy of any ejected electron depends only upon the frequency, ω, of the light particle with which it collided, and this energy is *independent* of light intensity. This is different from the classical case, which predicts that energy is proportional to light intensity (amplitude squared of the electromagnetic wave). In the particle picture, each light particle has energy $\hbar\omega$, and light intensity is given by the particle flux.

In 1905 Einstein explained the photoelectric effect by postulating that light behaves as a particle and that (in agreement with Planck's work) it is quantized in energy, so that

$$E = \hbar\omega \tag{2.15}$$

where $\hbar = 1.05492 \times 10^{-34}$ J s is *Planck's constant*. It is important to notice that the key result, $E = \hbar\omega$, comes directly from experiment. Also notice that for $\hbar\omega$ to have the dimensions of energy, $\hbar$ must have dimensions of J s. A quantity of this type is called an action. The units J s can also be expressed as kg m^2 s^{-1}.

The quantum of light is called a *photon*. A photon has zero mass and is an example of an elemental quantity in quantum mechanics. In quantum mechanics, one talks of light being quantized into particles called photons.

From classical electrodynamics we know that electromagnetic *plane waves carry momentum* of magnitude $p = U/c$, where U is the electromagnetic energy density. This means that if a photon has energy E then its momentum is $p = E/c$. Because experiment shows that light is quantized in energy so that $E = \hbar\omega$, it seems natural that momentum should also be quantized. Following this line of thought, we may write $p = \hbar\omega/c$ or, since $\omega = c2\pi/\lambda = ck$, photon momentum can be written as

$$\mathbf{p} = \hbar\mathbf{k} \tag{2.16}$$

This is another key result that will be important as we continue to develop our understanding of quantum mechanics. We could have guessed this result from dimensional analysis. Planck's constant $\hbar$ is measured in units of kg m^2 s^{-1}, which, if we divide by a length, gives units of momentum. Since, for a plane wave of wavelength λ, there is only one natural inverse length scale $k = 2\pi/\lambda$, it is reasonable to suggest that momentum is just $\mathbf{p} = \hbar\mathbf{k}$.

To get a feel for what this quantization of energy and momentum means in practice, let's put in some numbers. Using the relationship $E = \hbar\omega$, it follows that, if we know the energy of a photon measured in electron volts, then the photon wavelength λ in free space measured in nanometers is given by the expression

$$\lambda_{\text{photon}}(\text{nm}) = \frac{1240}{E(\text{eV})} \tag{2.17}$$

The energy of $\lambda = 1000$ nm wavelength light is $E = 1.24$ eV. Compared with room temperature thermal energy $k_B T = 25$ meV, the quantized energy of $\lambda = 1000$ nm wavelength light is large. Using the relationship $p = \hbar k$, the momentum of $\lambda = 1000$ nm wavelength light is just $p = 6.63 \times 10^{-28}$ kg m s^{-1}.

Before photons were introduced, we were very successful in describing the properties of electromagnetic phenomena using Maxwell's equations. It seems appropriate to ask under what circumstances one can still use the classical description rather than the quantum description. From our experience with classical mechanics, we can anticipate the answer. If there are a large number of incoherent photons associated with a particular electromagnetic field, we can expect the classical description to give accurate results. If there are very few photons or there are special conditions involving a coherent superposition of photons, then a quantum description will be more appropriate.

By way of a practical example, let's see what this means for wireless communication. To be specific, consider electromagnetic radiation emitted from the antenna of a cellular telephone. This particular cell phone operates at a center frequency $f_0 = 1$ GHz (wavelength $\lambda_0 = 0.3$ m), and the radiated power is $P_0 = 300$ mW. Assuming isotropic radiation, one can calculate the electric field, photon flux, and number of photons per cubic wavelength at a distance R from the antenna. To start with, we will consider the case $R = 1$ km. In the absence of absorption, at distance R the power per unit area is decreased by a factor $P_R = P_0/4\pi R^2$, which, when $R = 1$ km, gives $P_R = 0.3/(4\pi \times 10^6) = 24$ nW m^{-2}. The electric field is found from $P_R = S = |E_R|^2/Z_0$, so $E_R = (P_R Z_0)^{1/2} = (P_0 Z_0/R^2)^{1/2}$. For the example, we are interested in $R = 1$ km, and this gives an electric field $E_R = (24 \times 10^{-9} \times 377)^{1/2} = 3 \times 10^{-3}$ V m^{-1}. Since the photon energy $\hbar\omega = \hbar 2\pi f_0$, we have $\hbar\omega = 6.6 \times 10^{-25}$ J for each photon at frequency $f_0 = 1$ GHz. The number of photons per second per unit area is $n_{\text{photon}}^{\text{area}} = P_R/\hbar\omega$. In our case, this gives $n_{\text{photon}}^{\text{area}} = 3.6 \times 10^{16}$ m^{-2} s^{-1} or, in terms of the number of photons per cubic wavelength, $n_{\text{photon}}^{\lambda_0^3} = n_{\text{photon}}^{\text{area}} \lambda_0^3/c = 3.3 \times 10^6$, which is a large number. In fact, the number of photons is so large that effects due to the discrete nature of quantized photons are, for all practical purposes, washed out.

At another extreme, one may figure out the distance at which there is approximately only one photon per cubic wavelength. Under these conditions, one expects the quantized nature of the photon to play an important role in determining how information can be transmitted over the wireless channel. Using the expression $n_{\text{photon}}^{\lambda_0^3} = n_{\text{photon}}^{\text{area}} \lambda_0^2/f_0$, which for the case of one photon per cubic wavelength $n_{\text{photon}}^{\lambda_0^3} = 1$ gives a distance of about $R_1 = (\lambda_0^2 P_0/4\pi f_0 \hbar\omega)^{1/2}$, the result turns out to be $R_1 = 1.8 \times 10^6$ m $= 1800$ km for our example.[1] The power at this distance is a very small number, $P_{R_1} = c\hbar\omega/\lambda_0^3 = 7.4 \times 10^{-15}$ W m^{-2}.

It appears to be safe to state that any conventional cell phone will not need a quantum mechanical description of the electromagnetic fields. Further, one can assume that Maxwell's classical equations will be of use for many practical situations that require calculation of electromagnetic radiation fields.

---

[1] For comparison, the equatorial radius of the Earth is 6400 km.

We are left asking what the essential aspects of the quantum theory are and whether quantum theory can be applied to other situations. Obviously, quantization of energy and momentum, linear superposition, and interference are important. Starting with the last two, we may question why photon particles are relatively easy to prepare so that they show the key quantum effects of linear superposition and interference. The reason why photons often appear as waves is that they do not scatter strongly among themselves. This fact makes it rather simple to create photons that have a well-defined wavelength. Photons with a well-defined wavelength have a long coherence time and a long coherence length.

2.1.4   The link between quantization of photons and other particles

If photons are particles with energy $E = \hbar\omega$, wavelength λ, and quantized momentum $\mathbf{p} = \hbar\mathbf{k}_{\mathrm{photon}}$, then there may be other particles that are also characterized by energy, wavelength, and momentum. The essential link between quantization of photons and quantization of other particles such as electrons is momentum. In general, *interaction between particles involves exchange of momentum*. We already know that both the photon and the electron have momentum and that they can interact with each other. Such interaction must involve exchange of momentum. We now make the observation that if the photon momentum is quantized it is natural to assume that electron momentum is quantized. The uncomfortable alternative is to have two types of momentum, quantized momentum for photons and classical momentum for electrons!

Interaction between a photon and an electron causes momentum exchange. In a photoelectric-effect experiment a photon with quantized momentum $\hbar\mathbf{k}_{\mathrm{photon}}$ and energy $E = \hbar\omega$ collides with an electron in a metal. The photon energy is absorbed, and the electron is ejected from the metal. The collision process may be described using a diagram similar to Fig. 2.6.

Experimental verification that electron momentum is quantized in a way similar to that of photons would require showing that an electron is characterized by a wavelength. For example, measurement of an electron on the right time or length scale might exhibit wave-related interference effects similar to those seen for photons. Because

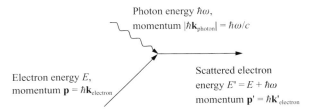

Photon energy $\hbar\omega$,
momentum $|\hbar\mathbf{k}_{\mathrm{photon}}| = \hbar\omega/c$

Electron energy E,
momentum $\mathbf{p} = \hbar\mathbf{k}_{\mathrm{electron}}$

Scattered electron
energy $E' = E + \hbar\omega$
momentum $\mathbf{p}' = \hbar\mathbf{k}'_{\mathrm{electron}}$

Fig. 2.6. The momentum and energy exchange between a photon and an electron may be described in a scattering diagram in which time flows from left to right. The electron has initial wave vector $\mathbf{k}_{\mathrm{electron}}$ and scattered wave vector $\mathbf{k}'_{\mathrm{electron}}$. The quantized momentum carried by the photon is $\mathbf{k}_{\mathrm{photon}}$.

electron kinetic energy is related to momentum and quantized momentum is related to wavelength, we can estimate the wavelength λ_e an electron with mass m_0 and energy E in free space would have. The answer is $\lambda_e = 2\pi\hbar/\sqrt{2m_0E}$, which for an electron with $E = 1\,\text{eV}$ gives a quite small wavelength $\lambda_e = 1.226\,\text{nm}$. In addition, unlike photons, electrons can interact quite strongly with themselves via the coulomb potential. If one wished to perform an interference experiment with electrons, one would need to prepare a beam of electrons with well-defined energy and then scatter the electrons with a coulomb potential analogous to the slits used by Young in his experiments with light. One may conclude that establishing the wave nature of electrons requires some careful experiments.

2.1.5 Diffraction and interference of electrons

In 1927 Davisson and Germer reported that an almost monoenergetic beam of electrons of kinetic energy $E = p^2/2m_0$ incident on a crystal of nickel gave rise to Bragg scattering peaks for electrons emerging from the metal. As illustrated in Fig. 2.7 the periodic array of atoms that forms a nickel crystal of lattice constant $L = 0.352\,\text{nm}$ creates a periodic coulomb potential from which the electrons scatter in a manner similar to light scattering from Young's slits. The observation of intensity maxima for electrons emerging from the crystal showed that electrons behave as waves. The electron waves exhibited the key features of diffraction, linear superposition, and interference. The experiment of Davisson and Germer supported the 1924 ideas of de Broglie for electron "waves" in atoms. An electron of momentum $\mathbf{p} = \hbar\mathbf{k}$ (where $|\mathbf{k}| = 2\pi/\lambda$) has wavelength

$$\lambda_e = 2\pi\hbar\frac{1}{p} = 2\pi\hbar\frac{1}{\sqrt{2m_0E}} \tag{2.18}$$

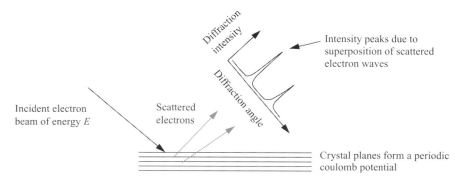

Fig. 2.7. A monoenergetic beam of electrons scattered from a metal crystal showing intensity maxima. The periodic array of atoms that forms the metal crystal creates a periodic coulomb potential from which electrons scatter. The observation of intensity maxima for electrons emerging from the crystal is evidence that electrons behave as waves.

Electrons of kinetic energy $E = p^2/2m_0$ behave as waves in such a way that

$$\psi(\mathbf{r}, t) \sim e^{-i(Et/h - \mathbf{k}\cdot\mathbf{r})} \tag{2.19}$$

where $\mathbf{p} = \hbar\mathbf{k}$.

2.1.6 When is a particle a wave?

From our discussion of the photoelectric effect in Section 2.1.3 it is clear that the electron and photon sometimes appear to behave as particles and sometimes appear to behave as waves. Other, atomic-scale entities such as neutrons and protons can also appear to behave either as particles or waves. There appears to be a *complementarity* in the way one treats their behavior. Neutrons, protons, and electrons can seem like particles, with a mass and momentum. However, if one looks on an appropriate length or time scale, they might exhibit the key characteristics of waves, such as superposition and interference. Deciding which is an appropriate description depends upon the details of the experiment. A quick analysis using waves usually, but not always, reveals the best approach. Isolated systems with little scattering can often exhibit wave-like behavior. This is because the states of the system are long-lived. It is easier to measure the consequences of wave-like behavior such as superposition and interference when states are long-lived. The reason for this is simply that coherent wave effects that last a relatively long time can be less spectrally broadened and hence easier to interpret. In addition, long-lived states give the experimenter more time to perform the measurement.

Obviously, when we are considering this apparent wave-particle duality, an important scale is set by the particle wavelength and its corresponding energy.

Photon energy is quantized as $E = \hbar\omega$, photon mass is zero, photon momentum is $\mathbf{p} = \hbar\mathbf{k}$, and photon wavelength is $\lambda = 2\pi/k$. The dispersion relationship for the photon moving at the speed of light in free space is $E = \hbar c k$ or, more simply, $\omega = ck$.

Electron momentum is quantized as $\mathbf{p} = \hbar\mathbf{k}$, electron mass is $m_0 = 9.109565 \times 10^{-31}$ kg, and electron energy is $E = p^2/2m_0$. The dispersion relationship for an isolated electron moving in free space is $E = \hbar^2 k^2/2m_0$. If we know the energy E of the electron measured in electron volts, then the electron wavelength λ_e in free space measured in units of nanometers is given by the expression

$$\lambda_e(\text{nm}) = \frac{1.226}{\sqrt{E(\text{eV})}} \tag{2.20}$$

This means that an electron with a kinetic energy of $E = 100\,\text{eV}$ would have a wavelength of $\lambda_e = 0.1226\,\text{nm}$. To measure this wavelength one needs to scatter electrons from a structure with a coulomb potential that has a similar characteristic length scale. A good example is a crystalline metal such as the nickel used in the Davisson and Germer experiment.

Similarly, other finite-mass particles, such as the neutron, have a wavelength that is inversely related to the square root of the particle's kinetic energy. For the neutron, we have a free space wavelength

$$\lambda_n(\text{nm}) = \frac{0.0286}{\sqrt{E(\text{eV})}} \qquad (2.21)$$

Clearly, a neutron of kinetic energy $E = 100\,\text{eV}$ has a very small wavelength $\lambda_n = 0.00286\,\text{nm}$, which is quite difficult to observe in an experiment.

Now that we know that experiments have shown that atomic-scale particles can exhibit wave-like behavior, it is time to investigate some of the consequences. A good starting point is to consider the properties of atoms since atoms make up solids. However, before we can hope to make any significant progress in that direction, we need to develop a theory powerful enough to provide the answers to our questions about atoms. Historically, Heisenberg introduced such a theory first. While this work, which was published in 1925, is insightful and interesting, it does not follow our approach in this book, and so we will not discuss it here. In 1926 Schrödinger introduced the theory that we will use.

2.2 The Schrödinger wave equation

There is a need to generalize what we have learned thus far about the wave properties of atomic-scale particles. On the one hand, the formalism needs to incorporate the wave nature observed in experiments; on the other hand, the approach should, in the appropriate limit, incorporate the results of classical physics.

We haven't yet learned how to derive an equation to describe the dynamics of particles with wavy character, so we will start by making some guesses. Based on our previous experience with classical mechanics and classical electrodynamics, we will assume that time, t, is a continuous, smooth parameter and that position, $\mathbf{r}$, is a continuous, smooth variable. To describe the dynamics of wavy particles, it seems reasonable to assume that we will wish to find quantities such as particle position $\mathbf{r}$ and momentum $\mathbf{p}$ as a function of time.

We know that waviness is associated with the particle, so let's introduce a wave function ψ that carries the appropriate information. Young's slits experiments suggest that such a wave function, which depends upon position and time, can be formed from a linear superposition of plane waves. Under these conditions, it seems reasonable to consider the special case of plane waves without loss of generality. So, now we have

$$\psi(\mathbf{r}, t) = A e^{i(\mathbf{k} \cdot \mathbf{r} - \omega t)} \qquad (2.22)$$

The fact that the wave function $\psi(\mathbf{r}, t)$ is a complex quantity presents a bit of a problem. The wave function cannot be treated like an ordinary vector in classical electrodynamics, where one only uses complex quantities as a mathematical convenience and measured quantities are the real component. As will become apparent, in quantum mechanics the wave function is a true complex quantity, and hence it cannot be measured directly. We cannot use $\psi(\mathbf{r}, t)$ to represent the particle directly, because it is a complex number, and this is at variance with our everyday experience that quantities such as particle position are real. The easiest way to guarantee a real value is to measure its intensity, $\psi(\mathbf{r}, t)^*\psi(\mathbf{r}, t) = |\psi(\mathbf{r}, t)|^2$. As with the Young's slits experiment, the intensity gives a measure of the probability of the entity's presence in different regions of space. The probability of finding the particle at position $\mathbf{r}$ in space at time t is proportional to $|\psi(\mathbf{r}, t)|^2$. Recognizing that the particle is definitely in some part of space, we can *normalize* the intensity $|\psi(\mathbf{r}, t)|^2$ so that integration over all space is unity. This defines a *probability density* for finding the particle at position $\mathbf{r}$ in space at time t. If we wish to find the most likely position of our wavy particle in space, we need to weight the probability distribution with position $\mathbf{r}$ to obtain the average position $\langle \mathbf{r} \rangle$. The way to do this is to perform an integral over all space so that

$$\langle \mathbf{r} \rangle = \int_{-\infty}^{\infty} \psi^*(\mathbf{r}, t)\mathbf{r}\psi(\mathbf{r}, t)d^3\mathbf{r} = \int_{-\infty}^{\infty} \mathbf{r}|\psi(\mathbf{r}, t)|^2d^3r \tag{2.23}$$

In quantum mechanics, the average value of position is $\langle \mathbf{r} \rangle$ and is called the *expectation value* of the position *operator*, $\mathbf{r}$.

To understand and track particle dynamics, we will need to know other quantities such as the particle momentum, $\mathbf{p}$. In quantum mechanics, particle momentum is quantized in such a way that $\mathbf{p} = \hbar\mathbf{k}$, so to find the average value of momentum $\langle \mathbf{p} \rangle$ we need to perform an integral over all space, so that

$$\langle \mathbf{p} \rangle = \int_{-\infty}^{\infty} \psi^*(\mathbf{k}, t)\hbar\mathbf{k}\,\psi(\mathbf{k}, t)d^3k \tag{2.24}$$

Of course, Eqn (2.24) requires that we find the function $\psi(\mathbf{k}, t)$. This can be done by taking the Fourier transform of $\psi(\mathbf{r}, t)$. By adopting this approach, we will be changing our description in terms of wave functions $\psi(\mathbf{k}, t)$ in k space to wave functions $\psi(\mathbf{r}, t)$ in real space. This change in space is accompanied by a change in the form of the momentum operator. To show how to find the momentum operator in real space, let's restrict ourselves to motion in the x direction only. Under these circumstances,

$$\langle \hat{p}_x \rangle = \int_{-\infty}^{\infty} \psi^*(k_x)\hbar k_x\,\psi(k_x)dk_x \tag{2.25}$$

The momentum operator $\hat{p}_x$ has a ˆ to remind us that it is a quantum operator. Notice that for convenience we ignore the time dependence $e^{-i\omega t}$ of the wave function when evaluating $\psi^*\psi$, since the time-dependent terms cancel. Taking the Fourier transform of $\psi(k_x)$ to obtain $\psi(x)$ gives

$$\langle \hat{p}_x \rangle = \frac{1}{2\pi} \int\limits_{-\infty}^{\infty} dk_x \left(\int\limits_{-\infty}^{\infty} dx' \psi^*(x') e^{ik_x x'} \right) \hbar k_x \left(\int\limits_{-\infty}^{\infty} dx \psi(x) e^{-ik_x x} \right) \tag{2.26}$$

Integrating the far right-hand term in the brackets by parts using $\int UV' dx = UV - \int U'V dx$ with $U = \psi(x)$ and $V' = e^{-ik_x x}$ gives

$$\int\limits_{-\infty}^{\infty} dx \psi(x) e^{-ik_x x} = \left[\frac{1}{-ik_x} e^{-ik_x x} \psi(x) \right]_{-\infty}^{\infty} + \int\limits_{-\infty}^{\infty} dx \frac{1}{ik_x} \cdot \frac{\partial}{\partial x} \psi(x) e^{-ik_x x} \tag{2.27}$$

The oscillatory function in the square brackets is zero in the limit $x \to \pm\infty$, so our expression for $\langle \hat{p}_x \rangle$ becomes

$$\langle \hat{p}_x \rangle = \frac{\hbar}{2i\pi} \int\limits_{-\infty}^{\infty} dk_x \int\limits_{-\infty}^{\infty} dx' \psi^*(x') e^{ik_x x'} \int\limits_{-\infty}^{\infty} dx e^{-ik_x x} \frac{\partial}{\partial x} \psi(x) \tag{2.28}$$

which we may rewrite as

$$\langle \hat{p}_x \rangle = -i\hbar \int\limits_{-\infty}^{\infty} dx' \int\limits_{-\infty}^{\infty} dx \psi^*(x') \cdot \frac{1}{2\pi} \int\limits_{-\infty}^{\infty} dk_x e^{-ik_x(x-x')} \cdot \frac{\partial}{\partial x} \psi(x) \tag{2.29}$$

Recognizing that

$$\frac{1}{2\pi} \int\limits_{\infty}^{-\infty} dk_x e^{-ik_x(x-x')} = \delta(x - x')$$

allows one to write

$$\langle \hat{p}_x \rangle = -i\hbar \int\limits_{-\infty}^{\infty} dx' \int\limits_{-\infty}^{\infty} dx \psi^*(x') \delta(x - x') \frac{\partial}{\partial x} \psi(x) \tag{2.30}$$

so that finally we have

$$\boxed{\langle \hat{p}_x \rangle = -i\hbar \int\limits_{-\infty}^{\infty} dx \psi^*(x) \frac{\partial}{\partial x} \psi(x)} \tag{2.31}$$

The important conclusion is that if $\hat{p}_x = \hbar k_x$ in k space (momentum space) then the momentum operator in real space is $-i\hbar \cdot \partial/\partial x$. The momentum operator in real space is a *spatial* derivative. The momentum operator and the position operator form a *conjugate pair* linked by a Fourier transform. In this sense there is a full symmetry between the position and momentum operator.

Table 2.4. *Classical variables and quantum operators for* $\psi(\mathbf{r}, t)$

Description	Classical theory	Quantum theory
Position	$\mathbf{r}$	$\mathbf{r}$
Potential	$V(\mathbf{r}, t)$	$V(\mathbf{r}, t)$
Momentum	p_x	$-i\hbar\dfrac{\partial}{\partial x}$
Energy	E	$i\hbar\dfrac{\partial}{\partial t}$

While position and momentum are important examples of operators in quantum mechanics, so is potential. If potential is a scalar function of position only, the quantum mechanical operator for potential is just $V(x)$. However, in the more general case, potential is also a function of time, so that $V(x, t)$.

Using what we know from classical mechanics and electrodynamics and considering the ramifications of the Young's slits experiment and the photoelectric effect forced us to create the concept of particles with wavy character. These wavy particles have quantized energy and quantized momentum. Further speculation led us to the concepts of a particle wave function, position and momentum operators, and expectation values.

Summarizing, in quantum mechanics, every particle can be described by using a wave function $\psi(\mathbf{r}, t)$, where $|\psi(\mathbf{r}, t)|^2$ is the probability of finding the particle in the volume d^3r at position $\mathbf{r}$ at time t. The wave function and its spatial derivative are continuous, finite, and single-valued.

The wave function $\psi(\mathbf{r}, t)$ is not physical, because it cannot be measured. The quantity that is measured, and hence physical, is $|\psi(\mathbf{r}, t)|^2$, or an expectation value of a quantum operator. Quantum operators are often associated with classical variables. Table 2.4 lists the classical variables along with the corresponding quantum operators we have discussed thus far.

Because time is a *parameter* and hence not an operator, it does not appear in Table 2.4. Time is not a dynamical variable, and so it does not have an expectation value. It is merely a parameter that we use to measure system evolution. This fact exposes a weakness in our theory. In Table 2.4 we have listed the energy operator for the wave function $\psi(x, t)$ as $i\hbar \cdot \partial/\partial t$. This cannot strictly be true since time is not an operator. This inconsistency is a hint that there exists a more complete theory for which time is not just a parameter. However, for the purposes of this book, we choose to ignore this issue.

The probability of finding a particle with wave function $\psi(\mathbf{r}, t)$ somewhere in space is unity. Hence, the wave function is normalized in such a way that

$$\int_{-\infty}^{\infty} \psi^* \psi \, d^3r = \int_{-\infty}^{\infty} |\psi|^2 \, d^3r = 1 \tag{2.32}$$

The average or *expectation* value of an operator $\hat{A}$ is

$$\langle \hat{A} \rangle = \int_{-\infty}^{\infty} \psi^* \hat{A} \psi \, d^3 r \tag{2.33}$$

We still need to advance our understanding if we are to describe the dynamics of our new wavy quantum mechanical particles. What is needed is a dynamical equation. Again, we will make use of our previous experience with classical mechanics, in which the total energy function or Hamiltonian of a particle mass m moving in potential V is

$$H = T + V = \frac{\hat{p}^2}{2m} + V \tag{2.34}$$

Now, all we have to do is substitute our expressions for quantum mechanical momentum and potential into the equation. Because we have quantum mechanical operators which act on wave functions in one dimension, this gives

$$H\psi(x, t) = \frac{-\hbar^2}{2m} \frac{\partial^2}{\partial x^2} \psi(x, t) + V(x, t)\psi(x, t) \tag{2.35}$$

In three dimensions, this equation is written

$$H\psi(\mathbf{r}, t) = \frac{-\hbar^2}{2m} \nabla^2 \psi(\mathbf{r}, t) + V(\mathbf{r}, t)\psi(\mathbf{r}, t) \tag{2.36}$$

where

$$\nabla^2 \psi(\mathbf{r}, t) = \frac{\partial^2 \psi}{\partial x^2} + \frac{\partial^2 \psi}{\partial y^2} + \frac{\partial^2 \psi}{\partial z^2} \tag{2.37}$$

Replacing the Hamiltonian with the energy operator, we have

$$H\psi(\mathbf{r}, t) = i\hbar \frac{\partial}{\partial t} \psi(\mathbf{r}, t) \tag{2.38}$$

where

$$H = \left(\frac{-\hbar^2}{2m_0} \nabla^2 + V(\mathbf{r}, t) \right) \tag{2.39}$$

is the *Hamiltonian operator*. The equation

$$\boxed{ \left(\frac{-\hbar^2}{2m} \nabla^2 + V(\mathbf{r}, t) \right) \psi(\mathbf{r}, t) = i\hbar \frac{\partial}{\partial t} \psi(\mathbf{r}, t) } \tag{2.40}$$

is called the *Schrödinger equation*. This equation can be used to describe the behavior of quantum mechanical particles in three-dimensional space. The fact that the Schrödinger equation is only first-order in the time derivative indicates that the wave function $\psi(x, t)$ evolves from a single initial condition.

We now consider the special case of a closed system in which energy is conserved and potential energy is time-independent in such a way that $V = V(\mathbf{r})$. For convenience, we return to the one-dimensional case to show how the time and spatial parts of the wave function may be separated out. This is done by using the method of separation of variables. If we assume the wave function can be expressed as a product $\psi(x, t) = \psi(x)\phi(t)$, then substitution into the one-dimensional Schrödinger equation gives

$$\frac{-\hbar^2}{2m}\frac{\partial^2}{\partial x^2}\psi(x)\phi(t) + V(x)\psi(x)\phi(t) = i\hbar\frac{\partial}{\partial t}\psi(x)\phi(t) \tag{2.41}$$

We then divide both sides by $\psi(x)\phi(t)$, so that the left-hand side is a function of x only and the right-hand side is a function of t only. This is true if both sides are equal to a constant E. It follows that

$$E\phi(t) = i\hbar\frac{\partial}{\partial t}\phi(t) \tag{2.42}$$

is the *time-dependent Schrödinger equation*,

$$\left(\frac{-\hbar^2}{2m}\frac{\partial^2}{\partial x^2} + V(x)\right)\psi(x) = E\psi(x) \tag{2.43}$$

is the *time-independent Schrödinger equation* in one dimension, and the constant E is just the *energy eigenvalue* of the particle described by the wave function.

It is important to remember that these two equations apply when the potential may be considered independent of time. Of course, the potential is not truly time-independent in the sense that it must have been created at some time in the past. However, we assume that the transients associated with this creation have negligible effect on the calculated results, so that the use of a static potential is an excellent approximation.

The solution to the time-dependent Schrödinger equation is of simple harmonic form

$$\phi(t) = e^{-i\omega t} \tag{2.44}$$

where $E = \hbar\omega$. Notice that Schrödinger's equation *requires* the exponential form $e^{-i\omega t}$. Alternatives such as $e^{+i\omega t}$ or trigonometric functions such as $\sin(\omega t)$ are not allowed. Unlike oscillatory solutions in classical mechanics, Schrödinger's equation gives no choice for the form of the time dependence appearing in the wave function.

The solution to the time-independent Schrödinger equation is $\psi(x)$, which is independent of time. The wave function $\psi(x, t) = \psi(x)\phi(t)$ is called a *stationary state* because the probability density $|\psi(x, t)|^2$ is independent of time. A particle in such a time-independent state will remain in that state until acted upon by some external entity that forces it out of that state.

Wave functions that are solutions to the time-independent Schrödinger equation are normalized in such a way that

$$\int_{-\infty}^{\infty} \psi_n^*(\mathbf{r})\psi_n(\mathbf{r})\, d^3r = 1 \tag{2.45}$$

Here we have adopted a notation in which the integer n labels the wave function associated with a given energy eigenvalue. n is called a quantum number. Wave functions with different quantum numbers, say n and m, have the mathematical property of orthogonality, so that

$$\int_{-\infty}^{\infty} \psi_n^*(\mathbf{r})\psi_m(\mathbf{r})\, d^3r = 0 \tag{2.46}$$

These facts may be summarized by stating that the wave functions are *orthonormal* or

$$\int_{-\infty}^{\infty} \psi_n^*(\mathbf{r})\psi_m(\mathbf{r})\, d^3r = \delta_{nm} \tag{2.47}$$

where δ_{nm} is the Kronecker delta function which has the value of unity when $n = m$ and otherwise is zero. We will find these properties useful as we find out more about the predictions of Schrödinger's equation.

2.2.1   The wave function description of an electron in free space

Perhaps the simplest application of the Schrödinger wave function description of a particle is that of an electron moving unimpeded through space. To find out what the predictions of Schrödinger's equation are, we start by writing down the time-independent Schrödinger equation for an electron mass m_0. The equation is

$$H\psi_n(\mathbf{r}) = E_n\psi(\mathbf{r}) \tag{2.48}$$

or

$$\frac{-\hbar^2}{2m_0}\nabla^2\psi_n(\mathbf{r}) + V(\mathbf{r})\psi_n(\mathbf{r}) = E_n\psi_n(\mathbf{r}) \tag{2.49}$$

where for the case of free space we set the potential $V(\mathbf{r}) = 0$. The *energy eigenvalues* are E_n and ψ_n are *eigenstates* so that

$$\psi_n(\mathbf{r}, t) = \psi_n(\mathbf{r})e^{-i\omega t} \tag{2.50}$$

The wave functions and corresponding energy eigenvalues are labeled by the quantum number n. For an electron in free space, $V(\mathbf{r}) = 0$, and so we have

$$E_n = \frac{\hat{p}_n^2}{2m_0} = \frac{\hbar^2 k_n^2}{2m_0} = \hbar\omega_n \tag{2.51}$$

and

$$\psi_n(\mathbf{r}, t) = (A e^{i\mathbf{k}\cdot\mathbf{r}} + B e^{-i\mathbf{k}\cdot\mathbf{r}}) \, e^{-i\omega_n t} \tag{2.52}$$

where the first term in the parentheses is a wave of amplitude A traveling in the $\mathbf{k}$ direction and the second term is a wave of amplitude B traveling in the $-\mathbf{k}$ direction. The term $e^{-i\omega_n t}$ gives the time dependence of the wave function. Selecting a boundary condition characterized by $B = 0$, and considering the case of motion in the x direction only, the wave function becomes

$$\psi_n(x, t) = A e^{i k_x x} e^{-i\omega_n t} \tag{2.53}$$

To find the momentum associated with this wave function, we apply the momentum operator to the wave function:

$$\hat{p}_x \psi_n(x, t) = -i\hbar \frac{\partial}{\partial x} \psi_n(x, t) = \hbar k_x \cdot A e^{i k_x x} e^{-i\omega_n t} = \hbar k_x \cdot \psi_n(x, t) \tag{2.54}$$

Since, in quantum mechanics, the real eigenvalue of an operator can, at least in principle, be measured, we may safely assume that the x-directed momentum of the particle is just $p_x = \hbar k_x$.

From these results, we conclude that an electron with mass m_0 *and* energy E in free space has momentum of magnitude $p = \hbar k = \sqrt{2m_0 E}$ and a nonlinear dispersion relation that gives $E_k = \hbar\omega = \hbar^2 k^2 / 2m_0$, or

$$\boxed{\omega(k) = \frac{\hbar k^2}{2m_0}} \tag{2.55}$$

This dispersion relation is illustrated in Fig. 2.8.

The Schrödinger equation does not allow an electron in free space to have just any energy and wavelength; rather, the electron is constrained to values given by the dispersion relation. Just as with the linear chain of harmonic oscillators discussed in Section 1.2.4, the underlying reason that the dispersion relation exists is intimately related to a type of symmetry in the structure of the equation of motion.

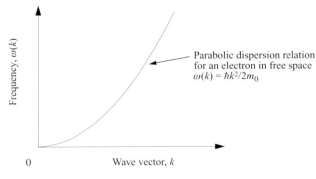

Parabolic dispersion relation
for an electron in free space
$\omega(k) = \hbar k^2 / 2m_0$

Frequency, $\omega(k)$

0   Wave vector, k

Fig. 2.8. Dispersion relation for an electron in free space.

2.2.2   The electron wave packet and dispersion

In the previous example, an electron in free space was described by traveling plane-wave states that extended over all space but were well-defined points in k space. Obviously, this is an extreme limit.

Suppose we wish to describe an electron at a particular average position in free space as a sum of a number of plane-wave eigenstates. We can force the electron to occupy a finite region of space by forming a *wave packet* from a continuum of plane-wave eigenstates. The wave packet consists of the superposition of many eigenstates that destructively interfere everywhere except in some localized region of space. For simplicity, we start with a plane wave of momentum $\hbar k_0$ in the x direction and create a Gaussian pulse from this plane wave in such a way that at time $t = 0$

$$\psi(x, t = 0) = A e^{ik_0 x} e^{-(x-x_0)^2/4\Delta x^2} \tag{2.56}$$

where the amplitude is $A = 1/(2\pi \Delta x^2)^{1/4}$, the mean position is $\langle x \rangle = x_0$, and a measure of the spatial spread in the wave function is given by the value of Δx. The probability density at time $t = 0$ is just a normalized Gaussian function of standard deviation Δx:

$$\psi^*(x, t = 0)\psi(x, t = 0) = |\psi(x, t = 0)|^2 = A^2 e^{-(x-x_0)^2/2\Delta x^2} \tag{2.57}$$

The Gaussian pulse contains a continuum of momentum components centered about the original plane-wave momentum $\hbar k_0$. To find the values of the momentum components in the Gaussian pulse, we take the Fourier transform of the wave function $\psi(x, t = 0)$. This gives

$$\psi(k, t = 0) = \frac{1}{A\sqrt{\pi}} e^{-i(k-k_0)x} e^{-(k-k_0)^2 \Delta x^2} \tag{2.58}$$

The corresponding probability density in k space (momentum space) is given by

$$|\psi(k, t = 0)|^2 = \frac{1}{A^2\pi} e^{-(k-k_0)^2 2\Delta x^2} = \frac{1}{A^2\pi} e^{-(k-k_0)^2/2\Delta k^2} \tag{2.59}$$

where k_0 is the average value of k, and a measure of the spread in the distribution of k is given by the standard deviation $\Delta k = 1/2\Delta x$. Because the product $\Delta k \Delta x = 1/2$ is a constant, this indicates that localizing the Gaussian pulse in real space will increase the width of the corresponding distribution in k space. Conversely, localizing the Gaussian pulse in k space increases the width of the pulse in real space. Recognizing that momentum $p = \hbar k$, we have

$$\Delta p \Delta x = \frac{\hbar}{2} \tag{2.60}$$

which is an example of the *uncertainty principle. Conjugate pairs of operators cannot be measured to arbitrary accuracy.* In this case, it is not possible to simultaneously know the exact position of a particle and its momentum.

To track the time evolution of the Gaussian wave packet, we need to follow the time dependence of each contributing plane wave. Since each plane wave has a time dependence of the form $e^{-i\omega_k t}$, we may write for time $t > 0$

$$\psi(k,t) = \frac{1}{A\sqrt{\pi}} e^{-i(k-k_0)x} e^{-(k-k_0)^2 \Delta x^2} e^{-i\omega_k t} \qquad (2.61)$$

The value of ω_k is given by the dispersion relation for a freely propagating electron of mass m_0 and energy $E_k = \hbar\omega_k = \hbar^2 k^2/2m_0$. The Taylor expansion about k_0 for the dispersion relation is

$$\omega(k) = \frac{\hbar k_0^2}{2m_0} + \frac{\hbar k_0(k-k_0)}{m_0} + \frac{\hbar(k-k_0)^2}{2m_0} \qquad (2.62)$$

To find the effect of dispersion on the Gaussian pulse as a function of time, we need to take the Fourier transform of $\psi(k,t)$ to obtain $\psi(x,t)$. The solution is

$$\psi(x,t) = \frac{1}{A\pi\sqrt{2}} e^{i(k_0 x - \omega_0 t)} \int\limits_{-\infty}^{\infty} e^{i(k-k_0)(x-x_0-(\hbar k_0 t/m_0))} e^{-(k-k_0)^2 \Delta x^2(1+i\hbar t/2m_0\Delta x^2)} dk \quad (2.63)$$

The prefactor is a plane wave oscillating at frequency $\omega_0 = \hbar k_0^2/2m_0$ and moving with momentum $\hbar k_0$ and a *phase velocity* $v_p = \hbar k_0/2m_0$. In the integral, the term $e^{i(k-k_0)(x-x_0-(\hbar k_0 t/m_0))}$ shows that the center of the wave packet moves a distance $\hbar k_0 t/m$ in time t, indicating a *group velocity* for the wave packet of $v_g = \hbar k_0/m_0$. If one describes a classical particle by a wave packet then the velocity of the particle is given by the group velocity of the wave packet.

The term

$$e^{-(k-k_0)^2 \Delta x^2(1+i\hbar t/2m_0\Delta x^2)} \qquad (2.64)$$

in the integral shows that the width of the wave packet increases with time. The width increases as

$$\Delta x(t) = \left(\Delta x^2 + \frac{\hbar^2 t^2}{4m_0^2 \Delta x^2} \right)^{1/2} \qquad (2.65)$$

The wave packet delocalizes as a function of time because of dispersion. Figure 2.9 illustrates the effect dispersion has, as a function of time, on the propagation of a Gaussian wave packet. The characteristic time $\Delta\tau_{\Delta x}$ for the width of the wave packet to double is $\Delta\tau_{\Delta x} = 2m_0\Delta x^2/\hbar$. We can use this time to illustrate why classical particles always appear particle-like.

Consider a classical particle of mass one gram ($m = 10^{-3}$ kg) the position of which is known to an accuracy of one micron $\Delta x = 10^{-6}$ m. Modeling this particle as a Gaussian wave packet gives a characteristic time

$$\Delta\tau_{\Delta x} = \frac{2m\Delta x^2}{\hbar} = \frac{2 \times 10^{-3} \times (10^{-6})^2}{1.05 \times 10^{-34}} = 2 \times 10^{19} \text{ s}$$

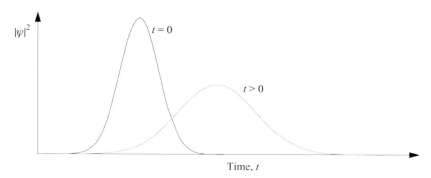

Fig. 2.9. Illustration of the time evolution of a Gaussian wave packet, showing the effect of dispersion.

This time is 6×10^{11} years, which is 30–60 times the age of the universe![2] The constant $\hbar$ sets an absolute scale that ensures that the macroscopic classical particle remains a classical particle for all time.

We may now contrast this result with an atomic-scale entity. Consider an electron of mass $m_0 = 9.1 \times 10^{-31}$ kg in a circular orbit of radius $a_B = 0.529177 \times 10^{-10}$ m around a proton. Here we are adopting the semiclassical picture of a hydrogen atom discussed in Section 2.2.3. The electron orbits in the coulomb potential of the proton with tangential speed 2.2×10^6 m s^{-1}. For purposes of illustration, we assume the electron is described by a Gaussian wave packet and that its position is know to an accuracy of $\Delta x = 10^{-11}$ m. In this case, one obtains a characteristic time that is

$$\Delta \tau_{\Delta x} = \frac{2 m_0 \Delta x^2}{\hbar} = \frac{2 \times 9.1 \times 10^{-31} \times (10^{-11})^2}{1.05 \times 10^{-34}} = 1.7 \times 10^{-18} \text{ s}$$

This time is significantly shorter than the time to complete one orbit ($\tau_{\text{orbit}} \sim 1.5 \times 10^{-17}$ s). Long before the electron wave packet can complete one orbit it has lost all of its particle character and has completely delocalized over the circular trajectory. The concept of a semiclassical wave packet describing a particle-like electron orbiting the proton of a hydrogen atom just does not make any sense.

From our discussion, it is clear that the effect of dispersion in the wave components of a Gaussian pulse is to increase pulse width as a function of time. This has a profound influence on our ability to assign wave or particle character to an object of mass, m. For a freely propagating pulse, the momenta of the plane-wave components do not change, so the product $\Delta p \Delta x$ must be increasing with time from its minimum value at time $t = 0$. We can now write the uncertainty relation for momentum and position more accurately as

---

[2] The age of the universe is thought to be in the range 1×10^{10}–2×10^{10} years.

$$\Delta p \Delta x \geq \frac{\hbar}{2} \tag{2.66}$$

This relationship controls the precision with which it is possible to simultaneously know the position of a particle and its momentum.

2.2.3   The hydrogen atom

In addition to the case of an electron moving freely through space, we can also consider an electron confined by a potential to motion in some local region. A very important confining potential for an electron comes from the positive charge on protons in the nucleus of an atom. We consider hydrogen because it consists of just one proton and one electron. The proton has positive charge and the electron has negative charge. In 1911, Rutherford showed experimentally that electrons appear to orbit the nucleus of atoms. With this in mind, we start by considering a single electron moving in the coulomb potential of the single proton in a hydrogen atom (see Fig. 2.10).

Note the use of a spherical coordinate system. The proton is at the origin and the electron is at position $\mathbf{r}$. The electron has charge $-e = -1.602176 \times 10^{-19}$ C and mass $m_0 = 9.109381 \times 10^{-31}$ kg. The proton charge is equal and opposite to that of the electron, and the proton mass is $m_\mathrm{p} = 1.672621 \times 10^{-27}$ kg. The ratio of proton mass to electron mass is $m_\mathrm{p}/m_0 = 1836.15$.

Because the ratio of proton mass to electron mass is more than 1800, it seems reasonable to think of the light electron as orbiting the center-of-mass of the heavy proton.

Classically, we think of the electron in a curved orbit similar to that sketched in Fig. 2.11. In such an orbit, the electron is accelerating and hence must, according to classical electrodynamics, radiate electromagnetic waves. As the electron energy radiates away, the radius of the orbit decreases until the electron collapses into the proton nucleus. This should all happen within a time of about 0.1 ns. Experiment shows that the classical theory does not work! Hydrogen is observed to be stable. In addition, the spectrum of hydrogen is observed to consist of discrete spectral lines – which, again, is a feature not predicted by classical models.

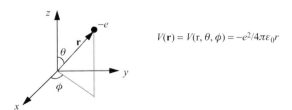

$$V(\mathbf{r}) = V(r, \theta, \phi) = -e^2/4\pi\varepsilon_0 r$$

Fig. 2.10. A hydrogen atom consists of an electron and a proton. It is natural to choose a spherical coordinate system to describe a single electron moving in the coulomb potential of the single proton.

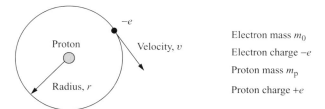

Fig. 2.11. Illustration of a classical circular orbit of an electron mass m_0 moving with velocity v in the coulomb potential of a proton mass m_p. This classical view predicts that hydrogen is unstable.

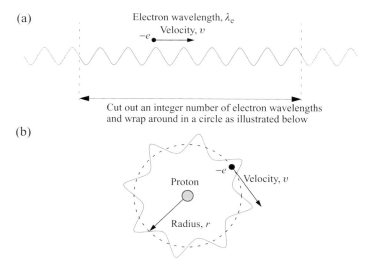

Fig. 2.12. (a) Illustration of electron wave propagating in free space with wavelength λ_e. (b) Illustration of an electron wave wrapped around in a circular orbit about a proton. Single-valuedness of the electron wave function suggests that only an integer number of electron wavelengths can fit into a circular orbit of radius r.

Having established the wavy nature of the electron when traveling in free space and when scattered, as shown in the Davisson and Germer experiment, it seems reasonable to insist that an electron in a circular orbit must also exhibit a wavy character. Imposing wavy character on the electron moving in a circular orbit around the proton is interesting, because the geometry forces the wave to fold back upon itself. Since we anticipate any wave function describing the electron to be single-valued, only integer wavelengths can be fit into a circular orbit of a given circumference. Figure 2.12 illustrates the idea. In Fig. 2.12(a) we imagine cutting out an integer number, n, wavelengths of an electron wave function moving in free space and, as shown in Fig. 2.12(b), wrapping it around a circular orbit of radius $r = n\lambda_e/2\pi$, where n is a nonzero positive integer.

It is possible to put together an *ad-hoc* explanation of the properties of the hydrogen atom using just a few postulates. In 1913, Bohr showed that the spectral properties of hydrogen may be described quite accurately if one adopts the following rules:

1. Electrons exist in stable circular orbits around the proton.
2. Electrons may make transitions between orbits by emission or absorption of a photon of energy $\hbar\omega$.
3. The angular momentum of the electron in a given orbit is quantized according to $p_\theta = n\hbar$, where n is a nonzero positive integer.

Postulate 3 admits the wavy nature of the electron. Stable, long-lived orbits only exist when an integer number of wavelengths can fit into the classical orbit. This is analogous to the resonance or long-lived state of a plucked guitar string.

Classically, the atom can radiate electromagnetic waves when there is a net oscillating dipole moment. This involves net oscillatory current flow due to the movement of charge. However, if one associates a wavelength with the electron, then a resonance can exist when an integer number of wavelengths can fit into the classical orbit. The resonance may be thought of as two counter-propagating waves whose associated currents exactly cancel. In this case, there is no net oscillatory current, and the electron state is long-lived.

The postulates of Bohr allowed many parameters to be calculated, such as the average radius of an electron orbit and the energy difference between orbits. The following examples demonstrate the calculations of such quantities.

2.2.3.1 Calculation of the average radius of an electron orbit in hydrogen

We start by equating the electrostatic force and the centripetal force:

$$\frac{-e^2}{4\pi\varepsilon_0 r^2} = -\frac{m_0 v^2}{r} \tag{2.67}$$

where the left-hand term is the electrostatic force and the right-hand term is the centripetal force. Because angular momentum is quantized, $p_\theta = n\hbar = m_0 v r_n$, in which we note that the subscript n on the radius r is due to n taking on integer values. We may now rewrite our equation for angular momentum as $m_0^2 v^2 = n^2\hbar^2/r_n^2$ or $m_0 v^2/r_n = (1/r_n m_0) \cdot (n^2\hbar^2/r_n^2)$. Substituting into our expression for centripetal force gives

$$\frac{e^2}{4\pi\varepsilon_0 r_n^2} = \frac{1}{r_n m_0} \cdot \frac{n^2\hbar^2}{r_n^2} \tag{2.68}$$

and hence

$$r_n = \frac{4\pi\varepsilon_0 n^2\hbar^2}{m_0 e^2} \tag{2.69}$$

is the radius of the n-th orbit. The radius of each orbit is quantized. The spatial scale is set by the radius for $n = 1$, which gives

$$r_1 = a_B = \frac{4\pi \varepsilon_0 \hbar^2}{m_0 e^2} = 0.529177 \times 10^{-10} \text{ m} \tag{2.70}$$

which is called the *Bohr radius*.

2.2.3.2 Calculation of energy difference between electron orbits in hydrogen

Calculation of the energy difference between orbits is important, since it will allow us to predict the optical spectra of excited hydrogen atoms.

We start by equating the classical momentum with the quantized momentum of the n-th electron orbit. Since the electron is assumed to exist in a stable circular orbit around the proton, as shown schematically in Fig. 2.11, the momentum of the electron is $m_0 v = n\hbar / r_n$, and we have

$$v = \frac{n\hbar}{m_0 r_n} = \frac{m_0 e^2}{4\pi \varepsilon_0 n^2 \hbar^2} \cdot \frac{n\hbar}{m_0} = \frac{e^2}{4\pi \varepsilon_0 n\hbar} \tag{2.71}$$

for the electron velocity of the n-th orbit. The value of v for $n = 1$ is $v = 2.2 \times 10^6$ m s^{-1}. We now obtain the kinetic energy of the electron:

$$T = \frac{1}{2} m_0 v^2 = \frac{1}{2} m_0 \frac{e^4}{(4\pi \varepsilon_0)^2 n^2 \hbar^2} \tag{2.72}$$

The potential energy is just the force times the distance between charges, so

$$V = \frac{-e^2}{4\pi \varepsilon_0 r_n} = -m_0 \frac{e^4}{(4\pi \varepsilon_0)^2 n^2 \hbar^2} \tag{2.73}$$

Total energy for the n-th orbit is

$$E_n = T + V = -\frac{1}{2} m_0 \frac{e^4}{(4\pi \varepsilon_0)^2 n^2 \hbar^2} \tag{2.74}$$

We note that $T = -V/2$. The result is a specific example of the more general "virial theorem".

The energy difference between orbits with quantum number n_1 and n_2 is

$$\boxed{E_{n_2} - E_{n_1} = \frac{m_0}{2} \frac{e^4}{(4\pi \varepsilon_0)^2 \hbar^2} \left(\frac{1}{n_1^2} - \frac{1}{n_2^2} \right)} \tag{2.75}$$

Clearly, the prefactor in this equation is a natural energy scale for the system. The numerical value of the prefactor corresponds to $n = 1$ and is the lowest energy state, E_1:

$$E_1 = Ry = \frac{-m_0}{2} \frac{e^4}{(4\pi \varepsilon_0)^2 \hbar^2} = -13.6058 \text{ eV} \tag{2.76}$$

Ry is called the Rydberg constant.

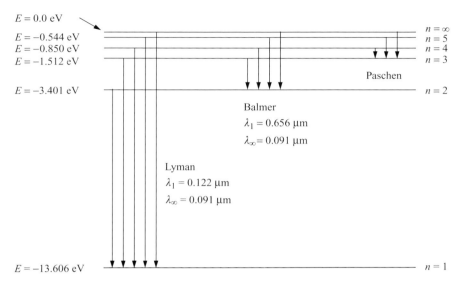

Fig. 2.13. Photon emission spectra of excited hydrogen consist of a discrete number of spectral lines corresponding to transitions from high energy levels to lower energy levels. Different groups of characteristic emission line spectra have been given the names of those who first observed them.

When the electron orbiting the proton of a hydrogen atom is excited to a high energy state, it can lose energy by emitting a photon of energy $\hbar\omega$. Because the energy levels of the hydrogen atom are quantized, photon emission spectra of excited hydrogen consist of a discrete number of spectral lines. In Fig. 2.13, the emission lines correspond to transitions from high energy levels to lower energy levels.

It is also possible for the reverse process to occur. In this case, photons with the correct energy can be absorbed, causing an electron to be excited from a low energy level to a higher energy level. This absorption process could be represented in Fig. 2.13 by changing the direction of the arrows on the vertical lines.

Different groups of energy transitions result in emission of photons of energy, $\hbar\omega$. The characteristic emission line spectra have been given the names (Lyman, Balmer, Paschen) of those who first observed them.

The Bohr model of the hydrogen atom is somewhat of a hybrid between classical and quantum ideas. What is needed is a model that derives directly from quantum mechanics. From what we know so far, the way to do this is to use the Schrödinger equation to describe an electron moving in the spherically symmetric coulomb potential of the proton charge.

In spherical coordinates, the time-independent solutions to the Schrödinger equation are of the form

$$\psi_{nlm}(r, \theta, \phi) = R_n(r)\Theta_l(\theta)\Phi_m(\phi) \tag{2.77}$$

where we have separated out r-, θ-, and ϕ-dependent parts to the wave equation. The resulting three equations have wave functions that are quantized with integer quantum numbers n, l, and m and are separately normalized. For example, it can be shown that the function $\Phi_m(\phi)$ must satisfy

$$\frac{\partial^2}{\partial\phi^2}\Phi_m(\phi) + m^2\Phi_m(\phi) = 0 \tag{2.78}$$

where m is the quantum number for the wave function

$$\Phi_m(\phi) = Ae^{im\phi} \tag{2.79}$$

The normalization constant A can be found from

$$\int_0^{2\pi} \Phi_m^*(\phi)\Phi_m(\phi)d\phi = 1 \tag{2.80}$$

$$A^2 \int_0^{2\pi} e^{-im\phi}e^{im\phi}d\phi = A^2 \int_0^{2\pi} d\phi = 2\pi A^2 = 1 \tag{2.81}$$

$$A = \frac{1}{\sqrt{2\pi}} \tag{2.82}$$

Hence,

$$\Phi_m(\phi) = \frac{1}{\sqrt{2\pi}}e^{im\phi} \tag{2.83}$$

It is reasonable to expect that $\Phi_m(\phi)$ should be single-valued, thereby forcing the function to repeat itself every 2π. This happens if m is an integer. The other quantum numbers are also integers with a special relationship to one another. We had $\psi_{nlm}(r, \theta, \phi) = R_n(r)\Theta_l(\theta)\Phi_m(\phi)$ with quantum numbers n, l, and m. The quantum numbers that specify the state of the electron are

$n = 1, 2, 3, \ldots$

$l = 0, 1, 2, \ldots (n-1)$

$m = \pm l, \ldots \pm 2, \pm 1, 0$

The principal quantum number n specifies the energy of the Bohr orbit. The quantum numbers l and m relate to the quantization of orbital angular momentum. The orbital angular momentum quantum number is l, and the azimuthal quantum number is m. Because, in this theory, the energy level for given n is independent of quantum numbers l and m, there is an n^2 degeneracy in states of energy E_n, since

$$\sum_{l=0}^{l=n-1} (2l+1) = n^2 \tag{2.84}$$

(see Exercise 2.9).

In addition to n, l, m, the electron has a spin quantum number $s = \pm 1/2$. Electron spin angular momentum, $s\hbar$, is an intrinsic property of the electron that arises due to the influence of special relativity on the behavior of the electron. In 1928, Dirac showed that electron spin emerges as a natural consequence of a relativistic treatment of the Schrödinger equation. We will not discuss Dirac's equations because, in practice, Schrödinger's nonrelativistic equation is extraordinarily successful in describing most of the phenomena in which we will be interested. We will, however, remember to include the intrinsic electron spin quantum number when appropriate.

2.2.4 Periodic table of elements

The hydrogen atom discussed in Section 2.2.3 is only one type of atom. There are over 100 other atoms, each with their own unique characteristics. The incredible richness of the world we live in can, in part, be thought of as due to the fact that there are many ways to form different combinations of atoms in the gas, liquid, and solid states. Chemistry sets out to introduce a methodology that predicts the behavior of different combinations of atoms. To this end, chemists find it convenient to use a periodic table in which atoms are grouped according to the similarities in their chemical behavior. For example, H, Li, Na and other atoms form the column IA elements of the periodic table because they all have a single electron available for chemical reaction with other atoms. It is the number of electrons a given atom has available for chemical reaction that is used by chemists to characterize groups of atoms. The rules of quantum mechanics can help us to understand why atoms behave the way they do and why chemists can group atoms according to the number of electrons available for chemical reaction.

2.2.4.1 The Pauli exclusion principle and the properties of atoms

It is an experimental fact that no two electrons in an interacting system can have the same quantum numbers n, l, m, s. Each electron state is assigned an unique set of values of n, l, m, s. This is the *Pauli exclusion principle*, which determines many properties of atoms in the periodic table, including the formation of electron shells. For a given n in an atom there are only a finite number of values of l, m, and s that an electron may have. If there is an electron assigned to each of these values, then a complete shell is formed. Completed shells occur in the chemically inert noble elements of the periodic table, which are He, Ne, Ar, Kr, Xe, and Rn. All other atoms apart from H use one of these atoms as a core and add additional electron states to incomplete subshell states. Electrons in these incomplete subshells are available for chemical reaction with other atoms and therefore dominate the chemical activity of the atom. This simple version of the shell model, summarized by Table 2.5, works quite well. There are, however, a few complications for high atomic number atoms, which we will not consider here.

One may now use what we know to figure out the lowest-energy electronic configurations for atoms. This lowest-energy state of an atom is also called the *ground state*. We still use quantum numbers n and m, but replace $l = 0$ with s, $l = 1$ with p, $l = 2$

Table 2.5. *Electron shell states*

n	l	m	$2s$	Allowable states in subshell	Allowable states in complete shell
1	0	0	± 1	2	2
2	0	0	± 1	2	8
	1	-1	± 1		
	1	0	± 1	6	
		1	± 1		
3	0	0	± 1	2	18
	1	-1	± 1		
		0	± 1	6	
		1	± 1		
	2	-2	± 1		
		-1	± 1		
		0	± 1	10	
		1	± 1		
		2	± 1		

with d, and $l = 3$ with f. This naming convention comes from early work that labeled spectroscopic lines as sharp, principal, diffuse, and fundamental.

For example, the electron configuration for Si($z = 14$) is $1s^2 2s^2 2p^6 3s^2 3p^2$, where the atomic number z is the number of electrons in the atom. In this notation, $2p^6$ means that $n = 2, l = 1$, and there are six electrons in the $2p$ shell. Note that there are four electrons in the outer $n = 3$ shell and that the $n = 1$ and $n = 2$ shells are completely full. Full shells are chemically inert, and in this case $1s^2 2s^2 2p^6$ is the Ne ground state, so we can write the ground-state configuration for Si as [Ne]$3s^2 3p^2$. Silicon consists of an inert (chemically inactive) neon core and four chemically active electrons in the $n = 3$ shell. In a crystal formed from Si atoms, it is the chemically active electrons in the $n = 3$ shell that interact to form the chemical bonds that hold the crystal together. These electrons are called *valence electrons*, and they occupy valence electron states of the crystal. Table 2.6 illustrates this classification method.

Technologically important examples of ground-state atomic configurations include:

Al	[Ne]$3s^2 3p^1$	group IIIB
Si	[Ne]$3s^2 3p^2$	group IVB
P	[Ne]$3s^2 3p^3$	group VB
Ga	[Ar]$3d^{10} 4s^2 p^1$	group IIIB
Ge	[Ar]$3d^{10} 4s^2 p^2$	group IVB
As	[Ar]$3d^{10} 4s^2 p^3$	group VB
In	[Kr]$4d^{10} 5s^2 p^1$	group IIIB

Table 2.6. *Electron ground-state for first 18 elements of the periodic table*

Atomic number	Element		$n = 1$ $l = 0$ $1s$	$n = 2$ $l = 0$ $2s$	$n = 2$ $l = 1$ $2p$	$n = 3$ $l = 0$ $3s$	$n = 3$ $l = 1$ $3p$	Shorthand notation
1	H		1					$1s^1$
2	He		2					$1s^2$
3	Li	[He] core		1				$1s^2 2s^1$
4	Be	2 electrons		2				$1s^2 2s^2$
5	B			2	1			$1s^2 2s^2 2p^1$
6	C			2	2			$1s^2 2s^2 2p^2$
7	N			2	3			$1s^2 2s^2 2p^3$
8	O			2	4			$1s^2 2s^2 2p^4$
9	F			2	5			$1s^2 2s^2 2p^5$
10	Ne			2	6			$1s^2 2s^2 2p^6$
11	Na	[Ne] core				1		[Ne] $3s^1$
12	Mg	10 electrons				2		[Ne] $3s^2$
13	Al					2	1	[Ne] $3s^2 3p^1$
14	Si					2	2	[Ne] $3s^2 3p^2$
15	P					2	3	[Ne] $3s^2 3p^3$
16	S					2	4	[Ne] $3s^2 3p^4$
17	Cl					2	5	[Ne] $3s^2 3p^5$
18	Ar					2	6	[Ne] $3s^2 3p^6$

Table 2.7 is the periodic table of elements giving the ground-state configuration for each element along with the atomic mass.

2.2.5   Crystal structure

Classical mechanics deals with the motion of macroscopic bodies, and classical electrodynamics deals with the motion of electromagnetic fields in the limit of large quantum numbers. As we apply our knowledge of quantum mechanics, we will be interested in, among other things, the interaction of solid matter with electromagnetic fields. Because of this, we need to understand something about the properties of solids. First, we will review how atoms are arranged spatially to form a crystalline solid. Following this, we will briefly discuss the properties of a single crystal semiconductor. Specifically, we will discuss a semiconductor, because of its importance for modern technology.

   Solids are made of atoms. As illustrated in Fig. 2.14, there are different ways of arranging atoms spatially to form a solid. Atoms in a crystalline solid, for example, are arranged in a spatially periodic fashion. Because this periodicity extends over many

Table 2.7. *The periodic table of elements*

IA	IIA	IIIB	IVB	VB	VIB	VIIB	VIII	VIII	VIII	IB	IIB	IIIA	IVA	VA	VIA	VIIA	Noble
Hydrogen **H** $1s^1$ 1.0079																	Helium **He** $1s^2$ 4.0026
Lithium **Li** $1s^2 2s^1$ 6.941	Beryllium **Be** $1s^2 2s^2$ 9.0122											Boron **B** $1s^2 2s^2 2p^1$ 10.81	Carbon **C** $1s^2 2s^2 2p^2$ 12.01	Nitrogen **N** $1s^2 2s^2 2p^3$ 14.007	Oxygen **O** $1s^2 2s^2 2p^4$ 15.999	Fluorine **F** $1s^2 2s^2 2p^5$ 18.998	Neon **Ne** $1s^2 2s^2 2p^6$ 20.18
Sodium **Na** $[\mathrm{Ne}]\,3s^1$ 22.9898	Magnesium **Mg** $[\mathrm{Ne}]\,3s^2$ 24.305											Aluminum **Al** $[\mathrm{Ne}]\,3s^2 3p^1$ 26.982	Silicon **Si** $[\mathrm{Ne}]\,3s^2 3p^2$ 28.086	Phosphorus **P** $[\mathrm{Ne}]\,3s^2 3p^3$ 30.974	Sulfur **S** $[\mathrm{Ne}]\,3s^2 3p^4$ 32.064	Chlorine **Cl** $[\mathrm{Ne}]\,3s^2 3p^5$ 35.453	Argon **Ar** $[\mathrm{Ne}]\,3s^2 3p^6$ 39.948
Potassium **K** $[\mathrm{Ar}]\,4s^1$ 39.09	Calcium **Ca** $[\mathrm{Ar}]\,4s^2$ 40.08	Scandium **Sc** $[\mathrm{Ar}]\,3d^1 4s^2$ 44.956	Titanium **Ti** $[\mathrm{Ar}]\,3d^2 4s^2$ 47.90	Vanadium **V** $[\mathrm{Ar}]\,3d^3 4s^2$ 50.942	Chromium **Cr** $[\mathrm{Ar}]\,3d^5 4s^1$ 52.00	Manganese **Mn** $[\mathrm{Ar}]\,3d^5 4s^2$ 54.938	Iron **Fe** $[\mathrm{Ar}]\,3d^6 4s^2$ 55.85	Cobalt **Co** $[\mathrm{Ar}]\,3d^7 4s^2$ 58.93	Nickel **Ni** $[\mathrm{Ar}]\,3d^8 4s^2$ 58.71	Copper **Cu** $[\mathrm{Ar}]\,3d^{10} 4s^1$ 63.55	Zinc **Zn** $[\mathrm{Ar}]\,3d^{10} 4s^2$ 65.38	Gallium **Ga** $[\mathrm{Ar}]\,3d^{10} 4s^2 4p^1$ 69.72	Germanium **Ge** $[\mathrm{Ar}]\,3d^{10} 4s^2 4p^2$ 72.59	Arsenic **As** $[\mathrm{Ar}]\,3d^{10} 4s^2 4p^3$ 74.922	Selenium **Se** $[\mathrm{Ar}]\,3d^{10} 4s^2 4p^4$ 78.96	Bromine **Br** $[\mathrm{Ar}]\,3d^{10} 4s^2 4p^5$ 79.91	Krypton **Kr** $[\mathrm{Ar}]\,3d^{10} 4s^2 4p^6$ 83.80
Rubidium **Rb** $[\mathrm{Kr}]\,5s^1$ 85.47	Strontium **Sr** $[\mathrm{Kr}]\,5s^2$ 87.62	Yttrium **Y** $[\mathrm{Kr}]\,4d^1 5s^2$ 88.91	Zirconium **Zr** $[\mathrm{Kr}]\,4d^2 5s^2$ 91.22	Niobium **Nb** $[\mathrm{Kr}]\,4d^4 5s^1$ 92.91	Molybdenum **Mo** $[\mathrm{Kr}]\,4d^5 5s^1$ 95.94	Technetium **Tc** $[\mathrm{Kr}]\,4d^5 5s^2$ 98.91	Ruthenium **Ru** $[\mathrm{Kr}]\,4d^7 5s^1$ 101.07	Rhodium **Rh** $[\mathrm{Kr}]\,4d^8 5s^1$ 102.90	Palladium **Pd** $[\mathrm{Kr}]\,4d^{10} 5s^0$ 106.40	Silver **Ag** $[\mathrm{Kr}]\,4d^{10} 5s^1$ 107.87	Cadmium **Cd** $[\mathrm{Kr}]\,4d^{10} 5s^2$ 112.40	Indium **In** $[\mathrm{Kr}]\,4d^{10} 5s^2 5p^1$ 114.82	Tin **Sn** $[\mathrm{Kr}]\,4d^{10} 5s^2 5p^2$ 118.69	Antimony **Sb** $[\mathrm{Kr}]\,4d^{10} 5s^2 5p^3$ 121.75	Tellurium **Te** $[\mathrm{Kr}]\,4d^{10} 5s^2 5p^4$ 127.60	Iodine **I** $[\mathrm{Kr}]\,4d^{10} 5s^2 5p^5$ 126.90	Xenon **Xe** $[\mathrm{Kr}]\,4d^{10} 5s^2 5p^6$ 131.30
Cesium **Cs** $[\mathrm{Xe}]\,6s^1$ 132.91	Barium **Ba** $[\mathrm{Xe}]\,6s^2$ 137.34	Lanthanum **La** $[\mathrm{Xe}]\,5d^1 6s^2$ 138.91	Hafnium **Hf** $[\mathrm{Xe}]\,4f^{14} 5d^2 6s^2$ 178.49	Tantalum **Ta** $[\mathrm{Xe}]\,4f^{14} 5d^3 6s^2$ 180.95	Tungsten **W** $[\mathrm{Xe}]\,4f^{14} 5d^4 6s^2$ 183.85	Rhenium **Re** $[\mathrm{Xe}]\,4f^{14} 5d^5 6s^2$ 186.2	Osmium **Os** $[\mathrm{Xe}]\,4f^{14} 5d^6 6s^2$ 190.20	Iridium **Ir** $[\mathrm{Xe}]\,4f^{14} 5d^7 6s^2$ 192.22	Platinum **Pt** $[\mathrm{Xe}]\,4f^{14} 5d^9 6s^1$ 195.09	Gold **Au** $[\mathrm{Xe}]\,4f^{14} 5d^{10} 6s^1$ 196.97	Mercury **Hg** $[\mathrm{Xe}]\,4f^{14} 5d^{10} 6s^2$ 200.59	Thallium **Tl** $[\mathrm{Xe}]\,4f^{14} 5d^{10} 6s^2 6p^1$ 204.37	Lead **Pb** $[\mathrm{Xe}]\,4f^{14} 5d^{10} 6s^2 6p^2$ 207.19	Bismuth **Bi** $[\mathrm{Xe}]\,4f^{14} 5d^{10} 6s^2 6p^3$ 208.98	Polonium **Po** $[\mathrm{Xe}]\,4f^{14} 5d^{10} 6s^2 6p^4$ 210	Astatine **At** $[\mathrm{Xe}]\,4f^{14} 5d^{10} 6s^2 6p^5$ 210	Radon **Rn** $[\mathrm{Xe}]\,4f^{14} 5d^{10} 6s^2 6p^6$ 222
Francium **Fr** $[\mathrm{Rn}]\,7s^1$ 223	Radium **Ra** $[\mathrm{Rn}]\,7s^2$ 226	Actinium **Ac** $[\mathrm{Rn}]\,6d^1 7s^2$ 227															

Rare earths

Lanthanides

Cerium **Ce** $[\mathrm{Xe}]\,4f^1 5d^1 6s^2$ 140.12	Praseodymium **Pr** $[\mathrm{Xe}]\,4f^3 6s^2$ 140.91	Neodymium **Nd** $[\mathrm{Xe}]\,4f^4 6s^2$ 144.24	Promethium **Pm** $[\mathrm{Xe}]\,4f^5 6s^2$ 145	Samarium **Sm** $[\mathrm{Xe}]\,4f^6 6s^2$ 150.35	Europium **Eu** $[\mathrm{Xe}]\,4f^7 6s^2$ 151.96	Gadolinium **Gd** $[\mathrm{Xe}]\,4f^7 5d^1 6s^2$ 157.25	Terbium **Tb** $[\mathrm{Xe}]\,4f^9 5d^0 6s^2$ 158.92	Dysprosium **Dy** $[\mathrm{Xe}]\,4f^{10} 5d^0 6s^2$ 162.50	Holmium **Ho** $[\mathrm{Xe}]\,4f^{11} 5d^0 6s^2$ 164.93	Erbium **Er** $[\mathrm{Xe}]\,4f^{12} 5d^0 6s^2$ 167.26	Thulium **Tm** $[\mathrm{Xe}]\,4f^{13} 5d^0 6s^2$ 168.93	Ytterbium **Yb** $[\mathrm{Xe}]\,4f^{14} 5d^0 6s^2$ 173.04	Lutetium **Lu** $[\mathrm{Xe}]\,4f^{14} 5d^1 6s^2$ 174.97

Actinides

Thorium **Th** $[\mathrm{Rn}]\,6d^2 7s^2$ 232.04	Protactinium **Pa** $[\mathrm{Rn}]\,5f^2 6d^1 7s^2$ 231	Uranium **U** $[\mathrm{Rn}]\,5f^3 6d^1 7s^2$ 238.03	Neptunium **Np** $[\mathrm{Rn}]\,5f^5 6d^0 7s^2$ 237.05	Plutonium **Pu** $[\mathrm{Rn}]\,5f^6 6d^0 7s^2$ 244	Americium **Am** $[\mathrm{Rn}]\,5f^7 6d^0 7s^2$ 243	Curium **Cm** $[\mathrm{Rn}]\,5f^7 6d^1 7s^2$ 247	Berkelium **Bk** $[\mathrm{Rn}]\,5f^9 6d^0 7s^2$ 247	Californium **Cf** $[\mathrm{Rn}]\,5f^{10} 7s^2$ 251	Einsteinium **Es**	Fermium **Fm**	Mendelevium **Md**	Nobelium **No**	Lawrencium **Lw**

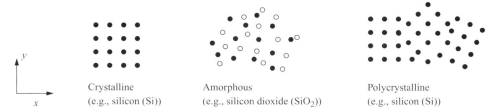

Fig. 2.14. Illustration of different types of solids according to atomic arrangement.

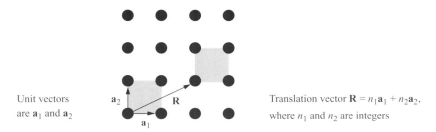

Fig. 2.15. A two-dimensional square lattice can be created by translating the unit vectors $\mathbf{a}_1$ and $\mathbf{a}_2$ through space according to $\mathbf{R} = n_1\mathbf{a}_1 + n_2\mathbf{a}_2$, where n_1 and n_2 are integers.

periods, crystalline solids are said to be characterized by long-range spatial order. The positions of atoms that form amorphous solids do not have long-range spatial order. Polycrystalline solids are made up of small regions of crystalline material with boundaries that break the spatial periodicity of atom positions.

2.2.5.1 Three types of solids classified according to atomic arrangement

Atoms in a crystalline solid are located in space on a *lattice*. The *unit cell* is a lattice volume, which is representative of the entire lattice and is repeated throughout the crystal. The smallest unit cell that can be used to form the lattice is called a *primitive cell*.

2.2.5.2 Two-dimensional square lattice

Figure 2.15 shows a square lattice of atoms and a square unit cell translated by vector $\mathbf{R}$. In general, the unit vectors $\mathbf{a}_i$ do not have to be in the $\mathbf{x}$ and $\mathbf{y}$ directions.

2.2.5.3 Three-dimensional crystals

Three-dimensional crystals are made up of a periodic array of atoms. For a given lattice there exists a basic unit cell that can be defined by the three vectors $\mathbf{a}_1$, $\mathbf{a}_2$, and $\mathbf{a}_3$.

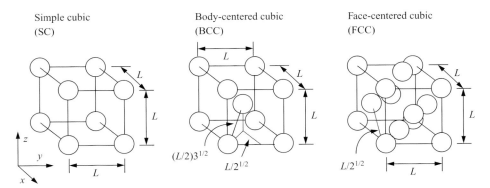

Fig. 2.16. Illustration of the indicated three-dimensional cubic unit cells, each of lattice constant L.

Crystal structure is defined as a real-space translation of basic points throughout space via

$$\mathbf{R} = n_1\mathbf{a}_1 + n_2\mathbf{a}_2 + n_3\mathbf{a}_3 \tag{2.85}$$

where n_1, n_2, and n_3 are integers. This complete real-space lattice is called the Bravais lattice. The volume of the basic unit cell (the primitive cell) is

$$\Omega_{\text{cell}} = \mathbf{a}_1 \cdot (\mathbf{a}_2 \times \mathbf{a}_3) \tag{2.86}$$

A good choice for the vectors $\mathbf{a}_1$, $\mathbf{a}_2$, and $\mathbf{a}_3$ that defines the primitive unit cell is due to Wigner and Seitz. The Wigner–Seitz cell about a lattice reference point is specified in such a way that any point of the cell is closer to that lattice point than any other. The Wigner–Seitz cell may be found by bisecting with perpendicular planes all vectors connecting a reference atom position to all atom positions in the crystal. The smallest volume enclosed is the Wigner–Seitz cell.

2.2.5.4 Cubic lattices in three dimensions

Possibly the simplest three-dimensional lattice to visualize is one in which the unit cell is cubic. In Fig. 2.16, L is called the *lattice constant*.

The simple cubic (SC) unit cell has an atom located on each corner of a cube side L. The body-centered cubic (BCC) unit cell is the same as the SC but with an additional atom in the center of the cube. Elements with the BCC crystal structure include Fe ($L = 0.287$ nm), Cr ($L = 0.288$ nm), and W ($L = 0.316$ nm). The face-centered cubic (FCC) unit cell is the same as the SC but with an additional atom on each face of the cube. Elements with the FCC crystal structure include Al ($L = 0.405$ nm), Ni ($L = 0.352$ nm), Au ($L = 0.408$ nm), and Cu ($L = 0.361$ nm). The diamond crystal structure shown in Fig. 2.17 consists of two interpenetrating FCC lattices off-set from each other by $L(\hat{\mathbf{x}} + \hat{\mathbf{y}} + \hat{\mathbf{z}})/4$. Elements with the *diamond crystal structure* include Si ($L = 0.543$ nm), Ge ($L = 0.566$ nm), and C ($L = 0.357$ nm). GaAs has the *zinc*

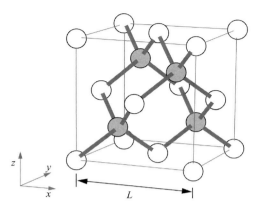

Fig. 2.17. Illustration of the diamond lattice cubic unit cell with lattice constant L. The tetrahedrally coordinated nearest-neighbor bonds are shown. GaAs is an example of a III-V compound semiconductor with the zinc blende crystal structure. This is the same as diamond, except that instead of one atom type populating the lattice the atom type alternates between Ga and As.

blende crystal structure which is the same as the diamond except that instead of one atom type populating the lattice the atom type alternates between Ga and As. Group III-V compound semiconductors with the zinc blende crystal structure include GaAs ($L = 0.565$ nm), AlAs ($L = 0.566$ nm), AlGaAs, InP ($L = 0.587$ nm), InAs ($L = 0.606$ nm), InGaAs, and InGaAsP.

2.2.5.5 The reciprocal lattice

Because the properties of crystals are often studied using wave-scattering experiments, it is important to consider the reciprocal lattice in reciprocal space (also known as wave vector space or k space). Given the basic unit cell defined by the vectors $\mathbf{a}_1$, $\mathbf{a}_2$, and $\mathbf{a}_3$ in real space, one may construct three fundamental reciprocal vectors, $\mathbf{g}_1$, $\mathbf{g}_2$, and $\mathbf{g}_3$, in reciprocal space defined by $\mathbf{a}_i \cdot \mathbf{g}_j = 2\pi \delta_{ij}$, so that $\mathbf{g}_1 = 2\pi(\mathbf{a}_2 \times \mathbf{a}_1)/\Omega_{\text{cell}}$, $\mathbf{g}_2 = 2\pi(\mathbf{a}_3 \times \mathbf{a}_1)/\Omega_{\text{cell}}$, and $\mathbf{g}_3 = 2\pi(\mathbf{a}_1 \times \mathbf{a}_2)/\Omega_{\text{cell}}$.

Crystal structure may be defined as a reciprocal-space translation of basic points throughout the space, in which

$$\boxed{\mathbf{G} = n_1\mathbf{g}_1 + n_2\mathbf{g}_2 + n_3\mathbf{g}_3} \tag{2.87}$$

where n_1, n_2, and n_3 are integers. The complete space spanned by $\mathbf{G}$ is called the *reciprocal lattice*. The volume of the reciprocal-space unit cell is

$$\Omega_k = \mathbf{g}_1 \cdot (\mathbf{g}_2 \times \mathbf{g}_3) = \frac{(2\pi)^3}{\Omega_{\text{cell}}} \tag{2.88}$$

The *Brillouin zone* of the reciprocal lattice has the same definition as the Wigner–Seitz cell in real space. The first Brillouin zone may be found by bisecting with perpendicular

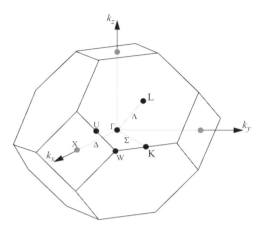

Fig. 2.18. Illustration of the Brillouin zone for the face-centered cubic lattice with lattice constant L. Some high-symmetry points are $\Gamma = (0, 0, 0)$, $X = (2\pi/L)(1, 0, 0)$, $L = (2\pi/L)(0.5, 0.5, 0.5)$, $W = (2\pi/L)(1, 0.5, 0)$. The high-symmetry line between the points Γ and X is labeled Δ, the line between the points Γ and L is Λ, and the line between Γ and K is Σ.

planes all reciprocal-lattice vectors. The smallest volume enclosed is the first Brillouin zone.

As an example, consider a face-centered cubic lattice in real space. To find the basic reciprocal lattice vectors for a face-centered cubic lattice we note that the basic unit cell vectors in real space are $\mathbf{a}_1 = (0, 1, 1)(L/2)$, $\mathbf{a}_2 = (1, 0, 1)(L/2)$, and $\mathbf{a}_3 = (1, 1, 0)(L/2)$, so that $\mathbf{g}_1 = 2\pi(-1, 1, 1)/L$, $\mathbf{g}_1 = 2\pi(1, -1, 1)/L$, and $\mathbf{g}_1 = 2\pi(1, 1, -1)/L$. Hence, the reciprocal lattice of a face-centered cubic lattice in real space is a body-centered cubic lattice.

Figure 2.18 is an illustration of the first Brillouin zone for the face-centered cubic lattice with lattice constant L. As may be seen, in this case the Brillouin zone is a truncated octahedron.

2.2.6   Electronic properties of bulk semiconductors and heterostructures

The energy states of an electron in a hydrogen atom are quantized and may only take on discrete values. The same is true for electrons in all atoms. There are discrete energy levels that an electron in a single atom may have, and all other energy values are not allowed.

In a single-crystal solid, electrons from the many atoms that make up the crystal can interact with one another. Under these circumstances, the discrete energy levels of single-atom electrons disappear, and instead there are finite and continuous ranges or bands of energy states with contributions from many individual atom electronic states.

The lowest-energy state of a semiconductor crystal exists when electrons occupy all available valence-band states. In a pure semiconductor crystal there is a range of energy states that are not allowed and this is called the energy band gap. A typical band-gap energy is $E_g = 1\,\text{eV}$. The band-gap energy separates valence-band states from conduction-band states by an energy of at least E_g. In this way the semiconductor retains a key feature of an atom – there are *allowed* energy states and *disallowed* energy states. The existence of an electron energy band gap in a semiconductor crystal may only be explained by quantum mechanics.

Single crystals of the group IV element Si or the group III–V binary compound GaAs are examples of technologically important semiconductors. Semiconductor crystals have characteristic electronic properties. Electrons are not allowed in the energy band gap that separates valence-band electronic states in energy from conduction-band states. A pure crystalline semiconductor is an electrical insulator at low temperature. In the lowest-energy state, or *ground state*, of a pure semiconductor, all electron states are occupied in the *valence band* and there are no electrons in the *conduction band*. The periodic array of atoms that forms the semiconductor crystal creates a periodic coulomb potential. If an electron is free to move in the material, its motion is influenced by the presence of the periodic potential. Typically, electrons with energy near the conduction-band minimum or energy near the valence-band maximum have an electron dispersion relation that may be characterized by a parabola, $\omega(k) \propto k^2$. The kinetic energy of the electron in the crystal may therefore be written as $E_k = \hbar\omega = \hbar^2 k^2 / 2m^*$, where m^* is called the *effective electron mass*. The value of m can be greater or less than the mass of a "bare" electron moving in free space. Because different semiconductors have different periodic coulomb potentials, different semiconductors have different values of effective electron mass. For example, the conduction-band effective electron mass in GaAs is approximately $m^* = 0.07\,m_0$, and in InAs it is near $m^* = 0.02\,m_0$, where m_0 is the bare electron mass.

As illustrated in Fig. 2.19, at finite temperatures the electrically insulating ground state of the semiconductor changes, and electrical conduction can take place due to thermal excitation of a few electrons from the valence band into the conduction band. Electrons excited into the conduction band and absences of electrons (or *holes*) in the valence band are free to move in response to an electric field, and they can thereby contribute to electrical conduction.

The thermal excitation process involves lattice vibrations that collide via the coulomb interaction with electrons. In addition, electrons may scatter among themselves. The various collision processes allow electrons to equilibrate to a temperature T, which is the same as the temperature of the lattice. Electrons in thermal equilibrium have a statistical distribution in energy that includes a small finite probability at relatively high energy.

The statistical energy distribution of electrons in thermal equilibrium at absolute temperature T is typically described by the Fermi–Dirac distribution function

$$f_{\mathbf{k}}(E_{\mathbf{k}}) = \frac{1}{e^{(E_{\mathbf{k}} - \mu)/k_B T} + 1} \tag{2.89}$$

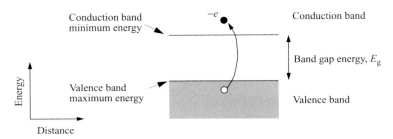

Fig. 2.19. Energy–distance diagram of a semiconductor showing valence band, conduction band, and band-gap energy E_g. Also shown is the excitation of an electron from the valence band into the conduction band.

In this expression, the chemical potential μ is defined as the energy to place an extra electron into the system of n electrons. $f_{\mathbf{k}}(E_{\mathbf{k}})$ is the probability of occupancy of a given electron state of energy $E_{\mathbf{k}}$, and so it has a value between 0 and 1. The Fermi–Dirac distribution is driven by the Pauli exclusion principle which states that *identical indistinguishable half-odd-integer spin particles cannot occupy the same state.* An almost equivalent statement is that the total n particle eigenfunction must be antisymmetric (i.e., must change sign) upon the permutation of any two particles.

The distribution function for electrons in the limit of low temperature ($T \rightarrow 0\,\text{K}$) becomes a step function, with electrons occupying all available states up to energy $\mu_{T=0}$ and no states with energy greater than $\mu_{T=0}$. This low-temperature limit is so important that it has a special name: it is called the Fermi energy, E_{F}. For electrons with effective mass m^*, one may write $E_{\text{F}} = \hbar^2 k_{\text{F}}^2 / 2m^*$, where k_{F} is called the Fermi wave vector.

At finite temperatures, and in the limit of electron energies that are large compared with the chemical potential, the distribution function takes on the Boltzmann form $f_{\mathbf{k}}(E \rightarrow \infty) = e^{-E/k_{\text{B}}T}$. This describes the high-energy tail of an electron distribution function at finite temperatures. We will now use this fact to gain insight into the electronic properties of a semiconductor.

As shown schematically in Fig. 2.19, a valence-band electron with enough energy to surmount the semiconductor band-gap energy can enter the conduction band. The promotion of an electron from the valence band to the conduction band is an example of an *excitation* of the system. In the presence of such excitations, the lattice vibrations can collide with electrons in such a way that a distribution of electrons exists in the conduction band and a distribution of *holes* (absences of electrons) exists in the valence band. The energy distribution of occupied states is usually described by the Fermi–Dirac distribution function given by Eqn (2.89), in which energy is measured from the conduction-band minimum for electrons and from the valence-band maximum for holes. The concept of electron and hole distribution functions to describe an excited semiconductor is illustrated in Fig. 2.20.

A typical value of the band-gap energy is $E_g = 1\,\text{eV}$. For example, at room temperature pure crystalline silicon has $E_g = 1.12\,\text{eV}$. A measure of conduction-band

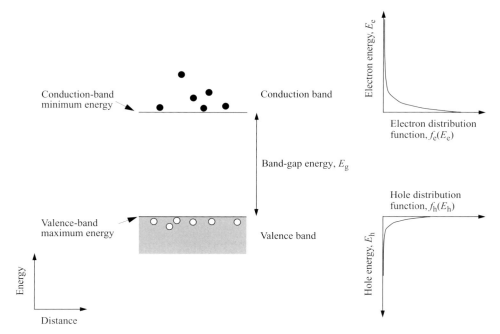

Fig. 2.20. Energy–distance diagram of a semiconductor showing valence band, conduction band, and band-gap energy E_g. The excitation of electrons from the valence band into the conduction band creates an electron distribution function $f_e(E_e)$ in the conduction band and a hole distribution function $f_h(E_h)$ in the valence band. Electron energy E_e is measured from the conduction-band minimum and hole energy E_h is measured from the valence-band maximum.

electron occupation probability at room-temperature energy $k_B T = 25\,\text{meV}$ can be obtained using the Boltzmann factor. For $E_g = 1\,\text{eV}$, this gives $prob \sim e^{-E_g/k_B T} = e^{-40} = 4 \times 10^{-18}$, and the resulting electrical conductivity of such *intrinsic* material is not very great. Semiconductor technology makes use of *extrinsic* methods to increase electrical conductivity to a carefully controlled and predetermined value.

Electrons that are free to move in the material may be introduced into the conduction band by adding impurity atoms. This extrinsic process, called *substitutional doping*, replaces atoms of the pure semiconductor with atoms the effect of which is to add mobile charge carriers. In the case of n-type doping, electrons are added to the conduction band. To maintain overall charge neutrality of the semiconductor, for every negatively charged electron added to the conduction band there is a positively charged impurity atom located at a substitutional doping site in the crystal. As the density of mobile charge carriers is increased, the electrical conductivity of the semiconductor can increase dramatically.

The exact amount of impurity concentration or doping level in a thin layer of semiconductor can be precisely controlled using modern crystal growth techniques such as molecular beam epitaxy (MBE) or metalorganic chemical vapor deposition (MOCVD). In addition to controlling the impurity concentration profile, the same

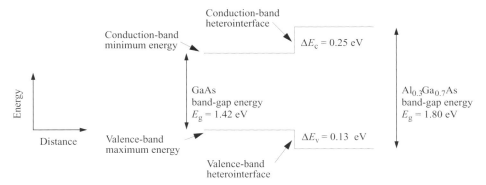

Fig. 2.21. Diagram to illustrate the conduction-band potential energy step ΔE_c and the valence-band potential energy step ΔE_v created at a heterointerface between GaAs and $Al_{0.3}Ga_{0.7}As$. The band-gap energy of the two semiconductors is indicated.

epitaxial semiconductor crystal growth techniques may be used to grow thin layers of different semiconductor materials on top of one another to form a *heterostructure*. Excellent single-crystal results may be achieved if the different semiconductors have the same crystal symmetry and lattice constant. In fact, even nonlattice-matched semiconductor crystal layers may be grown, as long as the strained layer is thin and the lattice mismatch is not too great. The interface between the two different materials is called a *heterointerface*. Semiconductor epitaxial layer thickness is controllable to within a monolayer of atoms, and quite complex structures can be grown with many heterointerfaces.

Of course, a basic question concerns the relative position of the band gaps or "band structure line up". At a heterointerface consisting of material with two different band-gap energies, part of the band-gap energy difference appears as a potential step in the conduction band. Figure 2.21 shows a conduction-band potential energy step ΔE_c and a valence-band potential energy step ΔE_v created at a heterointerface between GaAs and $Al_{0.3}Ga_{0.7}As$. By carefully designing a multi-layer heterostructure semiconductor it is possible to create a specific potential as a function of distance in the conduction band. Typically, atomically abrupt changes are possible at heterointerfaces and more gradual changes in potential may be achieved either by using changes in doping concentration or by forming semiconductor alloys the band gap of which changes as a function of position in the crystal growth direction. Figure 2.22 shows an electron microscope cross-section image of epitaxially grown single-crystal semiconductor consisting of a sequence of 4-monolayer-thick Ge/Si layers. The periodic layered heterostructure forms a superlattice that can be detected by electron diffraction measurements (Fig. 2.22, inset). The fact that a monoenergetic electron beam creates a diffraction pattern is, in itself, both a manifestation of the wavy nature of electrons and a measure of the quality of the superlattice.

The use of modern crystal growth techniques to define the composition of semiconductor layers with atomic precision allows new types of electronic and photonic

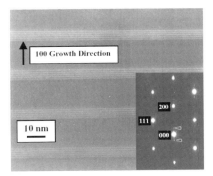

Fig. 2.22. High-resolution transmission electron microscope cross-section image of epitaxially grown single-crystal layers of Ge(100) interspersed with sequences of 4-monolayer Si and 4-monolayer Ge layers. The inset electron diffraction pattern confirms the high perfection of the superlattice periodicity. The image is taken in (400) bright field mode. Image courtesy of R. Hull and J. Bean, University of Virginia.

devices to be designed. As we will discover in the next chapter, when devices make use of electron motion through potentials that change rapidly on a length scale comparable to the wavelength associated with the electron, these devices will operate according to the rules of quantum mechanics.

2.2.6.1 The heterostructure diode

Epitaxial crystal growth techniques can be used to create a heterostructure diode. To illustrate the current–voltage characteristics of such a diode, we will consider the special case of a unipolar n-type device formed using the heterointerface between GaAs and $Al_{0.3}Ga_{0.7}As$. The GaAs is heavily n-type, and the wider band-gap $Al_{0.3}Ga_{0.7}As$ is more lightly doped n-type. The lightly doped $Al_{0.3}Ga_{0.7}As$ is depleted due to the presence of a conduction-band potential energy step ΔE_c at the heterointerface. As shown in Fig. 2.23, the depletion region exposes the positive charge of the substitutional dopant atoms. The potential energy profile $eV(x)$ and the value of the depletion width w in the $Al_{0.3}Ga_{0.7}As$ region with uniform impurity concentration n may be found by solving Poisson's equation $\nabla \cdot \mathbf{E} = \rho/\varepsilon_0\varepsilon_r$ (Eqn (1.37)). Considering an electric field $\mathbf{E}$ in one dimension and noting that the charge density is $\rho = en$, this gives

$$\frac{\partial E_x}{\partial x} = \frac{en}{\varepsilon_0\varepsilon_r} = -\frac{\partial^2}{\partial x^2}V(x) \tag{2.90}$$

where $V(x)$ is the potential profile of the conduction-band minimum. The potential profile is found by integrating Eqn (2.90), and the depletion region width for a built-in potential energy barrier $eV_0 = \Delta E_c - eV_n$ is approximately

$$w = \left(\frac{2\varepsilon_0\varepsilon_r}{en}V_0\right)^{1/2} \tag{2.91}$$

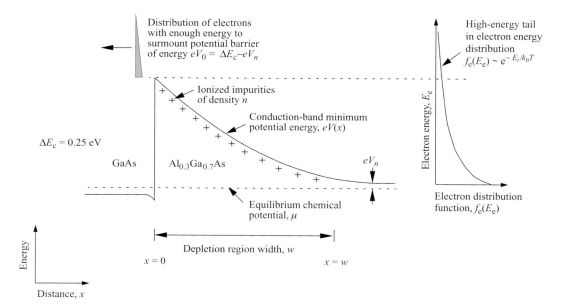

Fig. 2.23. Diagram of the conduction-band minimum of a unipolar n-type GaAs/Al$_{0.3}$Ga$_{0.7}$As heterostructure diode. The GaAs is heavily doped, and the Al$_{0.3}$Ga$_{0.7}$As is lightly doped. The conduction-band offset at the heterointerface is $\Delta E_c = 0.25\,\mathrm{eV}$. The depletion region width is w. The chemical potential μ used to describe the electron distribution under zero-bias equilibrium conditions is shown.

Under an applied external voltage bias of V_{ex}, Eqn (2.91) becomes

$$w = \left(\frac{2\varepsilon_0 \varepsilon_r}{en}(V_0 - V_{\mathrm{ex}}) \right)^{1/2} \tag{2.92}$$

In Fig. 2.23 an effective potential energy barrier of approximately eV_0 always exists for electrons trying to move left-to-right from the heavily doped GaAs into the less doped Al$_{0.3}$Ga$_{0.7}$As region. This limits the corresponding current flow to a constant, I_0. However, electrons of energy E_e moving right-to-left from Al$_{0.3}$Ga$_{0.7}$As to GaAs see an effective barrier $e(V_0 - V_{\mathrm{ex}})$, which depends upon the applied external voltage V_{ex}. In a simple thermionic emission model, this contribution to current flow can have an exponential dependence upon voltage bias because, as illustrated in Fig. 2.23, the high-energy tail of the Fermi–Dirac distribution function has an exponential Boltzmann form, $e^{-E_e/k_B T}$. Only those electrons in the undepleted Al$_{0.3}$Ga$_{0.7}$As region with enough energy to surmount the potential barrier can contribute to the right-to-left electron current. It follows that the voltage dependence of right-to-left electron current is $I_0 e^{-eV_{\mathrm{ex}}/k_B T}$.

The net current flow across the diode is just the sum of right-to-left and left-to-right current, which gives

$$I = I_0(e^{-eV_{\mathrm{ex}}/n_{\mathrm{id}}k_B T} - 1) \tag{2.93}$$

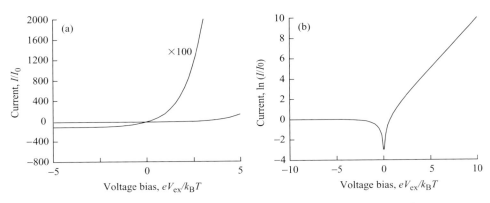

Fig. 2.24. (a) Current–voltage characteristics of an ideal diode plotted on a linear scale. (b) Current–voltage characteristics of an ideal diode. The natural logarithm of normalized current is plotted on the vertical axis.

In this expression we have included a phenomenological factor n_{id}, called the ideality factor, which takes into account the fact that there may be deviations from the predictions of the simple thermionic emission model we have used. A nonideal diode has $n_{id} > 1$. Figure 2.24 shows the predicted current–voltage characteristics of a typical diode for the ideal case when $n_{id} = 1$.

It is clear from Fig. 2.24 that a diode has a highly nonlinear current–voltage characteristic. The exponential increase in current with positive or forward voltage bias and the essentially constant current with negative or reverse voltage bias are very efficient mechanisms for controlling current flow. It is this characteristic *exponential sensitivity* of an output (current) to an input control signal (voltage) that makes the diode such an important device and a basic building block for constructing other, more complex, devices.

The concept of exponential sensitivity can often be used as a guide when we are trying to decide whether a new device or device concept is likely to find practical applications.

2.3 Example exercises

Exercise 2.1
(a) Given that Planck's radiative energy density spectrum for thermal light is

$$S(\omega) = \frac{\hbar\omega^3}{\pi^2 c^3} \frac{1}{e^{\hbar\omega/k_B T} - 1}$$

show that in the low-frequency (long-wavelength) limit this reduces to the Rayleigh–Jeans spectrum

$$S(\omega) = \frac{k_B T}{\pi^2 c^3} \omega^2$$

in which the Planck constant does not appear. This is an example of the *correspondence principle*, in which as $\hbar \to 0$ one obtains the result known from classical mechanics. Show that the high-frequency (short-wavelength) limit of Planck's radiative energy density spectrum reduces to the Wien spectrum

$$S(\omega) = \frac{\hbar\omega^3}{\pi^2 c^3} e^{-\hbar\omega/k_B T}$$

(b) Find the energy of *peak* radiative energy density, $\hbar\omega_{peak}$, in Planck's expression for $S(\omega)$, and compare the *average* value $\hbar\omega_{average}$ with $k_B T$.

(c) Show that the total radiative energy density for thermal light is

$$S_{total} = \frac{\pi^2 k_B^4 T^4}{15 c^3 \hbar^3}$$

(d) The Sun has a surface temperature of 5800 K and an average radius 6.96×10^8 m. Assuming that the mean Sun–Earth distance is 1.50×10^{11} m, what is the peak radiative power per unit area incident on the upper Earth atmosphere facing the Sun?

Exercise 2.2

Find the normalized autocorrelation function for a rectangular pulse of width t_0 in time. What is the coherence time of this pulse?

Exercise 2.3

Show that the de Broglie wavelength of an electron of kinetic energy $E(eV)$ is

$$\lambda_e = 1.23/\sqrt{E(eV)} \text{ nm}$$

(a) Calculate the wavelength and momentum associated with an electron of kinetic energy 1 eV.

(b) Calculate the wavelength and momentum associated with a photon that has the same energy.

(c) Show that constructive interference for a monochromatic plane wave of wavelength λ scattering from two planes separated by distance d occurs when $n\lambda = 2d\cos(\theta)$, where θ is the incident angle of the wave measured from the plane normal and n is an integer. What energy electrons and what energy photons would you use to observe Bragg scattering peaks when performing electron or photon scattering measurements on a nickel crystal with lattice constant $L = 0.352$ nm?

Exercise 2.4

An expression for the energy levels of a hydrogen atom is $E_n = -13.6/n^2$ eV, where n is an integer, such as $n = 1, 2, 3 \ldots$.

(a) Using this expression, draw an energy level diagram for the hydrogen atom.

(b) Derive the expression for the energy (in units of eV) and wavelength (in units of nm) of emitted light from transitions between energy levels.

(c) Calculate the three longest wavelengths (in units of nm) for transitions terminating at $n = 2$.

Exercise 2.5

Write down a Schrödinger equation for (a) a helium atom and (b) a simple one-dimensional harmonic oscillator with potential $V(x) = \kappa x^2 / 2$.

Exercise 2.6

In quantum mechanics the quantum of lattice vibration is called a phonon. Consider a longitudinal polar-optic phonon in GaAs that has energy $\hbar \omega = 36.3$ meV, where $\hbar$ is Planck's constant and ω is the angular frequency of the lattice vibration. Assuming a particle mass of $m = 72 \times m_p$, where m_p is the mass of a proton, estimate the amplitude of oscillation of this phonon relative to the nearest-neighbor spacing. The lattice constant of GaAs is $L = 5.65 \times 10^{-10}$ m.

Exercise 2.7

Consider a helium atom (He) with one electron missing. Estimate the energy difference between the ground state and the first excited state. Express the answer in units of eV. The binding energy of the electron in the hydrogen atom is 13.6 eV.

Exercise 2.8

If momentum $\hat{p}_x = \hbar k_x$, then the expectation value of momentum for a particle described by wave function $\psi(x)$ can be obtained from

$$\langle \hat{p}_x \rangle = \int \phi^*(k_x) \hbar k_x \phi(k_x) dk_x$$

where $\phi(k_x)$ is the Fourier transform of $\psi(x)$, so that

$$\langle \hat{p}_x \rangle = \frac{1}{2\pi} \int\limits_{-\infty}^{\infty} dk_x \left(\int\limits_{-\infty}^{\infty} dx' \psi^*(x') e^{ik_x x'} \right) \hbar k_x \left(\int\limits_{-\infty}^{\infty} dx \psi(x) e^{-ik_x x} \right)$$

(a) Integrating by parts and using the relation $\delta(x - x') = \frac{1}{2} \int_{-\infty}^{\infty} dk \, e^{ik(x-x')}$, show that

$$\langle \hat{p}_x \rangle = \int\limits_{-\infty}^{\infty} dx \psi^*(x) \left(-i\hbar \frac{\partial}{\partial x} \right) \psi(x)$$

so that one may conclude that if $\hat{p}_x = \hbar k_x$ in k space (momentum space) then the momentum operator in real space is $-i\hbar \cdot \partial/\partial x$.

(b) There is a full symmetry between the position operator and the momentum operator. They form a *conjugate pair*. In real space, momentum is a differential operator. Show that in k space position is a differential operator, $i\hbar \cdot \partial/\partial p_x$, by evaluating expectation value

$$\langle \hat{x} \rangle = \int\limits_{-\infty}^{\infty} dx\, \psi^*(x) x \psi(x)$$

in terms of $\phi(k_x)$, which is the Fourier transform of $\psi(x)$.

(c) The wave function for a particle in real space is $\psi(x, t)$. Usually, it is assumed that position x and time t are continuous and smoothly varying. Given that particle energy is quantized so that $E = \hbar\omega$, show that the energy operator for the wave function $\psi(x, t)$ is $i\hbar \cdot \partial/\partial t$.

Exercise 2.9

In Section 2.2.3.2 it was stated that the degeneracy of state ψ_{nlm} in a hydrogen atom is n^2. Show that this is so by proving

$$\sum_{l=0}^{l=n-1} (2l + 1) = n^2$$

SOLUTIONS

Solution 2.1

(a) Given that Planck's radiative energy density spectrum for thermal light is

$$S(\omega) = \frac{\hbar\omega^3}{\pi^2 c^3} \frac{1}{e^{\hbar\omega/k_B T} - 1}$$

we wish to show that in the low-frequency (long-wavelength) limit this reduces to the Rayleigh–Jeans spectrum

$$S(\omega) = \frac{k_B T}{\pi^2 c^3} \omega^2$$

in which the Planck constant does not appear.

Expanding the exponential in the expression for radiative energy density

$$S(\omega) = \frac{\hbar\omega^3}{\pi^2 c^3} \frac{1}{e^{\hbar\omega/k_B T} - 1}$$

gives

$$e^{\hbar\omega/k_B T} \sim 1 + \hbar\omega/k_B T + \cdots$$

in the low-frequency limit, so that

$$S(\omega) \sim \frac{\hbar\omega^3}{\pi^2 c^3} \frac{1}{1 + \hbar\omega/k_B T + \cdots - 1} = \frac{k_B T}{\pi^2 c^3}\omega^2$$

which is the Rayleigh–Jeans spectrum. This is an example of the *correspondence principle* in which as $\hbar \to 0$ one obtains the result known from classical mechanics.

In the high-frequency limit $\omega \to \infty$, so that the exponential term in $S(\omega)$ becomes $1/(e^{\hbar\omega/k_B T} - 1) \to e^{-\hbar\omega/k_B T}$. It follows that the high-frequency (short-wavelength) limit of Planck's radiative energy density spectrum reduces to the Wien spectrum

$$S(\omega) = \frac{\hbar\omega^3}{\pi^2 c^3} e^{-\hbar\omega/k_B T}$$

(b) To find the energy of *peak* radiative energy density, $\hbar\omega_{\text{peak}}$, in Planck's expression for $S(\omega)$, we seek

$$\frac{d}{d\omega}S(\omega) = \frac{d}{d\omega}\left(\frac{\hbar\omega^3}{\pi^2 c^3} \frac{1}{e^{\hbar\omega/k_B T} - 1}\right) = 0$$

$$\frac{3\hbar\omega^2}{\pi^2 c^3} \frac{1}{e^{\hbar\omega/k_B T} - 1} - \frac{\hbar\omega^2}{\pi^2 c^3}\left(\frac{\hbar}{k_B T}\right)\frac{e^{\hbar\omega/k_B T}}{(e^{\hbar\omega/k_B T} - 1)^2} = 0$$

$$\frac{3\hbar\omega^2}{\pi^2 c^3} - \frac{\hbar\omega^3}{\pi^2 c^3}\left(\frac{\hbar}{k_B T}\right)\frac{e^{\hbar\omega/k_B T}}{e^{\hbar\omega/k_B T} - 1} = 0$$

Dividing by $\hbar\omega^2/\pi^2 c^3$, letting $x = \hbar\omega/k_B T$, and multiplying by $(e^x - 1)$ gives

$$3(e^x - 1) - xe^x = 0$$

or

$$3(1 - e^{-x}) - x = 0$$

which we may solve numerically to give $x \sim 2.82$, so that $\hbar\omega_{\text{peak}} \sim 2.82 \times k_B T$. For the Sun with $T = 5800\,\text{K}$, this gives a peak in $S(\omega)$ at energy $\hbar\omega_{\text{peak}} = 1.41\,\text{eV}$, or, equivalently, frequency $\nu = 341\,\text{THz}$ ($\omega = 2.14 \times 10^{15}\,\text{rad s}^{-1}$).

To find an average value of z distributed as $f(z)$ we seek $\langle f(z)\rangle = \int zf(z)dz/\int f(z)dz$. In our case, we will be weighting the integral by the energy $\hbar\omega$, so

$$\hbar\omega_{\text{average}} = \int_0^\infty \hbar\omega S(\omega)d\omega \bigg/ \int_0^\infty S(\omega)d\omega$$

$$\hbar\omega_{\text{average}} = \int_0^\infty \hbar\omega \frac{\hbar\omega^3}{\pi^2 c^3} \frac{1}{e^{\hbar\omega/k_B T} - 1}d\omega \bigg/ \int_0^\infty \frac{\hbar\omega^3}{\pi^2 c^3} \frac{1}{e^{\hbar\omega/k_B T} - 1}d\omega$$

Introducing $x = \hbar\omega/k_B T$ so that $dx = (\hbar/k_B T)d\omega$, $\omega = x(k_B T/\hbar)$, and $d\omega = dx(k_B T/\hbar)$ gives

$$\hbar\omega_{\text{average}} = \frac{\dfrac{\hbar^2}{\pi^2 c^3} \cdot \dfrac{k_B T}{\hbar} \cdot \left(\dfrac{k_B T}{\hbar}\right)^4 \cdot \displaystyle\int_0^\infty \dfrac{x^4}{e^x - 1}dx}{\dfrac{\hbar}{\pi^2 c^3} \cdot \dfrac{k_B T}{\hbar} \cdot \left(\dfrac{k_B T}{\hbar}\right)^3 \cdot \displaystyle\int_0^\infty \dfrac{x^3}{e^x - 1}dx} = k_B T \cdot \frac{\displaystyle\int_0^\infty \dfrac{x^4}{e^x - 1}dx}{\displaystyle\int_0^\infty \dfrac{x^3}{e^x - 1}dx}$$

To solve the integrals we note that[3]

$$\int_0^\infty \frac{x^{p-1}}{e^{rx} - q}dx = \frac{1}{qr^p}\Gamma(p)\sum_{k=1}^\infty \frac{q^k}{k^p}$$

where $\Gamma(p)$ is the gamma function.[4]

For the numerator, $\Gamma(p = 5) = 24$, and the integral is approximately 25. For the denominator, we use $\Gamma(p = 4) = 6$, and the integral is approximately 6.5. Hence, $\hbar\omega_{\text{average}} \sim k_B T \cdot (25/6.5) = 3.85 k_B T$. We notice that the average value is greater than the peak value. This is expected, since the function $S(\omega)$ is not symmetric.

(c) To find the total radiative energy density for thermal light $S_{\text{total}}(\omega)$ we need to integrate the expression for $S(\omega)$ over all frequencies. This gives

$$S_{\text{total}} = \int_{\omega=0}^{\omega=\infty} S(\omega)d\omega = \int_{\omega=0}^{\omega=\infty} \frac{\hbar\omega^3}{\pi^2 c^3}\frac{1}{e^{\hbar\omega/k_B T} - 1}d\omega = \frac{k_B^4 T^4}{\pi^2 c^3 \hbar^3}\int_{x=0}^{x=\infty}\frac{x^3}{e^x - 1}dx$$

where $x = \hbar\omega/k_B T$. The integral on the right-hand side is standard:

$$\int_{x=0}^{x=\infty}\frac{x^3}{e^x - 1}dx = \frac{\pi^4}{15}$$

giving the final result

$$S_{\text{total}} = \frac{\pi^2 k_B^4 T^4}{15 c^3 \hbar^3}$$

The fact that the total energy density is proportional to T^4 is known as the Stefan–Boltzmann radiation law.

(d) The Sun has a surface temperature of 5800 K, so the total radiative energy density of the Sun is

[3] I. S. Gradshteyn and I. M. Ryzhik, *Table of Integrals, Series, and Products*, Academic Press, San Diego, 1980, p. 326 (ISBN 0 12 294760 6).

[4] M. Abramowitz and I. A. Stegun, *Handbook of Mathematical functions*, Dover, New York, 1974, pp. 267–273 (ISBN 0 486 612724 4).

$$S_{\text{total}} = \frac{\pi^2 k_B^4 T^4}{15c^3\hbar^3} = \frac{\pi^2 \times (1.3807 \times 10^{-23})^4 (5800)^4}{15 \times (3 \times 10^8)^3 (1.054 \times 10^{-34})^3} = 0.857 \text{ J m}^{-3}$$

The average radius of the Sun is $r_{\text{Sun}} = 6.96 \times 10^8$ m, and the mean Sun–Earth distance is $R_{\text{Sun–Earth}} = 1.50 \times 10^{11}$ m. The peak radiative power per unit area incident on the upper Earth atmosphere *facing* the Sun is given by

$$S_{\text{total}} \times c \times \left(\frac{r_{\text{Sun}}}{R_{\text{Sun–Earth}}}\right)^2 = 0.857 \times (3 \times 10^8) \times \left(\frac{6.96 \times 10^8}{1.50 \times 10^{11}}\right)^2 = 5.5 \text{ kW m}^{-2}$$

To obtain the *average* radiation per unit area (including the Earth surface not facing the Sun) we divide by 4 to give 1.4 kW m^{-2}.

Solution 2.2

In this exercise we wish to find the normalized autocorrelation function for a rectangular pulse of width t_0 in time. The pulse is described by the function $f(t)$, and the normalized autocorrelation function is defined as

$$g(\tau) = \frac{\langle f^*(t) f(t + \tau)\rangle}{\langle f^*(t) f(t)\rangle}$$

where τ is a time delay. In our case, $g(\tau) = 0$, except for delays in the range $-t_0 \leq \tau \leq t_0$, where

$$g(\tau) = \frac{1}{t_0}(t_0 + \tau) \qquad \text{for } -t_0 \leq \tau \leq 0$$

and

$$g(\tau) = \frac{1}{t_0}(t_0 - \tau) \qquad \text{for } 0 < \tau \leq t_0$$

The coherence time is found using

$$\tau_c = \int\limits_{\tau=-\infty}^{\tau=\infty} |g(\tau)|^2 d\tau = \int\limits_{\tau=-t_0}^{\tau=0} \frac{1}{t_0^2}(t_0 + \tau)^2 d\tau + \int\limits_{\tau=0}^{\tau=t_0} \frac{1}{t_0^2}(t_0 - \tau)^2 d\tau = -\frac{2}{3}\left[\frac{(t_0 - \tau)^3}{t_0^2}\right]_{\tau=0}^{\tau=t_0}$$

$$\tau_c = \frac{2}{3}t_0$$

Solution 2.3

The fact that the de Broglie wavelength of an electron of kinetic energy $E(\text{eV})$ is given by

$$\lambda_e = \frac{1.23}{\sqrt{E(\text{eV})}} \text{ nm}$$

follows directly from the electron energy dispersion relation $E = \hbar^2 k^2/2m$, where $k = 2\pi/\lambda_e$.

(a) Putting in the numbers for an electron of energy 1 eV gives the wavelength of the electron as λ_e (1 eV) = 1.23 nm and the momentum of the electron as

$$p_e(1\ \text{eV}) = \hbar k = \frac{h}{\lambda} = 5.40 \times 10^{-25}\ \text{kg m s}^{-1}$$

(b) For comparison, the wavelength and momentum associated with a photon that has the same energy may also be calculated. We start by noticing that the wavelength of a photon is $\lambda = c/f = 2\pi c/\omega$. Hence, for a photon of energy $E = \hbar\omega = 1$ eV we have

$$\frac{2\pi\hbar c}{\hbar\omega} = \frac{hc}{E} = \frac{6.626 \times 10^{-34} \times 2.998 \times 10^8}{1.602 \times 10^{-19}} = \lambda_{\text{photon}}(1\ \text{eV}) = 1241\ \text{nm}$$

The momentum of the photon is $p_{\text{photon}} = \hbar k$, so that

$$p_{\text{photon}}(1\ \text{eV}) = \frac{h}{\lambda} = 5.34 \times 10^{-28}\ \text{kg m s}^{-1}$$

The momentum associated with a photon of energy 1 eV is about 1000 times less than that of an electron of energy 1 eV.

(c) A monochromatic plane wave of wavelength λ scatters from two planes separated by distance d. As shown in the figure below, constructive interference occurs when the extra path length $n\lambda = 2d\cos(\theta)$, where θ is the incident angle of the wave measured from the plane-normal and n is an integer.

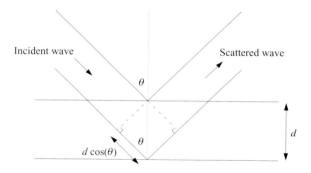

If $n = 1$ and $\theta = 0$, then $\lambda = 2d$. For a nickel crystal with lattice constant $L = 0.352$ nm one would use a monochromatic beam of electrons with kinetic energy greater than

$$E_{\text{electron}} = \left(\frac{1.23}{2 \times L}\right)^2 = 3.05\ \text{eV}$$

and photons with energy greater than

$$E_{\text{photon}} = \frac{1241}{2 \times L} = 1.76 \times 10^3\ \text{eV}$$

to observe Bragg scattering peaks from the crystal.

Solution 2.4

(a) Using the expression for the energy levels of a hydrogen atom $E_n = -13.6/n^2$ eV, where n is an integer such as $n = 1, 2, 3 \ldots$, we can draw the energy level diagram below for the hydrogen atom. The energy levels for $n = 1, 2, 3, 4, 5$, and ∞ are shown. The lowest energy (ground state) corresponds to $n = 1$.

(b) Electrons can make transitions between the energy levels of (a) by emitting light. The energy transitions are given by

$$\Delta E = E_{n_2} - E_{n_1} = -13.6 \left(\frac{1}{n_2^2} - \frac{1}{n_1^2} \right) \text{ eV}$$

Since $\lambda = 2\pi c/\omega$, the wavelength (in units of nm) of emitted light from transitions between energy levels is given by

$$\lambda = \frac{ch}{\Delta E} = \frac{ch}{13.6} \frac{1}{\dfrac{1}{n_1^2} - \dfrac{1}{n_2^2}} = \frac{91.163 \text{ nm}}{\dfrac{1}{n_1^2} - \dfrac{1}{n_2^2}}$$

$E = 0.0$ eV

$E = -0.544$ eV ·············· $n = \infty$ / $n = 5$

$E = -0.850$ eV ·············· $n = 4$

$E = -1.512$ eV ·············· $n = 3$

$E = -3.401$ eV ·············· $n = 2$

Energy

$E = -13.606$ eV ·············· $n = 1$

(c) We now calculate the three longest wavelengths (in nm) for transitions terminating at $n = 2$. We adopt the notation λ_{nm} for a transition from the quantum state labeled by quantum number n to the quantum state labeled by m. From our solution to (b) we have

$\lambda_{32} = 656.375$ nm

$\lambda_{42} = 486.204$ nm

$\lambda_{52} = 434.110$ nm

Solution 2.5

(a) Before finding the Schrödinger equation for a helium atom we remind ourselves of the general form of the Schrödinger equation for a time-independent potential:

$$\frac{-\hbar^2}{2m}\nabla^2\psi(\mathbf{r}, t) + V(\mathbf{r})\psi(\mathbf{r}, t) = H\psi(\mathbf{r}, t) = -\frac{\hbar}{i}\frac{\partial}{\partial t}\psi(\mathbf{r}, t)$$

The helium atom consists of two protons, two neutrons, and two electrons. For helium, we have two electrons, each of mass m_0, moving in the potential field of a nucleus that has charge $Z = 2e$ and mass M. The nucleus consists of the two protons and two neutrons and is assumed to behave as a single-point particle. Therefore, the Hamiltonian for the relative motion is

$$H = -\frac{\hbar^2}{2\mu}\left(\nabla_1^2 + \nabla_2^2\right) - \frac{Ze^2}{4\pi\varepsilon_0}\left(\frac{1}{r_1} + \frac{1}{r_2}\right) + \frac{e^2}{4\pi\varepsilon_0 r_{12}}$$

where μ is the reduced mass, so that $\mu^{-1} = m_0^{-1} + M^{-1}$. Electron 1 is a distance r_1 from the nucleus, and electron 2 is a distance r_2 from the nucleus. The distance between electron 1 and electron 2 is r_{12}. The first term on the right-hand side is the kinetic energy of the electrons. For the second term on the right-hand side we assume that each electron of charge $-e$ sees the coulomb potential due to the nucleus of charge $Ze = 2e$. The last term on the right-hand side is from coulomb repulsion between the two electrons.

(b) A particle mass m moving in a simple one-dimensional harmonic oscillator potential $V(x) = \kappa x^2/2$ has a Hamiltonian that is just the sum of kinetic energy and potential energy. This gives

$$H = -\frac{\hbar^2}{2m}\frac{d^2}{dx^2} + \frac{\kappa x^2}{2}$$

Solution 2.6

In GaAs, the total energy of the longitudinal polar-optic phonon is $\hbar\omega_{LO} = 36.3$ meV, where $\hbar$ is Planck's constant. Using the classical relationship between oscillator amplitude and energy developed in section 1.2.2, we have amplitude

$$A = \left(\frac{2\hbar\omega_{LO}}{m}\right)^{1/2}\frac{1}{\omega_{LO}}$$

$$\omega_{LO} = \frac{e\hbar\omega_{LO}}{\hbar} = \frac{1.60 \times 10^{-19} \times 36.3 \times 10^{-3}}{1.05 \times 10^{-34}} = 5.53 \times 10^{13} \text{ rad s}^{-1}$$

We now need to estimate the mass of the particle. We note that the relative atomic mass of Ga is 69.72 and the relative atomic mass of As is 74.92. Arbitrarily assuming a particle mass $m = 72 \times m_p = 72 \times 1.67 \times 10^{-27} = 1.20 \times 10^{-25}$ kg gives an oscillation amplitude

$$A = \left(\frac{2\hbar\omega_{LO}}{m}\right)^{1/2}\frac{1}{\omega_{LO}} = \left(\frac{2 \times 1.60 \times 10^{-19} \times 36.3 \times 10^{-3}}{1.20 \times 10^{-25}}\right)^{1/2}\frac{1}{5.53 \times 10^{13}}$$

$$A = 5.63 \times 10^{-12} \text{ m}$$

The lattice constant for GaAs, which has the zinc blende crystal structure, is $L = 5.65 \times 10^{-10}$ m. The spacing between the nearest Ga and As is $L\sqrt{3}/4 = 2.45 \times 10^{-10}$ m, so A is approximately $5.63 \times 10^{-2}/2.45 = 0.0229$ (about 2.3%) of the nearest neighbor atom spacing or about 1% of the lattice constant L.

Solution 2.7

We wish to estimate the energy difference between the ground state and the first excited state of a helium atom (He) with one electron missing. This is essentially the same system as the hydrogen atom except that the nucleus now has a charge $2e$.

We start by noting that the helium ion potential for an electron in the n-th orbit is

$$V = \frac{Ze^2}{4\pi \varepsilon_0 r_n} = \frac{-m}{n^2 \hbar^2} \left(\frac{Ze^2}{4\pi \varepsilon_0} \right)^2$$

where $Z = 2$ and $Ry = -\frac{1}{2}m(e^2/4\hbar\pi\varepsilon_0)^2 = -13.6$ eV, so that $E_n = (Z^2/n^2)Ry$. Hence, for $n_1 = 1$ and $n_2 = 2$ the energy difference is

$$E_{n_2} - E_{n_1} = 4Ry \left(\frac{1}{n_1^2} - \frac{1}{n_2^2} \right) = 4Ry \left(1 - \frac{1}{4} \right) = 3Ry = 40.8 \text{ eV}$$

This is four times the value of $E_2 - E_1$ for the hydrogen atom. The reason is that the energy levels are proportional to Z^2 for hydrogen-like ions, and in our case $Z = 2$.

Solution 2.8

(a) Given that linear momentum in the x direction is $\hat{p}_x = \hbar k_x$, the expectation value of momentum for a particle described by wave function $\psi(x)$ can be obtained from $\langle \hat{p}_x \rangle = \int \phi^*(k_x)\hbar k_x \phi(k_x)dk_x$, where $\phi(k_x)$ is the Fourier transform of $\psi(x)$, so that

$$\langle \hat{p}_x \rangle = \frac{1}{2\pi} \int_{-\infty}^{\infty} dk_x \left(\int_{-\infty}^{\infty} dx' \psi^*(x')e^{ik_x x'} \right) \hbar k_x \left(\int_{-\infty}^{\infty} dx \psi(x)e^{-ik_x x} \right)$$

Solving this integral one may show that if the momentum operator $\hat{p}_x = \hbar k_x$ in k space (momentum space) then the momentum operator in real space is $\hat{p}_x = -i\hbar \cdot \partial/\partial x$.

Integrating by parts, assuming evaluation at the limits is zero, gives

$$\langle \hat{p}_x \rangle = \frac{\hbar}{2i\pi} \int_{-\infty}^{\infty} dk_x \int_{-\infty}^{\infty} dx' \psi^*(x')e^{ik_x x'} \int_{-\infty}^{\infty} dx e^{-ik_x x} \frac{\partial}{\partial x} \psi(x)$$

Making use of the fact that a delta function can be expressed as

$$\frac{1}{2\pi} \int_{-\infty}^{\infty} dk_x e^{-ik_x(x-x')} = \delta(x - x')$$

gives

$$\langle \hat{p}_x \rangle = -i\hbar \int\limits_{-\infty}^{\infty} dx' \int\limits_{-\infty}^{\infty} dx \, \psi^*(x') \delta(x - x') \frac{\partial}{\partial x} \psi(x)$$

$$\langle \hat{p}_x \rangle = -i\hbar \int\limits_{-\infty}^{\infty} dx \, \psi^*(x) \frac{\partial}{\partial x} \psi(x)$$

from which it follows that the linear momentum operator in real space is $\hat{p}_x = -i\hbar \partial / \partial x$.

(b) and (c) involve the same idea as expressed in (a).

Solution 2.9

In Section 2.2.3.2 the energy of the hydrogen atom in state ψ_{nlm} depends upon the principal quantum number n but not on the orbital quantum number l or the azimuthal quantum number m. For a given n, the allowed values of l are $l = 0, 1, 2, \ldots, (n - 1)$ and the allowed values of m are $m = \pm l, \ldots, \pm 2, \pm 1, 0$. Hence, the degeneracy of a state with principal quantum number n is a sum over n terms:

$$\sum_{l=0}^{l=n-1} (2l + 1) = 1 + 3 + 5 + \cdots + (2n - 5) + (2n - 3) + (2n - 1)$$

Reordering the right-hand side as

$$\sum_{l=0}^{l=n-1} (2l + 1) = (2n - 1) + (2n - 3) + (2n + 5) + \cdots + 5 + 3 + 1$$

suggests one adds the two equations. Doing this gives

$$2 \sum_{l=0}^{l=n-1} (2l + 1) = 2n + 2n + 2n + \cdots$$

Since there are n terms on the right-hand side, this may be written

$$2 \sum_{l=0}^{l=n-1} (2l + 1) = n2n$$

and so we may conclude that the degeneracy of the state ψ_{nlm} is

$$\sum_{l=0}^{l=n-1} (2l + 1) = n^2$$

3 Using the Schrödinger wave equation

3.1 Introduction

The purpose of this chapter is to give some practice calculating what happens to an electron moving in a potential according to Schrödinger's equation. There is a remarkable richness in the type and variety of the predictions. In fact, to the uninitiated, specific solutions to Schrödinger's equations can be quite unexpected. For this reason alone, one should be motivated to explore the possibilities. Getting used to the behavior of waves can take some time, so in this chapter we want to carefully reveal some of the key features of Schrödinger's equation that describe the waviness of matter.

Let's start by considering some basics. According to our approach, a particle of mass m moves in space as a function of time in the presence of a potential. Time and space are assumed to be smooth and continuous. The potential can cause the electron motion to be localized to one region of space, forming what is called a *bound state*. The alternative one may consider is an electron able to move anywhere in space, in which case the electron is in an *unbound state* (sometimes called a scattering state).

In Section 2.2 we introduced the time-independent Schrödinger equation for a particle of mass m in a potential $V(\mathbf{r})$, which is a function of space only. The second-order differential equation is

$$H\psi_n(\mathbf{r}) = \left(-\frac{\hbar^2}{2m}\nabla^2 + V(\mathbf{r}) \right)\psi_n(\mathbf{r}) = E_n\psi_n(\mathbf{r}) \tag{3.1}$$

where H is the Hamiltonian operator, E_n are energy eigenvalues, and $\psi_n(\mathbf{r})$ are time-independent stationary states, resulting in

$$\psi_n(\mathbf{r}, t) = \psi_n(\mathbf{r})e^{-i\omega t} \tag{3.2}$$

The *eigenfunctions* $\psi_n(\mathbf{r})$ are found for a given potential by using the Schrödinger wave equation and applying *boundary conditions*. The fact that application of boundary conditions to the wave equation gives rise to eigenstates is a direct consequence of the mathematical structure of the equation. One may think of the wave equation as containing a vast number of possible solutions. The way to extract results for a

111

particular set of circumstances is to specify the potential $V(\mathbf{r})$ and apply boundary conditions.

When the potential is independent of time, solutions to Schrödinger's equation are called *stationary states*. This is because the probability density for such states $|\psi(\mathbf{r}, t)|^2$ is independent of time. It is a little unfortunate that solutions to the time-independent Schrödinger equation are called stationary states when, in fact, they have a time dependence. If the eigenvalue has energy $E = \hbar\omega$, then the time dependence is of the form $e^{-i\omega t}$. The simplest stationary state is a bound-state standing wave in space with energy $\hbar\omega$ and time dependence $e^{-i\omega t}$. This wave function is stationary in the sense that it does not propagate a flux through space. As an example, a standing wave can be constructed from two identical but counter-propagating traveling waves of the form $e^{i(kx-\omega t)}$ such that $\psi_n = (e^{i(kx-\omega t)} + e^{i(-kx-\omega t)})/2 = \cos(kx) \cdot e^{-\omega t}$. In contrast to a standing wave, a traveling wave of the form $e^{i(kx-\omega t)}$ carries a constant flux. Such functions can also be solutions to the time-independent Schrödinger equation. In this case the state is only stationary in the weaker sense that there is no time-varying flux component.

Proper solutions to the Schrödinger wave equation require that $\psi_n(\mathbf{r})$ and $\nabla\psi_n(\mathbf{r})$ are *continuous* everywhere. To give some practice with wave functions, in the next section we show the effect of not complying with this requirement.

3.1.1   The effect of discontinuity in the wave function and its slope

We want to show that the wave function and the spatial derivative of the wave function should be continuous. The argument we use relies on the notion that energy is a constant of the motion and hence a well-behaved quantity. For ease of discussion, we consider a one-dimensional wave function that is real, and we set the potential energy to zero.

First, it will be shown that a *discontinuity in a wave function* would result in a nonphysical infinite contribution to the expectation value of the particle's kinetic energy, T. Figure 3.1 illustrates a discontinuity in a wave function, $\psi(x)$, at position x_0.

We start our analysis by approximating $\psi(x)$ as a piece-wise function, and then we consider the contribution to the expectation value of kinetic energy in the limit $x \to x_0$.

Fig. 3.1. Illustration of a discontinuity in wave function $\psi(x)$ at position $x = x_0$. Notice that the discontinuity is characterized by an infinite slope in the wave function.

To first order, this gives

$$\Delta\langle\hat{T}\rangle = \lim_{\eta\to\infty} \frac{-\hbar^2}{2m} \int\limits_{x_0-\frac{1}{\eta}}^{x_0+\frac{1}{\eta}} \psi^*\frac{d^2\psi}{dx^2}dx \tag{3.3}$$

where η is a dummy variable. Since we have assumed $\psi(x)$ is real-valued, the complex conjugate is dropped. Integration by parts gives

$$\Delta\langle\hat{T}\rangle = \lim_{\eta\to\infty} \int\limits_{x_0-\frac{1}{\eta}}^{x_0+\frac{1}{\eta}} \psi\left(\frac{-\hbar^2}{2m}\frac{d^2\psi}{dx^2}\right)dx$$

$$= \frac{-\hbar^2}{2m}\lim_{\eta\to\infty}\left(\psi\frac{d\psi}{dx}\bigg|_{x_0-\frac{1}{\eta}}^{x_0+\frac{1}{\eta}} - \int\limits_{x_0-\frac{1}{\eta}}^{x_0+\frac{1}{\eta}}\left(\frac{d\psi}{dx}\right)^2 dx\right) \tag{3.4}$$

For $\psi(x)$ *discontinuous*, the term on the far right of Eqn (3.4) is infinite. To show this we use the median value theorem for functions $f(x)$ and $g(x)$:

$$\int\limits_a^b f(x)g(x)dx = f(\xi)\int\limits_a^b g(x)dx \tag{3.5}$$

where $a < \xi < b$ with $f(x) = g(x) = d\psi/dx$, so that

$$\int\limits_{x_0-\frac{1}{\eta}}^{x_0+\frac{1}{\eta}}\left(\frac{d\psi}{dx}\right)^2 dx = \frac{d\psi}{dx}\bigg|_\xi \int\limits_{x_0-\frac{1}{\eta}}^{x_0+\frac{1}{\eta}}\frac{d\psi}{dx}dx \tag{3.6}$$

$$\int\limits_{x_0-\frac{1}{\eta}}^{x_0+\frac{1}{\eta}}\left(\frac{d\psi}{dx}\right)^2 dx = \frac{d\psi}{dx}\bigg|_\xi \left[\psi\left(x_0+\frac{1}{\eta}\right) - \psi\left(x_0-\frac{1}{\eta}\right)\right]_{x_0-\frac{1}{\eta}}^{x_0+\frac{1}{\eta}} \tag{3.7}$$

for $x_0 - 1/\eta < \xi < x_0 + 1/\eta$. The term in the rectangular brackets is a non-zero constant and within the limit $\eta\to\infty$, $\xi\to x_0$, so that

$$\int\limits_{x_0-\frac{1}{\eta}}^{x_0+\frac{1}{\eta}}\left(\frac{d\psi}{dx}\right)^2 dx = \frac{d\psi}{dx}\bigg|_{x\to x_0}\left[\psi\left(x_0+\frac{1}{\eta}\right) - \psi\left(x_0-\frac{1}{\eta}\right)\right]_{x_0-\frac{1}{\eta}}^{x_0+\frac{1}{\eta}} = \infty \tag{3.8}$$

because $d\psi/dx|_{x\to x_0} = \psi'|_{x\to x_0} = \infty$.

Hence, we conclude that the contribution to the expectation value of kinetic energy due to a discontinuity in the wave function is infinite, $\Delta\langle\hat{T}\rangle\to\infty$. The infinite

$\psi(x)$

Distance, x

Kink at position x_0

Fig. 3.2. Illustration of a kink in wave function $\psi(x)$ at position $x = x_0$. Notice that the kink is characterized by a discontinuity in the slope of the wave function.

contribution to kinetic energy comes from the infinite slope of the wave function. In the physical world that we normally experience, there are no infinite contributions to energy, and so a solution that includes an infinite energy term is called "*unphysical*". To avoid such an unphysical result, one requires that the wave function be continuous.

Assuming that our real wave function is continuous, we now show that a discontinuity in the spatial derivative of the wave function makes a finite contribution to the expectation value of kinetic energy. A discontinuity in the spatial derivative corresponds to a kink in our wave function. We will show that the kink makes a contribution $\Delta\langle\hat{T}\rangle = -(\hbar^2/2m)\psi\,\Delta(\psi')$ to the expectation value of the kinetic energy, where ψ is the wave function at the kink and $\Delta(\psi')$ is the difference between the slopes of the wave function on the two sides of the kink. Figure 3.2 illustrates a wave function, $\psi(x)$, with a kink at position x_0.

From our previous work, the contribution of the kink at position x_0 to the expectation value of the particle's kinetic energy is

$$\Delta\langle\hat{T}\rangle = \lim_{\eta\to\infty} \int_{x_0-\frac{1}{\eta}}^{x_0+\frac{1}{\eta}} \psi\left(\frac{-\hbar^2}{2m}\frac{d^2\psi}{dx^2}\right)dx$$

$$= \frac{-\hbar^2}{2m}\lim_{\eta\to\infty}\left(\psi\frac{d\psi}{dx}\bigg|_{x_0-\frac{1}{\eta}}^{x_0+\frac{1}{\eta}} - \int_{x_0-\frac{1}{\eta}}^{x_0+\frac{1}{\eta}}\left(\frac{d\psi}{dx}\right)^2 dx\right) \quad (3.9)$$

Since this time we know that $\psi(x)$ is continuous, $d\psi/dx$ has finite values everywhere. In this case the integral on the far right-hand side of Eqn (3.9) vanishes as $\eta\to\infty$, and we are left with

$$\Delta\langle\hat{T}\rangle = \frac{-\hbar^2}{2m}\lim_{\eta\to\infty}\psi\frac{d\psi}{dx}\bigg|_{x_0-\frac{1}{\eta}}^{x_0+\frac{1}{\eta}} \quad (3.10)$$

$$\Delta\langle\hat{T}\rangle = \frac{-\hbar^2}{2m}\left(\psi\left(x_0+\frac{1}{\eta}\right)\psi'\left(x_0+\frac{1}{\eta}\right) - \psi\left(x_0-\frac{1}{\eta}\right)\psi'\left(x_0-\frac{1}{\eta}\right)\right) \quad (3.11)$$

Because $\psi(x)$ is continuous, it follows that

$$\psi\left(x_0+\frac{1}{\eta}\right) = \psi\left(x_0-\frac{1}{\eta}\right) = \psi(x_0) \quad (3.12)$$

If we let

$$\Delta \psi = \psi'\left(x_0 + \frac{1}{\eta}\right) - \psi'\left(x_0 - \frac{1}{\eta}\right) \tag{3.13}$$

then we may write

$$\Delta \langle \hat{T} \rangle = -\frac{\hbar^2}{2m} \psi(x_0) \Delta \psi \tag{3.14}$$

So we may conclude that a kink in a wave function at position x_0 makes a contribution to the expectation value of the kinetic energy that is proportional to the value of the wave function multiplied by the difference in slope of the wave function at x_0. Allowing solutions to the wave function that contain *arbitrary* kinks would add arbitrary values of kinetic energy. Rather than deal with this possibility explicitly, we avoid it and other complications by requiring that the spatial derivative of the wave function describing a particle moving in a well-behaved potential be continuous. This is a significant constraint on the number and type of wave functions that are allowed.

In the theory we have developed, solutions to Schrödinger's time-independent equation (Eqn (3.1)) must be consistent with the potential $V(\mathbf{r})$. If a potential is badly behaved and possesses a delta-function singularity, then the spatial derivative of the wave function is discontinuous. A delta-function potential creates a kink in the wave function. Delta-function potentials often show up in models of physical phenomena in which the characteristic spatial curvature of the wave function is small compared with the spatial extent of the potential (see Exercise 3.5).

It follows from our analysis that if ψ and ψ' are smooth and continuous, the expectation value of kinetic energy $\langle \hat{T} \rangle$ is proportional to the integral over all space of the value of the wave function multiplied by the wave function curvature. Hence, the energy of a wave function increases if it has more "wiggles" or, equivalently, more regions of high curvature (change in slope).

3.2 Wave function normalization and completeness

In the previous section we found some criteria for obtaining solutions to Schrödinger's equation. Because $|\psi_n(\mathbf{r})|^2 d^3r$ is simply the probability of finding a particle in state $\psi_n(\mathbf{r})$ in the volume element d^3r at position $\mathbf{r}$ in space, it is also convenient to require that eigenstates $\psi_n(\mathbf{r})$ be *normalized*. This normalization requirement means that if the particle is in state $\psi_n(\mathbf{r})$ then integration of $|\psi_n(\mathbf{r})|^2$ over all space yields unity. The probability of finding the particle somewhere in space is unity. In fact, the wave functions that appear in the time-independent Schrödinger equation

$$H\psi_n(\mathbf{r}) = \left(-\frac{\hbar^2}{2m}\nabla^2 + V(\mathbf{r})\right)\psi_n(\mathbf{r}) = E_n \psi_n(\mathbf{r}) \tag{3.15}$$

are *orthogonal* and *normalized*. This *orthonormal* property can be expressed mathematically as

$$\int\limits_{-\infty}^{\infty} \psi_n^*(\mathbf{r})\psi_m(\mathbf{r})d^3r = \delta_{nm} \qquad (3.16)$$

where the Kronecker delta function $\delta_{nm} = 0$ unless $n = m$, in which case $\delta_{nn} = 1$. In addition, the orthogonal wave functions ψ_n form a *complete set*, so that any arbitrary wave function $\psi(\mathbf{r})$ can be expressed as a sum of orthonormal wave functions weighted by coefficients a_n in such a way that

$$\psi(\mathbf{r}) = \sum_n a_n \psi_n(\mathbf{r}) \qquad (3.17)$$

When we normalize the wave function, we give it dimensions. For example, if a one-dimensional electron wave function confined to a region of space between $x = 0$ and $x = L$ is of the form $\psi_n(x) = A\sin(n\pi x/L)$, then normalization, which requires

$$\int\limits_{x=0}^{x=L} |\psi_n(x)|^2 dx = 1 \qquad (3.18)$$

means that $|A| = \sqrt{2/L}$. Thus, a one-dimensional wave function has dimensions of inverse square-root length. Because we choose to measure length in meters, this means $|A|$ has units $m^{-1/2}$. If the wave function exists in an infinite volume, we will need a different way to normalize it or will have to spend time figuring out how to get around integrals that diverge. A sensible approach is to compare ratios of wave functions in such a way that $|A|^2$ can be interpreted as a relative particle density.

3.3 Inversion symmetry in the potential

Continuing our exploration of solutions to Schrödinger's equation, we now look at the influence of symmetry. The time-independent Schrödinger equation is of the Hamilton form

$$\left(-\frac{\hbar^2}{2m}\nabla^2 + V(\mathbf{r})\right)\psi_n(\mathbf{r}) = H_n\psi_n(\mathbf{r}) = E_n\psi_n(\mathbf{r}) \qquad (3.19)$$

where $H = T + V$ is the Hamiltonian containing the kinetic energy term $-(\hbar^2/2m)\nabla^2$ and a potential energy term $V(\mathbf{r})$.

There are a number of simplifying concepts that may be used when solving problems involving the Schrödinger equation. One is the result of symmetry that may exist in the mathematical structure of the Hamiltonian. An important example is the situation in which there is a natural coordinate system such that $V(\mathbf{r})$ has *inversion symmetry*, where

$$V(\mathbf{r}) = V(-\mathbf{r}) \tag{3.20}$$

In this case, the symmetry in the potential forces eigenfunctions with different eigenvalues to have either *odd* or *even parity*, i.e. $\psi_n(\mathbf{r}) = \pm\psi_n(-\mathbf{r})$, where

$$\psi_n(\mathbf{r}) = +\psi_n(-\mathbf{r}) \tag{3.21}$$

has even parity, and

$$\psi_n(\mathbf{r}) = -\psi_n(-\mathbf{r}) \tag{3.22}$$

has odd parity. A simple symmetry in the potential energy forces wave functions to also have a definite symmetry.

Suppose there are a pair of eigenfunctions that have the same eigenvalues *and do not obey this parity requirement* – i.e., they are *linearly independent*. Let $\psi(\mathbf{r})$ be a function that does not obey the parity requirement. In this situation, $\psi(\mathbf{r})$ and $\psi(-\mathbf{r})$ are linearly independent, so one can always construct

$$\psi_+(\mathbf{r}) = \psi(\mathbf{r}) + \psi(-\mathbf{r}) \tag{3.23}$$

with even parity and

$$\psi_-(\mathbf{r}) = \psi(\mathbf{r}) - \psi(-\mathbf{r}) \tag{3.24}$$

with odd parity; these do obey the required parity.

Summarizing our discussion, we anticipate that inversion symmetry in the potential will result in wave functions with definite parity. To see what this means in practice, we now consider an example in which an electron is placed in a potential with inversion symmetry.

3.3.1 One-dimensional rectangular potential well with infinite barrier energy

A simple potential with inversion symmetry is a one-dimensional, rectangular potential well with infinite barrier energy of the type illustrated in Fig. 3.3. Before we proceed, it is necessary to decide on the natural coordinate system for the one-dimensional potential well. In this particular case, inversion symmetry in the potential such that $V(x) = V(-x)$ requires that $x = 0$ is in the middle of the potential well.

We assume a particle of mass m is described using Schrödinger's time-independent equation (Eqn (3.15)). We want to solve this differential equation for the wave function and energy eigenvalues subject to *boundary condition* $\psi = 0$ for $-L/2 > x > L/2$. Such a boundary condition requires that the wave function vanish at the edge of the potential well. Wave functions cannot penetrate into a potential barrier of infinite energy.

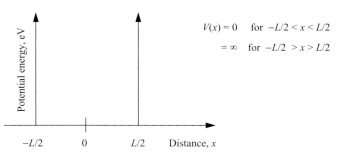

Fig. 3.3. Sketch of a one-dimensional rectangular potential well with infinite barrier energy. The width of the well is L and the potential is such that $V(x) = 0$ for $-L/2 < x < L/2$ and $V(x) = \infty$ for $-L/2 > x > L/2$.

Between $x = -L/2$ and $x = L/2$ the potential is zero and the particle wave function is of the form given by Eqn (2.52). Ignoring any time dependence, at the $x = -L/2$ boundary we require

$$\psi(x)|_{x=-L/2} = Ae^{-ikL/2} + Be^{ikL/2} = 0 \tag{3.25}$$

and at the $x = L/2$ boundary we require

$$\psi(x)|_{x=L/2} = Ae^{ikL/2} + Be^{-ikL/2} = 0 \tag{3.26}$$

Since there is only one solution for the normalized wave function, $\psi(x)$, the determinant of coefficients for the two equations (Eqn (3.25) and Eqn (3.26)) is zero:

$$\begin{vmatrix} e^{-ikL/2} & e^{ikL/2} \\ e^{ikL/2} & e^{-ikL/2} \end{vmatrix} = e^{-ikL} - e^{ikL} = 0 \tag{3.27}$$

It follows that the boundary conditions require solutions in which $\sin(k_n L) = 0$, where $k_n L = n\pi$ and n is a positive nonzero integer. The corresponding wave functions are sinusoidal and of wavelength

$$\lambda_n = \frac{2L}{n} \tag{3.28}$$

where n is a positive nonzero integer. Expressed in terms of the wave vector we may write

$$k_n = n\frac{2\pi}{2L} = \frac{n\pi}{L} \text{ for } n = 1, 2, 3, \ldots \tag{3.29}$$

and the energy eigenvalues are simply $E_n = \hbar^2 k_n^2/2m$. Hence,

$$E_n = \frac{\hbar^2}{2m} \cdot \frac{n^2\pi^2}{L^2} \tag{3.30}$$

The *lowest-energy state* or *ground state* of the system has energy eigenvalue

$$E_1 = \frac{\hbar^2 \pi^2}{2mL^2} \tag{3.31}$$

This is the *zero-point energy* for the bound state. There is no classical analog for zero-point energy. Zero-point energy arises from the wavy nature of particles. This forces the lowest-energy bound state in a confining potential to have a minimum curvature and hence a minimum energy. One may think of this energy as a measure of how confined the particle is in real space. In this sense, the wavy nature of particles in quantum mechanics imposes a relationship between energy and space. A particle confined to any region of space cannot sustain an average energy below the ground state or zero-point energy.

Another point worth discussing is the sinusoidal nature of the eigenfunctions in the region $-L/2 > x > L/2$. For a given periodicity, sinusoidal wave functions have the lowest curvature and hence the lowest eigenenergy. In fact, for the given boundary conditions, they are the only solution possible. We may think of Schrödinger's second-order time-independent differential equation and boundary conditions as excluding all but a few wave function solutions.

Wave functions in our simple one-dimensional potential well are of either *even* or *odd parity*. The energy levels and wave functions for the first three lowest-energy states are

$$E_1 = \frac{\hbar^2 \pi^2}{2mL^2} = \hbar \omega_1 \qquad \psi_1(x) = \left(\frac{2}{L}\right)^{1/2} \cos\left(\frac{\pi}{L} x\right) \qquad \text{(has even parity)} \tag{3.32}$$

$$E_2 = 4E_1 = \hbar \omega_2 \qquad \psi_2(x) = \left(\frac{2}{L}\right)^{1/2} \sin\left(\frac{2\pi}{L} x\right) \qquad \text{(has odd parity)} \tag{3.33}$$

$$E_3 = 9E_1 = \hbar \omega_3 \qquad \psi_3(x) = \left(\frac{2}{L}\right)^{1/2} \cos\left(\frac{3\pi}{L} x\right) \qquad \text{(has even parity)} \tag{3.34}$$

For an electron in the conduction band of GaAs in a potential well of width $L = 20$ nm and with an effective electron mass $m^* = 0.07 \times m_0$ the energy of the ground state is $E_1 = 13.4$ meV, corresponding to an angular frequency $\omega_1 = 2 \times 10^{13}$ rad s^{-1}.

Figure 3.4(a) is a sketch of a one-dimensional, rectangular potential well with infinite barrier energy showing energy eigenvalues E_1, E_2, and E_3. Figure 3.4(b) sketches the first three eigenfunctions ψ_1, ψ_2, and ψ_3 for the potential shown in Fig. 3.4(a).

These wave functions appear to be similar to standing waves. To find the time dependence of the eigenfunctions $\psi(x, t) = \psi(x)\phi(t)$, we need to solve the time-dependent Schrödinger equation (Eqn (2.42))

$$E\phi(t) = i\hbar \frac{\partial}{\partial t} \phi(t) \tag{3.35}$$

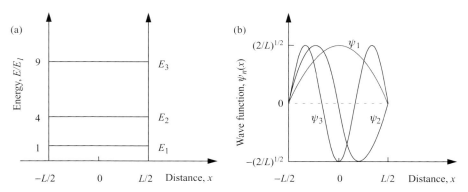

Fig. 3.4. (a) Sketch of a one-dimensional, rectangular potential well with infinite barrier energy showing the energy eigenvalues E_1, E_2, and E_3. (b) Sketch of the eigenfunctions ψ_1, ψ_2, and ψ_3 for the potential shown in (a).

The solutions are of the form $\phi_n(t) = e^{-iE_n t/\hbar} = e^{-i\omega_n t}$, where $E_n = \hbar\omega_n$. Hence, the solution for a wave function with quantum number n is

$$\psi_n(x, t) = \psi_n(x)e^{-i\omega_n t} \tag{3.36}$$

which can be written as

$$\psi_n(x, t) = \left(\frac{2}{L}\right)^{1/2} \sin\left(k_n\left(x + \frac{2}{L}\right)\right) \cdot e^{-i\omega_n t} \tag{3.37}$$

where $k_n = n\pi/L$ is the wave vector and $\omega_n = \hbar k_n^2/2m$ is the angular frequency of the n-th eigenstate. In Eqn (3.37) we see that the wave function $\psi_n(x, t)$ consists of a spatial part and an oscillatory time-dependent part. The *frequency of oscillation* of the time-dependent part is given by the energy eigenvalue $E_n = \hbar\omega$, which we found using the *spatial* solution to the time-independent Schrödinger equation. This link between spatial and temporal solutions can be traced back to the separation of variables in Schrödinger's equation (Eqn (2.42) and Eqn (2.43) in Chapter 2). Such separation of variables is possible when the potential is time independent.

3.4 Numerical solution of the Schrödinger equation

The time-independent Schrödinger equation describing a particle of mass m constrained to motion in a time-independent, one-dimensional potential $V(x)$ is (Eqn (3.1))

$$H\psi_n(x) = \left(-\frac{\hbar^2}{2m}\frac{d^2}{dx^2} + V(x)\right)\psi_n(x) = E_n\psi_n(x) \tag{3.38}$$

To solve this equation numerically, one must first discretize the functions. A sensible first approach samples the wave function and potential at a discrete set of $N + 1$ equally-spaced points in such a way that position $x_j = j \times h_0$, the index $j = 0, 1, 2, \ldots, N$,

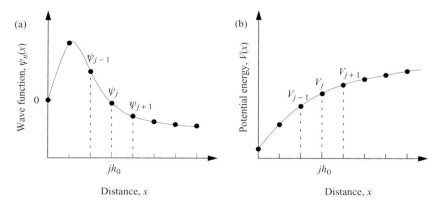

Fig. 3.5. (a) Sampling the wave function $\psi(x)$ in such a way that the interval between sampling points is h_0. (b) Sampling the potential $V(x)$ at equally-spaced intervals.

and h_0 is the interval between adjacent sampling points. Thus, one may define $x_{j+1} \equiv x_j + h_0$. Such sampling of a particle wave function and potential is illustrated in Fig. 3.5.

The region in which we wish to solve the Schrödinger equation is of length $L = N h_0$. At each sampling point the wave function has value $\psi_j = \psi(x_j)$ and the potential is $V_j = V(x_j)$. The first derivative of the discretized wave function $\psi(x_j)$ in the finite-difference approximation is

$$\frac{d}{dx}\psi(x_j) = \frac{\psi(x_j) - \psi(x_{j-1})}{h_0} \tag{3.39}$$

To find the second derivative of the discretized wave function, we use the three-point finite-difference approximation, which gives

$$\frac{d^2}{dx^2}\psi(x_j) = \frac{\psi(x_{j-1}) - 2\psi(x_j) + \psi(x_{j+1})}{h_0^2} \tag{3.40}$$

Substitution of Eqn (3.40) into Eqn (3.38) results in the matrix equation

$$H\psi(x_j) = -u_j\psi(x_{j-1}) + d_j\psi(x_j) - u_{j+1}\psi(x_{j+1}) = E\psi(x_j) \tag{3.41}$$

where the Hamiltonian is a symmetric tri-diagonal matrix. The diagonal matrix elements are

$$d_j = \frac{\hbar^2}{mh_0^2} + V_j \tag{3.42}$$

and the adjacent off-diagonal matrix elements are

$$u_j = \frac{\hbar^2}{2mh_0^2} \tag{3.43}$$

As discussed in Section 3.3.1, to find the eigenenergies and eigenstates of a particle of mass m in a one-dimensional, rectangular potential well with infinite barrier energy the boundary conditions require $\psi_0(x_0) = \psi_N(x_N) = 0$. Because the boundary conditions force the wave function to zero at positions $x = 0$ and $x = L$, Eqn (3.41) may be written as

$$(\mathbf{H} - E\mathbf{I})\psi$$

$$= \begin{bmatrix} (d_1 - E) & -u_2 & 0 & 0 & \cdots \\ -u_2 & (d_2 - E) & -u_3 & 0 & \cdots \\ 0 & -u_3 & (d_3 - E) & -u_4 & \cdots \\ \vdots & \vdots & \vdots & \ddots & \\ & & & -u_{N-1} & (d_{N-1} - E) \end{bmatrix} \begin{bmatrix} \psi_1 \\ \psi_2 \\ \vdots \\ \psi_{N-1} \end{bmatrix} = 0$$

$$(3.44)$$

where $\mathbf{H}$ is the Hamiltonian matrix and $\mathbf{I}$ the identity matrix. The solutions to this equation may be found using conventional numerical methods. Programming languages such as MATLAB also have routines that efficiently diagonalize the tri-diagonal matrix and solve for the eigenvalues and eigenfunctions.

In the preceding, we have considered the situation in which the particle is confined to one-dimensional motion in a region of length L. Outside this region, the potential is infinite and the wave function is zero. Particle motion is localized to one region of space, so only bound states can exist as solutions to the Schrödinger equation. Because the particle is not transmitted beyond the boundary positions $x = 0$ and $x = L$, this is a quantum nontransmitting boundary problem.

There are, of course, other situations in which we might be interested in a region of space of length L through which particles can enter and exit via the boundaries at position $x = 0$ and $x = L$. When there are transmitting boundaries at positions $x = 0$ and $x = L$, then $\psi_0(x_0) \neq 0$ and $\psi_N(x_N) \neq 0$. In this case there is the possibility of unbound particle states as well as sources and sinks of particle flux to consider. To deal with these and other extensions, the quantum transmitting boundary method may be used.[1]

3.5 Current flow

If one were to place an electron in the one-dimensional potential well with infinite barrier energy we have been discussing one might reasonably expect there to be circumstances under which it is able to move around. Of course, as a particle moves around there must be a corresponding current flow. To understand this and to quantify any current flow

[1] C. S. Lent and D. J. Kirkner, *J. Appl. Phys.* **67**, 6353 (1990), C. L. Fernando and W. R. Frensley, *J. Appl. Phys.* **76**, 2881 (1994), and Z. Shano, W. Porod, C. S. Lent, and D. J. Kirkner, *J. Appl. Phys.* **78**, 2177 (1995).

in quantum mechanics, we need to find the appropriate operator. To find the current density operator, we start by making the reasonable and simplifying assumption that the electron moves in a potential that is real.

From Maxwell's equations or by elementary consideration of current conservation, the change in charge density ρ is related to the divergence of current density $\mathbf{J}$ through

$$\frac{\partial}{\partial t}\rho(\mathbf{r}, t) = -\nabla \cdot \mathbf{J}(\mathbf{r}, t) \tag{3.45}$$

This is the classical expression for *current continuity*. The time dependence of charge density (the *temporal* dependence of a scalar field) is related to net current into or out of a region of space (the *spatial* dependence of a vector field). For a particle of charge e, we identify $\rho = e|\psi|^2$, so that

$$\frac{\partial \rho}{\partial t} = e\frac{\partial}{\partial t}(\psi^*\psi) = e\left(\psi^*\frac{\partial \psi}{\partial t} + \psi\frac{\partial \psi^*}{\partial t}\right) \tag{3.46}$$

There are two terms in the parentheses on the right-hand side of the equation that we wish to find. To obtain these terms one makes use of the fact that the time-dependent Schrödinger equation for a particle of mass m is

$$i\hbar\frac{\partial}{\partial t}\psi(\mathbf{r}, t) = \left(-\frac{\hbar^2}{2m}\nabla^2 + V(\mathbf{r})\right)\psi(\mathbf{r}, t) = H\psi(\mathbf{r}, t) \tag{3.47}$$

Multiplying both sides by $\psi^*(\mathbf{r}, t)$ gives

$$i\hbar\psi^*(\mathbf{r}, t)\frac{\partial}{\partial t}\psi(\mathbf{r}, t) = -\frac{\hbar^2}{2m}\psi^*(\mathbf{r}, t)\nabla^2\psi(\mathbf{r}, t) + \psi^*(\mathbf{r}, t)V(\mathbf{r})\psi(\mathbf{r}, t) \tag{3.48}$$

which, when multiplied by $e/i\hbar$, is the first term in our expression for $\partial\rho/\partial t$. To find the second term one takes the complex conjugate of Eqn (3.47) and multiplies both sides by $\psi(\mathbf{r}, t)$. In effect, we interchange $\psi(\mathbf{r}, t)$ and $\psi^*(\mathbf{r}, t)$ and change the sign of i in Eqn (3.48). Now, because $V(\mathbf{r})$ is real, one may write $\psi^*(\mathbf{r}, t)V(\mathbf{r})\psi(\mathbf{r}, t) = \psi(\mathbf{r}, t)V(\mathbf{r})\psi^*(\mathbf{r}, t)$, giving the rate of change of charge density

$$\frac{\partial \rho}{\partial t} = -\psi^*\frac{e\hbar}{i2m}\nabla^2\psi + \psi\frac{e\hbar}{i2m}\nabla^2\psi^* + \frac{e\psi^*\psi}{i\hbar}(V(\mathbf{r}) - V(\mathbf{r})) \tag{3.49}$$

$$\frac{\partial \rho}{\partial t} = \frac{ie\hbar}{2m}(\psi^*\nabla^2\psi - \psi\nabla^2\psi^*) = \frac{ie\hbar}{2m}\nabla \cdot (\psi^*\nabla\psi - \psi\nabla\psi^*) = -\nabla \cdot \mathbf{J} \tag{3.50}$$

Hence, the *current density* operator for a particle described by state ψ is

$$\boxed{\mathbf{J} = -\frac{ie\hbar}{2m}(\psi^*\nabla\psi - \psi\nabla\psi^*)} \tag{3.51}$$

or in one dimension

$$\mathbf{J} = -\frac{ie\hbar}{2m}\left(\psi^*(x)\frac{\partial}{\partial x}\psi(x) - \psi(x)\frac{\partial}{\partial x}\psi^*(x)\right) \tag{3.52}$$

The above derivation of the current density operator requires that the potential be real. If the potential is complex, it can have the effect of absorbing or creating particles. This is a useful feature that helps in solving some types of problems. However, we are not going to consider complex potentials in this book.

A point worth highlighting is the obvious symmetry in the expression for the current operator. This symmetry has an important influence on the type of wave functions that can carry current. We explore this in the next two sections.

3.5.1 Current in a rectangular potential well with infinite barrier energy

Initially, we are interested in finding the current carried by the lowest-energy state of a one-dimensional, rectangular potential with infinite barrier energy and well width L. In Section 3.3.1 we chose the one-dimensional, rectangular potential well with infinite barrier energy to be centered at $x = 0$. Our calculations identified the $n = 1$ ground-state energy as

$$E_1 = \hbar\omega_1 = \frac{\hbar^2\pi^2}{2mL^2} \tag{3.53}$$

and the ground-state wave function appeared as a standing wave

$$\psi_1(x, t) = \left(\frac{2}{L}\right)^{1/2} \sin\left(\frac{\pi}{L}\left(x + \frac{L}{2}\right)\right) \cdot e^{-i\omega_1 t} \tag{3.54}$$

To find the current due to this state, we substitute the wave function into the expression for the current density. The current density in this case is

$$\mathbf{J} = -\frac{ie\hbar}{2m}\left(\psi_1^*(x)\frac{\partial}{\partial x}\psi_1(x) - \psi_1(x)\frac{\partial}{\partial x}\psi_1^*(x)\right) = 0 \tag{3.55}$$

from which we conclude that current is not carried by a single standing wave. This should come as no surprise, since a standing wave can be thought of as a resonance consisting of two counter-propagating traveling waves whose individual contributions to current flow exactly cancel each other out.

We are still left with the question of how particles are transported in the one-dimensional potential well. The answer is that a superposition of electron states is required. This is easily illustrated by example.

Let us consider current flow in a linear superposition state consisting of a simple combination of the ground-state wave function, ψ_1, and the first excited-state wave function, ψ_2, so that

$$\psi(x, t) = \frac{1}{\sqrt{2}}(\psi_1(x, t) + \psi_2(x, t)) \tag{3.56}$$

Substituting this into our expression for current density gives the result

$$\mathbf{J} = -\frac{2e\pi\hbar}{mL^2}\left(\cos\left(\frac{\pi x}{L}\right)\cos\left(\frac{2\pi x}{L}\right) + \frac{1}{2}\sin\left(\frac{\pi x}{L}\right)\sin\left(\frac{2\pi x}{L}\right)\right)\sin((\omega_2 - \omega_1)t)$$

(3.57)

This equation shows that current is carried by a superposition of stationary bound states. The ability of a linear superposition of stationary states to carry current is due to the symmetry embedded in the form of the current operator. We can also see how current flows in the one-dimensional potential well as a function of time. The current has an oscillatory time dependence, which in our particular example is given by the difference frequency, $\omega_2 - \omega_1$, between the ψ_1 and ψ_2 eigenstates.

In the next section we consider current flow for an unbound state. In this situation the symmetry that gave zero current for a stationary bound state is broken, and we may evaluate the current in unbound, plane-wave traveling states.

3.5.2   Current flow due to a traveling wave

Consider the simple case of a particle with charge e and mass m that is in an unbound state. The particle is described by a wave function that is a plane wave traveling from left to right of the form $\psi(x) = e^{i(kx - \omega t)}$. To calculate the current flow associated with this state, we substitute into our expression for the current operator

$$J_x = -\frac{ie\hbar}{2m}\left(\psi^*(x)\frac{\partial}{\partial x}\psi(x) - \psi(x)\frac{\partial}{\partial x}\psi^*(x)\right)$$

(3.58)

$$J_x = -\frac{ie\hbar}{2m}(e^{-ikx} \cdot ik \cdot e^{ikx} + e^{ikx} \cdot ik \cdot e^{-ikx}) = -\frac{ie\hbar}{2m} \cdot 2ik = \frac{e\hbar k}{m}$$

(3.59)

Since momentum in the x direction is $p_x = mv_x = \hbar k$, the current associated with the traveling wave may be written in the familiar form $J_x = ev_x$, where e is the particle charge and v_x is the electron velocity.

If we were to construct an electron wave function consisting of a plane wave traveling from left to right and an identical but counter-propagating wave, then the individual contributions to current flow exactly cancel and there is zero net current flow. The symmetry that gives rise to zero net current flow for a superposition of two identical but counter-propagating traveling waves is the same symmetry that creates the standing-wave state or stationary state previously discussed in Section 3.1 and Section 3.5.1.

3.6 Degeneracy as a consequence of symmetry

Continuing our theme of symmetry, we investigate the effect symmetry in a potential has on the number of eigenstates with the same energy. The number of states with the same energy eigenvalue is called the *degeneracy* of the state. Degeneracy is most often a direct consequence of symmetry in the potential. In situations in which this is *not* the case, the degeneracy is said to be accidental. Possibly the best way to learn about degeneracy is by considering an example, so we will do this next.

3.6.1   Bound states in three dimensions and degeneracy of eigenvalues

To illustrate how degeneracy arises from symmetry in a potential we consider the case of a box-shaped potential of side L the interior of which has zero potential energy and the exterior of which has infinite potential energy.

For a potential box in three dimensions with infinite barrier energy we have

$$V(x, y, z) = 0 \text{ for } -\frac{L}{2} < x, y, z < \frac{L}{2} \tag{3.60}$$

$$V(x, y, z) = \infty \text{ for } -\frac{L}{2} > x, y, z > \frac{L}{2} \tag{3.61}$$

Using the results given in Section 3.3.1 for the one-dimensional, rectangular potential well with infinite barrier energy, the energy eigenvalues for the three-dimensional potential well are

$$E_{n_x, n_y, n_z} = \frac{\hbar^2 k^2}{2m} = \frac{\hbar^2}{2m}\left(k_x^2 + k_y^2 + k_z^2\right) \tag{3.62}$$

where $k^2 = k_x^2 + k_y^2 + k_z^2$. For the box-shaped potential, it follows that

$$k_x = \frac{n_x \pi}{L} \qquad k_y = \frac{n_y \pi}{L} \qquad k_z = \frac{n_z \pi}{L} \tag{3.63}$$

where n_x, n_y and n_z are nonzero positive integers. The corresponding energy eigenvalues are

$$E_{n_x, n_y, n_z} = \frac{\hbar^2 \pi^2}{2mL^2}\left(n_x^2 + n_y^2 + n_z^2\right) = n^2 E_{1,1,1} \tag{3.64}$$

where the lowest energy value is

$$E_{1,1,1} = \frac{3\hbar^2 \pi^2}{2mL^2} \tag{3.65}$$

The energy eigenvalues are labeled by three nonzero positive-integer quantum numbers that correspond to orthogonal x, y, and z eigenfunctions. Because n_x, n_y, and n_z label independent contributions to the total energy eigenvalue, many of these energy eigenvalues are *degenerate* in energy. Such degeneracy exists because *different* combinations of quantum numbers result in the same value of energy. For example, $E_{n_x=1, n_y=2, n_z=3}$ can be rearranged in a number of ways:

$(1, 2, 3), (1, 3, 2), (2, 1, 3), (2, 3, 1), (3, 1, 2), (3, 2, 1)$

All six states have the same energy $14E_{1,1,1}$, and so the state is said to be six-fold degenerate.

The lowest-energy state is the ground state, which has quantum numbers $n_x = n_y = n_z = 1$ and energy $E_{1,1,1}$. Only one combination of quantum numbers has this energy, and thus the state is nondegenerate.

The conclusion is that degeneracy of energy eigenvalues is another consequence of symmetry in the potential. If the potential is changed in such a way that the symmetry is broken, a new set of energy eigenvalues is created that will have a different degeneracy. Usually, reducing or breaking the level of symmetry will reduce the degeneracy.

3.7 Symmetric finite-barrier potential

So far we have discussed states bound by a rectangular potential well with infinite barrier energy. Bound states can also exist in a potential well with finite barrier energy. To find out more about this situation, we next consider the case of a particle of mass m in the presence of a simple symmetric, one-dimensional, rectangular potential well of total width $2 \times L$. The potential has finite barrier energy so that $V(x) = 0$ for $-L < x < L$ and $V(x) = V_0$ elsewhere. The value of V_0 is a finite, positive constant.

We proceed in the usual way by first sketching the potential and writing down the Schrödinger equation we intend to solve. It is clear from Fig. 3.6 that any bound state in the system must have energy $E < V_0$. A particle with energy $E > V_0$ will belong to the class of *continuum* unbound states. The *binding energy* of a particular bound state is the minimum energy required to promote the particle from that bound state to an unbound state. At present, we are only concerned with the properties of bound states of the system.

The time-independent Schrödinger equation for one dimension is

$$\left(-\frac{\hbar^2}{2m} \frac{d^2}{dx^2} + V(x) \right) \psi(x) = E\psi(x) \tag{3.66}$$

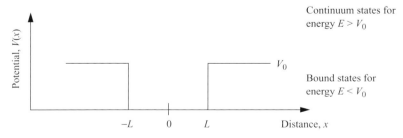

Fig. 3.6. Sketch of a simple symmetric one-dimensional, rectangular potential well of width $2L$. The potential is such that $V(x) = 0$ for $-L < x < L$ and $V(x) = V_0$ elsewhere. The value of V_0 is a finite, positive constant.

By symmetry we have $\psi(x) = \pm\psi(-x)$, and for $|x| < L$ the wave function either has even parity

$$\psi(x) = A\cos(kx) \tag{3.67}$$

or odd parity

$$\psi(x) = B\sin(kx) \tag{3.68}$$

The magnitude of the wave vector in Eqn (3.67) and Eqn (3.68) is

$$k = \sqrt{\frac{2mE}{\hbar^2}} \tag{3.69}$$

For $|x| > L$ the wave function is

$$\psi(x) = Ce^{-\kappa|x|} \tag{3.70}$$

and

$$\kappa = \sqrt{\frac{2m(V_0 - E)}{\hbar^2}} \tag{3.71}$$

for $E < V_0$.

The alternative solution for Eqn (3.70), $\psi = Ce^{\kappa x}$, is not allowed because the expectation value of potential energy $\langle V \rangle$ would have an infinite negative contribution. In addition, our probabilistic interpretation of the bound-state wave function squared requires $\int \psi^*(x)\psi(x)dx$ to be finite. Technically, one says ψ must be square-integrable.

From Eqn (3.69) $k^2 = 2mE/\hbar^2$ and from Eqn (3.71) $\kappa^2 = 2m(V_0 - E)/\hbar^2$, and introducing $L^2 K_0^2$, we have

$$L^2 K_0^2 = L^2 k^2 + L^2\kappa^2 = \frac{(2mE + 2mV_0 - 2mE)}{\hbar^2}L^2 = \frac{2mV_0L^2}{\hbar^2} \tag{3.72}$$

It is clear that this condition relating k and κ must be satisfied when we find solutions for $\psi(x)$. This is an equation for a circle in the $(L\kappa, Lk)$ plane of radius:

$$LK_0 = \sqrt{\frac{2mV_0L^2}{\hbar^2}} \tag{3.73}$$

Since $V(x)$ and $d^2\psi(x)/dx^2$ are finite everywhere, $\psi(x)$ and $d\psi(x)/dx$ must be continuous everywhere, including at $x = \pm L$. Hence, the boundary conditions

$$\psi(x)|_{x=L+\delta} = \psi(x)|_{x=L-\delta} \tag{3.74}$$

and

$$\frac{d}{dx}\psi(x)\bigg|_{x=L+\delta} = \frac{d}{dx}\psi(x)\bigg|_{x=L-\delta} \tag{3.75}$$

lead directly to solutions of $\psi(x)$ at $x = L$

Even-parity solutions must satisfy

$$A\cos(kL) = Ce^{-\kappa L} \tag{3.76}$$

from Eqn (3.74), and

$$-kA\sin(kL) = -\kappa Ce^{-\kappa L} \tag{3.77}$$

from Eqn (3.75), so that *even-parity* solutions require

$$kL \cdot \tan(kL) = \kappa L \tag{3.78}$$

and *odd-parity* solutions require

$$kL \cdot \cot(kL) = -\kappa L \tag{3.79}$$

Solutions that simultaneously satisfy the equation for a circle of radius $LK_0 = \sqrt{2mV_0L^2}/\hbar$ in the $(\kappa L, kL)$ plane *and* Eqn (3.78) or Eqn (3.79) can be found by graphical means. One plots the equations, and solutions exist wherever the curves intersect.

3.7.1   Calculation of bound states in a symmetric, finite-barrier potential

In this section we first consider the case of a potential well similar to that sketched in Fig. 3.6 for which $LK_0 = 1$. Solutions occur when the equation for a circle of radius LK_0 and Eqn (3.78) or Eqn (3.79) are simultaneously satisfied in the $(\kappa L, kL)$ plane.

When $LK_0 = 1$, then $V_0L^2 = \hbar^2/2m$, and, as shown in Fig. 3.7(a), there is one solution with even parity.

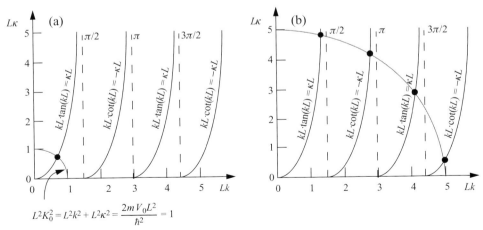

$$L^2 K_0^2 = L^2 k^2 + L^2 \kappa^2 = \frac{2mV_0L^2}{\hbar^2} = 1$$

Fig. 3.7. Illustration of the graphical method to find the solution for a simple symmetric, one-dimensional, rectangular potential well with finite barrier energy such that $V(x) = 0$ for $-L < x < L$ and $V(x) = V_0$ elsewhere. In (a) the value of V_0 results in $2mV_0L^2/\hbar^2 = 1$, and there is one solution. In (b) the value of V_0 results in $2mV_0L^2/\hbar^2 = 25$, and there are four solutions.

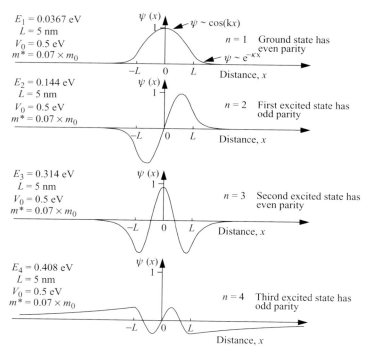

Fig. 3.8. Sketch of the eigenfunctions ψ_1, ψ_2, ψ_3, and ψ_4 for a simple symmetric one-dimensional rectangular potential well of total width $2 \times L$ with finite barrier energy such that $V(x) = 0$ for $-L < x < L$ and $V(x) = V_0$ elsewhere. Effective electron mass is $m^* = 0.07 \times m_0$, barrier energy is $V_0 = 0.5$ eV, and total well width is $2 \times L = 10$ nm. The four bound energy eigenvalues measured from the bottom of the potential well are $E_1 = 0.0367$ eV, $E_2 = 0.144$ eV, $E_3 = 0.314$ eV, and $E_4 = 0.408$ eV. Because E_4 is close in value to V_0 the wave function ψ_4 is not well confined by the potential and so extends well into the barrier region.

When $Lk_0 = 5$ there are four solutions, two with even parity and two with odd parity. See Fig. 3.7(b). In this case, $V_0 L^2 = 25\hbar^2/2m$. It is clear from Fig. 3.7(b) that there will always be at least four solutions if $2m V_0 L^2/\hbar^2 > (3\pi/2)^2$ or $2m V_0 L^2/\hbar^2 > 22.2$.

To illustrate the wave functions for the bound states of an electron in a finite potential well, they are sketched in Fig. 3.8 for the case in which $2m^* V_0 L^2/\hbar^2 = 23.0$. Here the effective electron mass is taken to be $m^* = 0.07 \times m_0$, which is the value for an electron in the conduction band of the semiconductor GaAs. In the figure, the boundaries of the potential at position $x = \pm L$ are indicated to help identify the penetration of the wave function into the barrier region.

We may now draw some conclusions concerning the nature of solutions one obtains for a particle in our simple rectangular potential well with finite-barrier energy. First, note that wave functions have alternating even and odd parity. This is a direct consequence of the symmetry of the potential. Second, there is always at least one lowest-energy state solution that is *always* of *even parity*. This result is obvious from our graphical solution, but again it is a consequence of the potential's symmetry. Third, for a finite potential well, the wave function always extends into the barrier region. This is particularly apparent for the $n = 4$ state shown in Fig. 3.8. In quantum mechanics, when a wave function penetrates a potential barrier one talks about *tunneling* into the barrier. For the case of the simple potential we have been considering, tunneling *always* has the effect of reducing the energy of the eigenvalue E_n compared with the infinite potential well case. This reduction in energy is easy to understand if we recognize that tunneling decreases the curvature of the wave function and hence the energy associated with it.

We complete our analysis of a rectangular potential well with finite barrier energy by making sure that in the limit $V_0 \to \infty$ we regain our previous expression for the rectangular potential well with infinite barrier energy.

In the limit $V_0 \to \infty$, notice that $L^2 K_0^2 = L^2 k^2 + L^2 \kappa^2 = 2mV_0L^2/\hbar^2 \to \infty$. The intersection of the circle of radius $LK_0 \to \infty$ with curves defined by $kL \cdot \tan(kL) = \kappa L$ and $kL \cdot \cot(kL) = -\kappa L$ now occurs at $Lk = n\pi/2$, where $n = 1, 2, 3, \ldots$. The energy eigenvalues are

$$E_n = \frac{\hbar^2}{2m} \cdot n^2 k^2 = \frac{\hbar^2}{2m} \cdot \frac{n^2 \pi^2}{4L^2} \qquad (3.80)$$

This is comforting, because it is the same result previously derived in Section 3.3.1 (Eqn (3.30)) for the energy levels of a particle in a rectangular potential well with infinite barrier energy. The only difference between Eqn (3.80) and Eqn (3.30) is the factor 4 in the denominator, which arises from the fact that the rectangular potential well of infinite barrier energy had a total well width of L and the finite barrier energy potential we have been considering has total well width of $2 \times L$.

While the value of E_1 given in Fig. 3.8 was 36.7 meV, the ground-state energy of the same electron with effective electron mass $m^* = 0.07 \times m_0$ but in a rectangular potential well of width $2 \times L = 10$ nm with infinite barrier energy is $E_1 = 53.7$ meV. As we have discussed previously, this reduction in eigenenergy for the particle in a rectangular potential well with finite barrier energy is due to the fact the wave function can reduce its curvature by tunneling into the barrier region.

3.8 Transmission and reflection of unbound states

In this section we will explore the transmission and reflection of a particle incident on a potential step. The classical result is that a particle of mass m is transmitted if its energy is greater than the potential step and is reflected if its energy is less than the potential step. The velocity of a classical particle with energy greater than the potential step changes as it passes the step. While the corresponding change in momentum exerts a force, the particle is still transmitted. In quantum mechanics, the situation is different,

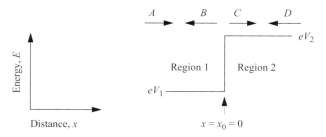

Fig. 3.9. Sketch of a one-dimensional potential step. In region 1 the potential energy is eV_1, and in region 2 the potential energy is eV_2. The coefficients A and C correspond to waves traveling left to right in regions 1 and 2, respectively. The coefficients B and D correspond to waves traveling right to left in regions 1 and 2, respectively. The transition between region 1 and region 2 occurs at position $x = x_0$. The incident particle is assumed to have energy $E > eV_2$.

and a particle with energy greater than the potential step is not necessarily transmitted. The wavy nature of an unbound state ensures that it feels the presence of changes in potential. Physically, this manifests itself as a scattering event with particle transmission and reflection probabilities. A simple situation to consider is transmission and reflection of unbound states from a potential step in one dimension.

Consider a particle of energy E and mass m incident from the left on a step potential of energy $V_0 = V_2 - V_1$ at position x_0. We will only consider the case $E > V_2$. Figure 3.9 is a sketch of the potential energy. The one-dimensional potential step occurs at the boundary between region 1 and region 2.

Solutions of the time-independent Schrödinger equation

$$((-\hbar^2/2m)\nabla^2 + V(x))\psi(x) = E\psi(x)$$

are of the form

$$\psi_1(x) = Ae^{ik_1x} + Be^{-ik_1x} \qquad \text{for } x < x_0 \tag{3.81}$$

and

$$\psi_2(x) = Ce^{ik_2x} + De^{-ik_2x} \qquad \text{for } x > x_0 \tag{3.82}$$

These equations describe left- and right-traveling waves in the two regions. The energy of the particle in region 1 (with $V_1 = 0$) is

$$E = \frac{\hbar^2 k_1^2}{2m} \tag{3.83}$$

so that

$$k_1 = \frac{(2mE)^{1/2}}{\hbar} \tag{3.84}$$

When $E < V_2$ for the potential step shown in Fig. 3.9, k_2 is imaginary. This results in $\psi_2(x \to \infty) = Ce^{-k_2x}|_{x\to\infty} = 0$ and $|A|^2 = |B|^2 = 1$, which has the physical meaning that the particle is completely reflected at the potential step (see Exercise 3.1).

In general, a particle of energy E moving in a one-dimensional potential with value V_j in the j-th region has a wave vector

$$k_j = \frac{(2m_j(E - eV_j))^{1/2}}{\hbar} \tag{3.85}$$

For $E > V_j$, the wave vector k_j is real, and for $E < V_j$ the wave vector k_j is imaginary.

We now have the means to calculate the scattering probability of a particle moving from region 1 to region 2 for the potential shown in Fig. 3.9. Often we will be interested in applying our knowledge to situations involving electron transport in semiconductors. A complication we need to consider is that in a semiconductor the electron scatters from the periodic potential created by crystal atoms and that the resulting dispersion relation gives the electron an effective mass that is different from that of a free electron. In the case of electrons moving across a semiconductor heterostructure interface there exists the possibility that the effective electron mass can change. Since, in some models, it is possible for the electron to have a different effective mass in each region, we need to consider two situations, one where $m_1 = m_2$ and one where $m_1 \neq m_2$.

3.8.1   Scattering from a potential step when $m_1 = m_2$

In this case, we assume that m_j is constant through all regions.

At the boundary between regions 1 and 2 at x_0 the wave functions are linked by the constraint that the wave function ψ and the derivative $d\psi/dx$ are continuous, so that

$$\psi_1|_{x_0} = \psi_2|_{x_0} \tag{3.86}$$

$$\frac{d}{dx}\psi_1\bigg|_{x_0} = \frac{d}{dx}\psi_2\bigg|_{x_0} \tag{3.87}$$

These boundary conditions and Eqn (3.81) and Eqn (3.82) give, for $x_0 = 0$,

$$A + B = C + D \tag{3.88}$$

$$A - B = \frac{k_2}{k_1}C - \frac{k_2}{k_1}D \tag{3.89}$$

To find simple expressions for the probability amplitude in both regions we apply initial conditions. Suppose we know that the particle is incident from the left. Then $|A|^2 = 1$ and $|D|^2 = 0$, since there is no left-traveling wave in region 2. Substituting $A = 1$ and $D = 0$ into Eqn (3.88) and Eqn (3.89) gives

$$1 + B = C \tag{3.90}$$

$$1 - B = \frac{k_2}{k_1}C \tag{3.91}$$

Adding Eqn (3.90) and Eqn (3.91) gives

$$2 = (1 + k_2/k_1)C \tag{3.92}$$

$$C = \frac{2}{(1 + k_2/k_1)} \tag{3.93}$$

and from Eqn (3.90)

$$B = C - 1 = \frac{2}{(1 + k_2/k_1)} - 1 = \frac{(2 - 1 - k_2/k_1)}{(1 + k_2/k_1)} \tag{3.94}$$

$$B = \frac{(1 - k_2/k_1)}{(1 + k_2/k_1)} \tag{3.95}$$

Identifying electron velocity in the j-th region as $v_j = \hbar k_j/m$, the transmission probability $|C|^2$ and reflection probability $|B|^2$ may be written as

$$|C|^2 = \frac{4}{(1 + k_2/k_1)^2} = \frac{4}{(1 + v_2/v_1)^2} \tag{3.96}$$

$$|B|^2 = \frac{(1 - k_2/k_1)^2}{(1 + k_2/k_1)^2} = \left(\frac{v_1 - v_2}{v_1 + v_2}\right)^2 \tag{3.97}$$

Because $V_2 > V_1$, it follows that $k_2 < k_1$, so that $k_2/k_1 < 1$. It can easily be seen that $|C|^2$ is always bigger than 1 (hence bigger than $|A|^2$), which means the probability amplitude for finding the particle anywhere in region 2 is greater than the probability of finding it anywhere in region 1. To understand this, we should consider the velocity of the particle in each region $v_j = \hbar k_j/m$, where $k_j = \sqrt{2m(E - V_j)}/\hbar$ so $v_1 > v_2$, because $V_2 > V_1$. We know that, classically, when a particle moves faster it spends less time at any given position and the probability of finding it in each infinitesimal part of space dx is smaller. However, we should be very careful about this conclusion and not confuse it with the concept of transmission and reflection that is meaningful only when we consider probability current density. In fact, when we are dealing with traveling waves the probability current is a more useful concept than probability itself – as opposed to bound states, in which the opposite is true.

Before considering probability current density, it is helpful to find solutions for transmission and reflection probability in the situation in which $m_1 \neq m_2$.

3.8.2   Scattering from a potential step when $m_1 \neq m_2$

In this case, we assume that m_j varies from region to region. At the boundary between regions 1 and 2 we require continuity in the wave function ψ and the derivative $(1/m_j) \cdot d\psi/dx$, so that

$$\psi_1|_{x_0} = \psi_2|_{x_0} \tag{3.98}$$

$$\frac{1}{m_1}\frac{d}{dx}\psi_1\bigg|_{x_0} = \frac{1}{m_2}\frac{d}{dx}\psi_2\bigg|_{x_0} \tag{3.99}$$

Inadequacies in our model force us to choose a boundary condition that ensures conservation of current $j_x \propto e p_x / m$ rather than $d\psi_1/dx|_{x_0} = d\psi_2/dx|_{x_0}$ (more accurate models satisfy both of these conditions).

These conditions and Eqn (3.81) and Eqn (3.82) give, for $x_0 = 0$,

$$A + B = C + D \tag{3.100}$$

$$A - B = \frac{m_1 k_2}{m_2 k_1} C - \frac{m_1 k_2}{m_2 k_1} D \tag{3.101}$$

If we know that the particle is incident from the left, then $A = 1$ and $D = 0$, giving

$$1 + B = C \tag{3.102}$$

$$1 - B = \frac{m_1 k_2}{m_2 k_1} C \tag{3.103}$$

We now solve for the transmission probability $|C|^2$ and the reflection probability $|B|^2$. The result is

$$|C|^2 = \frac{4}{\left(1 + \dfrac{m_1 k_2}{m_2 k_1}\right)^2} \tag{3.104}$$

$$|B|^2 = \frac{\left(1 - \dfrac{m_1 k_2}{m_2 k_1}\right)^2}{\left(1 + \dfrac{m_1 k_2}{m_2 k_1}\right)^2} \tag{3.105}$$

Compared with Eqn (3.96) and Eqn (3.97), the ratio m_1/m_2 appearing in Eqn (3.104) and Eqn (3.105) gives an extra degree of freedom in determining transmission and reflection probability. It is this extra degree of freedom that will allow us to engineer the transmission and reflection probability in device design. Having established this, we now proceed to calculate probability current density for an electron scattering from a potential step.

3.8.3   Probability current density for scattering at a step

Probability current density for transmission and reflection is different from transmission and reflection probability. We will be interested in calculating the incident current $\mathbf{J}_\mathrm{I}$, reflected current $\mathbf{J}_\mathrm{R}$, and transmitted current $\mathbf{J}_\mathrm{T}$, shown schematically in Fig. 3.10.

From our work in Section 3.8.2, the solution for the wave function will be of the form

$$\psi_1 = A e^{ik_1 x} + B e^{-ik_1 x} \tag{3.106}$$

$$\psi_2 = C e^{ik_2 x} + D e^{-ik_2 x} \tag{3.107}$$

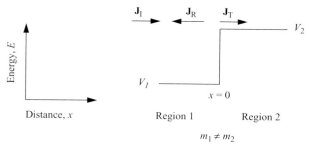

Fig. 3.10. Sketch of a one-dimensional, rectangular potential step. In region 1 the potential energy is V_1 and particle mass is m_1. In region 2 the potential energy is V_2 and particle mass is m_2. The transition between region 1 and region 2 occurs at position $x = x_{12}$. Incident probability current density $\mathbf{J}_I$, reflected probability current density $\mathbf{J}_R$, and transmitted probability current density $\mathbf{J}_T$ are indicated.

For a particle incident from the left, we had $|A|^2 = 1$, $|D|^2 = 0$. Adopting the boundary conditions

$$\psi_1|_{x=0} = \psi_2|_{x=0} \tag{3.108}$$

$$\left. \frac{1}{m_1} \frac{d}{dx} \psi_1 \right|_{x=0} = \left. \frac{1}{m_2} \frac{d}{dx} \psi_2 \right|_{x=0} \tag{3.109}$$

gives reflection probability

$$|B|^2 = \frac{(1 - m_1 k_2 / m_2 k_1)^2}{(1 + m_1 k_2 / m_2 k_1)^2} \tag{3.110}$$

and transmission probability

$$|C|^2 = \frac{4}{(1 + m_1 k_2 / m_2 k_1)^2} \tag{3.111}$$

We now calculate current using the current operator

$$\mathbf{J} = \frac{-ie\hbar}{2m} (\psi^* \nabla \psi - \psi \nabla \psi^*)$$

The incident current is

$$\mathbf{J}_I = \frac{e\hbar k_1}{m_1} |A|^2 \tag{3.112}$$

The reflected current is

$$\mathbf{J}_R = \frac{-e\hbar k_1}{m_1} |B|^2 \tag{3.113}$$

and the transmitted current is

$$\mathbf{J}_T = \frac{e\hbar k_2}{m_2} |C|^2 \tag{3.114}$$

The reflection coefficient for the particle flux is

$$Refl = -\frac{\mathbf{J}_R}{\mathbf{J}_I} = \left|\frac{B}{A}\right|^2 = \left(\frac{(1 - m_1 k_2/m_2 k_1)}{(1 + m_1 k_2/m_2 k_1)}\right)^2 \tag{3.115}$$

where the minus sign indicates current flowing in the negative x direction. This is the same as the reflection probability given by Eqn (3.105), because the ratio of velocity terms that contribute to particle flux is unity. The transmission coefficient for the particle flux is

$$Trans = \frac{\mathbf{J}_T}{\mathbf{J}_I} = \frac{m_1 k_2}{m_2 k_1}\left|\frac{C}{A}\right|^2 = \frac{4 k_1 k_2/m_1 m_2}{\left(\dfrac{k_1}{m_1} + \dfrac{k_2}{m_2}\right)^2} = 1 - Refl \tag{3.116}$$

where we note that $Trans + Refl = 1$. The fact that $Trans + Refl = 1$ is expected since current conservation requires that the incident current must equal the sum of the transmitted and reflected current.

3.8.4   Impedance matching for unity transmission across a potential step

In this section we continue our discussion of particle scattering at the potential step shown schematically in Fig. 3.10. Suppose we want a flux transmission probability of unity for a particle of energy $E > V_2$ approaching the potential step from the left. Since momentum $p = \hbar k = mv$ we can identify velocity $v_j = \hbar k_j/m_j$ as the physically significant quantity in the expression for the transmission coefficient. Substituting v_j into Eqn (3.116) gives

$$Trans = \frac{m_1 k_2}{m_2 k_1}\left|\frac{C}{A}\right|^2 = \frac{m_1 k_2}{m_2 k_1}\frac{4}{(1 + v_2/v_1)^2} = \frac{v_2}{v_1}\frac{4}{(1 + v_2/v_1)^2} \tag{3.117}$$

If $Trans = 1$, then Eqn (3.117) can be rewritten

$$1 = \frac{v_2}{v_1}\frac{4}{(1 + v_2/v_1)^2} \tag{3.118}$$

which shows that unity transmission occurs when the velocity of the particle in the two regions is matched[2] in such a way that $v_2/v_1 = 1$. In microwave transmission line theory, this is called an impedance matching condition. To figure out when impedance matching occurs as a function of particle energy, we start with

$$\frac{v_2}{v_1} = \frac{m_1 k_2}{m_2 k_1} = \frac{m_1}{m_2} \cdot \left(\frac{2 m_2 \hbar^2 (E - V_2)}{2 m_1 \hbar^2 (E - V_1)}\right)^{1/2} = \left(\frac{m_1(E - V_2)}{m_2(E - V_1)}\right)^{1/2} \tag{3.119}$$

[2] J. F. Müller, A. F. J. Levi, and S. Schmitt-Rink, *Phys. Rev.* **B38**, 9843 (1988) and T. H. Chiu and A. F. J. Levi, *Appl. Phys. Lett.* **55**, 1891 (1989).

so that impedance matching ($v_2/v_1 = 1$) will occur when

$$1 = \frac{m_1}{m_2} \cdot \frac{E - V_2}{E - V_1} \tag{3.120}$$

or

$$\frac{m_2}{m_1} = \frac{E - V_2}{E - V_1} \tag{3.121}$$

Clearly, for an electron incident on the potential step with energy E, the value of E for which $Trans = 1$ depends upon the ratio of effective electron mass in the two regions and the difference in potential energy between the steps. To see what this means in practice, we now consider a specific example.

For a potential step of 1 eV we set $V_1 = 0$ eV and $V_2 = 1$ eV. We assume that the electron mass is such that $m_1 = 10 \times m_2$, so that the particle flux transmission coefficient $Trans = 1$ when

$$\frac{m_2}{m_1} = \frac{E - V_2}{E - V_1} = \frac{E - 1}{E} = \frac{1}{10} \tag{3.122}$$

Hence the particle energy when $Trans = 1$ is

$$E = \frac{10}{9} = 1.11 \text{ eV} \tag{3.123}$$

When $m_1 = 2 \times m_2$, the particle flux transmission coefficient $Trans = 1$ when

$$\frac{m_2}{m_1} = \frac{E - V_2}{E - V_1} = \frac{E - 1}{E} = \frac{1}{2} \tag{3.124}$$

so that the particle energy is $E = 2.00$ eV.

We can also calculate $Trans$ in the limit when energy E goes to infinity. In this case

$$v_2/v_1|_{E \to \infty} = \sqrt{\frac{m_1}{m_2}} \tag{3.125}$$

and

$$Trans|_{E \to \infty} = \lim_{E \to \infty} (v_2/v_1) \frac{4}{(1 + v_2/v_1)^2} \tag{3.126}$$

For the case $m_1/m_2 = 10$

$$Trans|_{E \to \infty} = (\sqrt{10}) \frac{4}{(1 + \sqrt{10})^2} = 0.73 \tag{3.127}$$

and when $m_1/m_2 = 2$ one finds

$$Trans|_{E \to \infty} = (\sqrt{2}) \frac{4}{(1 + \sqrt{2})^2} = 0.97 \tag{3.128}$$

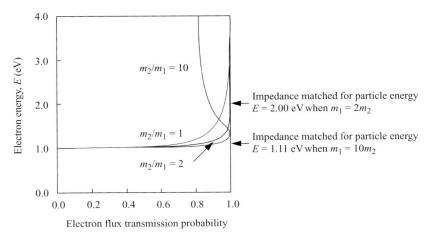

Fig. 3.11. Electron flux transmission probability as a function of energy for electron motion across the potential step illustrated in Fig. 3.10, with $V_2 - V_1 = 1$ eV. In region 1 the electron has mass m_1, and in region 2 the electron has mass m_2.

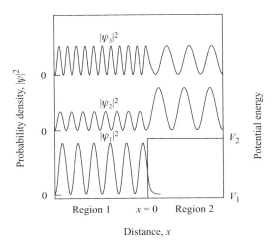

Fig. 3.12. Sketch of probability density $|\psi|^2$ for three types of wave function. Wave function ψ_1 corresponds to a particle with energy $E < V_2$, and so it is essentially zero in region 2 far from $x = 0$. Wave function ψ_2 corresponds to a particle with energy $E > V_2$. In this case, because particle velocity is lower in region 2, the probability density in region 2 is greater than in region 1. Wave function ψ_3 corresponds to a particle of energy $E > V_2$ that is impedance matched for transmission across the potential step. In this situation, the velocity on either side of the potential step is the same, as is the amplitude of ψ_3.

The result of calculating electron flux transmission probability as a function of energy for electron motion across the 1 eV potential step is shown in Fig. 3.11 for the cases when $m_2 = m_1, m_2 = 2 \times m_1$, and $m_2 = 10 \times m_1$.

Impedance matching particle transmission at a potential step also changes the nature of the wave function and hence the probability density. To illustrate this, Fig. 3.12 is a sketch of probability density $|\psi|^2$ for three types of wave function. Wave function ψ_1 corresponds to a particle with energy $E < V_2$, and so it is essentially zero in region 2 apart from a small exponentially decaying contribution in the potential barrier near $x = 0$. Wave function ψ_2 corresponds to a particle with energy $E > V_2$, which has finite transmission probability. In this case, because particle velocity is lower in region 2, the probability density in region 2 is greater than in region 1. Wave function ψ_3 corresponds to a particle of energy $E > V_2$ that is impedance matched for transmission across the potential step. In this situation, the velocity of the particle and the amplitude of ψ_3 are the same on either side of the potential step.

3.9 Particle tunneling

If electrons or other particles of mass m and kinetic energy E impinge on a potential barrier of energy V_0 and thickness L in such a way that $E < V_0$, they penetrate into the barrier, just as with the potential step considered above. If the potential barrier is thin, there is a significant chance that the particle can be transmitted through the barrier. This is called quantum mechanical tunneling. Tunneling of a finite mass particle is *fundamentally quantum mechanical.*

Analogies are often made between tunneling in quantum mechanics and effects that occur in electromagnetism. However, evanescent field coupling in classical electromagnetism is merely the large photon number limit of quantum mechanical photon particle tunneling. Tunneling is a purely quantum mechanical effect.

For electrons, the tunnel current density depends exponentially upon the barrier thickness, L, and barrier energy, V_0, and so it tends not to be important when $V_0 L$ is large.

For a particle of energy $E < V_0$, the solution for the wave function in regions 1 and 2 contains right- and left-propagating terms with the coefficients indicated in Fig. 3.13. In region 1, the wave function is of the form $\psi_1(x) = Ae^{ikx} + Be^{-ikx}$, and in region 2 the wave function is of the form $\psi_2(x) = Ce^{\kappa x} + De^{-\kappa x}$, where $k = \sqrt{2mE}/\hbar$ and $\kappa = \sqrt{2m(V_0 - E)}/\hbar$. Notice that the particle wave function has left- and right-propagating exponentially decaying solutions inside the barrier. This is required if tunnel current is to flow through the barrier. Because we will be considering a particle incident from the left, in region 3 we need only consider a right-propagating wave. To find the general solution, we require that the wave function and its derivative be continuous at $x = 0$ and $x = L$. This condition gives four equations:

$$A + B = C + D \tag{3.129}$$

$$(A - B) = \frac{i\kappa}{k}(D - C) \tag{3.130}$$

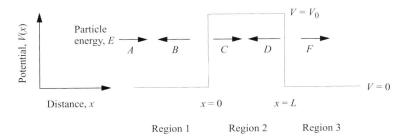

Fig. 3.13. Illustration of a particle of energy E and mass m incident on a one-dimensional, rectangular potential barrier of energy $V = V_0$ and thickness L. The incident wave has amplitude $A = 1$, and the transmitted wave has amplitude F. Wave reflection at positions $x = 0$ and $x = L$ can give rise to resonances in transmission as a function of particle energy.

$$C e^{\kappa L} + D e^{-\kappa L} = F e^{ikL} \tag{3.131}$$

$$C e^{\kappa L} - D e^{-\kappa L} = \frac{ik}{\kappa} F e^{ikL} \tag{3.132}$$

Since the tunneling transmission probability is $|F/A|^2$, we proceed to eliminate the other coefficients. Obtaining the solution we seek requires some manipulation of equations which, to save the reader time and effort, will be reproduced here.

Adding Eqn (3.129) and Eqn (3.130) gives

$$2A = \left(1 - \frac{i\kappa}{k}\right)C + \left(1 + \frac{i\kappa}{k}\right)D \tag{3.133}$$

Adding Eqn (3.131) and Eqn (3.132) gives

$$2C e^{\kappa L} = \left(1 + \frac{ik}{\kappa}\right)F e^{ikL} \tag{3.134}$$

Subtracting Eqn (3.131) and Eqn (3.132) gives

$$2D e^{-\kappa L} = \left(1 - \frac{ik}{\kappa}\right)F e^{ikL} \tag{3.135}$$

Using Eqn (3.134) and Eqn (3.135) to eliminate C and D from Eqn (3.133) leads to

$$2A = \left(1 - \frac{i\kappa}{k}\right)\left(1 + \frac{ik}{\kappa}\right)\frac{F}{2}e^{ikL-\kappa L} + \left(1 + \frac{i\kappa}{k}\right)\left(1 - \frac{ik}{\kappa}\right)\frac{F}{2}e^{ikL+\kappa L} \tag{3.136}$$

$$\frac{F}{A} = \frac{4k\kappa e^{-ikL}}{(k - i\kappa)(\kappa + ik)e^{-\kappa L} + (k + i\kappa)(\kappa - ik)e^{\kappa L}} \tag{3.137}$$

$$\frac{F}{A} = \frac{4ik\kappa}{(-(\kappa - ik)^2 e^{\kappa L} + (\kappa + ik)^2 e^{-\kappa L})} \cdot e^{-ikL} \tag{3.138}$$

Dividing the top and bottom of Eqn (3.138) by κ^2 gives

$$\frac{F}{A} = \frac{4ik/\kappa}{(-(1 - ik/\kappa)^2 e^{\kappa L} + (1 + ik/\kappa)^2 e^{-\kappa L})} \cdot e^{-ikL} \tag{3.139}$$

The magnitude squared of Eqn (3.139) is

$$\frac{F^* F}{A^* A} = \frac{16(k/\kappa)^2}{(1 + ik/\kappa)^2((1 + ik/\kappa)^*)^2(e^{-2\kappa L} + e^{2\kappa L}) - ((1 + ik/\kappa)^*)^4 - (1 + ik/\kappa)^4}$$

(3.140)

$$\left|\frac{F}{A}\right|^2 = \frac{16(k/\kappa)^2}{(1 - ik/\kappa)^2(1 + ik/\kappa)^2((e^{-\kappa L} - e^{\kappa L})^2 + 2) - (1 - ik/\kappa)^4 - (1 + ik/\kappa)^4}$$

(3.141)

$$\left|\frac{F}{A}\right|^2 = \frac{16(k/\kappa)^2}{(1 - ik/\kappa)^2(1 + ik/\kappa)^2(4\sinh^2(\kappa L) + 2) - (1 - ik/\kappa)^4 - (1 + ik/\kappa)^4}$$

(3.142)

Dealing with the last two terms in the parentheses in the denominator, it is easy to check that

$$(1 - ik/\kappa)^4 + (1 + ik/\kappa)^4 = -16(k/\kappa)^2 + 2(1 + ik/\kappa)^2(1 - ik/\kappa)^2$$ (3.143)

We now use this by substituting Eqn (3.143) into Eqn (3.142) to give

$$\left|\frac{F}{A}\right|^2 = \frac{(k/\kappa)^2}{\frac{1}{4}((1 + ik/\kappa)^2(1 - ik/\kappa)^2)\sinh^2(\kappa L) + (k/\kappa)^2}$$

(3.144)

$$\left|\frac{F}{A}\right|^2 = \frac{(k/\kappa)^2}{\frac{1}{4}(1 + (k/\kappa)^2)^2\sinh^2(\kappa L) + (k/\kappa)^2}$$

(3.145)

So the tunneling transmission probability is

$$\boxed{\left|\frac{F}{A}\right|^2 = \frac{1}{1 + \left(\dfrac{k^2 + \kappa^2}{2k\kappa}\right)^2 \sinh^2(\kappa L)}}$$

(3.146)

Equation (3.146) will be of use in Chapter 4 when we derive the same result using a different method. However, our immediate goal is to plot the probability density distribution for an electron of energy E traveling left to right and incident on a rectangular potential barrier of energy V_0 in a way that $E < V_0$. To obtain this probability distribution, we also need to find C, D, and B. Rewriting Eqn (3.134) and Eqn (3.135) gives

$$C = \left(1 + \frac{i\kappa}{k}\right)\frac{F}{2}e^{ikL}e^{-\kappa L}$$

(3.147)

$$D = \left(1 - \frac{i\kappa}{k}\right)\frac{F}{2}e^{ikL}e^{\kappa L}$$

(3.148)

Using Eqn (3.129), Eqn (3.130), Eqn (3.147), and Eqn (3.148) one may obtain an expression of B that is

$$B = -i\frac{1}{2}Fe^{ikL} \times \frac{k^2 + \kappa^2}{k\kappa}\sinh(\kappa L) \qquad (3.149)$$

The calculation is now complete except for normalization of the wave function. Unfortunately, because we are dealing with traveling waves, the wave function cannot be normalized. To understand why, consider the expression for the absolute value squared of the wave function in regions 1, 2, and 3, which is given by the following three equations:

$$|\psi_1|^2 = |A|^2 + |B|^2 + 2\text{Re}(AB^*e^{2ikx}) \qquad (3.150)$$
$$|\psi_2|^2 = |C|^2e^{2\kappa x} + |D|^2e^{-2\kappa x} + 2\text{Re}(CD^*) \qquad (3.151)$$
$$|\psi_3|^2 = |F|^2 \qquad (3.152)$$

Normalization of the total wave function ψ requires evaluation of the integral, as follows:

$$\int_{-\infty}^{\infty} |\psi|^2 dx = \int_{-\infty}^{0} (|A|^2 + |B|^2 + 2\text{Re}(AB^*e^{2ikx}))dx$$
$$+ \int_{0}^{L} (|C|^2e^{2\kappa x} + |D|^2e^{-2\kappa x} + 2\text{Re}(CD^*))dx + \int_{L}^{\infty} |F|^2 dx \qquad (3.153)$$

This integral is infinite, so we cannot find a value for A that satisfies the normalization condition $\int |\psi|^2 dx = 1$ (Eqn (2.32)). However, one may still evaluate the *relative* probability $|\psi|^2/|A|^2$ by keeping A as an unknown constant.

As an example, for an electron of energy $E = 0.9$ eV incident on a potential barrier of energy $V_0 = 1$ eV located between $x = 0$ nm and $x = 0.5$ nm, we calculate $|\psi|^2/|A|^2$. Figure 3.14 shows results of doing this.

In Fig. 3.14 we have set $A = 1$ for convenience and have plotted: (a) the real part of ψ, and (b) probability $|\psi|^2$ with distance. Using Eqn (3.146) and plugging in the numbers for k, κ, and L, we will find that $|F/A| = |F| \cong 0.548$, so that the electron transmission probability is $|\psi_3|^2 = |F|^2 \cong 0.3$. The probability in region 3 is a constant with distance because the electron is described using a simple traveling wave moving from left to right.

The description of the electron in region 1 is a little more complex. Rewriting Eqn (3.150),

$$|\psi_1|^2 = (|A|^2 + |B|^2 + 2\text{Re}(AB^*e^{2ikx})) = 1 + |B|^2 + 2|B|\cos(2kx - \theta) \qquad (3.154)$$

where we have set $A = 1$ and $B = |B|e^{i\theta}$, with θ being a phase determined by the boundary conditions. To obtain a numerical value for $|B|$, we use Eqn (3.149) and

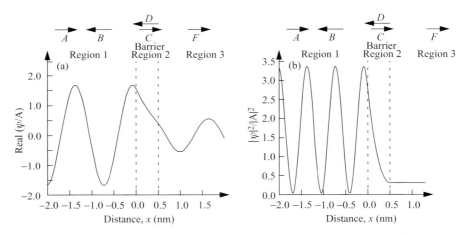

Fig. 3.14. (a) Real part of wave function as a function of distance for an electron of energy $E = 0.9$ eV traveling from left to right and incident on a rectangular potential barrier of energy $V_0 = 1$ eV and width $L = 0.5$ nm. (b) Probability for electron in (a).

obtain $|B| \cong 0.837$. It is clear from Eqn (3.154) that $|\psi_1|^2$ will oscillate with distance between $(1 + |B|)^2$ and $(1 - |B|)^2$ or 3.375 and 0.027 in our numerical example. The reason for the oscillation in electron probability in region 1 is that the incoming right-traveling component of the wave function ψ_1 with amplitude A interferes with the left-traveling component of the wave function with amplitude B. The left-traveling component was created when the incoming right-traveling wave function was reflected from the potential barrier.

Region 2 has left- and right-propagating, exponentially decaying solutions of amplitudes C and D, respectively. These two components must exist for current to flow through the potential barrier (see Exercise 3.1(c)).

3.9.1   Electron tunneling limit to reduction in size of CMOS transistors

Tunneling of electrons is not an abstract theoretical concept of no practical significance. It can dominate performance of many components, and it is of importance for many electronic devices. For example, tunneling of electrons through oxide barriers contributes directly to leakage current in metal-oxide-semiconductor (MOS) field-effect transistors (FETs). Because of this, it is a *fundamental limit* to the continued reduction in size (*scaling*) of CMOS transistors. As one reduces gate length and minimum feature size of such a transistor (see Fig. 3.15), one must also reduce the gate oxide thickness, t_x, to maintain device performance. This is because in simple models transconductance is proportional to gate capacitance. For very small minimum feature sizes, the transistor can no longer function because t_x is so small that electrons efficiently tunnel from the metal gate into the source drain channel.

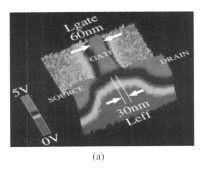  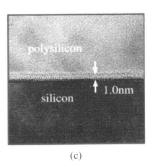

(a)   (b)   (c)

Fig. 3.15. (a) Scanning capacitance micrograph of the two-dimensional doping profile in a 60-nm gate length n-type MOSFET. The effective channel length is measured to be only $L_{\text{eff}} = 30$ nm. (b) Transmission electron microscope cross-section through a 35-nm gate length MOSFET. The channel length is only about 100 silicon lattice sites long. An enlargement of the channel region delineated is shown in (c). The gate oxide thickness estimated from the image is only about 1.0 nm. Images courtesy of G. Timp, University of Illinois.

SiO_2, which is used as the oxide for CMOS transistors, also has a minimum thickness due to the fact that the constituents of the oxide are discrete particles (atoms). According to the SIA roadmap, by the year 2012 transistors will have a minimum feature size of 50 nm and a gate oxide thickness of $t_x = 1.3$ nm, which is so thin that on average it contains only five Si-atom-containing layers. The following generation of transistors, with a minimum feature size of 30 nm, will have a gate oxide that is only three Si-atom layers thick, corresponding to $t_x = 0.8$ nm. Because there are two interfaces – one for the metal gate and one for the channel – that are not completely oxidized, the actual oxide may be thought of as only one atomic layer thick. This leaves a physically thin tunnel barrier for electrons. The absence of a single monolayer of oxide atoms over a significant area under the transistor gate could lead to undesirable leakage current due to electron tunneling between the metal gate and the source drain channel.

The fact that SiO_2 is made up of discrete atoms is a *structural limit*. One cannot vary the thickness of the oxide by, say, one quarter of a monolayer because the minimum layer thickness is, by definition, one atomic layer. These observations allow one to draw the trivial conclusion that, in addition to tunneling, there are *structural* limits to scaling conventional devices such as MOS transistors.[3]

To circumvent these difficulties, there is interest in developing a gate insulator for CMOS that would allow a thicker dielectric insulator layer to be used in transistor fabrication. For a constant gate capacitance, the way to achieve this is to increase the value of the dielectric's relative permittivity, ε_r. To see this, consider a parallel-plate capacitor of area A. Capacitance is

$$C = \frac{\varepsilon_0 \varepsilon_r A}{t_x} \tag{3.155}$$

[3] M. Schulz, *Nature* **399**, 729 (1999).

So, for constant C and area A, increasing thickness t_x requires an increase in ε_r. Unfortunately, using known material properties, there are limits to how far this approach may be exploited. Nevertheless, there is interest in establishing a so-called *high-k* (corresponding to a high value of ε_r) dielectric technology[4] to carry CMOS technology beyond 2012.

3.10 The nonequilibrium electron transistor

What we know about quantum mechanical transmission and reflection of electrons at potential-energy steps can be applied to the design of a new type of semiconductor transistor. The device we are going to design may not be of tremendous use today, but it does allow us to practice the quantum mechanics we have learned so for. We treat it, therefore, as a prototype device.

We can use the techniques described in Section 2.2.6 to create a conduction-band potential for our semiconductor transistor that operates by injecting electrons from a lightly n-type doped, wide band-gap emitter in such a way that they have high velocity while traversing a thin, heavily n-type doped base region. Electrons that successfully traverse the base of thickness x_B without scattering emerge as current flowing in the lightly n-type doped collector. Figure 3.16(a) is a schematic diagram of a potential for this double heterojunction unipolar transistor. Indicated in the figure are the emitter current I_E, base current I_B, collector current I_C, and voltages V_{BE} and V_{CE} for the transistor biased in the common emitter configuration.

In fact, we are trying to design a nonequilibrium electron transistor (NET) that makes use of relatively high-velocity electron transmission through the base with little or no scattering. This type of electron motion is called extreme nonequilibrium electron transport. If the distance a nonequilibrium electron travels before scattering is on average λ_B, then the probability of electron transmission through the base region without scattering is e^{-x_B/λ_B}. Based on these considerations alone, it is clearly advantageous to adopt a design with a very thin base region.

To illustrate some other design considerations, Fig. 3.16(b) shows a schematic diagram of the conduction band edge potential (CB_{min}) of an $AlSb_{0.92}As_{0.08}$/InAs/GaSb double heterojunction unipolar NET under bias. The transistor base consists of a 10-nm-thick epilayer of InAs with a high carrier concentration of two-dimensionally confined electrons, $n \sim 2 \times 10^{12}$ cm^{-2}. The collector arm is a 350-nm-thick layer of Te doped ($n \sim 1 \times 10^{16}$cm^{-3}) GaSb, and the entire epilayer structure is grown by molecular beam epitaxy on a (001)-oriented n-type GaSb substrate.

The potential energy step at the emitter–base heterointerface is $\phi_{EB} = 1.3$ eV, and the potential energy step at the base–collector interface is $\phi_{BC} = 0.8$ eV. The device can

---

[4] D. Buchanan, *IBM J. Res. Develop.* **43**, 245 (1999).

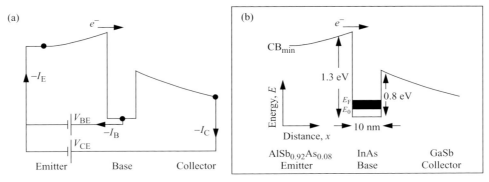

Fig. 3.16. (a) Schematic diagram of a potential double heterojunction unipolar transistor. Indicated in the figure are the emitter current I_E, base current I_B, collector current I_C, and voltages V_{BE} and V_{CE} for the transistor biased in the common emitter configuration. The solid dots indicate electrical contacts between the transistor and the leads that connect to the batteries. (b) Schematic diagram of the conduction band of an $AlSb_{0.92}As_{0.08}$/InAs/GaSb double heterojunction unipolar NET under bias. The conduction-band minimum CB_{min} is indicated, as are the confinement energy E_0 and the Fermi energy E_F of the occupied two-dimensional electron states in the InAs base. Electrons indicated by e^- are injected from the forward-biased $AlSb_{0.92}As_{0.08}$ emitter into the InAs base region with a large excess kinetic energy. The effective electron mass near CB_{min} is $m_{InAs} = 0.021 \times m_0$ in the base and $m_{GaSb} = 0.048 \times m_0$ in the collector.

operate at room temperature because both the emitter barrier energy ϕ_{EB} and collector barrier energy ϕ_{BC} are much greater than the ambient thermal energy $k_B T \sim 0.025$ eV, so that reverse (leakage) currents are small.

The lowest-energy bound-electron state in the finite rectangular potential well of the transistor base is indicated as E_0. At low temperatures, electrons in the potential well occupy quantum states up to a Fermi energy indicated by $E_F = \hbar^2 k_F^2 / 2m$, where k_F is the Fermi wave vector.

Because emitter electrons injected into the base have high kinetic energy $E_k \sim \phi_{EB}$, they also have a high velocity. A typical nonequilibrium electron velocity is near 10^8 cm s^{-1}, so the majority of injected electrons with a large component of momentum in the x direction traverse the base of thickness $x_B = 10$ nm in the very short time of about 10 fs (assuming, of course, that one may describe an electron as a point particle with a well-defined *local* velocity).

It is important to inject electrons in a narrow range of energies close to E_k. Optimum device performance occurs when quantum reflections from ϕ_{BC} are minimized, and this may only be achieved for a limited range of injection energies. Reflections from the abrupt change in potential at ϕ_{BC} are minimized when the nonequilibrium electron group velocity (the slope $\partial \omega / \partial k$ at energy E_k in Fig. 3.17) is the same on either side of the base–collector junction. As discussed in Section 3.8, this impedance-matching condition is $m_{InAs}/m_{GaSb} = E/(E - \phi_{BC})$, where m_{InAs} and m_{GaSb} are the effective electron masses in the base and collector respectively. Therefore, by choosing E_k, ϕ_{BC},

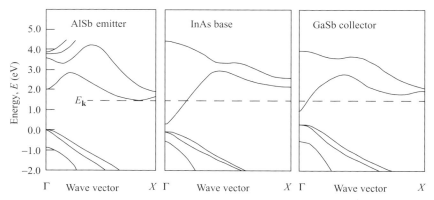

Fig. 3.17. Dispersion curves for electron motion through the three materials forming the AlSb/InAs/GaSb double heterostructure unipolar NET. The dashed line indicates the approximate value of energy for an electron moving through the device. Electron states used in transmission of an electron energy E through (100)-oriented semiconductor layers are near the points where the dashed line intersects the solid curves. The value of effective electron mass near the conduction band minimum (CB_{min}) in the ΓX direction in the AlSb emitter is close to the free-electron mass, m_0. At an energy E_k, which is above CB_{min} in the base and collector, the effective electron masses are greater than their respective values near CB_{min}.

base and collector materials quantum reflections from ϕ_{BC} may be eliminated for a small range ($\sim$0.5 eV) of E_k. It is worth mentioning that at a real heterojunction interface, impedance matching also requires that the symmetry of the electron wave function in the base and collector be similar.[5]

Although we know that degradation in device performance due to quantum mechanical reflection can be designed out of the device, there are other difficulties that require additional thought. For example, an injected electron, with energy E_k and wave vector **k**, moving in the x direction is able to scatter via the coulomb interaction into high-energy states in the base, changing energy by an amount $\hbar\omega$ and changing wave vector by **q**. Inelastic processes of this type include the electron scattering from lattice vibrations of the semiconductor crystal. Another possible inelastic process involves the high-energy injected electron scattering from low-energy (thermal) electron in the transistor base. Because these scattering processes can change the direction of travel of the electron, they can also dramatically reduce the number of electrons that traverse the base and flow as current in the collector. Such dynamical constraints imposed by inelastic scattering are quite difficult to deal with and thus make NET design a significant task. Nevertheless, some NETs do have quite good device characteristics.

Figure 3.18 shows measured common base current gain $\alpha = I_C/I_E$ and common emitter current gain $\beta = I_C/I_B$ for a device maintained at a temperature $T = 300$ K.[6]

[5] M. D. Stiles and D. R. Hamann, *Phys. Rev.* **B38**, 2021 (1988). If the symmetry of the electron wave function is not matched on either side of the interface, the electron can suffer *complete* quantum mechanical reflection *even* if the electron velocity is precisely matched!

[6] A. F. J. Levi and T. H. Chiu, *Appl. Phys. Lett.* **51**, 984 (1987).

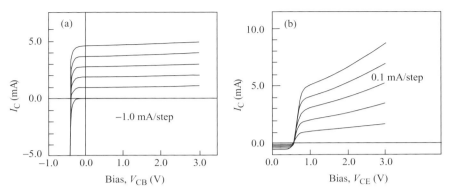

Fig. 3.18. (a) Room-temperature ($T = 300$ K) common base current gain characteristics of the device shown schematically in Fig. 3.16(b). Curves were taken in steps beginning with an injected emitter current of zero. The emitter area is 7.8×10^{-5} cm^2, I_C is collector current, and V_{CB} is collector base voltage bias. (b) Room-temperature common emitter current gain characteristics of the device in (a). Curves were taken in steps of $I_B = 0.1$ mA, beginning with an injected base current of zero. V_{CE} is collector emitter voltage bias. Current gain $\beta \sim 10$. Measurement performed with sample at temperature $T = 200$ K.

As may be seen in Fig. 3.18(b), the room-temperature value of β increases from $\beta = 10$ at $V_{CE} = 1.0$ V to $\beta = 17$ at $V_{CE} = 3.0$ V. At temperature $T = 77$ K (the temperature at which nitrogen gas becomes liquid), the current gain $\beta = 12$ at $V_{CE} = 1.0$ V and $\beta = 40$ at $V_{CE} = 3.0$ V. We note that the reduction in base–collector potential barrier ϕ_{BC}, with increasing collector voltage bias V_{CE}, improves collector efficiency for incoming nonequilibrium electrons and, in agreement with simple calculations, contributes to the observed slope in the common emitter saturation characteristics.

It is possible to devise experiments that further explore the role of quantum mechanical reflection from the abrupt change in potential ϕ_{BC} seen by a nonequilibrium electron approaching the base collector heterostructure. Figure 3.19(a) shows the band diagram for a NET device with a 15-nm-thick AlSb tunnel emitter that may be used to inject an essentially monoenergetic beam of electrons perpendicular to the heterointerface. After traversing the 10-nm-thick base, electrons impinge on the GaSb collector. If the injection energy E_k is less than the base–collector barrier energy, $\phi_{BC} \sim 0.8$ eV, no electrons are collected, and all the injected current I_E flows in the base. For $E_k > \phi_{BC}$, some electrons traverse the base and subsequently contribute to the collector current I_C.

An important scattering mechanism determining collector efficiency is quantum mechanical reflection from the step change in potential ϕ_{BC}. As discussed previously in Section 3.8, this reflection is determined in part by electron velocity mismatch across the abrupt InAs/GaSb heterointerface. For electron velocities v_{InAs} in InAs and v_{GaSb} in GaSb, quantum mechanical reflection is $Refl = ((v_{InAs} - v_{GaSb})/(v_{InAs} + v_{GaSb}))^2$. In this simplified example, we assume that there is no contribution to $Refl$ from a mismatch in the character (symmetry) of the electron wave function across the interface. Since, in our structure, $E_k \propto V_{BE}$ to within ± 0.1 eV, we may explore electron collection

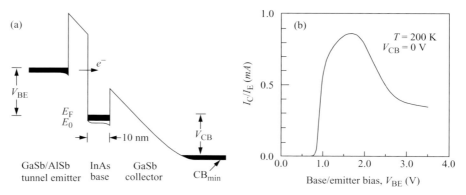

Fig. 3.19. (a) Conduction band diagram of an $AlSb_{0.92}As_{0.08}$/InAs/GaSb tunnel emitter heterojunction unipolar NET under base emitter bias V_{BE} and collector base bias V_{CB}. The conduction band minimum CB_{min} is indicated, as are the confinement energy E_0 and the Fermi energy E_F of the occupied two-dimensional electron states in the 10-nm-thick InAs base. The energy of a beam of electrons traversing the base and impinging on the base collector heterostructure is controlled by varying V_{BE}. (b) Measured ratio of the collector and emitter currents, I_C/I_E, as functions of base emitter bias, V_{BE}. The ratio I_C/I_E is related to the electron transmission probability across the base–collector heterostructure.

efficiency as a function of injection energy by plotting the ratio I_C/I_E with base–emitter voltage bias V_{BE}. Typical results of such a measurement at a temperature $T = 200$ K are shown in Fig. 3.19(b).[7] As may be seen, for injection energy E_k less than ϕ_{BC}, no electrons are collected. For $E_k > \phi_{BC}$, the ratio I_C/I_E increases rapidly with decreasing velocity mismatch at either side of the heterointerface. Maximum base transport efficiency occurs for $E_k \sim 2.5$ eV. With further increase in E_k, collection efficiency decreases, and finally for $E_k \sim 2.5$ eV less than 50% of electrons are collected. At energies greater than 2.5 eV, electrons are injected into electronic states above the lowest conduction band in which both velocity and wave function character mismatch creating high quantum mechanical reflection. Hence, for $E_k \geq 2.5$ eV, electron collection efficiency decreases dramatically.

If we summarize what we have achieved so far, it becomes apparent that we now know enough about quantum mechanics to understand much of what goes into the design of a new type of transistor. The device operates by injecting nonequilibrium electrons that are transported through a region of semiconductor only 10 nm thick. The existence of extreme nonequilibrium electron transport is a necessary condition for successful device operation. Transistor performance is improved by carefully matching the velocity across the base–collector semiconductor heterojunction. In fact, we were able to explore the velocity-matching condition by using quantum mechanical tunneling to inject electrons at different energies into the base–collector semiconductor heterojunction region.

Upon reflection, you may conclude that it is quite surprising that we have been so successful in understanding key elements of this transistor design without the need to

---

[7] T. H. Chiu and A. F. J. Levi, *Appl. Phys. Lett.* **55**, 1891 (1989).

develop a more sophisticated approach. In fact, our understanding is quite superficial. There are a number of complex and subtle issues concerning the description of nonequilibrium electron motion in the presence of strong elastic and inelastic scattering that you may wish to investigate at some later time. A significantly deeper understanding of nonequilibrium electron transport is beyond the scope of this book, and so we will not pursue that topic here.

3.11 Example exercises

Exercise 3.1

(a) Classically (i.e., from Maxwell's equations), change in charge density, ρ, is related to divergence of current density, $\mathbf{J}$. Use this fact and the time-dependent Schrödinger wave equation for particles of mass m and charge e to derive the current operator

$$\mathbf{J} = -\frac{ie\hbar}{2m}\left(\psi^* \nabla \psi - \psi \nabla \psi^*\right)$$

(b) If a wave function can be expressed as $\psi(x,t) = Ae^{i(kx-\omega t)} + Be^{i(-kx-\omega t)}$, show that particle flux is proportional to $A^2 - B^2$.

(c) For current to flow through a tunnel barrier, the wave function must contain *both* left- and right-propagating, exponentially decaying solutions with a phase difference. If $\psi(x,t) = Ae^{\kappa x - i\omega t} + Be^{-\kappa x - i\omega t}$, show that particle flux is proportional to $\mathrm{Im}(AB^*)$. Hence, show that if $\psi(x,t) = Be^{-\kappa x - i\omega t}$ particle flux is zero.

(d) Show that a particle of energy E and mass m moving from left to right in a one-dimensional potential in such a way that $V(x) = 0$ for $x < 0$ and $V(x) = V_0$ for $x \geq 0$ has unity reflection probability if $E < eV_0$.

Exercise 3.2

(a) An electron in the conduction band of GaAs has an effective electron mass $m_e^* = 0.07m_0$. Find the values of the first three energy eigenvalues, assuming a rectangular, infinite, one-dimensional potential well of width $L = 10$ nm and $L = 20$ nm. Find an expression for the difference in energy levels, and compare it with room-temperature thermal energy $k_B T$.

(b) For a quantum box in GaAs with no occupied electron states, estimate the size below which the conduction band quantum energy level spacing becomes larger than the classical coulomb blockade energy. Assume that the confining potential for the electron may be approximated by a potential barrier of infinite energy for $|x| > L/2$, $|y| > L/2$, $|z| > L/2$, and zero energy elsewhere.

Exercise 3.3

Electrons in the conduction band of different semiconductors have differing effective electron mass. In addition, the effective electron mass, m_j, is almost never the same as the free electron mass, m_0. These facts are usually accommodated in simple models of semiconductor heterojunction barrier transmission and reflection by requiring $\psi_1|_{0-\delta} = \psi_2|_{0+\delta}$ and $(1/m_1) \cdot \nabla \psi_1|_{0-\delta} = (1/m_2) \cdot \nabla \psi_2|_{0+\delta}$ at the heterojunction interface. Why are these boundary conditions used? Calculate the transmission and reflection coefficients for an electron of energy E, moving from left to right, impinging normally to the plane of a semiconductor heterojunction potential barrier of energy V_0. The effective electron mass on the left-hand side is m_1, and the effective electron mass on the right-hand side is m_2. Under what conditions will there be no reflection, even for a step of finite barrier energy? Express this condition in terms of electron velocities on either side of the interface.

Exercise 3.4

An asymmetric one-dimensional potential well of width L and zero potential energy has an infinite potential energy barrier on the left-hand side and a finite constant potential of energy V_0 on the right-hand side. Find the minimum value of L for which an electron has at least one bound state when $V_0 = 1$ eV.

Exercise 3.5

(a) Consider a one-dimensional potential well approximated by a delta function in space so that $V(x) = -b\delta(x = 0)$. Show that there is one bound state for a particle of mass m, and find its energy and eigenstate.

   (b) Show that any one-dimensional delta-function potential with $V(x) = \pm b\delta(x = 0)$ always introduces a kink in the wave function describing a particle of mass m.

Exercise 3.6

A nonequilibrium electron transistor (NET) of the type shown schematically in Fig. 3.16 is to be designed for use in a high-speed switching application. In this situation the NET must have a high current drive capability. To achieve this one needs to ensure that space-charging effects in the emitter and collector barriers are avoided. How would you modify the design of the NET to support current densities in excess of 10^5 A cm^{-2}?

Exercise 3.7

Using the method outlined in Section 3.4, write a computer program to solve the Schrödinger wave equation for the first four eigenvalues and eigenstates of an electron with effective mass $m_e = 0.07 \times m_0$ confined to a rectangular potential well of width $L = 10$ nm bounded by infinite barrier potential energy.

SOLUTIONS

Solution 3.1

(a) We will use the classical expression relating the divergence of charge density to current density to obtain an expression for the quantum mechanical current operator.

The change in charge density ρ is related to the divergence of current density **J** through

$$\frac{\partial}{\partial t}\rho(\mathbf{r}, t) = -\nabla \cdot \mathbf{J}(\mathbf{r}, t)$$

In quantum mechanics, we assume that the charge density of a particle with charge e and wave function ψ can be written $\rho = e|\psi|^2$, so that

$$\frac{\partial \rho}{\partial t} = e\frac{\partial}{\partial t}(\psi^*\psi) = e\left(\psi^*\frac{\partial \psi}{\partial t} + \psi\frac{\partial \psi^*}{\partial t}\right)$$

We now make use of the fact that the time-dependent Schrödinger equation for a particle of mass m moving in a real potential is

$$i\hbar\frac{\partial}{\partial t}\psi(\mathbf{r}, t) = \left(-\frac{\hbar^2}{2m}\nabla^2 + V(\mathbf{r})\right)\psi(\mathbf{r}, t) = H\psi(\mathbf{r}, t)$$

Multiplying both sides by $\psi^*(\mathbf{r}, t)$ gives

$$i\hbar\psi^*(\mathbf{r}, t)\frac{\partial}{\partial t}\psi(\mathbf{r}, t) = -\frac{\hbar^2}{2m}\psi^*(\mathbf{r}, t)\nabla^2\psi(\mathbf{r}, t) + \psi^*(\mathbf{r}, t)V(\mathbf{r})\psi(\mathbf{r}, t)$$

which, when multiplied by $e/i\hbar$, is the first term on the right-hand side in our expression for $\partial\rho/\partial t$. To find the second term, one takes the complex conjugate of the time-dependent Schrödinger equation and multiplies both sides by $\psi(\mathbf{r}, t)$. Because $V(\mathbf{r})$ is taken to be real, we have $\psi^*(\mathbf{r}, t)V(\mathbf{r})\psi(\mathbf{r}, t) = \psi(\mathbf{r}, t)V(\mathbf{r})\psi^*(\mathbf{r}, t)$, giving the rate of change of charge density

$$\frac{\partial\rho}{\partial t} = -\psi^*\frac{e\hbar}{i2m}\nabla^2\psi + \psi\frac{e\hbar}{i2m}\nabla^2\psi^* + \frac{e\psi^*\psi}{i\hbar}(V(\mathbf{r}) - V(\mathbf{r}))$$

$$\frac{\partial\rho}{\partial t} = \frac{ie\hbar}{2m}(\psi^*\nabla^2\psi - \psi\nabla^2\psi^*)$$

$$\frac{\partial\rho}{\partial t} = \frac{ie\hbar}{2m}\nabla \cdot (\psi^*\nabla\psi - \psi\nabla\psi^*) = -\nabla \cdot \mathbf{J}$$

Hence, an expression for the current density operator for a particle charge e described by state ψ is

$$\mathbf{J} = -\frac{ie\hbar}{2m}(\psi^*\nabla\psi - \psi\nabla\psi^*)$$

This expression has an obvious symmetry, some consequences of which we will now explore.

(b) In this exercise we will work with a particle of mass m and charge e that is described by a wave function $\psi(x, t) = Ae^{i(kx-\omega t)} + Be^{i(-kx-\omega t)}$, where k is real. This aim is to show that particle flux is proportional to $A^2 - B^2$.

Starting with the current density operator in one dimension and substituting in the expression for the wave function gives

$$J_x = -\frac{ie\hbar}{2m}\left(\psi^*\frac{\partial}{\partial x}\psi - \psi\frac{\partial\psi^*}{\partial x}\right)$$

$$\psi^*\frac{\partial\psi}{\partial x} = \left(A^*e^{-ikx} + B^*e^{ikx}\right)\left(ikAe^{ikx} - ikBe^{-ikx}\right)$$

$$\psi^*\frac{\partial\psi}{\partial x} = ik|A|^2 - ikA^*Be^{-2ikx} + ikB^*Ae^{2ikx} - ik|B|^2$$

and

$$\psi\frac{\partial\psi^*}{\partial x} = \left(Ae^{ikx} + Be^{-ikx}\right)\left(-ikA^*e^{-ikx} + ikB^*e^{ikx}\right)$$

$$\psi\frac{\partial\psi^*}{\partial x} = -ik|A|^2 - ikBA^*e^{-2ikx} + ikAB^*e^{2ikx} + ik|B|^2$$

so that, finally,

$$J_x = -\frac{ie\hbar}{m}\left(ik\left(|A|^2 - |B|^2\right)\right) = \frac{e\hbar k}{m}\left(|A|^2 - |B|^2\right)$$

This is not an unexpected result, since the wave function $\psi(x, t)$ is made up of two traveling waves. The first traveling wave Ae^{ikx} might consist of an electron probability density $\rho = |A|^2$ moving from left to right at velocity $v = \hbar k/m$ carrying current density $j_+ = ev_+\rho = e\hbar k|A|^2/m$. The second travelling wave Be^{-ikx} consists of an electron probability density $\rho = |B|^2$ moving from right to left at velocity $v = -\hbar k/m$ carrying current density $j_- = ev_-\rho = -e\hbar k|B|^2/m$. The net current density is just the sum of the currents $j_x = j_+ + j_- = e\hbar k(|A|^2 - |B|^2)/m$.

(c) A particle mass of m and charge e is described by the wave function $\psi(x, t) = Ae^{\kappa x - i\omega t} + Be^{-\kappa x - i\omega t}$, where κ is real. We wish to show that for current to flow through a tunnel barrier the wave function must contain both left- and right-propagating, exponentially decaying solutions with a phase difference.

As in part (b) we start with the current density operator in one dimension and substitute in the expression for the wave function. This gives

$$J_x = -\frac{ie\hbar}{2m}\left(\psi^*\frac{\partial}{\partial x}\psi - \psi\frac{\partial\psi^*}{\partial x}\right)$$

$$\psi^*\frac{\partial\psi}{\partial x} = \left(A^*e^{\kappa x} + B^*e^{-\kappa x}\right)\left(A\kappa e^{\kappa x} - B\kappa e^{-\kappa x}\right)$$

$$\psi^*\frac{\partial\psi}{\partial x} = \kappa|A^2|e^{2\kappa x} + \kappa B^*A - \kappa A^*B - \kappa|B|^2e^{-2\kappa x}$$

and

$$\psi\frac{\partial\psi^*}{\partial x} = \left(Ae^{\kappa x} + Be^{-\kappa x}\right)\left(A^*\kappa e^{\kappa x} - B^*\kappa e^{-\kappa x}\right)$$

$$\psi\frac{\partial\psi^*}{\partial x} = \kappa|A|^2e^{2\kappa x} - \kappa AB^* + \kappa BA^* - \kappa|B|^2e^{-2\kappa x}$$

so that

$$J_x = \frac{ie\hbar}{2m} \cdot 2\kappa \left(A^*B - B^*A\right) = \frac{2e\hbar\kappa}{m}\left(\mathrm{Im}(AB^*)\right)$$

To show this last equality, one starts by explicitly writing the real and imaginary parts of the terms in the brackets

$$\left(A^*B - B^*A\right) = (A_{\mathrm{Re}} - iA_{\mathrm{Im}})(B_{\mathrm{Re}} + iB_{\mathrm{Im}}) - (B_{\mathrm{Re}} - iB_{\mathrm{Im}})(A_{\mathrm{Re}} + iA_{\mathrm{Im}})$$

Multiplying out the terms on the right-hand side gives

$$\left(A^*B - B^*A\right) = A_{\mathrm{Re}}B_{\mathrm{Re}} + iA_{\mathrm{Re}}B_{\mathrm{Im}} - iA_{\mathrm{Im}}B_{\mathrm{Re}} + A_{\mathrm{Im}}B_{\mathrm{Im}} - A_{\mathrm{Re}}B_{\mathrm{Re}} - iB_{\mathrm{Re}}A_{\mathrm{Im}}$$
$$+ iB_{\mathrm{Im}}A_{\mathrm{Re}} - A_{\mathrm{Im}}B_{\mathrm{Im}}$$

This simplifies to

$$\left(A^*B - B^*A\right) = 2iA_{\mathrm{Re}}B_{\mathrm{Im}} - 2iA_{\mathrm{Im}}B_{\mathrm{Re}} = 2i(A_{\mathrm{Re}}B_{\mathrm{Im}} - A_{\mathrm{Im}}B_{\mathrm{Re}})$$

Taking the imaginary part of the terms on the right-hand side allows one to add the real terms $A_{\mathrm{Re}}B_{\mathrm{Re}}$ and $A_{\mathrm{Im}}B_{\mathrm{Im}}$, so that the expression may be factored

$$\left(A^*B - B^*A\right) = -2i(\mathrm{Im}(A_{\mathrm{Re}}B_{\mathrm{Re}} - iA_{\mathrm{Re}}B_{\mathrm{Im}} + iA_{\mathrm{Im}}B_{\mathrm{Re}} + A_{\mathrm{Im}}B_{\mathrm{Im}}))$$
$$\left(A^*B - B^*A\right) = -2i(\mathrm{Im}((A_{\mathrm{Re}} + iA_{\mathrm{Im}})(B_{\mathrm{Re}} - iB_{\mathrm{Im}}))) = 2i\left(\mathrm{Im}(AB^*)\right)$$
$$J_x = \frac{ie\hbar}{2m} \cdot 2\kappa\left(A^*B - B^*A\right) = \frac{2e\hbar\kappa}{m}\left(\mathrm{Im}(AB^*)\right)$$

The significance of this result is that current can only flow through a one-dimensional tunnel barrier in the presence of *both* exponentially decaying terms in the wave function $\psi(x,t) = Ae^{\kappa x} + Be^{-\kappa x}$. In addition, the complex coefficients A and B must differ in phase. The right-propagating term $Ae^{\kappa x}$ can only carry current if the left-propagating term $Be^{-\kappa x}$ exists. This second term indicates that the tunnel barrier is not opaque and that transmission of current via tunneling is possible.

Suppose a particle of mass m and charge e is described by wave function $\psi(x,t) = Be^{-\kappa x - i\omega t}$. Substitution into the expression for the one-dimensional current density operator gives

$$J_x = -\frac{ie\hbar}{2m}\left(\psi^*\frac{\partial}{\partial x}\psi - \psi\frac{\partial \psi^*}{\partial x}\right) = -\frac{ie\hbar}{2m}\left(B^*e^{-\kappa x}(-B\kappa e^{-\kappa x}) - Be^{-\kappa x}\left(-B^*\kappa e^{-\kappa x}\right)\right)$$
$$J_x = -\frac{ie\hbar}{2m}\left(-B^*B\kappa e^{-2\kappa x} + BB^*\kappa e^{-2\kappa x}\right) = 0$$

(d) To show that a particle of energy E and mass m moving in a one-dimensional potential in such a way that $V(x) = 0$ for $x < 0$ in region 1 and $V(x) = V_0$ for $x \geq 0$ in region 2 has unity reflection probability if $E < eV_0$, we write down solutions of the time-independent Schrödinger equation $(-\hbar^2\nabla^2/2m + V(x))\psi(x) = E\psi(x)$. The solutions that describe left- and right-traveling waves in the two regions are

$$\psi_1(x) = Ae^{ik_1 x} + Be^{-ik_1 x} \qquad \text{for } x < 0$$

and

$$\psi_2(x) = Ce^{ik_2x} + De^{-ik_2x} \qquad \text{for } x > 0$$

When $E < V_0$, the value of k_2 is imaginary, and so $k_2 \to ik_2$. The value of D is zero to avoid a wave function with infinite value as $x \to \infty$. The value of C is finite, but the contribution to the wave function $\psi_2(x \to \infty) = Ce^{-k_2x}|_{x\to\infty} = 0$. Hence, we may conclude that the transmission probability in the limit $x \to \infty$ must be zero, since $|\psi_2(x \to \infty)|^2 = 0$. From this it follows that $|A|^2 = |B|^2 = 1$, which has the physical meaning that the particle is completely reflected at the potential step when $E < V_0$.

Alternatively, one could calculate the reflection probability from the square of the amplitude coefficient for a particle of energy E and mass m incident on a one-dimension potential step

$$B = \left(\frac{1 - k_2/k_1}{1 + k_2/k_1} \right)$$

When particle energy $E < V_0$, the value of k_2 is imaginary, so that $k_2 \to ik_2$, which gives

$$|B|^2 = B^*B = \left(\frac{1 + ik_2/k_1}{1 - ik_2/k_1} \right) \left(\frac{1 - ik_2/k_1}{1 + ik_2/k_1} \right) = 1$$

From this one may conclude that the particle has a unity reflection probability.

Solution 3.2

(a) In this exercise we are to find the differences in energy eigenvalues for an electron in the conduction band of GaAs that has an effective electron mass $m_e^* = 0.07m_0$ and is confined by a rectangular infinite one-dimensional potential well of width L with value either $L = 10$ nm or $L = 20$ nm.

We start by recalling our expression for energy eigenvalues of an electron confined in a one-dimensional, infinite, rectangular potential of width L:

$$E_n = \frac{\hbar^2}{2m_e^*} \cdot \frac{n^2\pi^2}{L^2}$$

For $m_e^* = 0.07 \times m_0$, this gives

$$E_n = \frac{n^2}{L^2} \times 8.607 \times 10^{-37} \text{ J} = \frac{n^2}{L^2} \times 5.37 \times 10^{-18} \text{ eV}$$

so that the first three energy eigenvalues for $L = 10$ nm are

$$E_1 = 54 \text{ meV}$$

$$E_2 = 215 \text{ meV}$$

$$E_3 = 484 \text{ meV}$$

and the differences in energy between adjacent energy eigenvalues are

$\Delta E_{12} = 161$ meV

$\Delta E_{23} = 269$ meV

Increasing the value of the potential well width by a factor of 2 to $L = 20$ nm reduces the energy eigenvalues by a factor of 4. The energy eigenvalues scale as $1/L^2$, and so they are quite sensitive to the size of the potential well. For practical systems, we would like ΔE_{nm} to be large, typically $\Delta E_{nm} \gg k_B T$. However, although E_n scales as n^2/L^2, the *difference* in energy levels *scales linearly* with increasing eigenvalue, n. For adjacent energy levels, the difference in energy increases linearly with increasing eigenvalue, as

$$\Delta E_{n+1,n} = \frac{\hbar^2}{2m_e} \cdot \frac{\pi^2}{L^2}\left((n+1)^2 - n^2\right) = \frac{\hbar^2}{2m_e} \cdot \frac{\pi^2}{L^2}(2n+1)$$

At room temperature $k_B T = 25$ meV. Thus for the case $n = 1$ we have the condition

$$L^2 \ll \frac{\hbar^2}{2m_e} \cdot \frac{\pi^2}{k_B T}(2n+1) = \frac{3\hbar^2}{2m_e} \cdot \frac{\pi^2}{k_B T} = \frac{3}{0.025} \times 5.37 \times 10^{-18} = 6.44 \times 10^{-16}\,\text{m}^2$$

So, for the situation we are considering, $L \ll 25$ nm.

(b) We now use the results of (a) to estimate the size of a quantum box in GaAs with no occupied electron states at which there is a cross-over from classical coulomb blockade energy to quantum energy eigenvalues dominating electron energy levels.

For a one-dimensional, rectangular potential well of width L and infinite barrier energy the eigenenergy levels are given by

$$E_{n_x} = \frac{\hbar^2}{2m} \cdot \frac{n_x^2 \pi^2}{L^2}$$

and the differences in energy levels are

$$\Delta E_{n_x+1,n_x} = \frac{\hbar^2}{2m} \cdot \frac{\pi^2}{L^2}(2n_x + 1)$$

Adding the energy contributions from the x, y, and z directions of the quantum box gives energy

$$E_n = \frac{\hbar^2}{2m} \cdot \frac{\left(n_x^2 + n_y^2 + n_z^2\right)\pi^2}{L^2}$$

so that we expect the difference in quantum energy to scale as $1/L^2$. Because the classical coulomb blockade energy $\Delta E = e^2/2C$ scales as $1/L$, with decreasing size there must be a cross-over to quantum energy eigenvalues dominating electron dynamics.

For a quantum box in GaAs with no occupied electron states, a simple estimate for the cross-over size is determined by $E_{n_x=1,n_y=1,n_z=1} = e^2/2C$. If we approximate the capacitance as $C = 4\pi\varepsilon_0\varepsilon_r(L/2)$, then

$$L = \frac{3\pi^2 \hbar^2}{2m_e^*} \cdot \frac{4\pi \varepsilon_0 \varepsilon_r}{e^2}$$

For GaAs, using $m_e^* = 0.07 \times m_0$ and $\varepsilon_r = 13.1$ gives a characteristic size for the crossover of $L = 147$ nm and a ground-state energy $E_1 = 6.75$ meV. Clearly, for quantum dots in GaAs characterized by a size $L \ll 150$ nm, quantization in energy states will dominate. However, if necessary, one should also be careful to include the physics behind the coulomb blockade by calculating a *quantum capacitance* using electron wave functions and a quantum coulomb blockade by self-consistently solving for the wave functions and Maxwell's equations.

Solution 3.3

To accommodate the fact that electrons in the conduction band of different semiconductors labeled 1 and 2 have differing effective electron mass, simple models of semiconductor heterojunction barrier transmission and reflection require that the electron wave function satisfy $\psi_1|_{0-\delta} = \psi_2|_{0+\delta}$ and $(1/m_1) \cdot \nabla \psi_1|_{0-\delta} = (1/m_2) \cdot \nabla \psi_2|_{0+\delta}$ at the heterojunction interface. This ensures conservation of current across the heterointerface. In this limited model, current conservation is taken to be more important than ensuring continuity in the first spatial derivative of the wave function.

We now calculate the transmission and reflection coefficient for an electron of energy E, moving from left to right, impinging normally to the plane of a semiconductor heterojunction potential barrier of energy V_0, where the effective electron mass on the left-hand side is m_1 and the effective electron mass on the right-hand side is m_2.

At the boundary between regions 1 and 2 at position x_0, we require continuity in ψ and $\nabla \psi / m_j$, so that

$$\psi_1|_{x_0} = \psi_2|_{x_0}$$

and

$$\left. \frac{1}{m_1} \frac{d}{dx} \psi_1 \right|_{x_0} = \left. \frac{1}{m_2} \frac{d}{dx} \psi_2 \right|_{x_0}$$

This leads to

$$A + B = C + D$$

and

$$A - B = \frac{m_1 k_2}{m_2 k_1} C - \frac{m_1 k_2}{m_2 k_1} D$$

If we know that the particle is incident from the left, then $A = 1$ and $D = 0$, giving

$$1 + B = C$$
$$1 - B = \frac{m_1 k_2}{m_2 k_1} C$$

Recognizing velocity $v_1 = \hbar k_1 / m_1$ and $v_2 = \hbar k_2 / m_2$ and solving for the transmission probability $|C|^2$ and the reflection probability $|B|^2$ gives

$$|C|^2 = \frac{4}{\left(1 + \dfrac{m_1 k_2}{m_2 k_1}\right)^2} = \frac{4}{\left(1 + \dfrac{v_2}{v_1}\right)^2}$$

$$|B|^2 = \frac{\left(1 - \dfrac{m_1 k_2}{m_2 k_1}\right)^2}{\left(1 + \dfrac{m_1 k_2}{m_2 k_1}\right)^2} = \frac{\left(1 - \dfrac{v_2}{v_1}\right)^2}{\left(1 + \dfrac{v_2}{v_1}\right)^2}$$

From this it is clear that there is no reflection when particle velocity is the same in each of the two regions. This is an example of an impedance-matching condition that, in this case, will occur when particle energy $E = V_0/(1 - m_2/m_1)$.

It is worth making a few comments about our approach to solving this exercise. We selected two boundary conditions to guarantee current conservation and no discontinuity in the electron wave function. However, with this choice, we were unable to avoid the possibility of kinks in the wave function. Such an approach was first discussed by Bastard.[8] Unfortunately, the Schrödinger equation cannot be solved correctly using different effective electron masses for electrons on either side of an interface. A more complex, but correct, way to proceed is to use the atomic potentials of atoms forming the heterointerface and solve using the bare electron mass and the usual boundary conditions

$$\psi|_{x_0 - \delta} = \psi|_{x_0 + \delta}$$

and

$$\frac{d\psi}{dx}\bigg|_{x_0 - \delta} = \frac{d\psi}{dx}\bigg|_{x_0 + \delta}$$

A simplified version of this situation might use the Kronig–Penney potential, which will be discussed in Chapter 4.

There are a number of additional reasons why the model we have used should only be considered elementary. For example, the electron can be subject to other types of scattering. Such scattering may destroy either momentum or both momentum and energy conservation near the interface. This may take the form of diffuse electron scattering due to interface roughness or high phonon emission probability. The presence of nonradiative electron recombination due to impurities and traps, or, typically in direct band-gap semiconductors, radiative processes may be important. These effects may result in a nonunity sum of transmission and reflection coefficients ($R + T \neq 1$).

Situations often arise in which an effective electron mass cannot be used to describe electron motion. In this case the electron dispersion relation is nonparabolic over the

---

[8] G. Bastard, *Phys. Rev.* **B24**, 5693 (1981).

energy range of interest. In addition, when the character (or s, p, d, etc., symmetry) of electron wave function is very different on both sides of potential step, the simple approach used above is inappropriate and can lead to incorrect results.

Solution 3.4

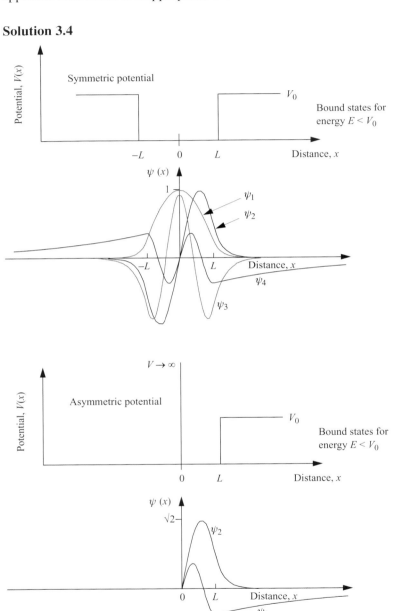

In this exercise, an *asymmetric* one-dimensional potential well of width L has an infinite potential energy barrier on the left-hand side and a finite constant potential of energy V_0 on the right-hand side. We aim to find the minimum value of L for which a particle of mass m has at least one bound state. A quick way to a solution is to compare with the results for the symmetric potential considered in Section 3.7. The above figure sketches four bound-state wave functions for a particle of mass m in a symmetric potential for which $V_0 L^2 = 25\hbar^2/2m$, so that $LK_0 = L\sqrt{2mV_0}/\hbar = 5$. As expected, the wave functions have alternating even and odd parity, with even parity for the lowest energy state, ψ_1.

The lower part of the above figure shows the asymmetric potential that arises when the potential for $x < 0$ is infinite. Because the wave function must be zero for $x < 0$, the only allowed solutions for bound states are of the forms ψ_2 and ψ_4 previously obtained for the symmetric case and $x > 0$. These correspond to the odd-parity wave functions that are found using the graphical method sketched below when $V_0 L^2 = 25\hbar^2/2m$, so that $LK_0 = L\sqrt{2mV_0}/\hbar = 5$.

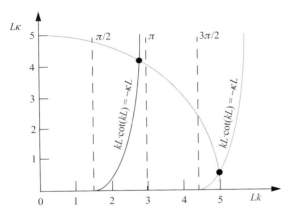

The lowest energy solution exists where the radius of the arc $LK_0 = L\sqrt{2mV_0}/\hbar$ intersects the function $kL \cdot \cot(kL) = -\kappa L$. Vanishing binding energy will occur when the particle is in a potential so that $LK_0 \to \pi/2$ and $k \ll k$. This situation is sketched below.

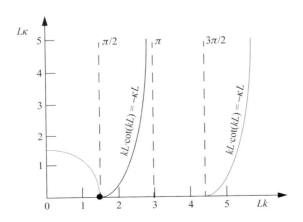

Since $LK_0 = L\sqrt{2mV_0}/\hbar$, in the limit of vanishing binding energy $L\sqrt{2mV_0}/\hbar \geq \pi/2$, so that

$$L \geq \pi\hbar/2\sqrt{2mV_0}$$

To find the minimum value of L, we need to know the values of V_0 and m. For the case in which the step potential energy is $V_0 = 1$ eV, the minimum value of L for an electron is $L_{\min} \geq \pi\hbar/2(2mV_0)^{1/2} \sim 0.3$ nm.

Solution 3.5

(a) Here we seek the eigenfunction and eigenvalue of a particle in a delta-function potential well, $V(x) = -b\delta(x = 0)$, where b is a measure of the strength of the potential. Due to the symmetry of the potential we expect the wave function to be an even function. The delta function will introduce a kink in the wave function. Schrödinger's equation for a particle subject to a delta-function potential is

$$\left(\frac{-\hbar^2}{2m}\frac{d^2}{dx^2} - b\delta(x = 0) \right)\psi(x) = E\psi(x)$$

Integrating between $x = 0-$ and $x = 0+$

$$\int_{x=0-}^{x=0+} \left(\frac{d^2}{dx^2}\psi(x) \right)dx = \int_{x=0-}^{x=0+} \left(\frac{-2m}{\hbar^2}(E + b\delta(x = 0))\psi(x) \right)dx$$

results in

$$\psi'(0+) - \psi'(0-) = \frac{-2mb}{\hbar^2}\psi(0)$$

where we have defined $d\psi/dx = \psi'$. One now notes that for an even function the spatial derivative is odd, so that $\psi'(0+) = -\psi'(0-)$, and one may write

$$\psi'(0+) - \psi'(0-) = 2\psi'(0+) = \frac{-2mb}{\hbar^2}\psi(0)$$

This is our boundary condition.

The Schrödinger equation is $(-\hbar^2/2m)d^2\psi(x)/dx^2 = E\psi(x)$ for $x > 0$ and $x < 0$. For a bound state we require $\psi(x) \to 0$ as $x \to \infty$ and $\psi(x) \to 0$, as $x \to -\infty$ so the only possibility is

$$\psi(x) = Ae^{-k|x|}$$

where $k = \sqrt{2m|E|/\hbar^2}$. Substituting this wave function into Schrödinger's equation gives $-\hbar^2k^2/2m = E$, but from our previous work

$$2\psi'(0+) = -2Ake^{-kx}|_{x=0} = \frac{-2mb}{\hbar^2}\psi(0) = \frac{-2mb}{\hbar^2}Ae^{-kx}|_{x=0}$$

so that $k = mb/\hbar^2$. Hence, we may conclude that

$$E = \frac{-mb^2}{2\hbar^2}$$

is the energy of the bound state.

To find the complete expression for the wave function we need to evaluate A. Normalization of the wave function requires

$$\int\limits_{x=-\infty}^{x=\infty} |\psi(x)|^2 dx = 1$$

and this gives $A = \sqrt{k}$ so that $\psi(x) = \sqrt{mb/\hbar^2}\, e^{-\frac{mb}{\hbar^2}|x|}$.

The spatial decay (the e^{-1} distance) for this wave function is just $\Delta x = 1/k = \hbar^2/mb$. Physically, it is reasonable to use the delta-function potential approximation when the spatial extent of the potential is much smaller than the spatial decay of the wave function. This might occur, for example, in the description of certain single-atom defects in a crystal.

(b) It follows from (a) that any delta function in the potential of the form $V(x) = \pm b\delta(x = 0)$ introduces a kink in the wave function. To show this one integrates Schrödinger's equation for a particle subject to a delta-function potential

$$\left(\frac{-\hbar^2}{2m} \frac{d^2}{dx^2} \pm b\delta(x = 0) \right) \psi(x) = E\psi(x)$$

between $x = 0-$ and $x = 0+$ in such a way that

$$\int\limits_{x=0-}^{x=0+} \left(\frac{d^2}{dx^2} \psi(x) \right) dx = \int\limits_{x=0-}^{x=0+} \left(\frac{-2m}{\hbar^2} (E \mp b\delta(x))\psi(x) \right) dx$$

$$\frac{d}{dx}\psi(0+) - \frac{d}{dx}\psi(0-) = \frac{\pm 2mb}{\hbar^2} \psi(0)$$

So, for finite $\psi(0)$ and b there is always a difference in the slope of the wave functions at $\psi(0+)$ and $\psi(0-)$, and hence a kink, in the wave function about $x = 0$. We may conclude that a delta-function potential introduces a kink in the wave function.

Solution 3.6

To avoid space-charging effects in the emitter and collector barriers of a NET, it is necessary to dope the emitter and collector barriers to an n-type impurity concentration density $n > j/ev_{av}$, where v_{nv} is an appropriate average electron velocity in the barrier and j is the current density. For typical values $v_{av} = 10^7$ cm s^{-1} and $n > 10^{17}$ cm^{-3} this gives a maximum current density near $j = 1.6 \times 10^5$ A cm^{-2}.

Solution 3.7

We would like to use the method outlined in Section 3.4 to numerically solve the one-dimensional Schrödinger wave equation for an electron of effective mass $m_e^* = 0.07 \times m_0$ in a rectangular potential well of width $L = 10$ nm bounded by infinite barrier energy.

Our solution is a MATLAB computer program which consists of two parts. The first part deals with input parameters such as the length L, effective electron mass, the number of discretization points, N, and the plotting routine. It is important to choose a high enough value of N so that the wave function does not vary too much between adjacent discretization points and so that the three-point finite-difference approximation used in Eqn (3.40) is reasonably accurate.

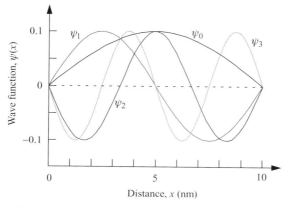

The second part of the computer program solves the discretized Hamiltonian matrix (Eqn (3.44)) and is a function, solve_schM, called from the main routine, Chapt3Exercise7. The diagonal matrix element given by Eqn (3.42) has potential values $V_j = 0$. The adjacent off-diagonal matrix elements are given by Eqn (3.43).

In this particular exercise the first four energy eigenvalues are $E_0 = 0.0537$ eV, $E_1 = 0.2149$ eV, $E_2 = 0.4834$ eV, and $E_3 = 0.8592$ eV. The eigenfunctions generated by the program and plotted in the above figure are not normalized.

The following lists an example program written in the MATLAB language. The main program is called Chapt3Exercise7.

Listing of MATLAB program for Exercise 3.7

```
%Chapt3Exercise7.m
%numerical solution to Schroedinger equation for
%rectangular potential well with infinite barrier energy
clear
clf;

length = 10;                              %length of well (nm)
npoints=200;                              %number of sample points
```

```
x=0:length/npoints:length;                          %position of sample points
mass=0.07;                                          %effective electron mass
num_sol=4;                                          %number of solutions sought

for i=1:npoints+1; v(i)=0; end                      %potential (eV)

[energy,phi]=solve_schM(length,npoints,v,mass,num_sol);   %call solve_schM

for i=1:num_sol
sprintf(['eigenenergy (',num2str(i),') = ',num2str(energy(i)),'eV'])%energy eigenvalues
end

figure(1);
plot(x,v,'b');xlabel('Distance (nm)'),ylabel('Potential energy, (eV)');
ttl=['Chapt3Exercise7, m* = ',num2str(mass),'m0, Length = ',num2str(length),'nm'];
title(ttl);

s=char('y','k','r','g','b','m','c');               %plot curves in different colors

figure(2);
for i=1:num_sol
  j=1+mod(i,7);
    plot(x,phi(:,i),s(j));                         %plot eigenfunctions
    hold on;
  end
xlabel('Distance (nm)'),ylabel('Wave function');
title(ttl);
hold off;
```

Listing of solve_schM function for MATLAB program used in Exercise 3.7

```
function [e,phi]=solve_schM(length,n,v,mass,num_sol)
% Solve Schrodinger equation for energy eigenvalues and eigenfunctions.
% [energy, phi_fun]=solve_sch(length,number,potential,mass,sol_num)
% solves time-independent Schrodinger equation in region bounded by 0<=x<=length.
% Potential is infinite outside this region.
%
% length                length of region (nm)
% n                     number of sample points
% v                     potential inside region (eV)
% mass                  effective electron mass
% sol_num               number of solutions sought
%
% e                     energy eigenvalue (eV)
% phi                   eigenfunction with eigenvalue = e
%
hbar=1.054571596;                      %Planck's constant (x10^34 J s)
hbar2=hbar^2;
echarge=1.602176462;                   %electron charge (x10^19 C)
baremass=9.10938188;                   %bare electron mass (x10^31 kg)
```

```
const=hbar2/baremass/echarge;        %(hbar^2)/(echarge*1nm^2*m0)
const=const/mass;                    %/m-effective
deltax=length/n;                     %x-increment = length(nm)/n
deltax2=deltax^2;
const=const/deltax2;

for i=2:n
    d(i-1)=v(i)+const;               %diagonal matrix element
    offd(i-1)=const/2;               %off-diagonal matrix element
end

t1=-offd(2:n-1);
Hmatrix=diag(t1,1);                  %upper diagonal of Hamiltonian matrix

Hmatrix=Hmatrix+diag(t1,-1);         %add lower diagonal of Hamiltonian matrix

t2=d(1:n-1);
Hmatrix=Hmatrix+diag(t2,0);          %add diagonal of Hamiltonian matrix
```

[phi,te]=eigs(Hmatrix,num_sol,'SM');%use matlab function eigs to find num_sol eigenfunctions and eigenvalues

```
for i=1:num_sol
    e(i)=te(i,i);                    %return energy eigenvalues in vector e
end
phi=[zeros(1,num_sol);phi;zeros(1,num_sol)]; %wave function is zero at x = 0 and x = length
return
```

4 The propagation matrix

4.1 Introduction

In the first two chapters of this book we learned about the way a particle moves in a potential. Because in quantum mechanics particles have wavy character, this modifies how they scatter from a change in potential compared with the classical case. In Section 3.8 we calculated transmission and reflection of an unbound particle from a one-dimensional potential step of energy eV_0. The particle was incident from the left and impinged on the potential barrier with energy $E > eV_0$. Significant quantum mechanical reflection probability for the particle occurred because the change in particle velocity at the potential step was large. This result is in stark contrast to the predictions of classical mechanics in which the particle velocity changes but there is no reflection.

In Section 3.10 we applied our knowledge of electron scattering from a step potential to the design of a new type of transistor. The analytic expressions developed were very successful in focusing our attention on the concept of matching electron velocities as a means of reducing quantum mechanical reflection that can occur at a semiconductor heterointerface. In this particular case, it is obvious that we could benefit from a model that is capable of taking into account more details of the potential. Such a model would be a next step in developing an accurate picture of transistor operation over a wide range of voltage bias conditions. We need a method for finding solutions to complicated potential structures for which analytic expressions are unmanageable.

Putting in the time and effort to develop a way of calculating electron transmission in one dimension will also be of benefit for those wishing to understand and apply quantum mechanics to a wide range of situations of practical importance. In this chapter, we will introduce our approach and apply it to situations that illustrate the interference effects of "wavy" electrons caused by scattering off changes in potential.

In the following, we would like to extend our calculation to go beyond a simple step change in potential to a one-dimensional potential of arbitrary shape. We will do this by approximating the arbitrary potential as a series of potential steps. The transmission and reflection coefficients are calculated at the first potential step for a particle of energy E

incident from the right. One then imagines the transmitted particle propagating to the next potential step, where it again has a probability of being transmitted or reflected. Associated with every potential step and free propagation to the next potential step is a 2×2 matrix which carries all of the amplitude and phase information on transmission and reflection at each potential step and propagation to the next potential step. The total one-dimensional propagation probability for a potential consisting of a number of potential steps may be calculated by multiplying together each 2×2 matrix associated with transmission and reflection at each potential step.

Before developing the method, it is useful to consider the conditions under which this approach will not introduce significant errors. Obviously it will work very well for a series of step potentials. The validity of the approach is a little less clear if we extend the idea and use a series of step changes in potential to approximate a smoothly varying potential. The errors introduced by the steps are a little difficult to quantify, but it is usually safe to say that errors are small if spacing between steps is small compared with the shortest particle wavelength being considered.

4.2 The propagation matrix method

Suppose an electron of energy E and mass m is incident from the left on an arbitrarily shaped, one-dimensional, smooth, continuous potential, $V(x)$. We can solve for a particle moving in an arbitrary potential by dividing the potential into a number of potential steps.

One may use the propagation matrix method to calculate the probability of the electron emerging on the right-hand side of the barrier. The problem is best approached by dividing it into small, easy-to-understand, logical parts. In the following, the four basic elements needed are summarized and then described sequentially in greater detail. In the next section, a step-by-step approach to writing a computer program to calculate transmission probability is given.

Part I summary: Calculate the propagation matrix $\hat{p}_{\text{step}}$ for transmission and reflection of the wave function representing a particle of energy E impinging on a single potential step. The potential step we consider is at position x_{j+1} in Fig. 4.1.

Part II summary: Calculate the propagation matrix $\hat{p}_{\text{free}}$ for propagation of the wave function between steps. The free propagation we consider is between positions x_j and x_{j+1} in Fig. 4.1. The distance of this free propagation is L_j.

Part III summary: Calculate the propagation matrix for the j-th region in Fig. 4.1. This is achieved if we multiply $\hat{p}_{\text{step}}$ and $\hat{p}_{\text{free}}$ to obtain the propagation matrix p_j for the j-th region of the discretized potential.

Part IV summary: Calculate the total propagation matrix $\hat{p}$ for the complete potential by multiplying together the propagation matrices for each region of the discretized potential.

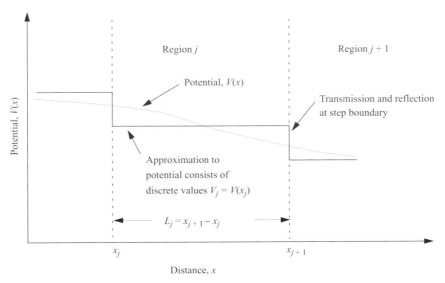

Fig. 4.1. Diagram illustrating approximation of a smoothly varying one-dimensional potential $V(x)$ with a series of potential steps. In this approach, the potential between positions x_j and x_{j+1} in region j is approximated by a value V_j. Associated with the potential step at x_j and free propagation distance $L_j = x_{j+1} - x_j$ is a 2×2 matrix which carries all of the amplitude and phase information on the particle.

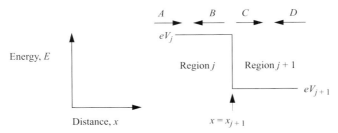

Fig. 4.2. Sketch of a one-dimensional potential step. In region j the potential energy is eV_j and in region $j+1$ the potential energy is eV_{j+1}. The coefficients A and C correspond to waves traveling left to right in regions j and $j+1$, respectively. The coefficients B and D correspond to waves traveling right to left in regions j and $j+1$, respectively. The transition between region j and region $j+1$ occurs at position $x = x_{j+1}$.

We now proceed to describe in more detail each of the parts summarized above that contribute to the propagation method.

Part I: The step propagation matrix

Figure 4.2 shows the detail of the potential step at position $j + 1$. The electron (or particle) has wave vector

$$k_j = \frac{(2m(E - eV_j))^{1/2}}{\hbar} \tag{4.1}$$

in region j, and the wave functions in regions j and $j + 1$ are

$$\psi_j = A_j e^{ik_j x} + B_j e^{-ik_j x} \tag{4.2}$$

$$\psi_{j+1} = C_{j+1} e^{ik_{j+1} x} + D_{j+1} e^{-ik_{j+1} x} \tag{4.3}$$

Following the convention we have adopted in this book, A and C are coefficients for the wave function traveling left to right in regions j and $j + 1$, respectively. B and D are the corresponding right-to-left traveling-wave coefficients.

The two wave functions given by Eqn (4.2) and Eqn (4.3) are related to each other by the constraint that ψ and $d\psi/dx$ must be continuous. This means that at the potential step that occurs at the boundary between regions j and $j + 1$ we require

$$\psi_j|_{x=x_{j+1}} = \psi_{j+1}|_{x=x_{j+1}} \tag{4.4}$$

and

$$\frac{d\psi_j}{dx}\bigg|_{x=x_{j+1}} = \frac{d\psi_{j+1}}{dx}\bigg|_{x=x_{j+1}} \tag{4.5}$$

Substituting Eqn (4.2) and Eqn (4.3) into Eqn (4.4) and Eqn (4.5) gives the two equations

$$A_j e^{ik_j x} + B_j e^{-ik_j x} = C_{j+1} e^{ik_{j+1} x} + D_{j+1} e^{-ik_{j+1} x} \tag{4.6}$$

$$A_j e^{ik_j x} - B_j e^{-ik_j x} = \frac{k_{j+1}}{k_j} C_{j+1} e^{ik_{j+1} x} - \frac{k_{j+1}}{k_j} D_{j+1} e^{-ik_{j+1} x} \tag{4.7}$$

It is worth mentioning that, as we have already discussed in Section 3.8.2, for semiconductor heterostructures with different effective electron mass, current continuity requires that all factors (k_{j+1}/k_j) in Eqn (4.7) be replaced with $(m_j k_{j+1}/m_{j+1} k_j)$.

By organizing into rows and columns the terms that contain left-to-right traveling waves of the form e^{ikx} and right-to-left traveling waves of the form e^{-ikx}, one may write Eqn (4.6) and Eqn (4.7) for a potential step at position $x_{j+1} = 0$ as a *matrix equation*:

$$\begin{bmatrix} 1 & 1 \\ 1 & -1 \end{bmatrix} \begin{bmatrix} A_j \\ B_j \end{bmatrix} = \begin{bmatrix} 1 & 1 \\ \dfrac{k_{j+1}}{k_j} & -\dfrac{k_{j+1}}{k_j} \end{bmatrix} \begin{bmatrix} C_{j+1} \\ D_{j+1} \end{bmatrix} \tag{4.8}$$

Unfortunately, as it stands, this expression is not in a very useful form. We would much prefer a simple equation of the type

$$\begin{bmatrix} A_j \\ B_j \end{bmatrix} = \hat{p}_{j_{\text{step}}} \begin{bmatrix} C_{j+1} \\ D_{j+1} \end{bmatrix} \tag{4.9}$$

where $\hat{p}_{j_{\text{step}}}$ is the 2×2 matrix describing wave propagation at a potential step. To obtain this expression, we need to eliminate the 2×2 matrix on the left-hand side of Eqn (4.8). This is not difficult to do. We simply recall from basic linear algebra that the inverse of a 2×2 matrix

$$\hat{A} = \begin{bmatrix} a_{11} & a_{12} \\ a_{21} & a_{22} \end{bmatrix} \tag{4.10}$$

is

$$\hat{A}^{-1} = \frac{1}{det} \begin{bmatrix} a_{22} & -a_{12} \\ -a_{21} & a_{11} \end{bmatrix} \tag{4.11}$$

where the determinant of $\hat{A}$ is given by

$$det = a_{11}a_{22} - a_{12}a_{21} \tag{4.12}$$

Hence, the inverse of $\begin{bmatrix} 1 & 1 \\ 1 & -1 \end{bmatrix}$ is $\frac{-1}{2}\begin{bmatrix} -1 & -1 \\ -1 & 1 \end{bmatrix} = \frac{1}{2}\begin{bmatrix} 1 & 1 \\ 1 & -1 \end{bmatrix}$, so that we may now write

$$\begin{bmatrix} A_j \\ B_j \end{bmatrix} = \frac{1}{2}\begin{bmatrix} 1 & 1 \\ 1 & -1 \end{bmatrix}\begin{bmatrix} 1 & \dfrac{1}{k_{j+1}} \\ k_j & -\dfrac{1}{k_{j+1}} \\ \end{bmatrix}\begin{bmatrix} C_{j+1} \\ D_{j+1} \end{bmatrix} = \hat{p}_{\text{step}}\begin{bmatrix} C_{j+1} \\ D_{j+1} \end{bmatrix} \tag{4.13}$$

where the step matrix is

$$\hat{p}_{j_{\text{step}}} = \frac{1}{2}\begin{bmatrix} 1 + \dfrac{k_{j+1}}{k_j} & 1 - \dfrac{k_{j+1}}{k_j} \\ 1 - \dfrac{k_{j+1}}{k_j} & 1 + \dfrac{k_{j+1}}{k_j} \end{bmatrix} \tag{4.14}$$

This is our result for the step potential that will be used later. We continue the development of the matrix method by considering the propagation between steps.

Part II: The propagation between steps

Propagation between potential steps separated by distance L_j carries phase information only so that $\psi_{A_j}e^{ik_jL_j} = \psi_{C_j}$ and $\psi_{B_j}e^{-ik_jL_j} = \psi_{D_j}$. This may be expressed in matrix form as

$$\begin{bmatrix} e^{ik_jL_j} & 0 \\ 0 & e^{-ik_jL_j} \end{bmatrix}\begin{bmatrix} A_j \\ B_j \end{bmatrix} = \begin{bmatrix} C_{j+1} \\ D_{j+1} \end{bmatrix} \tag{4.15}$$

or, alternatively,

$$\begin{bmatrix} A_j \\ B_j \end{bmatrix} = \hat{p}_{j_{\text{free}}}\begin{bmatrix} C_{j+1} \\ D_{j+1} \end{bmatrix} \tag{4.16}$$

where

$$
\hat{p}_{j_{\text{free}}} = \begin{bmatrix} e^{-ik_j L_j} & 0 \\ 0 & e^{ik_j L_j} \end{bmatrix}
$$

(4.17)

Part III: The propagation matrix p_j for the j-th region

Because we are dealing with probabilities, in order to find the combined effect of $\hat{p}_{j_{\text{free}}}$ and $\hat{p}_{j_{\text{step}}}$ we simply multiply the two matrices together. Hence, propagation across the complete j-th element consisting of a free propagation region and a step is

$$
\hat{p}_j = \hat{p}_{j_{\text{free}}} \hat{p}_{j_{\text{step}}} = \begin{bmatrix} p_{11} & p_{12} \\ p_{21} & p_{22} \end{bmatrix}
$$

(4.18)

When we multiply out the matrices $\hat{p}_{j_{\text{free}}} \hat{p}_{j_{\text{step}}}$ given by Eqn (4.17) and Eqn (4.14), respectively, it gives us the propagation matrix for the j-th region:

$$
p_j = \frac{1}{2} \begin{bmatrix} \left(1 + \dfrac{k_{j+1}}{k_j}\right) e^{-ik_j L_j} & \left(1 - \dfrac{k_{j+1}}{k_j}\right) e^{-ik_j L_j} \\ \left(1 - \dfrac{k_{j+1}}{k_j}\right) e^{ik_j L_j} & \left(1 + \dfrac{k_{j+1}}{k_j}\right) e^{ik_j L_j} \end{bmatrix}
$$

(4.19)

Notice that $p_{11} = p_{22}^*$ and $p_{21} = p_{12}^*$. This interesting symmetry, which is embedded in the matrix, allows us to write

$$
\hat{p}_j = \begin{bmatrix} p_{11} & p_{12} \\ p_{12}^* & p_{11}^* \end{bmatrix}
$$

(4.20)

Because we will make use of this fact later, it is worth spending some time investigating the origin of this symmetry. In Section 4.4 the concept of time-reversal symmetry is introduced. It is this that allows us to write Eqn (4.20).

Part IV: Propagation through an arbitrary series of step potentials

For the general case of N potential steps, we write down the propagation matrix for each region and multiply out to obtain the total propagation matrix,

$$
\hat{P} = \hat{p}_1 \hat{p}_2 \ldots \hat{p}_j \ldots \hat{p}_N = \prod_{j=1}^{j=N} \hat{p}_j
$$

(4.21)

Since the particle is introduced from the left, we know that $A = 1$, and if there is no reflection at the far right then $D = 0$. We may then rewrite

$$
\begin{bmatrix} A \\ B \end{bmatrix} = \left(\prod_{j=1}^{j=N} \hat{p}_j\right) \begin{bmatrix} C \\ D \end{bmatrix} = \hat{P} \begin{bmatrix} C \\ D \end{bmatrix}
$$

(4.22)

as

$$\begin{bmatrix} 1 \\ B \end{bmatrix} = \begin{bmatrix} p_{11} & p_{12} \\ p_{21} & p_{22} \end{bmatrix} \begin{bmatrix} C \\ 0 \end{bmatrix} \qquad (4.23)$$

In this case, the transmission probability is simply $|C|^2$, or

$$|C|^2 = \left| \frac{1}{p_{11}} \right|^2 \qquad (4.24)$$

Equation (4.24) is a particularly simple result. We will make use of this when we calculate the transmission probability of a particle through an essentially arbitrary one-dimensional potential.

4.3 Program to calculate transmission probability

To apply what we know so far about propagation of a particle through an arbitrary potential, we need to write a computer program or algorithm. In addition, because the exercises at the end of this chapter require a computer program, it is worth spending time becoming familiar with the basic approach.

Since you may choose to write the program code in one of a number of possible languages such as c, c++, f77, f90, MATLAB, and so on, in the following we will mainly be concerned with describing how to put the building blocks together in a natural flow.

The program calculates transmission probability for a particle of mass m incident from the left on a one-dimensional potential profile $V(x)$. The program uses the propagation matrix method and plots the results as a function of particle energy E. To illustrate how to put the code together we will refer to MATLAB in part because of both its popularity and the ease with which it handles matrix multiplication.

The following is the simple six-step flow of the propagation matrix algorithm.

STEP 1. Define values of useful constants such as the square root of minus one (eye), Planck's constant (hbar), electron charge (echarge) and particle mass (m). As a starting point, you may want to set your array size to 200 elements each for storing energy values, transmission coefficient values, and potential values. In some computer languages such as f77 you will have to dimension the size of arrays.

STEP 2. Set up the potential profile by discretizing $V(x)$ into, for example, N = 200 values, V(i), where i is an integer index for the array.

STEP 3. For fixed-particle energy E(j), calculate wave number k(i) for each position in the potential V(i). The wave number, which may be calculated using Eqn (4.1), can be complex. Then create and multiply through each propagation matrix, p, for each increment in x propagation to obtain total propagation matrix, bigP.

MATLAB example

```
bigP=[1,0;0,1];                              %default value of matrix bigP
  for i=1:N
    k(i)=sqrt(2*e*m*(E(j)-V(i)))/hbar;       %wave number at each position in potential V(j)
  end
  for n=1:(N-1)
    p(1,1)=0.5*(1+k(n+1)/k(n))*exp(-eye*k(n)*L(n));
    p(1,2)=0.5*(1-k(n+1)/k(n))*exp(-eye*k(n)*L(n));
    p(2,1)=0.5*(1-k(n+1)/k(n))*exp(eye*k(n)*L(n));
    p(2,2)=0.5*(1+k(n+1)/k(n))*exp(eye*k(n)*L(n));
    bigP=bigP*p;
  end
```

STEP 4. Calculate the transmission coefficient, Trans, for the current value of particle energy, E(j). It is convenient to use a negative *natural* logarithmic scale, since transmission probability often varies exponentially with particle energy E(j). In this approach, we use Trans $= -\log(|\text{bigP}(1)|^2)$, so that unity (maximum) transmission corresponds to Trans $= 0$ and low transmission corresponds to a large numerical value.

STEP 5. Increment to the next value of particle energy E(j). If E(j) isn't greater than the maximum energy you wish to consider, repeat the calculation from step 3 to obtain the next transmission coefficient for the next energy value.

STEP 6. Plot transmission coefficient data versus energy x(j) $=$ E(j) using any convenient graphing application.

For those interested in learning more about numerical solutions using the propagation matrix method and its application to electron transport across semiconductor heterostructures, the footnote gives a few useful references.[1]

4.4 Time-reversal symmetry

So far we have adopted the convention that a wavy particle incident on a potential step at position $x = x_{j+1}$ is initially traveling from left to right. As illustrated in Fig. 4.2, the incident wave has a coefficient A, and the forward scattered wave has coefficient C. The wave reflected at the potential step has a coefficient B, and a wave incident traveling from right to left has coefficient D. The corresponding wave functions on either side of the potential step are

$$\psi_j(x, t) = \left(A_j e^{ik_j x} + B_j e^{-ik_j x}\right)e^{-i\omega t} \tag{4.25}$$

[1] M. Steslicka and R. Kucharczyk, *Vacuum* **45**, 211 (1994), B. Jonson and S. T. Eng, *IEEE J. Quant. Electron.* **26**, 2025 (1990), M. O.Vassell, J. Lee, and H. F. Lockwood, *J. Appl. Phys.* **54**, 5206 (1983), and G. Bastard, *Phys. Rev.* **B24**, 5693 (1981).

and

$$\psi_{j+1}(x, t) = \left(C_{j+1}e^{ik_{j+1}x} + D_{j+1}e^{-ik_{j+1}x}\right)e^{-i\omega t} \tag{4.26}$$

If, as we have in fact assumed, the potential step is real and does not change with time, then there is no way we can admit the existence of a solution based on a left-to-right propagating incident wave without also allowing an associated time-reversed solution. To see why this is so, we start by writing the time-dependent Schrödinger equation

$$H\psi(x, t) = i\hbar\frac{\partial}{\partial t}\psi(x, t) \tag{4.27}$$

Reversing time requires changing the sign of the parameter $t \to -t$. Performing the sign change and taking the complex conjugate gives

$$H\psi^*(x, -t) = i\hbar\frac{\partial}{\partial t}\psi^*(x, -t) \tag{4.28}$$

The significance of this is that Eqn (4.27) and Eqn (4.28) are the same except for the replacement of $\psi(x, t)$ with $\psi(x, -t)$. This means that if $\psi(x, t)$ is a solution then the time-reversed solution $\psi^*(x, -t)$ also applies. To see what effect this has, one needs to compare the regular propagation matrix with the time-reversed case.

We start by recalling that the propagation matrix links coefficients A and B to C and D via a matrix

$$\begin{bmatrix} A \\ B \end{bmatrix} = \begin{bmatrix} p_{11} & p_{12} \\ p_{21} & p_{22} \end{bmatrix} \begin{bmatrix} C \\ D \end{bmatrix} \tag{4.29}$$

where we have arranged into rows and columns terms that contain left-to-right traveling waves of the form e^{ikx} and right-to-left traveling waves of the form e^{-ikx}. Since time reversal of the wave function $\psi_j(x, t) = (A_j e^{ik_j x} + B_j e^{-ik_j x})e^{-i\omega t}$ gives $\psi_j^*(x, -t) = (A_j^* e^{-ik_j x} + B_j^* e^{ik_j x})e^{-i\omega t}$ and does not change the energy of the system, we may write the time-reversed form of Eqn (4.29) as

$$\begin{bmatrix} B^* \\ A^* \end{bmatrix} = \begin{bmatrix} p_{11} & p_{12} \\ p_{21} & p_{22} \end{bmatrix} \begin{bmatrix} D^* \\ C^* \end{bmatrix} \tag{4.30}$$

The trick now is to take the complex conjugate of both sides and interchange rows so that a direct comparison can be made with Eqn (4.29). This gives our time-reversed solution as

$$\begin{bmatrix} A \\ B \end{bmatrix} = \begin{bmatrix} p_{22}^* & p_{21}^* \\ p_{12}^* & p_{11}^* \end{bmatrix} \begin{bmatrix} C \\ D \end{bmatrix} \tag{4.31}$$

Comparison of Eqn (4.29) and Eqn (4.31) confirms our previous assertion

$$p_{11} = p_{22}^* \tag{4.32}$$

and

$$p_{21} = p_{12}^* \tag{4.33}$$

This relationship may be thought of as a direct consequence of the time-reversal symmetry built into the Schrödinger equation when the potential is real and does not change with time.

There are situations in which time-reversal symmetry is broken. An example is when there is dissipation in the system. Often, dissipation can be modeled by introducing an imaginary part to the potential in Schrödinger's equation. However, we do not intend to explore this here. Rather, we proceed to make use of our result to find the condition for current conservation in a system with a real and time-independent potential.

4.5 Current conservation and the propagation matrix

Classically, the temporal change in charge density ρ is related to the spatial divergence of current density **J**. The relationship between charge density and current density is determined by conservation of charge and current. In Chapter 3 we showed that the current density operator for an electron of charge e is given by

$$\mathbf{J} = -\frac{ie\hbar}{2m}\left(\psi^*\nabla\psi - \psi\nabla\psi^*\right) \tag{4.34}$$

Our initial focus is electron propagation in the x direction across a one-dimensional potential step shown schematically in Fig. 4.3. We first note that the transmission coefficient $|C/A|^2$ and the reflection coefficients $|B/A|^2$ are ratios, so that absolute normalization of wave functions is not important. From Section 4.4 we know that if the spatial part of the wave function $\psi(x)$ is a traveling wave, then $\psi^*(x)$ is a time-reversed solution. The wave vector k_j describes a plane wave and is real because we are restricting our present discussion to simple traveling waves and because particle energy $E > eV_2$. Later, we will consider the case when $E < eV_2$ and k_j is imaginary.

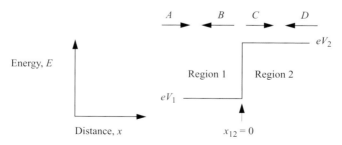

Fig. 4.3. Sketch of a one-dimensional potential step. In region 1 the potential energy is eV_1, and in region 2 the potential energy is eV_2. The transition between region 1 and region 2 occurs at position $x = x_{12}$.

For k_j real, the wave function and its complex conjugate in region 1 and in region 2 may be expressed as

$$\psi_1 = \frac{1}{\sqrt{k_1}}\left(Ae^{ik_1x} + Be^{-ik_1x}\right) \qquad \psi_1^* = \frac{1}{\sqrt{k_1}}\left(A^*e^{-ik_1x} + B^*e^{ik_1x}\right) \tag{4.35}$$

$$\psi_2 = \frac{1}{\sqrt{k_2}}\left(Ce^{ik_2x} + De^{-ik_2x}\right) \qquad \psi_2^* = \frac{1}{\sqrt{k_2}}\left(C^*e^{-ik_2x} + D^*e^{ik_2x}\right) \tag{4.36}$$

Because we assume a potential energy step between regions 1 and 2, the wave vectors k_1 and k_2 are, in general, different. The factor $1/\sqrt{k_j}$ appears in the equation because if we normalize for unit flux then $|A|^2 \to |A|^2/v_1$, so that $A \to A/\sqrt{k_1}$.

Applying the current density operator (Eqn (4.34)) when particle energy $E > eV_0$ gives

$$J(x) = \frac{e\hbar}{m}(|A|^2 - |B|^2) \text{ for } x < 0 \tag{4.37}$$

and

$$J(x) = \frac{e\hbar}{m}(|C|^2 - |D|^2) \text{ for } x > 0 \tag{4.38}$$

Equations (4.35) and (4.36), when used with the current density operator given by Eqn (4.34), allow us to state that current conservation requires

$$A^*A - B^*B = C^*C - D^*D \tag{4.39}$$

In words, the net current flow left to right in region 1 must be equal to the net flow in region 2.

The coefficients are related to each other by the 2×2 propagation matrix

$$\begin{bmatrix} A \\ B \end{bmatrix} = \begin{bmatrix} p_{11} & p_{12} \\ p_{21} & p_{22} \end{bmatrix} \begin{bmatrix} C \\ D \end{bmatrix} \tag{4.40}$$

so that

$$A = p_{11}C + p_{12}D \qquad A^* = p_{11}^*C^* + p_{12}^*D^* \tag{4.41}$$
$$B = p_{21}C + p_{22}D \qquad B^* = p_{21}^*C^* + p_{22}^*D^* \tag{4.42}$$

For no incoming wave from the right $D = 0$, we can make use of Eqn (4.41) and Eqn (4.42) and rewrite the left-hand side of Eqn (4.39) as

$$A^*A - B^*B = (p_{11}^*p_{11} - p_{21}^*p_{21})C^*C \tag{4.43}$$

However, we recall that for $D = 0$ current conservation (Eqn (4.39)) requires

$$A^*A - B^*B = C^*C \tag{4.44}$$

The only nontrivial way to simultaneously satisfy Eqn (4.43) and Eqn (4.44) is if

$$p_{11}^* p_{11} - p_{21}^* p_{21} = 1 \tag{4.45}$$

We now make use of the fact that $p_{11} = p_{22}^*$ and $p_{21} = p_{12}^*$, so that

$$p_{11} p_{22} - p_{12} p_{21} = 1 \tag{4.46}$$

or, in more compact form,

$$\boxed{\det(\hat{P}) = 1} \tag{4.47}$$

This, then, is an expression of *current conservation* when k_j is *real*. For imaginary k_j, Eqn (4.47) is modified to $\det(\hat{P}) = \pm i$, where the $\pm$ depends upon whether the incoming wave is from the left or the right.

4.6 The rectangular potential barrier

In Section 3.9 we discussed transmission and reflection of a particle incident on a rectangular potential barrier. A number of interesting results were obtained, including the quantum mechanical effect of tunneling. In this section, one goal is to explore the properties of the rectangular potential barrier using the propagation matrix method. This serves two purposes. First, we can get some practice using the propagation matrix method. Second, we can check our results against the solutions derived in Chapter 3 and in this way verify the validity of the propagation matrix method.

Figure 4.4 is a sketch of the rectangular potential barrier we will consider. A wavy particle incident on the barrier from the left with amplitude A sees a potential step-up in energy of eV_0 at $x = 0$, a barrier propagation region of length L, and a potential

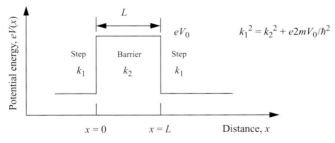

Fig. 4.4. Sketch of the potential of a one-dimensional rectangular barrier of energy eV_0. The thickness of the barrier is L. A particle mass m incident from the left of energy E has wave vector k_1. In the barrier region, the wave vector is k_2. The wave vectors k_1 and k_2 are related through $k_1^2 = k_2^2 + e2mV_0/\hbar^2$.

step-down at $x = L$. A particle of energy E, mass m, and charge e has wave number k_1 outside the barrier and k_2 in the barrier region $0 < x < L$.

4.6.1   Transmission probability for a rectangular potential barrier

We start by considering a particle impinging on a step change in potential between two regions in which the wave vector changes from k_1 to k_2 due to the potential step-up shown in Fig. 4.3. The corresponding wave function changes from ψ_1 to ψ_2. Solutions of the Schrödinger equation for a step change in potential are (Eqn (4.35) and Eqn (4.36))

$$\psi_1 = \frac{A}{\sqrt{k_1}}e^{ik_1x} + \frac{B}{\sqrt{k_1}}e^{-ik_1x} \tag{4.48}$$

$$\psi_2 = \frac{C}{\sqrt{k_2}}e^{ik_2x} + \frac{D}{\sqrt{k_2}}e^{-ik_2x} \tag{4.49}$$

Of course, the wave functions ψ_1 and ψ_2 are related to each other by the constraint that the wave function and its derivative must be continuous. Applying the condition that the wave function is continuous at the potential step

$$\psi_1|_\text{step} = \psi_2|_\text{step} \tag{4.50}$$

and that the derivative of the wave function is continuous

$$\frac{d\psi_1}{dx}\bigg|_\text{step} = \frac{d\psi_2}{dx}\bigg|_\text{step} \tag{4.51}$$

gives

$$\frac{A}{\sqrt{k_1}} + \frac{B}{\sqrt{k_1}} = \frac{C}{\sqrt{k_2}} + \frac{D}{\sqrt{k_2}} \tag{4.52}$$

$$\frac{A}{\sqrt{k_1}} - \frac{B}{\sqrt{k_1}} = \frac{k_2}{k_1}\frac{C}{\sqrt{k_2}} - \frac{k_2}{k_1}\frac{D}{\sqrt{k_2}} \tag{4.53}$$

Rewritten in matrix form, these equations become

$$\frac{1}{\sqrt{k_1}}\begin{bmatrix} 1 & 1 \\ 1 & -1 \end{bmatrix}\begin{bmatrix} A \\ B \end{bmatrix} = \frac{1}{\sqrt{k_2}}\begin{bmatrix} 1 & 1 \\ \dfrac{k_2}{k_1} & -\dfrac{k_2}{k_1} \end{bmatrix}\begin{bmatrix} C \\ D \end{bmatrix} \tag{4.54}$$

To eliminate the 2×2 matrix on the left-hand side of this equation, we must find its inverse matrix and multiply the equation by it. The determinant of the left-hand matrix is $(-1 - 1)/k_1 = -2/k_1$, so the inverse of the left-hand matrix is

$$\frac{k_1}{2\sqrt{k_1}}\begin{bmatrix} 1 & 1 \\ 1 & -1 \end{bmatrix} \tag{4.55}$$

Hence, we may rewrite Eqn (4.54) as

$$\begin{bmatrix} A \\ B \end{bmatrix} = \frac{k_1}{2} \frac{1}{\sqrt{k_1}} \begin{bmatrix} 1 & 1 \\ 1 & -1 \end{bmatrix} \begin{bmatrix} 1 & 1 \\ \dfrac{k_2}{k_1} & -\dfrac{k_2}{k_1} \end{bmatrix} \frac{1}{\sqrt{k_2}} \begin{bmatrix} C \\ D \end{bmatrix}$$

$$= \frac{1}{2\sqrt{k_1 k_2}} \begin{bmatrix} 1 & 1 \\ 1 & -1 \end{bmatrix} \begin{bmatrix} k_1 & k_1 \\ k_2 & -k_2 \end{bmatrix} \begin{bmatrix} C \\ D \end{bmatrix} \tag{4.56}$$

Multiplying out the two square matrices gives the 2×2 matrix describing propagation at the step-up in potential

$$\begin{bmatrix} A \\ B \end{bmatrix} = \frac{1}{2\sqrt{k_1 k_2}} \begin{bmatrix} k_1 + k_2 & k_1 - k_2 \\ k_1 - k_2 & k_1 + k_2 \end{bmatrix} \begin{bmatrix} C \\ D \end{bmatrix} \tag{4.57}$$

Since the rectangular potential barrier consists of a step-up and a step-down, we can make use of this symmetry and immediately calculate the 2×2 matrix for the step-down by simply interchanging k_1 and k_2. The total propagation matrix for the rectangular potential barrier of thickness L consists of the step-up 2×2 matrix multiplied by the propagation matrix from the barrier thickness L multiplied by the step-down matrix. Hence, our propagation matrix becomes,

$$\hat{P} = \frac{1}{2\sqrt{k_1 k_2}} \begin{bmatrix} k_1 + k_2 & k_1 - k_2 \\ k_1 - k_2 & k_1 + k_2 \end{bmatrix} \begin{bmatrix} e^{-ik_2 L} & 0 \\ 0 & e^{ik_2 L} \end{bmatrix} \frac{1}{2\sqrt{k_1 k_2}} \begin{bmatrix} k_2 + k_1 & k_2 - k_1 \\ k_2 - k_1 & k_2 + k_1 \end{bmatrix}$$

$$\tag{4.58}$$

$$\hat{P} = \frac{1}{4k_1 k_2} \begin{bmatrix} (k_1 + k_2)e^{-ik_2 L} & (k_1 - k_2)e^{ik_2 L} \\ (k_1 - k_2)e^{-ik_2 L} & (k_1 + k_2)e^{ik_2 L} \end{bmatrix} \begin{bmatrix} k_2 + k_1 & k_2 - k_1 \\ k_2 - k_1 & k_2 + k_1 \end{bmatrix} \tag{4.59}$$

To find the matrix elements of $\hat{P}$, we just multiply out the matrices in Eqn (4.59). For example p_{12} becomes

$$p_{12} = -i\frac{(k_2^2 + k_1^2)}{2k_1 k_2} \sin(k_2 L) \tag{4.60}$$

The next step we want to take is to calculate the transmission probability for a particle incident on the barrier. We already know that the transmission of a particle incident from the left is given by $|1/p_{11}|^2$, so that we will be interested in obtaining p_{11} from Eqn (4.59)

$$p_{11} = \frac{(k_2 + k_1)(k_1 + k_2)e^{-ik_2 L} + (k_1 - k_2)(k_2 - k_1)e^{ik_2 L}}{4k_1 k_2} \tag{4.61}$$

$$p_{11} = \frac{\left(k_2^2 + k_1^2 + 2k_1 k_2\right)e^{-ik_2 L} - \left(k_1^2 + k_2^2 - 2k_1 k_2\right)e^{ik_2 L}}{4k_1 k_2} \tag{4.62}$$

$$p_{11} = \frac{\left(k_2^2 + k_1^2\right)(e^{-ik_2L} - e^{ik_2L})}{4k_1k_2} + \frac{2k_1k_2(e^{-ik_2L} + e^{ik_2L})}{4k_1k_2} \tag{4.63}$$

$$p_{11} = -\frac{\left(k_2^2 + k_1^2\right)(e^{ik_2L} - e^{-ik_2L})}{2 \cdot 2k_1k_2} + \frac{1}{2}(e^{-ik_2L} + e^{ik_2L}) \tag{4.64}$$

In the following we will use this result to calculate transmission probability.

4.6.1.1 Transmission when $E \geq eV_0$

To take this further, we now specialize to the case in which the energy of the incident particle is greater than the potential barrier energy. In this case $E \geq eV_0$, and k_2 is real, so that Eqn (4.64) becomes

$$p_{11} = -i\frac{\left(k_2^2 + k_1^2\right)}{2k_1k_2} \sin(k_2L) + \cos(k_2L) \tag{4.65}$$

Hence the transmission probability $Trans = 1/(p_{11} \cdot p_{11}^*)$ is

$$Trans = \frac{1}{|p_{11}|^2} = \left(\left(\frac{\left(k_2^2 + k_1^2\right)}{2k_1k_2}\right)^2 \sin^2(k_2L) + \cos^2(k_2L)\right)^{-1} \tag{4.66}$$

This equation can be rearranged a little by first noting that $\cos^2(\theta) = 1 - \sin^2(\theta)$:

$$Trans = \left(\left(\frac{\left(k_2^2 + k_1^2\right)}{2k_1k_2}\right)^2 \sin^2(k_2L) + 1 - \sin^2(k_2L)\right)^{-1} \tag{4.67}$$

$$Trans = \left(1 + \left(\left(\frac{k_2^2 + k_1^2}{2k_1k_2}\right)^2 - 1\right) \sin^2(k_2L)\right)^{-1} \tag{4.68}$$

$$Trans = \left(1 + \frac{\left(k_1^4 + k_2^4 + 2k_1^2k_2^2 - 4k_1^2k_2^2\right)}{(2k_1k_2)^2} \sin^2(k_2L)\right)^{-1} \tag{4.69}$$

$$\boxed{Trans = \left(1 + \left(\frac{k_1^2 - k_2^2}{2k_1k_2}\right)^2 \sin^2(k_2L)\right)^{-1}} \qquad \text{for } E \geq eV_0 \tag{4.70}$$

For particle energy $E \geq eV_0$, this equation predicts unity transmission when $\sin^2(k_2L) = 0$. In this case, $Trans_{max} = 1$ and $k_2L = n\pi$ for $n = 1, 2, 3, \ldots$. Such resonances in transmission correspond to the situation in which the scattered waves originating from $x = 0$ and $x = L$ interfere and *exactly cancel* any reflection from the potential barrier.

4.6.1.2 Transmission when $E < eV_0$

For $E < eV_0$ the wave number k_2 becomes imaginary. So $k_2 \rightarrow ik_2$ with $k_2 = (2m(E - eV_0)/\hbar^2)^{1/2}$, and we can substitute into our expression for *Trans* to obtain

$$Trans = \left(1 + \left(\frac{k_1^2 + k_2^2}{2k_1k_2} \right)^2 \sinh^2(k_2 L) \right)^{-1} \qquad \text{for } E < eV_0 \qquad (4.71)$$

where we have made use of the fact that $\sin(k_j x) = (e^{ik_j x} - e^{-ik_j x})/2i$, $\sinh(k_j x) = (e^{k_j x} - e^{-k_j x})/2$, $\sin(ik_j x) = -i\sinh(k_j x)$, and $\sin^2(ik_j x) = -\sinh^2(k_j x)$.

In this case, the particle with energy less than eV_0 can only be transmitted through the barrier by tunneling. The concept of particle tunneling is purely quantum mechanical. Particles do not tunnel through potential energy barriers according to classical mechanics.

Now we can compare the results of Eqn (4.71), which we derived using the propagation matrix method, with those obtained previously in Chapter 3 and summarized by Eqn (3.146). As anticipated, the equations are identical, so we can proceed with increased confidence in our propagation matrix approach.

There is a great deal more that can be learned about transmission and reflection of a particle incident on a rectangular potential barrier of energy eV_0 by studying the properties of Eqn (4.70) and Eqn (4.71). Our approach is to investigate the analytic properties of these equations as a function of particle energy.

4.6.2   Transmission as a function of energy

Because the incident particle is often characterized by its energy, our first task is to obtain the *Trans* function in terms of energy. This involves no more than finding k_1 and k_2 expressed in terms of particle energy. In the barrier region, when $E \geq eV_0$, we have $k_2 = (2m(E - eV_0)/\hbar^2)^{1/2}$, so we may write $k_2^2 = 2m(E - eV_0)/\hbar^2$. In the nonbarrier region, $k_1^2 = 2mE/\hbar^2$.

4.6.2.1 Transmission when $E \geq eV_0$

We now recall that for an incoming particle energy greater than the barrier energy $E \geq eV_0$, and we had a transmission function given by Eqn (4.70)

$$Trans = \left(1 + \left(\frac{k_1^2 - k_2^2}{2k_1k_2} \right)^2 \sin^2(k_2 L) \right)^{-1} \qquad (4.72)$$

Using the relations $k_1^2 = 2mE/\hbar^2$ and $k_2^2 = 2m(E - eV_0)/\hbar^2$ gives

$$Trans = \left(1 + \frac{1}{4} \left(\frac{E - (E - eV_0)}{E^{1/2}(E - eV_0)^{1/2}} \right)^2 \sin^2(k_2 L) \right)^{-1} \qquad (4.73)$$

which allows us to write

$$Trans(E \geq eV_0) = \left(1 + \frac{1}{4}\left(\frac{e^2 V_0^2}{E(E - eV_0)}\right)\sin^2(k_2 L)\right)^{-1} \tag{4.74}$$

4.6.2.2 Transmission when $E < eV_0$

For the case when the incoming particle has energy less than the barrier energy $E < eV_0$, the wave vector k_2 is related to particle energy E by

$$k_2^2 = \frac{2m(eV_0 - E)}{\hbar^2} \tag{4.75}$$

The transmission function given by Eqn (4.71)

$$Trans = \left(1 + \left(\frac{k_1^2 + k_2^2}{2k_1 k_2}\right)^2 \sinh^2(k_2 L)\right)^{-1} \tag{4.76}$$

can now be expressed as

$$Trans(E < eV_0) = \left(1 + \frac{1}{4}\left(\frac{e^2 V_0^2}{E(eV_0 - E)}\right)\sinh^2(k_2 L)\right)^{-1} \tag{4.77}$$

4.6.3　Transmission resonances

The behavior of particle transmission as a function of wave vector or particle energy is particularly interesting. Previously, when $E \geq eV_0$, we had

$$Trans(E \geq eV_0) = \left(1 + \frac{1}{4}\left(\frac{e^2 V_0^2}{E(E - eV_0)}\right)\sin^2(k_2 L)\right)^{-1} \tag{4.78}$$

It may be shown (Exercise 4.1(a)) that Eqn (4.78) may be written as

$$Trans = \left(1 + \frac{1}{4} \cdot \frac{b^2}{(b + k_2^2 L^2)} \cdot \frac{\sin^2(k_2 L)}{k_2^2 L^2}\right)^{-1} \tag{4.79}$$

where the parameter $b = 2k_0 L = 2meV_0 L^2/\hbar^2$ is a measure of the "strength" of the potential barrier.

It is also possible to calculate maxima and minima in *Trans* as a function of $k_2 L$ or as a function of energy E. We start by analyzing for $k_2 L$. It is clear from Eqn (4.79) that *Trans* is a maximum when $\sin^2(k_2 L) = 0$. When this happens, $Trans_{max} = 1$ and $k_2 L = n\pi$ for $n = 1, 2, 3, \ldots$. $Trans_{max} = 1$ corresponds to resonances in transmission that occur when particle-waves back-scattered from the step change in barrier potential at positions $x = 0$ and $x = L$ interfere and exactly cancel each other, resulting in zero reflection from the potential barrier. A resonance with unity transmission requires exact cancellation in amplitude and phase of the coherent sum of all back-scattered waves.

The minimum of the *Trans* function occurs when $\sin^2(k_2 L) = 1$. This happens when $k_2 L = (2n - 1)\pi/2$ for $n = 1, 2, 3, \ldots$. The minimum of the *Trans* function may be written (Exercise 4.1(b)) as

$$Trans_{min} = 1 - \frac{b^2}{(2k_2^2 L^2 + b)^2} \tag{4.80}$$

or equivalently as

$$Trans_{min} = 1 - \frac{e^2 V_0^2}{(2E - eV_0)^2} \tag{4.81}$$

Another limiting case to consider is the situation in which $E \to eV_0$ or $k_2 L \to 0$. In this situation, one may expand $\sin^2(k_2 L) = k_2^2 L^2 + \cdots$ and substitute into our expression for *Trans*

$$Trans = \left(1 + \frac{1}{4} \cdot \frac{b^2}{\left(b + k_2^2 L^2\right)} \cdot \frac{\sin^2(k_2 L)}{k_2^2 L^2} \right)^{-1} \tag{4.82}$$

$$Trans(E \to eV_0) = \left(1 + \frac{1}{4} \cdot \frac{b^2}{\left(b + k_2^2 L^2\right)} \cdot \frac{k_2^2 L^2}{k_2^2 L^2} \right)^{-1} \Bigg|_{k_2 L \to 0} \tag{4.83}$$

$$Trans(E \to eV_0) = \left(1 + \frac{b}{4} \right)^{-1} = \frac{4}{4 + b} = \left(1 + \frac{em V_0 L^2}{2\hbar^2} \right)^{-1} \tag{4.84}$$

The results we have obtained are neatly summarized in Fig. 4.5. In this figure *Trans*, *Trans*$_{max}$, and *Trans*$_{min}$ are plotted as functions of $k_2 L$ for an electron incident on a rectangular potential barrier of energy $V_0 = 1.0$ eV and width $L = 1$ nm. The incident electron has energy $E \geq eV_0$.

One way of summarizing all the results of Section 4.6 is shown in Fig. 4.6. Figure 4.6(a) shows the potential energy of a one-dimensional rectangular barrier with energy eV_0 and thickness L. Figure 4.6(b) shows the transmission coefficient as a function of energy for an electron of energy E incident on the potential barrier. In this example, the rectangular potential barrier has energy $eV_0 = 1.0$ eV and width $L = 1$ nm. For electron energy less than eV_0, there is an exponential decrease in transmission probability with decreasing E, because transmission is dominated by tunneling. For electron energy greater than eV_0, there are resonances in transmission which occur when $2m(E - eV_0)L/\hbar = n\pi$, where $n = 1, 2, 3, \ldots$. At the peak of these resonances the transmission coefficient is unity.

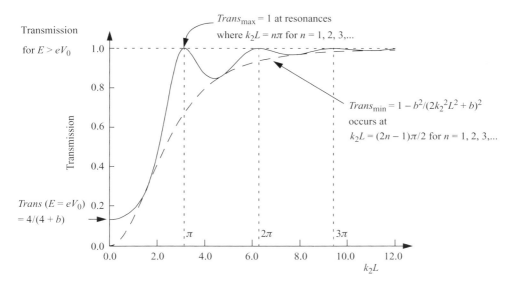

Fig. 4.5. Illustration of the behavior of the functions $Trans$ (solid line) and $Trans_{min}$ (broken line) showing the limiting values $Trans_{max}$ and $Trans(E = eV_0)$ when an incident electron has energy $E \geq eV_0$. In this example, the rectangular potential barrier has energy $eV_0 = 1.0$ eV and width $L = 1$ nm.

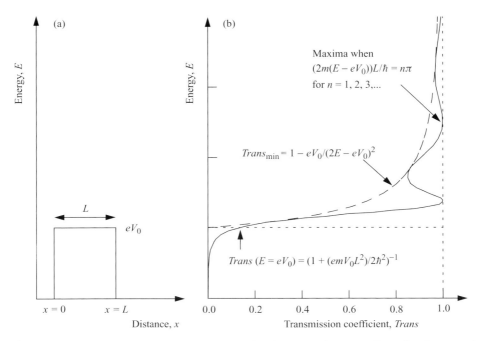

Fig. 4.6. (a) Plot of potential energy as a function of position showing a one-dimensional rectangular barrier of energy eV_0 and thickness L. (b) Solid line is plot of transmission coefficient as a function of energy for an electron of energy E incident on the potential barrier shown in (a). The broken line is $Trans_{min}$ and the dotted line is a guide to the eye for unity transmission and energy eV_0. In this example, the rectangular potential barrier has energy $eV_0 = 1.0$ eV and width $L = 1$ nm.

4.7 Resonant tunneling

In the previous section, resonances with unity peak transmission probability were found for a particle of energy $E > eV_0$ traversing a rectangular potential barrier of energy eV_0 and width L. This occurs because on resonance a coherent superposition of back-scattered particle-waves from step changes in potential at $x = 0$ and $x = L$ exactly cancel each other to give zero reflection from the potential barrier. A resonance with unity transmission requires exact cancellation in amplitude and phase of the coherent sum of all contributions to the back-scattered wave.

For particle energy $E < eV_0$, no resonances occur and transmission is dominated by simple quantum mechanical tunneling through the rectangular potential barrier. This fact may be explained as being due to the exponentially reduced contribution of the back-scattered wave from the step-change in potential at position $x = L$.

To obtain unity transmission probability for particle energy $E < eV_0$, a different type of potential must be considered. A simple example is the double rectangular potential barrier sketched in Fig. 4.7(a). For each barrier, width is 0.4 nm, well width is 0.6 nm, and barrier energy is 1 eV. As with the finite potential well of Section 3.7, there is *always* a symmetric lowest-energy resonance associated with this symmetric potential well. For the case we are considering, this lowest-energy resonance has energy E_0 and is indicated by a broken line in Fig. 4.7(a). The wave function associated with the resonance at energy E_0 is symmetric because the potential is symmetric. Due to the finite width and energy of the potential barrier on either side of the potential well, there is a finite probability that a particle can tunnel out of the potential well. Because of this, there is a finite lifetime associated with the resonance.

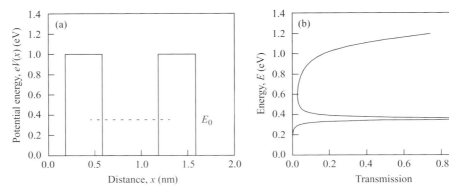

Fig. 4.7. (a) Symmetric double rectangular potential barrier with barrier width 0.4 nm, well width 0.6 nm, and barrier energy 1 eV. There is a resonance at energy E_0 for an electron mass m_0 in such a potential. This resonance energy is indicated by the broken line. (b) Transmission probability of an electron mass m_0 incident from the left on the potential given in (a). There is an overall increase in background transmission with increasing energy, and a unity transmission resonance occurs at energy $E = E_0 = 357.9$ meV with $\Gamma_{\text{FWHM}} = 30.7$ meV.

Figure 4.7(b) shows the transmission probability of an electron of mass m_0 incident from the left on the potential given in Fig. 4.7(a). With increasing electron energy there is an overall smooth increase in background transmission because the effective potential barrier seen by the electron decreases. At an incident energy $E = E_0 = 0.358$ eV there is a unity peak in transmission probability that stands out above the smoothly varying electron transmission background. This peak exists because an electron incident from the left tunnels via the resonance in the potential well at energy E_0. At the resonance peak, a coherent superposition of back-scattered particle-waves from step changes in the potential exactly cancel to give zero reflection and unity transmission.

The resonance has a finite width in energy, because if we place a particle in the potential well it can tunnel out. It seems natural to discuss this process in terms of a characteristic time or lifetime for the particle. The lifetime of the resonance at energy E_0 may be calculated approximately as $\tau = \hbar / \Gamma_{FWHM}$, where Γ_{FWHM} is the full width at half maximum in the transmission peak when fit to a Lorentzian line shape.[2] For the situation depicted in Fig. 4.7(b), Γ_{FWHM} is near 31 meV, giving a lifetime $\tau = 21$ fs. The lifetime, τ, is the characteristic response time of the system. The calculated lifetime assumes that the resonance may be described by a Lorentzian line shape. This, however, is always an approximation for a symmetric potential because, for such a case, the transmission line shape is always asymmetric. The asymmetry exists because of the overall increase in background transmission with increasing energy.

One may use modern crystal growth techniques to make a double-barrier potential similar to that shown in Fig. 4.7(a). The potential might be formed in the conduction band of a semiconductor heterostructure. For example, epitaxially grown AlAs may be used to form the two tunnel barriers, and GaAs may be used to form the potential well of an n-type unipolar diode. One expects the resonant tunneling heterostructure diode current–voltage characteristics of such a device to be proportional to electron transmission probability. While energy is not identical to applied voltage and current is

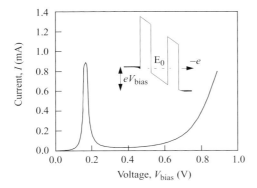

Fig. 4.8. Illustration of resonant tunnel diode current–voltage characteristics. There is a peak in current when the diode is voltage biased in such a way that electrons can resonantly tunnel through the potential well. A region of negative differential resistance also exists for some values of V_{bias}.

[2] See Table 2.3.

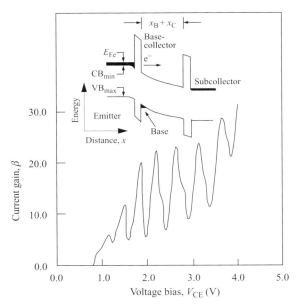

Fig. 4.9. Measured current gain, β, with bias, V_{CE}, for the AlAs/GaAs HBT sketched in the inset. Lattice temperature is $T = 4.2$ K, emitter area is 7.8×10^{-5} cm^2, and base current is $I_B = 0.1$ mA. The AlAs tunnel emitter is 8 nm thick, the p-type base is delta doped with a Be sheet concentration of 6×10^{13} cm^{-2}, the AlAs collector barrier is 5 nm thick, and emitter–collector barrier separation is $x_B + x_C = 40$ nm. The conduction band minimum, CB$_{min}$, the valence band maximum, VB$_{max}$, and the emitter Fermi energy, E_{Fe}, are indicated.

not identical to the transmission coefficient shown in Fig. 4.7(b), the trends are similar. In particular, a region of negative differential resistance exists that is of interest as an element in the design of electronic circuits. Figure 4.8 illustrates the current–voltage characteristics of a resonant tunneling heterojunction diode. The inset is the conduction band minimum profile under voltage bias V_{bias}.

In the next section we consider a heterostructure bipolar transistor with a tunnel barrier at the emitter–base junction and a tunnel barrier at the collector–subcollector boundary.

4.7.1   Heterostructure bipolar transistor with resonant tunnel barrier

It is possible to gain more insight into electron transport in nanoscale structures by combining what we know about resonant tunnel devices with a heterostructure bipolar transistor (HBT). The inset in Fig. 4.9 shows the conduction and valence band profiles of an n–p–n AlAs/GaAs HBT with an AlAs tunnel emitter and an AlAs tunnel barrier at the collector–subcollector interface.[3] The separation between the two tunnel barriers is

---

[3] A. F. J. Levi, unpublished.

$x_B + x_C = 40$ nm, where x_B is the thickness of the transistor base and x_C is the thickness of the collector. The p-type GaAs base region of the transistor consists of a thin two-dimensional sheet of substitutional Be atoms of average concentration 6×10^{13} cm^{-2}. The value of x_B is less than 10 nm and is determined by the spatial extent of the wave function in the x direction associated with the charge carriers. The emitter and subcollector are heavily doped n-type so that they have a Fermi energy E_{Fe} at low temperature.

The n–p–n HBT shown in Fig. 4.9 may be viewed as a three-terminal resonant tunnel device. The third terminal is the p-type base, which allows us to independently vary the energy of conduction band electrons that are tunnel-injected from the emitter into the base collector region, where resonant electron interference effects take place. Because this is a transistor, it is possible to fix the base current and measure common-emitter current gain β as a function of collector–emitter voltage bias V_{CE}.

To help obtain a clear understanding of electron motion in the device without complications introduced by thermal effects, it is useful to perform measurements at low temperatures. Figure 4.9 shows the results of measuring β at a temperature $T = 4.2$ K. As might be expected for a resonant tunnel structure, there are strong resonances in β associated with quantization of energy levels in the 40-nm-thick base–collector region. This confirms that quantum mechanical reflection and resonance phenomena play an important role in determining device performance. However, with a little thought, limitations to both device operation and modeling soon become apparent.

It is reasonable to consider why the minima in β do not approach closer to zero. One reason is the presence of elastic and inelastic electron scattering in the collector and p-type base. This has the effect of breaking the coherence of the wave function associated with resonant transport of an electron through the structure. The resulting decoherence reduces resonant interference and introduces an incoherent background contribution to current flow between emitter and collector.

It turns out that inelastic electron scattering presents a *fundamental* limit to transistor performance. Thus, in an effort to quantify this limit as a step to designing a better device, we might be interested in calculating the appropriate electron scattering rates. However, in addition to various other complications, such scattering rates cannot be evaluated independently of either V_{CE} or V_{CB}. This is obvious because the resonances depend upon the potential profile and hence the voltage bias across the device.

The existence of resonant effects illustrates that base and collector may no longer be separated, and so we cannot calculate one scattering rate for an electron traversing the base and a separate scattering rate for the electron traversing the collector. For this nanoscale transistor, the electron cannot be considered as a classical point-particle; rather it is a *nonlocal*, truly quantum mechanical object. The electron is nonlocal in the sense that it exists in the base and collector at the same time.

Another complication arises from the fact that any description of electron transport in nanoscale structures should not violate Maxwell's equations. Thus, one needs to be able to deal with the possibility of large dynamic space-charging effects due to quantum mechanical reflection as, for example, occurs in a resonant tunnel diode. Not only can elastic quantum mechanical reflections from an abrupt change in potential strongly influence inelastic scattering rates, but dissipative processes, such as scattering from lattice vibrations, can also, in turn, modify quantum reflection and transmission rates in a type of scattering-driven feedback. Ultimately, what is needed is a new approach to describing electron transport in nanoscale structures. Conventional techniques such as classical Boltzmann and Monte Carlo methods are neither adequate nor valid ways to model nanoscale devices of the type we have been considering.

While it is important to know something about limitations to understanding electron transport in nanoscale devices, a detailed investigation of these and related issues is beyond the scope of this book and so will not be considered further. Rather, in the next section we return to the study of quantum mechanical tunneling through a potential barrier and the phenomenon of resonant tunneling.

4.7.2   Resonant tunneling between two quantum wells

If the potential barrier separating two identical one-dimensional potential wells is of sufficient strength to isolate the ground-state eigenfunctions, then the ground-state eigenenergies are degenerate. This situation is depicted schematically in Fig. 4.10(a)

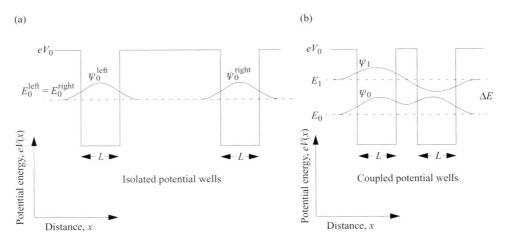

Fig. 4.10. (a) Two identical isolated potential wells with width L and barrier potential energy eV_0 have identical (degenerate) ground-state eigenenergies (indicated by the dotted line). The wave functions ψ_0^{left} and ψ_0^{right} associated with the left and right potential wells are symmetric and centered about their respective potential wells. (b) The lowest-energy wave functions for two identical potential wells coupled by a thin tunnel barrier are ψ_0 and ψ_1. The wave function ψ_0 is symmetric and has an eigenenergy E_0. The wave function ψ_1 is antisymmetric and has an eigenenergy E_1, which is greater in value than E_0.

for the case of rectangular potential wells each of width L and barrier energy eV_0. Each ground-state wave function ψ_0 is symmetric about its potential well and has an associated eigenenergy, $E_0^{\text{left}} = E_0^{\text{right}}$. The wave functions ψ_0^{left} and ψ_0^{right} are identical apart from a spatial translation.

Figure 4.10(b) is a sketch of what happens when the rectangular potential wells are no longer isolated by a strong potential barrier. Particle states originally associated with one well can interact via quantum mechanical tunneling with states in the other well. Now, the lowest-energy wave functions for two identical potential wells coupled by a thin tunnel barrier are ψ_0 and ψ_1. The wave function ψ_0 is symmetric and has an eigenenergy E_0. The wave function ψ_1 is antisymmetric and has an eigenenergy E_1, which is greater in value than E_0. The reason $E_0 < E_1$ is that the symmetric wave function ψ_0 has less integrated change in slope multiplied by the value of the wave function than the wave function ψ_1 (see Section 3.1.1).

A particle in eigenstate ψ_0 or ψ_1 can be found in either quantum well because the wave functions are spread out or *delocalized* across both potential wells. Tunneling through the central barrier is responsible for this.

The coupling between the original states via tunneling breaks degeneracy and causes splitting in energy levels $\Delta E = E_1 - E_0$. This effect is quite general in quantum mechanics and so is worth studying further.

We next consider the three barriers and two potential wells sketched in Fig. 4.11(a). As shown in Fig. 4.11(b), there are two resonances separated in energy by ΔE, one associated with an antisymmetric wave function and one with a symmetric wave function.

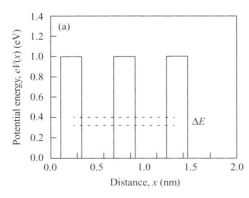 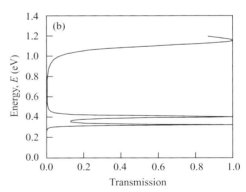

Fig. 4.11. (a) Symmetric triple rectangular potential barriers with barrier width 0.4 nm, well width 0.6 nm, and barrier energy 1 eV. For an electron of mass m_0 in such a potential, there are two resonant states separated in energy by energy ΔE. The energy of each resonance is indicated by the broken line. (b) Transmission probability of an electron mass m_0 incident from the left on the potential given in (a). There is an overall increase in background transmission with increasing energy, and unity transmission resonances occur at energy $E = 321.5$ meV ($\Gamma_{\text{FWHM}} = 12.2$ meV) and $E = 401.5$ meV ($\Gamma_{\text{FWHM}} = 20.8$ meV). The splitting in energy is $\Delta E = 80$ meV.

Each wave function may be viewed as formed from a linear combination of two lowest-energy symmetric wave function contributions for a single potential well, one from the left and one from the right potential well. These individual contributions are initially considered as degenerate bound states subsequently coupled via tunneling through the central potential barrier. The separation in energy between the two resonances ΔE depends upon the coupling strength through the central potential barrier. Again, the lowest-energy resonance in this double potential well is associated with a symmetric wave function, and the upper-energy resonance is associated with an antisymmetric wave function. The lowest-energy resonance has a smaller full width at half maximum Γ_{FWHM}, because the tunnel barrier seen by a particle is ΔE greater than the higher energy resonance.

In Fig. 4.11, the lowest-energy resonance peak is at 0.321 eV, and the upper-energy resonance is at 0.401 eV. The splitting in energy is $\Delta E = 80$ meV.

We now explore what happens to ΔE as the width of the central barrier decreases. This breaks a symmetry of the potential. According to our previous discussion, we expect coupling between the left and right potential wells to increase as the width of the central barrier decreases, and we also expect ΔE to increase. Indeed, as shown in Fig. 4.12, this is what happens. The lowest-energy resonance peak is at 0.272 eV, and the higher-energy resonance is at 0.456 eV. The energy splitting is now $\Delta E = 184$ meV. Notice that the energy splitting is greater than the previous example. Ultimately, as the thickness of the middle barrier vanishes, one anticipates obtaining the result for a single potential well of twice the original individual well width.

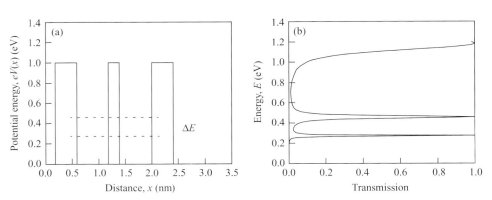

Fig. 4.12. (a) Symmetric triple rectangular potential barriers with outer barrier width 0.4 nm, central barrier width 0.2 nm, well width 0.6 nm, and barrier energy 1 eV. For an electron of mass m_0 in such a potential, there are two resonances separated in energy by energy ΔE. The energy of each resonance is indicated by the broken line. (b) Transmission probability of an electron of mass m_0 incident from the left on the potential given in (a). There is an overall increase in background transmission with increasing energy, and unity transmission resonances occur at energy $E = 271.9$ meV ($\Gamma_{FWHM} = 8.9$ meV) and $E = 455.9$ meV ($\Gamma_{FWHM} = 27.5$ meV). The splitting in energy is $\Delta E = 184$ meV.

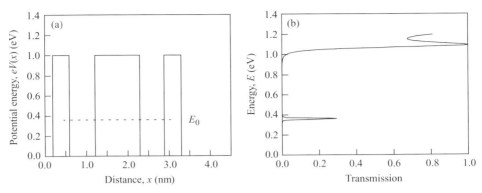

Fig. 4.13. (a) Symmetric triple rectangular potential barriers with outer barrier width 0.4 nm, central barrier width 1.1 nm, well width 0.6 nm, and barrier energy 1 eV. For an electron of mass m_0 in such a potential, there is a resonance at energy E_0, indicated by the broken line. (b) Transmission probability of an electron mass m_0 incident from the left on the potential given in (a). There is an overall increase in background transmission with increasing energy and nonunity transmission resonance at energy $E = E_0 = 360.2$ meV ($\Gamma_{FWHM} = 10.6$ meV). The value of transmission probability at energy E_0 is 0.308.

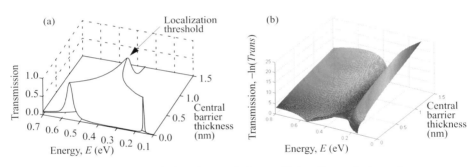

Fig. 4.14. (a) Three-dimensional plot of transmission coefficient for an electron of mass m_0 as a function of incident particle energy, E, and central barrier thickness for symmetric triple rectangular potential barriers. The outer barrier width is 0.4 nm, well width is 0.6 nm, and barrier energy is 1 eV. The localization threshold occurs when the central barrier thickness is 0.8 nm. (b) Three-dimensional plot of (a) with transmission plotted on a negative natural logarithmic scale.

Figure 4.13(b) is a plot of the transmission probability of an electron mass m_0 incident from the left on the symmetric triple rectangular potential barriers given in Fig. 4.11(a). There is an overall increase in background transmission with increasing energy and *nonunity* transmission resonance at energy $E = E_0 = 360.2$ meV.

Figure 4.14(a) is a three-dimensional plot of the transmission coefficient on a linear scale for an electron of mass m_0 as a function of both incident particle energy, E, and central barrier thickness for symmetric triple rectangular potential barriers. In Fig. 4.14(b) particle transmission is plotted on a negative natural logarithm scale. This means that unity transmission corresponds to zero and low transmission corresponds to a large number. More details in the transmission function can be seen using a logarithmic scale.

With increasing central barrier thickness, the coupling between degenerate bound states of the two wells decreases and the value of the energy splitting also decreases. Eventually, the energy splitting is so small that it disappears because it becomes smaller than the line width of an individual resonance ($\Delta E \leq \Gamma_{\mathrm{FWHM}}$). In the case we are considering, this happens when the central barrier is 0.8 nm thick. For central barriers thicker than this, the remaining resonant transmission peak has a value less than unity. Physically, one may think of the particle trapped or *localized* in one of the two potential wells with a line width determined by the escape probability via transmission through the thin barrier. The particle escapes by tunneling through the thin barrier much more readily than it is able to tunnel through the thick barrier. However, since transmission through the thick barrier is required to achieve resonant transmission via the degenerate bound states of the two potential wells, we may expect a *localization threshold* to occur when the central barrier is so thick that the energy splitting is less than the width in energy of the resonance. For central barrier thicknesses above the localization threshold, conduction of electrons by a transmission resonance is suppressed.

One can also break the symmetry of the potential by changing barrier energy. Figure 4.15(a) shows nonsymmetric triple rectangular potential barriers with barrier width 0.4 nm, well width 0.6 nm, left and central barrier energies 1 eV, and right barrier energy 0.6 eV. As indicated in Fig. 4.15(b), there is still a splitting in the resonance, but unity transmission is *not* obtained.

Figure 4.16 is a detail of the transmission peak showing that neither resonant peak of Fig. 4.15(b) has unity transmission.

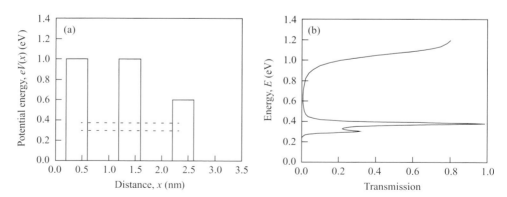

Fig. 4.15. (a) Nonsymmetric triple rectangular potential barriers with barrier width 0.4 nm, well width 0.6 nm, left and central barrier energies 1 eV, and right barrier energy 0.6 eV. (b) Transmission probability of an electron of mass m_0 incident from the left on the potential given in (a). There is an overall increase in background transmission with increasing energy and nonunity transmission resonance at two energies indicated by the broken lines.

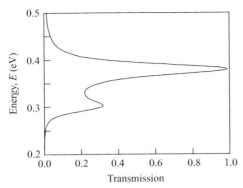

Fig. 4.16. Detail of transmission probability of an electron of mass m_0 incident from the left on the potential given in Fig. 4.15(a). There are nonunity transmission resonance peaks at two energies.

We now know quite a lot about resonant tunneling of particles such as electrons through various potentials. It is natural to ask if such structures can be applied for practical use. In fact, it is relatively straightforward to form band edge potential profiles similar to Fig. 4.7(a) using semiconductor heterostructures. Called resonant tunnel diodes, such devices were investigated in some detail during the late 1980s and early 1990s.[4]

Practical implementations using semiconductor structures require consideration of asymmetry in potential when voltage bias is applied across the device. This reduces the peak resonance transmission. In addition, for a device that passes electrical current, we need to consider space charging effects. If electron density in the potential well is great enough, it will distort the potential seen by an electron impinging on the barrier. For this reason, one must often calculate transmission probability self-consistently by simultaneously solving Schrödinger's equation for the electron wave functions and Poisson's equation for charge density. Further difficulties with design include the fact that resonant tunnel diodes must have low capacitance and low series resistance if they are to operate at high frequencies. This is particularly difficult to achieve in semiconductor heterostructures where tunnel barrier thickness is typically on a nanometer scale. For these and other reasons, interest in the resonant tunnel diode as a practical device has declined. However, one can still learn more about the nature of electron transport by studying such devices.

4.8 The potential barrier in the delta-function limit

In Section 4.6 the transmission and reflection of a particle incident on a rectangular potential were investigated. This type of potential barrier was used in Section 4.7

---

[4] For a survey see *Heterostructures and Quantum Devices*, eds. N. G. Einspruch and W. R. Frensley, Academic Press, San Diego, 1994 (ISBN 0 12 234124 4).

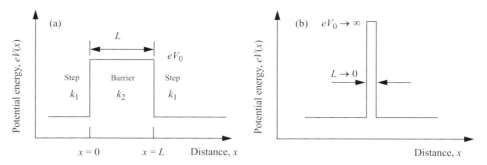

Fig. 4.17. (a) Sketch of a one-dimensional rectangular barrier of energy eV_0. The thickness of the barrier is L. A particle of mass m incident from the left of energy E has wave vector k_1. In the barrier region, the wave vector is k_2. The wave vectors k_1 and k_2 are related via $k_1^2 = k_2^2 + e2mV_0/\hbar^2$. (b) Potential energy as a function of position illustrating a one-dimensional rectangular potential barrier in the delta-function limit. An incident particle with wave vector k_1 and energy $E = \hbar^2 k_1^2/2m$ has wave vector ik_2 in the barrier.

to study resonant tunneling. Another important barrier we will explore is the delta-function potential. A delta-function potential energy barrier may be considered as the limit of a simultaneously infinite barrier energy and zero barrier width. The reason we are interested in this is that the results are quite simple and very easy to use. Our application will be to illustrate the behavior of electrons in periodic potentials that occur in crystals.

One way to obtain the delta-function limit is to consider the properties of the rectangular potential barrier of energy eV_0 and width L that we have been analyzing in this chapter. The delta-function limit can be reached if we allow $eV_0 \rightarrow \infty$ while $L \rightarrow 0$. This idea is illustrated in Fig. 4.17.

We have $eV_0 \rightarrow \infty$, $L \rightarrow 0$, so that $eV_0L \rightarrow constant$. Because $eV_0 \rightarrow \infty$, this means that $k_2 \rightarrow ik_2$ for all incident particle energies $E = \hbar^2 k_1^2/2m$.

Now $eV_0 = E - \hbar^2 k_2^2/2m$, and since $eV_0 \gg E$ one may write $eV_0 = -\hbar^2 k_2^2/2m$, or

$$k_2^2 L = -\frac{e2mV_0L}{\hbar^2} \tag{4.85}$$

From our previous work (Eqn (4.65))

$$p_{11} = \cos(k_2 L) - i\frac{\left(k_1^2 + k_2^2\right)}{2k_1 k_2} \sin(k_2 L) \tag{4.86}$$

which we can simplify in the limit $k_2 L \rightarrow 0$, so that $\cos(k_2 L) = 1 - k_2^2 L^2/2 + \cdots = 1$ and $\sin(k_2 L) = k_2 L - k_2^3 L^3/6 + \cdots = k_2 L$. This expansion allows us to rewrite Eqn (4.86) in the limit $k_2 L \rightarrow 0$ as

$$p_{11} = 1 - i\left(\frac{k_1^2}{2k_1 k_2} \cdot k_2 L + \frac{k_2^2}{2k_1 k_2} \cdot k_2 L\right) = 1 - i\left(\frac{k_1 L}{2} + \frac{k_2^2 L}{2k_1}\right) \tag{4.87}$$

Remembering that we are interested in $k_2 L \to 0$ and using Eqn (4.85)

$$p_{11} = 1 - i \left(\frac{k_1 L}{2} + \frac{k_2^2 L}{2k_1} \right) = 1 - i \left(0 - \frac{2em V_0 L}{2\hbar^2 k_1} \right) = 1 + i \frac{em V_0 L}{\hbar^2 k_1} \tag{4.88}$$

So that in the limit $k_2 L \to 0$ the expression for p_{11} is

$$\boxed{p_{11} = 1 + i \frac{k_0}{k_1}} \tag{4.89}$$

where $k_0 = me V_0 L / \hbar^2$ is a constant.

The transmission coefficient for a particle of mass m incident from the left on a delta function potential barrier is of a simple form that does *not* depend exponentially on the energy, E, of the incident particle

$$|C|^2 = \left| \frac{1}{p_{11}} \right|^2 = \left| \frac{1}{1 + i \dfrac{k_0}{k_1}} \right|^2 = \frac{E}{E + \dfrac{\hbar^2 k_0^2}{2m}} \tag{4.90}$$

where the electron wave vector $k_1 = \sqrt{2mE}/\hbar$. Equation (4.90) is plotted as a function of energy, E, in Fig. 4.18. In this particular case, the parameter k_0 is chosen so that $E_0 = \hbar^2 k_0^2 / 2m = 1$ eV. As may be seen in the figure, under these circumstances the particle has a transmission probability of 0.5 when particle energy is $E = E_0 = 1$ eV.

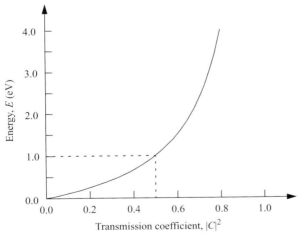

Fig. 4.18. Particle transmission probability for a delta-function barrier given by Eqn (4.90) plotted as a function of energy E. In this example, the value of k_0 is chosen so that $\hbar^2 k_0^2 / 2m = 1$ eV. In this case, the value of $|C|^2 = 0.5$ when $E = 1$ eV.

The other matrix element we need to find is p_{12}. We recall that previously we had from Eqn (4.60)

$$p_{12} = -i \frac{\left(k_2^2 + k_1^2\right)}{2k_1 k_2} \cdot \sin(k_2 L) \qquad (4.91)$$

so that in the limit $k_2 L \rightarrow 0$

$$\boxed{p_{12} = i \frac{k_0}{k_1}} \qquad (4.92)$$

Because we know from Eqn (4.32) and Eqn (4.33) that $p_{11} = p_{22}^*$ and $p_{12} = p_{21}^*$, we now have the complete expression for the propagation matrix $\hat{p}$ for a potential energy barrier in the delta-function limit.

4.9 Energy bands: the Kronig–Penney potential

So far, we have considered a single potential step, potential well, potential barrier, and resonant tunneling. In this section, we aim to use the propagation matrix method to explore what happens to electron transmission and reflection in a periodic delta-function potential. We will use this to approximate the potential seen by an electron in a crystal.

We proceed by first considering the impact the symmetry of a periodic potential has on electron wave functions. The first result we wish to introduce is Bloch's theorem. This theorem is important because it allows us to impose constraints on the type of wave functions that are allowed to exist in periodic potentials. Following this, in Section 4.9.2, we will show, using a model due to Kronig and Penney,[5] how transmission bands and band gaps are natural results of the presence of a periodic potential. We complete our study of periodic potentials by showing in Section 4.10 and Exercise 4.5 how the methods developed to describe electron motion in a crystal potential can be applied to other seemingly unrelated problems in engineering, such as the design of multi-layer dielectric coatings for optical mirrors.

4.9.1   Bloch's theorem

For simplicity, let's start by considering one-dimensional motion of an electron moving in free space. The electron is described by a wave function $\psi_{k_1}(x) = Ae^{i(k_1 x - \omega t)}$ and, as expected, the probability of finding the particle at any position in space is uniform. No particular location in space is more or less significant than any other location. The

[5] R. de L. Kronig and W. J. Penney, *Proc. Royal. Soc.* **A130**, 499 (1930).

underlying symmetry can be expressed by saying that probability is translationally invariant over all space.

We now introduce a periodic potential $V(x)$ in such a way that $V(x) = V(x + nL)$, where L is the minimum spatial period of the potential and n is an integer. Under these circumstances, it seems reasonable to expect that electron probability is modulated spatially by the same periodicity as the potential. The isotropic electron probability symmetry of free space is broken and replaced by a new probability symmetry which is translationally invariant over a space spanned by a unit cell size of L. We might describe electron probability that is the same in each unit cell by the function $|U(x)|^2$. Given this, it is clear that one is free to choose an electron wave function that is identical to $U_k(x)$ to within a phase factor e^{ikx}. To find the phase factor, we need to solve for the eigenstates of the one-electron Hamiltonian

$$H = \frac{-\hbar^2}{2m} \frac{d^2}{dx^2} + V(x) \tag{4.93}$$

where $V(x) = V(x + nL)$ for integer n. The electron wave function must be a *Bloch function* of the form

$$\boxed{\psi_k(x) = U_k(x)e^{ikx}} \tag{4.94}$$

where $U_k(x + nL) = U_k(x)$ has the same periodicity as the potential and n is an integer. The term e^{ikx} carries the phase information between unit cells via what is called the *Bloch wave vector k*.

If we wish to know how the value of an electron wave function changes from position x to position $x + L$, then we need to evaluate

$$\psi_k(x + L) = U_k(x + L)e^{ik(x+L)} \tag{4.95}$$

$$\psi_k(x + L) = U_k(x)e^{ikx} \cdot e^{ikL} \tag{4.96}$$

$$\boxed{\psi_k(x + L) = \psi_k(x) \cdot e^{ikL}} \tag{4.97}$$

Equations (4.94) and (4.97) are different ways of writing Bloch's theorem. In words, Bloch's theorem states that a potential with period L has wave functions that can be separated into a part with the same period as the potential and a plane-wave term e^{ikL}.

Equation (4.94) shows that electron probability $|\psi_k(x)|^2 = |U_k(x)|^2$ depends upon k but only contains the cell-periodic part of the wave function.

While one can see intuitively that Bloch's theorem must be correct, the consequences are quite dramatic when it comes to describing electron motion. Such motion in a periodic potential must involve propagation of the phase through the factor kx. Associated with wave vector k is a crystal momentum $\hbar k$. This momentum is not the same as the momentum of an actual electron because the $U_k(x)$ term in Eqn (4.94) means that the electron has a wide range of momenta. The Bloch wave function $\psi_k(x) = U_k(x)e^{ikx}$ is not an eigenfunction of the momentum operator $\hat{p} = -i\hbar\partial/\partial x$. Crystal momentum $\hbar k$ is an effective momentum of the electron and is an extremely useful way of describing electron motion in a periodic potential.

4.9.2   The propagation matrix applied to a periodic potential

We now move on to discuss electron motion in a periodic potential. The model we adopt makes use of the delta-function potential and Bloch's theorem which we developed in Section 4.8 and Section 4.9.1, respectively. Our model can be used as an approximation to the periodic coulomb potential which may be found in a crystal. Figure 4.19(a) is a sketch which represents the periodic potential seen by an electron due to a periodic array of atomic potentials. Figure 4.19(b) shows a schematic of the system we will consider. It is a one-dimensional periodic array of delta-function potential energy barriers with nearest-neighbor separation of L. This separation defines the unit cell of the periodic potential. Also shown in the figure are wave amplitudes A, B, C, and D for an electron scattering from a unit cell.

It follows from Bloch's theorem (Eqn (4.97)) that the coefficients A, B, C, and D are related to each other by a phase factor kL in such a way that

$$C = Ae^{ikL} \tag{4.98}$$

and

$$D = Be^{ikL} \tag{4.99}$$

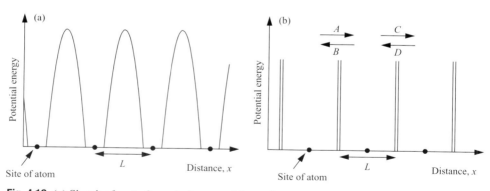

Fig. 4.19. (a) Sketch of part of a periodic potential seen by an electron due to an array of atomic potentials. The atom sites are indicated. The nearest-neighbor separation between barriers is L. (b) Plot of potential energy as a function of position, showing a periodic array of one-dimensional delta-function potential energy barriers. The nearest-neighbor separation between barriers is L. This defines the unit cell of the periodic potential. The potential is such that $V(x) = V(x + L)$, where L is the spatial period. This symmetry has a direct impact on the type of allowed eigenfunctions. The coefficients A, B, C, and D are the wave amplitudes for an electron scattering into and out of a given cell.

where k is the Bloch wave vector. Equations (4.98) and (4.99) may be expressed in matrix form as

$$\begin{bmatrix} A \\ B \end{bmatrix} = e^{-ikL} \begin{bmatrix} C \\ D \end{bmatrix} = \hat{p} \begin{bmatrix} C \\ D \end{bmatrix} \tag{4.100}$$

so that

$$\begin{bmatrix} p_{11} - e^{-ikL} & p_{12} \\ p_{21} & p_{22} - e^{-ikL} \end{bmatrix} \begin{bmatrix} C \\ D \end{bmatrix} = 0 \tag{4.101}$$

There is an interesting (nontrivial) solution to these linear homogeneous equations if the determinant of the 2×2 matrix in Eqn (4.101) vanishes – i.e.,

$$(p_{11} - e^{-ikL})(p_{22} - e^{-ikL}) - p_{12}p_{21} = 0 \tag{4.102}$$

$$p_{11}p_{22} + e^{-2ikL} - p_{22}e^{-ikL} - p_{11}e^{-ikL} - p_{12}p_{21} = 0 \tag{4.103}$$

Since $p_{11} = p_{22}^*$ and $p_{12} = p_{21}^*$, this can be rewritten

$$e^{-2ikL} - 2p_{11}e^{-ikL} = -p_{11}p_{22} + p_{12}p_{21} = -det(\hat{p}) \tag{4.104}$$

From current continuity $det(\hat{p}) = 1$ for real k (Eqn (4.47)). We now take the real part of both sides for terms in p:

$$e^{-2ikL} - 2\text{Re}(p_{11})e^{-ikL} = -1 \tag{4.105}$$

Noting that $e^{-ix} = \cos x - i \sin x$ allows us to write

$$\cos(2kL) - i\sin(2kL) - 2\text{Re}(p_{11})(\cos(kL) - i\sin(kL)) = -1 \tag{4.106}$$

Taking the imaginary part of this expression gives

$$\sin(2kL) - 2\text{Re}(p_{11})\sin(kL) = 0 \tag{4.107}$$

and, noting that $2\sin(x)\cos(y) = \sin(x+y) + \sin(x-y)$, if $x = y$ we can write $2\sin x \cos x = \sin(2x)$. Equation (4.107) then becomes

$$2\sin(kL)(\cos(kL) - \text{Re}(p_{11})) = 0 \tag{4.108}$$

Equations (4.107) and (4.108) require $\cos(kL) = \text{Re}(p_{11})$, and so k can only be real if

$$\boxed{|\text{Re}(p_{11})| \leq 1} \tag{4.109}$$

In general, this gives rise to bands of allowed real values of Bloch wave number k.

In the delta-function limit we had for a single barrier, $p_{11} = (1 + i(k_0/k_1))$ (Eqn (4.89)), where $k_0 = meV_0L/\hbar^2$ and k_1 is the wave vector outside the delta-function barrier.

In the Kronig–Penney model of a periodic potential, we have delta-function barriers separated by free-propagation regions of length L. The total propagation matrix for a cell of length L becomes

$$\hat{p} = \hat{p}_{\delta\text{-barrier}} \hat{p}_{\text{free}} = \begin{bmatrix} 1 + i\dfrac{k_0}{k_1} & i\dfrac{k_0}{k_1} \\ -i\dfrac{k_0}{k_1} & 1 - i\dfrac{k_0}{k_1} \end{bmatrix} \begin{bmatrix} e^{-ik_1 L} & 0 \\ 0 & e^{ik_1 L} \end{bmatrix} = \begin{bmatrix} p_{11} & p_{12} \\ p_{21} & p_{22} \end{bmatrix} \quad (4.110)$$

so the new p_{11} for the cell is

$$p_{11} = \left(1 + i\frac{k_0}{k_1}\right) \cdot e^{-ik_1 L} \quad (4.111)$$

We note that $e^{-ix} = \cos(x) - i\sin(x)$, so

$$p_{11} = \cos(k_1 L) - i\sin(k_1 L) + i\frac{k_0}{k_1}\cos(k_1 L) + \frac{k_0}{k_1}\sin(k_1 L) \quad (4.112)$$

Taking the real part gives

$$\text{Re}(p_{11}) = \cos(k_1 L) + \frac{k_0 L}{k_1 L} \cdot \sin(k_1 L) \quad (4.113)$$

Since we have the condition $|\text{Re}(p_{11})| \leq 1$ for allowed real values of k_1 (Eqn (4.108)) we may conclude that

$$-1 \leq \cos(k_1 L) + k_0 L \frac{\sin(k_1 L)}{k_1 L} \leq 1 \quad (4.114)$$

It is now possible to plot the function in Eqn (4.114) and establish regions of allowed real values of wave number k_1. This has been done in Fig. 4.20 for the case when $k_0 L = 20$.

The allowed bands are indicated by shaded regions. Notice that regions of nonpropagating states, which give rise to energy band gaps, become smaller with increasing values of $k_1 L$. The fact that the widths of forbidden bands decrease and the widths of the allowed bands increase with increasing $k_1 L$ is due mathematically to the decrease in the amplitude of the sine term. Also apparent in Fig. 4.20 is the fact that the boundary between the upper edge of allowed bands and the lower edge of forbidden ones occurs at values $k_1 L = n\pi$, where n is an integer $n = 1, 2, 3, \ldots$.

Previously we found that current continuity and real Bloch wave vector k in a periodic lattice expressed as Eqn (4.108) require

$$\cos(kL) = \text{Re}(p_{11}) \quad (4.115)$$

For the Kronig–Penney model, p_{11} is given by Eqn (4.111), and the real part of p_{11} is given by Eqn (4.113), so for real k_1 one may relate k to k_1 using the equation

$$\cos(kL) = \cos(k_1 L) + \frac{k_0}{k_1}\sin(k_1 L) \quad (4.116)$$

Figure 4.21 shows the dispersion relation $E_k(k)$ as a function of normalized Bloch wave vector kL for a Kronig–Penney model when $k_0 L = 80$ and $L = 0.15$ nm.

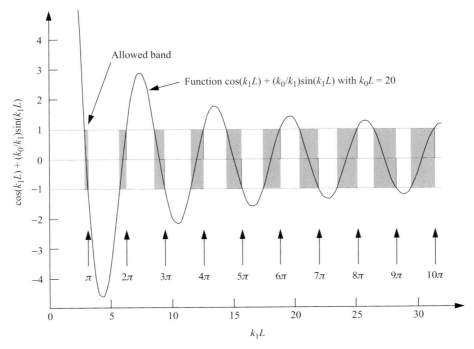

Fig. 4.20. Plot of the function $\cos(k_1 L) + (k_0/k_1) \sin(k_1 L)$ for $k_0 L = 20$. The allowed bands are indicated by shaded regions. Notice that regions of nonpropagating states (band gaps) get smaller with increasing values of $k_1 L$.

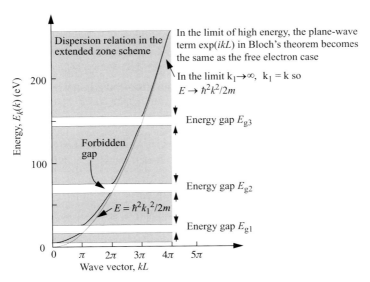

Fig. 4.21. Plot of dispersion relation in the extended zone representation. In this case, $k_0 L = 80$ with $L = 0.15$ nm.

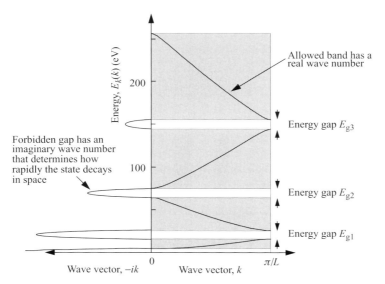

Fig. 4.22. Plot of dispersion relation in the reduced zone representation. Real wave vectors k have values from $k = 0$ to the Brillouin zone at $k = \pi/L$. The portions of energy and wave-vector space in which the wave vector can take on real values are indicated by the shaded regions. In regions where there is an energy gap, E_g, wave vectors take imaginary values. In this particular case, $k_0 L = 80$ with $L = 0.15$ nm.

Because the potential is periodic in L, we can redraw $E_k(k)$ as a function of Bloch wave vector k in the reduced zone. This is shown in Fig. 4.22.

The allowed energy bands of Fig. 4.21 may be thought of as arising from individual resonances of all the potential wells (of width L bounded by a delta-function potential energy barrier) that are coupled through tunneling. The finite lifetime of each resonance due to tunneling broadens the resonance energy. There are a large number of such resonances that overlap in energy to form a continuous allowed energy band.

Forbidden energy gap regions form where there are no resonances. At the boundary between the allowed and forbidden gap regions, wave functions have definite symmetry. Consider the energy gap region labelled E_{g1} in Fig. 4.21 which occurs at the Brillouin zone boundary $k = \pi/L$. The lower-energy wave function $\psi_{k=\pi/L}^{lower}$ with energy at the boundary between the allowed band and the forbidden band gap is an even function standing wave with respect to the delta-function potential. As illustrated in Fig. 4.23, the electron density associated with this wave function is a maximum between delta-function potential barriers. The upper-energy wave function $\psi_{k=\pi/L}^{upper}$ with energy at the boundary between the allowed band and forbidden band gap is an odd function standing wave with respect to the delta-function potential. The electron density associated with this wave function is a maximum at the delta-function potential barriers. The reason for the difference in energy of these states is easy to understand. $\psi_{k=\pi/L}^{upper}(x)$ has a significant portion of the wave function overlap with the delta-function potential, while $\psi_{k=\pi/L}^{lower}(x)$

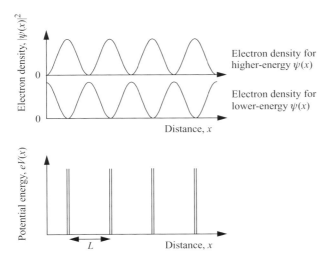

Fig. 4.23. Illustration of electron density at the Brillouin zone boundary $k = \pi/L$ of a periodic delta-function potential $V(x)$ showing states with high and low energy.

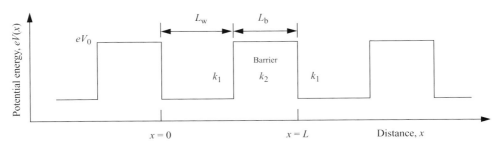

Fig. 4.24. Periodic array of rectangular potential barriers with energy eV_0 and width L_b. The potential wells have width L_w, and the cell repeats in distance $L = L_b + L_w$.

does not. Hence, the electron energy associated with $\psi_{k=\pi/L}^{\text{upper}}(x)$ picks up more energy from this overlap.

The presence of periodic potential barriers changes electron energy. If the electron state is such that it has a node at the position of the delta-function potential, it does not feel the presence of the potential and should have an energy that is the same as that of a free electron. This is the case at the bottom of the band gap ($\psi_{k=\pi/L}^{\text{lower}}(x)$), which is why in Fig. 4.21 the allowed band touches the parabolic free-electron dispersion relation at this point. The opposite is true for $\psi_{k=\pi/L}^{\text{upper}}(x)$, which has an antinode at the delta-function potential and so has energy greater than predicted by free-electron dispersion.

As a slight extension of the model we have used to describe particle motion in a periodic potential, we can consider the case of a periodic array of rectangular potential barriers of the type sketched in Fig. 4.24. This is the same as our previous Kronig–Penney model, except that the delta-function potential barriers have been substituted for rectangular potential barriers. As before, all we need to do to obtain the dispersion relation for particle motion in such a potential is to find propagation matrix $\hat{P}$ for the cell, evaluate the real part of p_{11}, and apply the conditions given by Eqn (4.109) and Eqn (4.115).

In this case, the total propagation matrix $\hat{P}$ for a cell of length L with barrier width L_b and well width L_w is the same as Eqn (4.58) but with an extra 2×2 matrix to describe free propagation in the well. Hence,

$$\hat{P} = \frac{1}{4k_1 k_2} \begin{bmatrix} e^{-ik_1 L_w} & 0 \\ 0 & e^{ik_1 L_w} \end{bmatrix} \begin{bmatrix} k_1 + k_2 & k_1 - k_2 \\ k_1 - k_2 & k_1 + k_2 \end{bmatrix} \begin{bmatrix} e^{-ik_2 L_b} & 0 \\ 0 & e^{-ik_2 L_b} \end{bmatrix}$$
$$\times \begin{bmatrix} k_2 + k_1 & k_2 - k_1 \\ k_2 - k_1 & k_2 + k_1 \end{bmatrix} \tag{4.117}$$

For particle energy $E > eV_0$, this gives

$$p_{11} = \left(\cos(k_2 L_b) - i \frac{(k_2^2 + k_1^2)}{2k_1 k_2} \sin(k_2 L_b) \right) e^{-ik_1 L_w} \tag{4.118}$$

and since $e^{-ik_1 L_w} = \cos(k_1 L_w) - i \sin(k_1 L_w)$, it follows that

$$\text{Re}(p_{11}) = \cos(k_2 L_b) \cos(k_1 L_w) - \frac{(k_2^2 + k_1^2)}{2k_1 k_2} \sin(k_2 L_b) \sin(k_1 L_w) \tag{4.119}$$

When $E < eV_0$, then $k_2 \to ik_2$, and Eqn (4.119) becomes

$$\text{Re}(p_{11}) = \cosh(k_2 L_b) \cos(k_1 L_w) + \frac{(k_2^2 - k_1^2)}{2k_1 k_2} \sinh(k_2 L_b) \sin(k_1 L_w) \tag{4.120}$$

The dispersion relation can then be found using Eqn (4.115).

As expected, in the limit $L_b \to 0$, as $eV_0 \to \infty$ (while keeping $k_0 = meV_0 L_b / \hbar$ a constant), Eqn (4.120) reduces to the result for the delta-function potential barrier given by Eqn (4.113).

4.9.3   Crystal momentum and effective electron mass

The presence of a periodic potential changes the way an electron moves. There are bands of energy described by the dispersion relation $\omega = \omega(k)$ in which an electron wave packet can propagate freely at velocity

$$v_g = \frac{\partial}{\partial k} \omega(k) \tag{4.121}$$

There are also forbidden bands of energy, or energy band gaps, where the electron cannot propagate freely. Clearly, the presence of a periodic potential has a dramatic influence on the way in which one describes the motion of electrons in a crystal.

The influence a periodic potential $V(x)$ has on the response of an electron to an external force requires the introduction of *crystal momentum*. To show this, we consider the effect of an external electric field $\mathbf{E}_x$ on the x component of electron motion. Our familiarity with classical mechanics suggests that an electron subject to an external force $-e\mathbf{E}_x$ causes a change in its momentum. To solve for electron motion in quantum mechanics, we start by writing down the total Hamiltonian:

$$H = \frac{\hat{p}_x^2}{2m} + V(x) + e\mathbf{E}_x x \tag{4.122}$$

The time-dependent Schrödinger equation (Eqn (2.39) and Eqn (2.40)) describing the electron wave function $\psi(x, t)$ is just

$$H\psi(x, t) = i\hbar \frac{\partial}{\partial k} \psi(x, t) \tag{4.123}$$

For an electron in state $\psi_0(x, t = 0)$ at time $t = 0$ we may rewrite Eqn (4.123) as

$$\psi(x, t) = e^{-iHt/\hbar} \psi_0(x, t = 0) \tag{4.124}$$

which for the Hamiltonian we are considering is

$$\psi(x, t) = e^{-i(\hat{p}_x^2/2m + V(x) + e\mathbf{E}_x x)t/\hbar} \psi_0(x, t = 0) \tag{4.125}$$

Replacing x with $(x + L)$ gives

$$\psi((x + L), t) = e^{-i(\hat{p}_x^2/2m + V(x+L) + e\mathbf{E}_x(x+L))t/\hbar} \psi_0((x + L), t = 0) \tag{4.126}$$

Using the fact that $V(x) = V(x + L)$ for a periodic potential, and using Bloch's theorem for an electron in a Bloch state, we can rewrite this as

$$\psi((x + L), t) = e^{-i(\hat{p}_x^2/2m + V(x) + e\mathbf{E}_x x)t/\hbar} e^{-ie\mathbf{E}_x Lt/\hbar} e^{ik(t=0)L} \psi_0(x, t = 0) \tag{4.127}$$

and hence

$$\psi((x + L), t) = e^{-ik(t)L} \psi_0(x, t = 0) \tag{4.128}$$

where

$$k(t) = -\frac{e\mathbf{E}_x t}{\hbar} + k(t = 0) \tag{4.129}$$

The time derivative of Eqn (4.129) can be written as

$$\frac{d}{dt}(\hbar k) = -e\mathbf{E}_x \tag{4.130}$$

One may now conclude that the effect of an external force is to change a quantity $\hbar k$. This is called the *crystal momentum*. Electrons move according to the rate of change of crystal momentum in the periodic potential.

In analogy with classical mechanics, we expect the acceleration of a particle in a given allowed band due to an external force to be described by

$$\frac{d}{dt}\left(\frac{d}{dt}x\right) = \frac{d}{dt}\left(\frac{\partial}{\partial k}\omega(k)\right) = \frac{\partial^2}{\partial k^2}\omega(k)\frac{dk}{dt} = \frac{1}{\hbar}\left(\frac{\partial^2}{\partial k^2}\omega(k)\right)\frac{d}{dt}(\hbar k) \tag{4.131}$$

$$\frac{d}{dt}\left(\frac{d}{dt}x\right) = \frac{1}{\hbar}\left(\frac{\partial^2}{\partial k^2}\omega(k)\right)(-e\mathbf{E}_x) \tag{4.132}$$

where we have used Eqn (4.121) and Eqn (4.130).

The result given by Eqn (4.131) can be expressed in the usual Newtonian form of force equals mass times acceleration if we introduce an effective mass $m^*(k)$ for the particle so that

$$m^*(k) = \frac{\hbar}{\dfrac{\partial^2}{\partial k^2}\omega(k)} \qquad\qquad (4.133)$$

Equation (4.133) indicates that effective mass is inversely proportional to the curvature of the dispersion relation for a given allowed band.

Usually, when discussing effective electron mass, a quantity $m_{\text{eff}}(k)$ is used, which is the effective electron mass normalized with respect to the bare electron mass m_0, so that

$$m^*(k) = m_{\text{eff}}(k) \times m_0 \qquad\qquad (4.134)$$

For an electron in the conduction band of GaAs, one often uses $m_{\text{eff}} = 0.07$, so the effective electron mass is $m^* = 0.07 \times m_0$.

To illustrate how a dispersion relation influences particle group velocity and effective particle mass, let's assume that a dispersion relation for a particular allowed band is described by a cosine function of wave vector k in such a way that

$$\omega(k) = \frac{E_{\text{b}}}{2\hbar}(1 - \cos(kL)) = \frac{E_{\text{b}}}{\hbar}\sin^2\left(\frac{kL}{2}\right) \qquad\qquad (4.135)$$

where E_{b} is the energy band width. Using this prototype dispersion relation one obtains a group velocity

$$v_{\text{g}}(k) = \frac{\partial}{\partial k}\omega(k) = \frac{E_{\text{b}}L}{2\hbar}\sin(kL) \qquad\qquad (4.136)$$

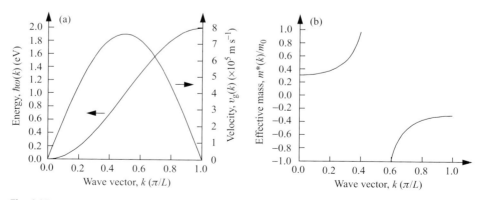

Fig. 4.25. (a) Electron dispersion relation $\omega(k) = E_{\text{b}}(1 - \cos(kL))/2\hbar$ and group velocity $v_{\text{g}}(k)$ as a function of wave vector k. Band width $E_{\text{b}} = 2$ eV, and lattice constant $L = 0.5$ nm. (b) Effective electron mass $m^*(k)$ as a function of wave vector k. The free electron mass is m_0.

and an effective particle mass

$$m^*(k) = \cfrac{\hbar}{m_0 \cfrac{\partial^2}{\partial k^2}\omega(k)} = \frac{2\hbar^2}{m_0 E_b L^2 \cos(kL)} \tag{4.137}$$

Equations (4.135), (4.136), and (4.137) are plotted in Fig. 4.25, assuming an electron band width $E_b = 2$ eV and a periodic potential lattice constant $L = 0.5$ nm. A quite remarkable prediction evident in Fig. 4.25(b) is the negative effective electron mass $m^*(k)$ for certain values of wave vector k. The physical meaning of this can be seen in Fig. 4.25(a) – electron velocity $v_g(k)$ can decrease with increasing wave vector k.

In fact, there are many interesting aspects to electron motion in a periodic potential. We have only touched on a few of them here.

4.9.3.1 The band structure of GaAs

So far we have considered a relatively simple model of a periodic potential. While the electron dispersion relations of actual crystalline solids are more complex, they do retain the important features, such as band gaps, established by our model. Figure 4.26

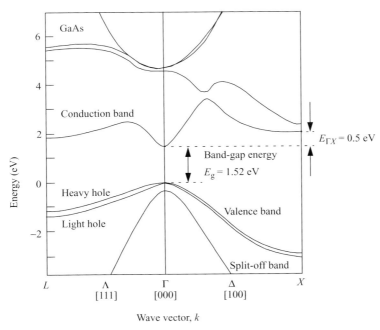

Fig. 4.26. Calculated low-temperature band structure of GaAs from the Γ symmetry point toward the L and X crystal symmetry points. The conduction band, heavy-hole band, light-hole band, and split-off band are indicated. The effective electron mass near the Γ symmetry point is approximately $m_e = 0.07 \times m_0$. The heavy-hole mass is $m_{hh} = 0.5 \times m_0$, and the light-hole mass is $m_{lh} = 0.08 \times m_0$. The band gap energy $E_g = 1.52$ eV is indicated, as is the energy $E_{\Gamma X}$ separating the minimum of the conduction band at Γ and the subsidiary minimum near the X symmetry point.

shows part of the calculated electron dispersion relation of GaAs along two of the principal symmetry directions, $\Gamma-X$ (or [100] direction, indicated by Δ) and $\Gamma-L$ (or [111] direction, indicated by Λ). GaAs has the zinc blende crystal structure with a low-temperature lattice constant of $L = 0.565$ nm. There is interest in GaAs and other III-V compound semiconductors[6] because they can be used to make laser diodes and high-speed transistors.

As may be seen in the dispersion relation of Fig. 4.26, there is a conduction band and a valence band separated in energy by a band gap. At temperatures close to 0 K, the band-gap energy is $E_g^{0\,K} = 1.52$ eV. At room temperature, this value reduces to $E_g^{300\,K} = 1.42$ eV, in part because the average spacing between the atoms in the crystal increases with elevated temperature.

4.10 Other engineering applications

The propagation matrix method can be used in a number of other engineering applications. By way of example, the problem to which we are going to apply the matrix method involves the propagation of a one-dimensional electromagnetic wave through an inhomogeneous dielectric medium.

At the beginning of this chapter we defined the propagation matrix via

$$\begin{bmatrix} A \\ B \end{bmatrix} = \hat{P} \begin{bmatrix} C \\ D \end{bmatrix}$$

where $\hat{P} = \hat{p}_1 \hat{p}_2 \ldots \hat{p}_i \ldots \hat{p}_N$ and

$$\hat{p}_i = \frac{1}{2} \begin{bmatrix} \left(1 + \dfrac{k_{j+1}}{k_j}\right)e^{-ik_j L_j} & \left(1 - \dfrac{k_{j+1}}{k_j}\right)e^{-ik_j L_j} \\ \left(1 - \dfrac{k_{j+1}}{k_j}\right)e^{ik_j L_j} & \left(1 + \dfrac{k_{j+1}}{k_j}\right)e^{ik_j L_j} \end{bmatrix} \tag{4.138}$$

For the optical problem $V(x) \to n_r(x)$, where $n_r(x)$ is the refractive index. As an example, consider an electromagnetic wave of wavelength $\lambda_{vac} = 1$ μm propagating in free space and incident on a loss-less dielectric with relative permittivity $\varepsilon_r > 1$ and relative permeability $\mu_r = 1$. The refractive index for vacuum is $n_{vac} = 1$, and for the dielectric it is $n_r = \sqrt{\varepsilon_r} > 1$. Example values for a dielectric refractive index are 1.45 for SiO_2 and 3.55 for Si. To find the transmission and reflection of the electromagnetic wave, we may use the propagation matrix of Eqn (4.138) directly, only now $k_j \to 2\pi/\lambda_j$,

---

[6] For additional information on the electronic band structure of semiconductors see M. L. Cohen and J. R. Chelikowsky, *Electronic Structure and Optical Properties of Semiconductors*, Spinger Series in Solid-State Science **75**, ed. M. Cardona, Springer-Verlag, New York, 1989 (ISBN 0 387 51391 4).

where $\lambda_j = \lambda_{\mathrm{vac}}/n_{r,j}$ and λ_{vac} is the wavelength in vacuum. Hence,

$$
\hat{p}_i = \frac{1}{2}
\begin{bmatrix}
\left(1 + \dfrac{\lambda_j}{\lambda_{j+1}}\right) e^{-i(2\pi L_j)/\lambda_j} & \left(1 - \dfrac{\lambda_j}{\lambda_{j+1}}\right) e^{-i(2\pi L_j)/\lambda_j} \\[2ex]
\left(1 - \dfrac{\lambda_j}{\lambda_{j+1}}\right) e^{i(2\pi L_j)/\lambda_j} & \left(1 + \dfrac{\lambda_j}{\lambda_{j+1}}\right) e^{i(2\pi L_j)/\lambda_j}
\end{bmatrix}
\tag{4.139}
$$

This may be rewritten as

$$
\hat{p}_i = \frac{1}{2}
\begin{bmatrix}
\left(1 + \dfrac{n_{r,j+1}}{n_{r,j}}\right) e^{-i(2\pi n_{r,j} L_j)/\lambda_{\mathrm{vac}}} & \left(1 - \dfrac{n_{r,j+1}}{n_{r,j}}\right) e^{-i(2\pi n_{r,j} L_j)/\lambda_{\mathrm{vac}}} \\[2ex]
\left(1 - \dfrac{n_{r,j+1}}{n_{r,j}}\right) e^{i(2\pi n_{r,j} L_j)/\lambda_{\mathrm{vac}}} & \left(1 + \dfrac{n_{r,j+1}}{n_{r,j}}\right) e^{i(2\pi n_{r,j} L_j)/\lambda_{\mathrm{vac}}}
\end{bmatrix}
\tag{4.140}
$$

If the electromagnetic wave is incident from the left, we know that $A = 1$, and, assuming no reflection at the far right, this gives $D = 0$. Thus,

$$
\begin{bmatrix} A \\ B \end{bmatrix} = \hat{P} \begin{bmatrix} C \\ D \end{bmatrix}
$$

becomes

$$
\begin{bmatrix} 1 \\ B \end{bmatrix} = \begin{bmatrix} p_{11} & p_{12} \\ p_{21} & p_{22} \end{bmatrix} \begin{bmatrix} C \\ 0 \end{bmatrix}
$$

The transmission coefficient is obtained from $1 = p_{11}C$ and the reflection coefficient from $B = p_{21}C$. Hence, transmission intensity is $|C|^2 = |1/p_{11}|^2$ and reflected intensity is $|B|^2 = |p_{21}/p_{11}|^2$.

From the preceding, it follows that reflected electromagnetic power from a step change in refractive index for a plane wave incident from vacuum is

$$
r = |B|^2 = \left| \frac{k_{\mathrm{vac}} - k_1}{k_{\mathrm{vac}} + k_1} \right|^2 = \left| \frac{1 - k_1/k_{\mathrm{vac}}}{1 + k_1/k_{\mathrm{vac}}} \right|^2 = \left(\frac{1 - n_r}{1 + n_r} \right)^2
\tag{4.141}
$$

This is just the well-known result from standard classical optics. The reflection in optical power r may be thought of as due to a velocity mismatch. To show this, we recall that the velocity of light v_j in an isotropic medium of refractive index $n_{r,j}$ is just $v_j = c/n_{r,j}$. Hence, reflection due to a step change in refractive index from $n_{r,j}$ to $n_{r,j+1}$ is

$$
r = \left| \frac{k_j - k_{j+1}}{k_j + k_{j+1}} \right|^2 = \left| \frac{n_{r,j} - n_{r,j+1}}{n_{r,j} + n_{r,j+1}} \right|^2 = \left| \frac{\dfrac{1}{v_j} - \dfrac{1}{v_{j+1}}}{\dfrac{1}{v_j} + \dfrac{1}{v_{j+1}}} \right|^2 = \left| \frac{v_{j+1} - v_j}{v_{j+1} + v_j} \right|^2
\tag{4.142}
$$

This is the same result we obtained in Section 3.8 for transmission of an electron energy E over a potential step V_0 where $E > eV_0$.

We conclude this section by noting that one may use the propagation matrix method to calculate not only transmission and reflection from a single dielectric step but also much more complex spatial variation in refractive index, such as occurs in multi-layer dielectric optical coatings. Exercise 4.5 makes use of the propagation matrix method to design multi-layer dielectric mirrors for a laser.

4.11 The WKB approximation

The method used in this chapter is not the only approach to solving transmission and reflection of particles due to spatial changes in potential. By way of example, in this section we describe what is known as the WKB approximation. In their papers published in 1926, Wentzel, Kramers, and Brillouin introduced into quantum mechanics what is in essence a semiclassical method.[7] In fact, their basic approach to solving differential equations had already been introduced years earlier, in 1837, by Liouville.[8]

The underlying idea may be explained by noting that a particle of energy $E = \hbar\omega$ moving in free space has a wave function $\psi(\mathbf{r}, t) = Ae^{i(\mathbf{k}\cdot\mathbf{r}-\omega t)}$ for which the wave vector $\mathbf{k}$ does not vary. If the particle now encounters a potential and the potential is slowly varying, then one should be able to obtain a *local* value of $\mathbf{k}$ from the local kinetic energy. For a potential that is varying slowly enough, this is usually a good approximation, and for $E > eV$ one may write $k(\mathbf{r}) = \sqrt{2m(E - V(\mathbf{r}))}/\hbar$. Of course, the phase of the wave function must also change from $\mathbf{k}\cdot\mathbf{r}$ to an integral $\int \mathbf{k}(\mathbf{r})\cdot d\mathbf{r}$ to take into account the fact that we assume a local wave vector that is a function of space.

To see how this works in practice, consider a particle moving in one dimension. We assume a slowly and smoothly varying potential $V(x)$ so that we can make the approximation $k = k(x)$ and integrate in the exponent of the wave function. In this semiclassical approximation, the spatial part of the wave function is of the form

$$\psi(x) = \frac{a}{\sqrt{k(x)}} \cdot e^{i\int^x k(x')dx'} \tag{4.143}$$

and

$$k(x) = \sqrt{\frac{2m(E - eV(x))}{\hbar}} \qquad \text{for } E > eV \tag{4.144}$$

$$k(x) = i\sqrt{\frac{2m(eV(x) - E)}{\hbar}} \qquad \text{for } E < eV \tag{4.145}$$

---

[7] G. Wentzel, *Z. Physik.* **38**, 518 (1926), H. A. Kramers, *Z. Physik.* **39**, 828 (1926), and L. Brillouin, *Compt. Rend.* **183**, 24 (1926).

[8] J. Liouville, *J. de Math.* **2**, 16, 418 (1837).

In general, there are left- and right-propagating waves so that

$$\psi_{WKB}(x) = \frac{1}{\sqrt{k(x)}}\left(Ae^{i\int^x k(x')dx'} + Be^{-i\int^x k(x')dx'}\right) \tag{4.146}$$

The accuracy of the WKB method relies on the fractional change in wave vector k_x being very much less than unity over the distance $\lambda/4\pi$. Obviously, at the classical turning points, where $E = eV(x)$ and $k(x) \to 0$, the particle wavelength is infinite. The way around this problem typically is achieved by using an appropriate connection formula.

To understand more about the WKB approach, we next explore the limiting case of low tunneling probability through an almost-opaque high-energy potential barrier.

4.11.1   Tunneling through a high-energy barrier of finite width

Consider the potential for a rectangular barrier of energy eV_0 and width L depicted in Fig. 4.27. Previously, we had a transmission coefficient for a particle of energy E, mass m, through a rectangular barrier energy V_0, which is given by Eqn (4.77).

$$Trans(E < eV_0) = \frac{1}{1 + \frac{1}{4} \cdot \frac{eV_0^2}{E(eV_0 - E)}\sinh^2(k_2 L)} \tag{4.147}$$

where $k_2^2 = (2m(eV_0 - E))/\hbar^2$ for a rectangular barrier.

Suppose the barrier is of high energy and finite width L, so that it is almost opaque to particle transmission. Then one may write

$$k_2 L = \sqrt{\frac{2m(eV_0 - E)}{\hbar^2}} \cdot L \gg 1 \tag{4.148}$$

We can now approximate the sinh term in Eqn (4.147) by

$$\sinh^2(k_2 L) = \left(\frac{1}{2}\left(e^{k_2 L} - e^{-k_2 L}\right)\right)^2 \sim \frac{1}{4}e^{2k_2 L} \tag{4.149}$$

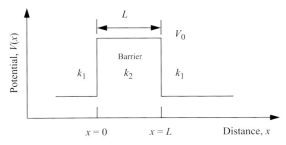

Fig. 4.27. Sketch of the potential of a one-dimensional, rectangular barrier of energy eV_0. The thickness of the barrier is L. A particle of mass m, incident from the left, of energy E, has wave vector k_1. In the barrier region the wave vector is k_2.

so that the transmission probability becomes

$$Trans \cong \cfrac{1}{1 + \cfrac{1}{16} \cdot \cfrac{eV_0^2}{E(eV_0 - E)} e^{2k_2 L}} \tag{4.150}$$

$$\boxed{Trans \cong 16 \cdot \frac{E(eV_0 - E)}{eV_0} e^{-2k_2 L}} \tag{4.151}$$

Notice that our approximations led us to an expression for transmission probability that has no backward-traveling component, thereby violating conditions for current flow established in Exercise 3.1(c).

The result given by Eqn (4.151) was obtained for the nearly opaque, rectangular potential barrier. Generalizing to a nearly opaque potential barrier of arbitrary shape, such as that shown schematically in Fig. 4.28, we adopt the WKB approximation. In this case, the expression for transmission becomes

$$\boxed{Trans \sim e^{-2 \int_{x_1}^{x_2} \sqrt{\frac{2m(eV(x) - E)}{\hbar^2}} dx}} \tag{4.152}$$

The physical meaning of this is that eV_0 has been replaced by a weighted "average" barrier energy eV_{av}, so that

$$k_2 L = \int_{x_1}^{x_2} \sqrt{\frac{2m(eV(x) - E)}{\hbar^2}} dx = \sqrt{\frac{2m}{\hbar^2}(eV_{av} - E)} \cdot L \tag{4.153}$$

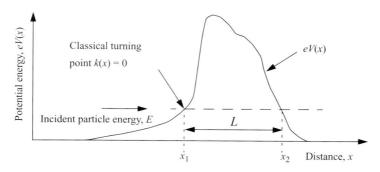

Fig. 4.28. Plot of potential energy as a function of position showing a one-dimensional barrier of energy $eV(x)$. An electron of energy E, mass m, incident on the potential barrier has classical turning points at positions x_1 and x_2. The distance an electron tunnels through such a barrier is $L = x_2 - x_1$.

4.12 Example exercises

Exercise 4.1

(a) Show for a rectangular potential barrier of energy eV_0 and thickness L that transmission probability for a particle of mass m and energy $E \geq eV_0$ moving normally to the barrier

$$Trans(E \geq eV_0) = \left(1 + \frac{1}{4}\left(\frac{e^2 V_0^2}{E(E - eV_0)} \right) \sin^2(k_2 L) \right)^{-1}$$

may be written as

$$Trans = \left(1 + \frac{1}{4} \cdot \frac{b^2}{\left(b + k_2^2 L^2 \right)} \cdot \frac{\sin^2(k_2 L)}{k_2^2 L^2} \right)^{-1}$$

where parameter $b = 2k_0 L = 2m e V_0 L^2 / \hbar^2$.

(b) Show that the function $Trans$ in (a) has a minimum when

$$Trans_{min} = 1 - \frac{b^2}{\left(2k_2^2 L^2 + b \right)^2}$$

and that this may be rewritten as

$$Trans_{min} = 1 - \frac{e^2 V_0^2}{(2E - eV_0)^2}$$

Exercise 4.2

Write a computer program in f77, MATLAB, c^{++}, or similar software that uses the propagation matrix method to find the transmission resonances of a particle of mass $m = 0.07 \times m_0$ (where m_0 is the bare electron mass) in the indicated one-dimensional potentials.

(a) A double barrier potential with the indicated barrier energy and widths.

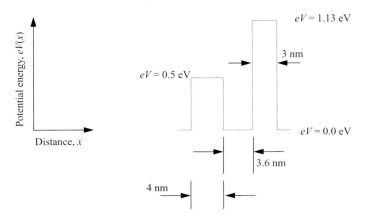

(b) A parabolic potential with $V(x) = (x^2 / L^2)$ eV for $|x| \leq L = 5$ nm and $V(x) = 0$ eV for $|x| > L$.

What happens to the energy levels in problems (a) and (b) if $m = 0.14 \, m_0$?

Your results should include: (i) a printout of the computer program you used; (ii) a computer-generated plot of the potential; (iii) a computer-generated plot of particle transmission as a function of incident energy for a particle incident from the left (you may find it useful to plot this on a natural log scale); (iv) a list of energy level values and resonant line widths.

Exercise 4.3

(a) Use your computer program developed in Exercise 4.2 to find transmission as a function of energy for a particle mass m_0 through 12 identical one-dimensional potential barriers each of energy 10 eV, width 0.1 nm, sequentially placed every 0.5 nm (so that the potential well between each barrier has width 0.4 nm). What are the allowed (band) and disallowed (band-gap) ranges of energy for particle transmission through the structure? How do you expect the velocity of the transmitted particle to vary as a function of energy?

(b) How do these bands compare with the situation in which there are only three barriers, each with 10 eV barrier energy, 0.1 nm barrier width, and 0.4 nm well width?

Exercise 4.4

Analyze the following problem: A particle of mass $m^* = 0.07 \times m_0$, where m_0 is the free-electron mass, has energy in the range $0 < E \leq 3$ eV and is incident on a potential $V(x) = 2 \sin^2(n 2\pi x / L)$ eV, for $0 \leq x \leq L$ and $V(x) = 0$ eV elsewhere. In the expression for the potential, n is a nonzero positive integer, and we decide to set $L = 20$ nm. What is the transmission probability of a particle incident from the left for the case in which: (a) $n = 1$, and (b) $n = 4$? How are your results altered if $m^* = 0.14 \, m_0$? What happens if the potential is distorted by an additional term $V'(x)$, where $V'(x) = V_0(1 - ((2x/L) - 1)^2)$ eV for $0 \leq x \leq L$ and $V'(x) = 0$ eV elsewhere, so that the total potential is now $V(x) = 2 \sin^2(n 2\pi x / L) eV + V'(x)$? Analyze this problem for different values of the prefactor, V_0. In particular, solve for the cases $V_0 = 0.1$ eV and $V_0 = 2.0$ eV when $n = 4$.

Your results should include: (i) a printout of the computer program you used; (ii) a computer-generated plot of the potential; (iii) a computer-generated plot of particle transmission as a function of incident energy for a particle incident from the left.

Exercise 4.5

(a) Use the propagation matrix to design a high-reflectivity Bragg mirror for electromagnetic radiation with center wavelength $\lambda_0 = 980$ nm incident normal to the surface of an AlAs/GaAs periodic dielectric layer stack consisting of 25 indentical layer-pairs. See following figure. Each individual dielectric layer has a thickness $\lambda/4n$, with n being the refractive index of the dielectric. Use $n_{AlAs} = 3.0$ for the refractive index of AlAs and $n_{GaAs} = 3.5$ for that of GaAs. Calculate and plot optical reflectivity in the wavelength range 900 nm $< \lambda <$ 1100 nm.

(b) Extend the design of your Bragg reflector to a two-mirror structure similar to that used in the design of a vertical-cavity surface-emitting laser (VCSEL). See following figure. This may be achieved by increasing the number of pairs to 50 and making the thickness of the central GaAs layer one wavelength long. Recalculate and plot the reflectivity over the same wavelength range as in (a). Use high wavelength resolution to find the band width of this optical pass band filter near $\lambda = 980$ nm.

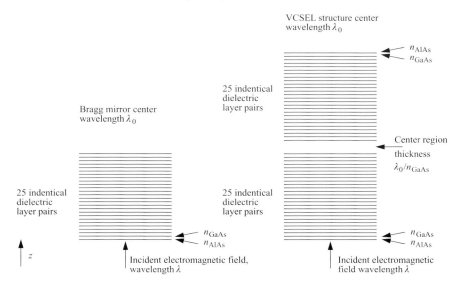

Your results should include a printout of the computer program you used and a computer-generated plot of particle transmission as a function of incident wavelength.

Exercise 4.6

Show that the Bloch wave function $\psi_k(x) = U_k(x)e^{ikx}$ is not an eigenfunction of the momentum operator $\hat{p} = -i\hbar \cdot \partial/\partial x$.

Exercise 4.7

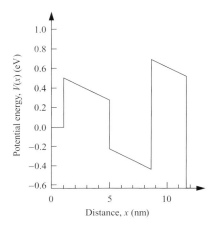

(a) Repeat the calculation in Exercise 4.2, but now apply a uniform electric field that falls across the double-barrier and single-well structure only, as shown in the above figure. The right-hand edge of the 3-nm-thick barrier is at a potential -0.63 eV below the left-hand edge of the 4-nm-thick barrier. Comment on the changes in transmission you observe.

(b) Rewrite your program to calculate transmission of a particle as a function of potential drop caused by the application of an electric field across the structure. Calculate the specific case of initial particle energy $E = 0.025$ eV with the particle incident on the structure from the left-hand side. Comment on how you might improve your model to calculate the current–voltage characteristics of a real semiconductor device with the same barrier structure.

Exercise 4.8
Using the method outlined in Section 3.4, write a computer program to solve the Schrödinger wave equation for the first two eigenvalues and eigenstates of an electron of mass m_0 confined to a double rectangular potential well sketched in the figure below. Each well is of width 0.6 nm and they are separated by 0.4 nm. The barrier potential energy is 1 eV.

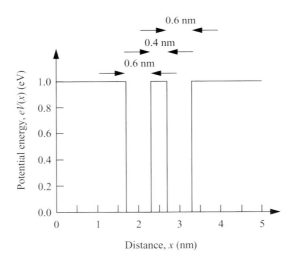

Exercise 4.9
Using the method outlined in Section 3.4, write a computer program to solve the Schrödinger wave equation for the first three eigenvalues and eigenstates of an electron with effective mass $m_e^* = 0.07 \times m_0$ confined to the periodic potential sketched in the figure below. Each of the eight quantum wells is of width 6.25 nm. Each quantum well is separated by a potential barrier of thickness 3.75 nm. The barrier potential energy is 1 eV.

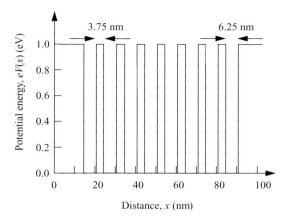

Relate the eigenfunctions to Bloch's theorem (Eqn (4.94)), and identify the value of the Bloch wave vector, k.

SOLUTIONS

Solution 4.1

(a) We start with the expression

$$Trans(E \geq V_0) = \left(1 + \frac{1}{4}\left(\frac{e^2 V_0^2}{E(E - eV_0)}\right)\sin^2(k_2 L)\right)^{-1}$$

Multiplying the terms in k out,

$$\left(k_1^2 = 2mE/\hbar^2, k_2^2 = 2m(E - eV_0)/\hbar^2, eV_0 = \left(k_1^2 - k_2^2\right)\hbar^2/2m\right)$$

we get

$$Trans = \left(1 + \frac{1}{4}\left(\frac{k_1^4 + k_2^4 - 2k_1^2 k_2^2}{k_1^2 k_2^2}\right)\sin^2(k_2 L)\right)^{-1}$$

and substituting for $k_1^2 = k_2^2 + 2meV_0/\hbar^2$ gives

$$Trans = \left(1 + \frac{1}{4} \cdot \frac{\left(k_2^2 + \frac{2meV_0}{\hbar^2}\right)^2 + k_2^4 - 2\left(k_2^2 + \frac{2meV_0}{\hbar^2}\right)k_2^2}{k_2^2\left(k_2^2 + \frac{2meV_0}{\hbar^2}\right)} \cdot \sin^2(k_2 L)\right)^{-1}$$

Expanding the terms in the numerator gives

$$2k_2^4 + \left(\frac{2meV_0}{\hbar^2}\right) + 2k_2^2\left(\frac{2meV_0}{\hbar^2}\right) - 2k_2^4 - 2k_2^2\left(\frac{2meV_0}{\hbar^2}\right)$$

so that

$$Trans = \left(1 + \frac{1}{4} \cdot \frac{\left(\dfrac{2meV_0}{\hbar^2}\right)^2}{\left(\dfrac{2meV_0}{\hbar^2} + k_2^2\right)k_2^2} \cdot \sin^2(k_2L)\right)^{-1}$$

Multiplying the second term both top and bottom by L^4, so that the terms in k_2^4 are dimensionless, gives

$$Trans = \left(1 + \frac{1}{4} \cdot \frac{\left(\dfrac{2meV_0L^2}{\hbar^2}\right)^2}{\left(\dfrac{2meV_0L^2}{\hbar^2} + k_2^2L^2\right)k_2^2L^2} \cdot \sin^2(k_2L)\right)^{-1}$$

Substituting in the parameter $b = 2k_0L = 2meV_0L^2/\hbar^2$, which is a measure of the "strength" of the potential barrier, gives the desired result:

$$Trans = \left(1 + \frac{1}{4} \cdot \frac{b^2}{\left(b + k_2^2L^2\right)} \cdot \frac{\sin^2(k_2L)}{k_2^2L^2}\right)^{-1}$$

(b) We now calculate the minimum of *Trans* as a function of k_2L or as a function of energy E. The minimum of the *Trans* function occurs when $\sin^2(k_2L) = 1$. When this happens, $k_2L = (2n - 1)\pi/2$ for $n = 1, 2, 3, \ldots$. Substituting this value into the expression obtained in (a) gives

$$Trans_{min} = \left(1 + \frac{1}{4} \cdot \frac{b^2}{\left(b + k_2^2L^2\right)} \cdot \frac{1}{k_2^2L^2}\right)^{-1}$$

$$Trans_{min} = \left(\frac{4\left(b + k_2^2L^2\right)k_2^2L^2 + b^2}{4\left(b + k_2^2L^2\right)k_2^2L^2}\right)^{-1} = \frac{4bk_2^2L^2 + 4k_2^4L^4}{b^2 + 4\left(bk_2^2L^2 + k_2^4L^4\right)}$$

$$Trans_{min} = \frac{4bk_2^2L^2 + 4k_2^4L^4 + b^2 - b^2}{\left(2k_2^2L^2 + b\right)^2} = \frac{\left(2k_2^2L^2 + b\right)^2 - b^2}{\left(2k_2^2L^2 + b\right)^2} = 1 - \frac{b^2}{\left(2k_2^2L^2 + b\right)^2}$$

One may rewrite this in terms of energy by substituting for $b = e2mV_0L^2/\hbar^2$ and $k_2^2 = 2m((E - eV_0)/\hbar^2)$ to give

$$Trans_{\min} = 1 - \frac{\left(\dfrac{2meV_0L^2}{\hbar^2}\right)^2}{\left(2 \cdot \dfrac{2m(E - eV_0)}{\hbar^2} \cdot L^2 + \dfrac{2meV_0L^2}{\hbar^2}\right)^2} = 1 - \frac{e^2V_0^2}{(2(E - eV_0) + eV_0)^2}$$

$$Trans_{\min} = 1 - \frac{e^2V_0^2}{(2E - eV_0)^2}$$

Solution 4.2

(a) We are to use the propagation matrix method to find the transmission resonances of a particle of mass $m = 0.07 \times m_0$ (where m_0 is the bare electron mass) in a one-dimensional potential consisting of two potential energy barriers. Since the particle mass is the same as the effective electron mass in the conduction band of GaAs, we might imagine that the potential could be created using heterostructures in the AlGaAs material system.

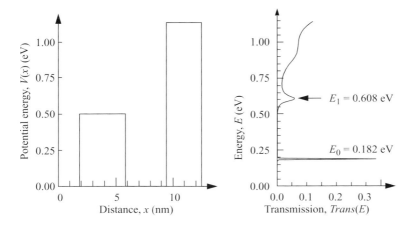

The above figure shows the one-dimensional potential as a function of distance, x, and calculated particle transmission as a function of particle energy, E. Particle transmission is plotted on a linear scale. There are two well-defined resonances, one at energy E_0 and one at energy E_1. To learn more about the resonances, it is a good idea to plot transmission on a logarithmic scale.

The figure below shows the one-dimensional potential as a function of distance, x, and calculated particle transmission as a function of particle energy, E. Particle transmission is plotted on a negative natural logarithm scale. This means that unity transmission is zero and low transmission corresponds to a large number.

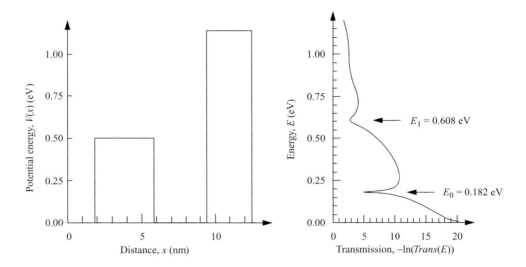

As indicated in the above figure, there are peaks in transmission at energy $E_0 = 0.182$ eV and $E_1 = 0.608$ eV. The resonance at energy E_0 has a value less than either potential barrier energy and is quite narrow in energy. This suggests that for the particle with energy E_0 the resonant state is reasonably well localized by the two potential barriers. A measure of the lifetime of such a localized state is given by the inverse of the width of the resonance. However, care must be taken in using a computer program to calculate the detailed line shape of the resonance. The program we used calculates transmission as a function of particle energy in approximately 6 meV energy increments. For this reason, features smaller than this in energy are not accurately calculated. Using a smaller energy increment gives more accurate results.

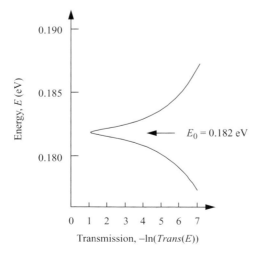

In the above figure the E_0 line shape has been calculated more accurately using 0.05 eV energy steps. The peak transmission is now near $-\ln(Trans(E_0)) = 1$, and the width of the peak is near $\Gamma_{\text{FWHM}} = 0.5$ meV. The lifetime of this particular state may be calculated approximately as $\tau = \hbar / \Gamma = 1.3$ ps. Transmission on resonance is not unity (corresponding to $-\ln(Trans(E_0)) = 0$), because the potential barriers used in this exercise are not symmetric.

You may wish to use the computer program to confirm that symmetrical potential barriers give unity transmission on resonance. If you do so, it is worth knowing that the exact position in energy of resonances depends upon both potential barrier width and energy.

The $E_1 = 0.608$ eV resonance in this exercise occurs at an energy greater than the lowest potential barrier energy. For this reason, the particle is not well localized by the two potential barriers, the resonance is broad in energy, and the resonance lifetime is much shorter than the E_0 resonance.

The following is an example of a computer program that can be used to calculate transmission as a function of energy. The program uses a simple approach. More sophisticated routines could be developed to optimize solutions for different applications.

Listing of MATLAB program for Exercise 4.2(a)

```
%Chapt4Exercise2a.m
%transmission through asymmetric double barrier
clear
clf;                        %set up potential profile position L(nm), potential V(eV)

N=6;                        %number of samples of potential

meff=0.07;                  %effective electron mass (m/m0)
vb1=0.5;                    %first potential barrier energy (eV)
vb2=1.13;                   %second potential barrier energy (eV)
bx1=4.0;                    %first potential barrier width (nm)
bx2=3.0;                    %second potential barrier width (nm)
wx1=3.6;                    %potential well width (nm)

L=[1,1,bx1,wx1,bx2,2]*1e-9; %distance array (nm)
V=[0,0,vb1,0,vb2,0];        %potential array

Emin=pi*1e-5;               %add (pi*1.0e-5) to energy to avoid divide by zero
Emax=1.4;                   %maximum particle energy (eV)
npoints=400;                %number of points in energy plot
dE=Emax/npoints;            %energy increment (eV)
hbar=1.0545715e-34;         %Planck's constant (Js)
eye=complex(0.,1.);         %square root of -1
m0=9.109382e-31;            %bare electron mass (kg)
m=meff*m0;                  %effective electron mass (kg)
echarge=1.6021764e-19;      %electron charge (C)
```

```
for j=1:npoints
    E(j)=dE*j+Emin;
    bigP=[1,0;0,1];                                    %default value of matrix bigP
for i=1:N
    k(i)=sqrt(2*echarge*m*(E(j)-V(i)))/hbar;    %wave number at each position in potential V(j)
end
for n=1:(N-1)
    p(1,1)=0.5*(1+k(n+1)/k(n))*exp(-eye*k(n)*L(n));
    p(1,2)=0.5*(1-k(n+1)/k(n))*exp(-eye*k(n)*L(n));
    p(2,1)=0.5*(1-k(n+1)/k(n))*exp(eye*k(n)*L(n));
    p(2,2)=0.5*(1+k(n+1)/k(n))*exp(eye*k(n)*L(n));
    bigP=bigP*p;
end
    Trans(j)=(abs(1/bigP(1,1)))^2;                 %transmission probability
end

figure(1);                                          %plot potential and transmission coefficient
Vp=[V;V];Vp=Vp(:);
dx=1e-12;                                            %small distance increment used in potential plot
Lx(1)=1.e-9;
for i=1:N
    for j=2:i
        Lx(i)=L(j)+Lx(j-1);                         %distance, x
    end
end
xp=[0,Lx(1)-dx,Lx(1),Lx(2)-dx,Lx(2),Lx(3)-dx,Lx(3),Lx(4)-dx,Lx(4),Lx(5)-dx,Lx(5),
Lx(6)]*1e9;
    subplot(1,2,1),plot(xp,Vp),axis([0,xp(12),0,1.4]),xlabel('Distance, x (nm)'),ylabel('Potential
energy, V(x) (eV)');
    ttl = sprintf('Chapt4Exercise2a, bx1=%3.1f bx2=%3.1f w1=%3.1f',bx1,bx2,wx1);
    title (ttl);
    subplot(1,2,2),plot(Trans,E),axis([0,1,0,1.4]),xlabel('Transmission coefficient'),ylabel('Energy,
E (eV)');
    ttl2 = sprintf('v1=%4.2f v2=%4.2f meff=%4.2f',vb1,vb2,meff);
    title (ttl2);

figure(2);
    subplot(1,2,1),plot(xp,Vp),axis([0,xp(12),0,1.4]),xlabel('Distance, x (nm)'),ylabel('Potential
energy, V(x) (eV)');
    title (ttl);
    subplot(1,2,2),plot(-log(Trans),E),xlabel('-ln trans. coeff.'),ylabel('Energy, E (eV)');
    title (ttl2);
```

(b) We are asked to calculate transmission through a one-dimensional potential that is parabolic with $V(x) = (x^2/L^2)$ eV for $|x| \leq L = 5$ nm and $V(x) = 0$ eV for $|x| > L$. We recognize the potential $V(x)$ for $|x| \ll L$ as that of a one-dimensional harmonic oscillator. The figure below shows the result of plotting the potential and the transmission for a particle mass $m = 0.07 \times m_0$, where m_0 is the bare electron mass.

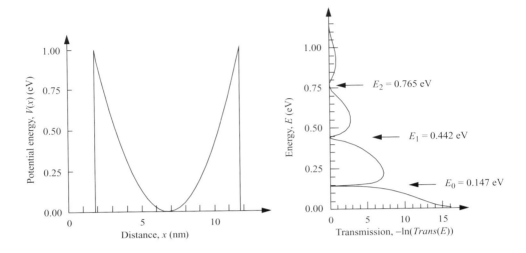

The well-defined transmission peak resonance at energy E_0 is unity (corresponding to $-\ln(Trans(E_0))$). As in Exercise 4.2(a), this may be confirmed by using a small energy increment when calculating transmission. The values of the two low-energy transmission peak resonances are in the ratio $E_1 = 3 \times E_0$, which is characteristic of quantized bound-state energy levels of a particle in a one-dimensional harmonic potential. Notice that the width in energy Γ_{FWHM} of each transmission peak increases with increasing resonance energy. The associated decrease in resonant-state lifetime $\tau = \hbar / \Gamma_{FWHM}$ arises because at greater values of resonance energy the particle finds it easier to tunnel through the potential barrier.

If mass m is increased by a factor 2 the energy levels and the energy-level spacing will decrease. For a rectangular potential well with infinite barrier energy the energy levels will decrease by a factor 2 since $E_n = \hbar^2 k^2 / 2m$. For a parabolic potential well the energy levels will decrease by a factor $1/\sqrt{2}$ (see Chapter 6 for more information on the parabolic harmonic oscillator potential).

Solution 4.3

(a) In this exercise we modify the computer program developed as part of Exercise 4.2 to find transmission as a function of energy for an electron of mass m_0 through 12 identical one-dimensional potential barriers each of energy 10 eV, width 0.1 nm, and sequentially placed every 0.5 nm so that the potential well between each barrier has width 0.4 nm.

The figure below shows the one-dimensional potential as a function of distance, x, and calculated particle transmission as a function of particle energy, E. Particle transmission is plotted on a negative natural logarithm scale. This means that unity transmission is zero and low transmission corresponds to a large number.

As is clear from the figure, there are bands in energy of near unity transmission and bands of energy where transmission is suppressed. The former correspond to conduction bands and the latter to band gaps in a semiconductor crystal. The approximate allowed (band, E_b) and disallowed (band gap, E_g) ranges of energy for particle transmission through the structure are indicated in the figure. One expects particle velocity to be slow near an allowed band edge, because one doesn't expect a discontinuous change in velocity from allowed to disallowed states. Disallowed states exist in the band gap where there are no propagating states and velocity is zero. Particle velocity should be fastest for energies near the middle of the allowed band.

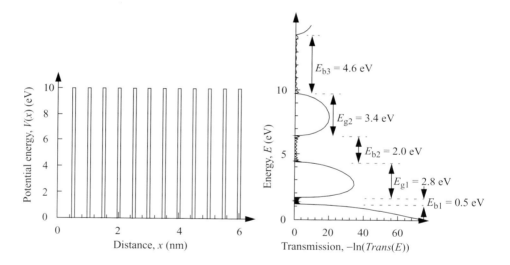

For 11 potential wells there are 11 transmission resonances, and associated with each resonance is a width in energy. When the finite width resonances overlap, they form a continuum transmission band.

(b) We now compare these results with the situation in which there are only three barriers, each with 10 eV barrier energy, 0.1 nm barrier width, and 0.4 nm well width. As may be seen in the figure below even with three potential barriers (and two potential wells), the basic structure of allowed and disallowed energy ranges has formed. For two potential wells, there are two transmission resonances. The resonances at higher energy are broader because the potential barrier seen by an electron at high energy is smaller.

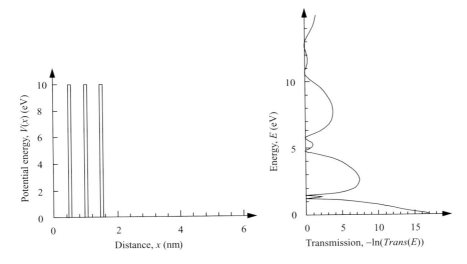

Solution 4.4

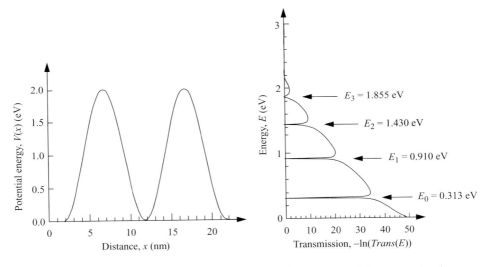

We are asked to consider a particle mass $m = 0.07 \times m_0$ with energy in the range $0 < E \le 3$ eV incident on a potential $V(x) = 2\sin^2(n2\pi x/L)$ eV, for $0 \le x \le L$ and $V(x) = 0$ eV elsewhere. In the expression for the potential, n is a nonzero positive integer, and we decide to set $L = 20$ nm. For part (a) of this exercise we set $n = 1$ and calculate the transmission probability of a particle incident from the left.

We anticipate unity transmission for well-defined resonances with energy less than the peak potential energy of 2 eV. In addition, because the potential minimum is of the form $V(x) \sim x^2$, we expect low-energy transmission resonances to occur at energies similar to those characteristic of quantized bound-state energy levels of a particle in a one-dimensional harmonic potential.

The above figure shows the one-dimensional potential as a function of distance, x, and calculated particle transmission as a function of particle energy, E. Particle transmission is plotted on a negative natural logarithm scale.

For part (b) of this exercise, we set $n = 4$ and calculate the transmission probability of a particle incident from the left.

Since L remains fixed at $L = 20$ nm but $n = 4$, the width of each well is decreased by a factor n. The reduction in well width increases the energy of the lowest-energy transmission resonance. For $n = 4$, there are $2n = 8$ peaks in the potential and $2n - 1 = 7$ potential wells. Instead of there being one distinct lowest-energy transmission resonance, as occurred in part (a) when $n = 1$, we now expect a transmission band formed from seven overlapping resonances.

The figure below shows the one-dimensional potential as a function of distance, x, for the case when $n = 4$. Also shown is the calculated particle transmission as a function of particle energy, E. Particle transmission is plotted on a negative natural logarithm scale.

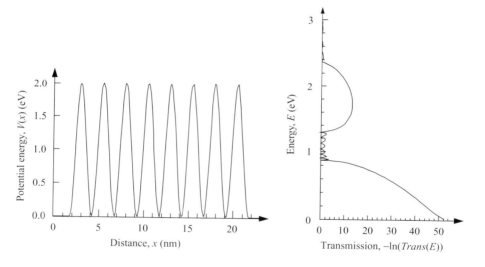

If the mass of the particle were increased by a factor of 2 from $m = 0.07 \times m_0$ to a value $m = 0.14\, m_0$, we would expect the energy of transmission resonances to be lowered and the width in energy of the transmission band to be reduced.

We now consider the case in which the potential is distorted by an additional term $V'(x)$, where $V'(x) = V_0(1 - ((2x/L) - 1)^2)$ eV for $0 \le x \le L$ and $V'(x) = 0$ eV elsewhere, so that the total potential is now $V(x) = 2\sin^2(n2\pi x/L)eV + V'(x)$. We begin our analysis by considering the case in which $V_0 = 0.1$ eV and $n = 4$.

The effect of this particular distortion on the total potential is relatively small, and so it may be thought of as a perturbation. It adds a wide and low potential barrier of peak energy $V_0 = 0.1$ eV. Particles with energy less than V_0 have a very low probability of transmission. For this reason, we may expect the effect of the additional term $V'(x)$ to be a rigid shift of the original potential $V(x) = 2\sin^2(n2\pi x/L)$ by approximately V_0. In this situation, the solution is the same as for the original potential, except that now all the energy levels are shifted up by V_0.

The figure below shows the one-dimensional potential $V(x) = 2\sin^2(n2\pi x/L) + V'(x)$ for the case in which $n = 4$ and distortion parameter $V_0 = 0.1$ eV. As may also be seen in the figure, the calculated particle transmission as a function of particle energy, E, is indeed shifted up in energy by approximately eV_0.

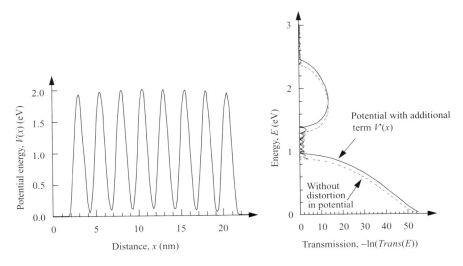

When the distortion in the potential is large, it is harder to guess the solution. Consider the case when $V_0 = 2.0$ eV in the expression $V'(x) = V_0(1 - ((2x/L) - 1)^2)$ eV, which applies when $0 \le x \le L$. Now the peak in the potential-energy distortion has the same value as the maximum potential energy appearing in the original potential energy $V(x) = 2\sin^2(n2\pi x/L)$ eV.

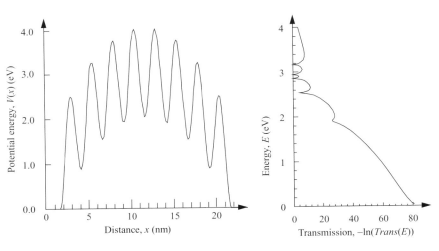

The above figure shows the one-dimensional potential $V(x) = 2\sin^2(n2\pi x/L) + V'(x)$ for the case in which $n = 4$ and distortion parameter $V_0 = 2.0\,\text{eV}$. The calculated particle transmission as a function of particle energy E is plotted on a negative natural logarithm scale.

Some transmission resonances do not have a maximum value of unity. The distortion in the periodic potential by the function $V'(x)$ breaks the exact cancellation in amplitude and phase of the coherent sum of all back-scattered waves for some resonances. For the same reasons discussed in Section 4.7, a *localization threshold* occurs when the central barrier is so thick that the energy splitting is less than the width in energy of the resonance. The particle escapes because it can tunnel through a thin effective barrier much more readily than it can through the thicker central barrier that couples degenerate bound states. The middle barrier is so thick, and tunneling is so suppressed, that there is minimal coupling between degenerate bound states, resulting in suppressed resonant transmission.

Solution 4.5

We wish to use the propagation matrix to design a high-reflectivity Bragg mirror for electromagnetic radiation with center wavelength $\lambda_0 = 980\,\text{nm}$ incident normally to the surface of an AlAs/GaAs periodic dielectric layer stack consisting of 25 indentical layer pairs. Each individual dielectric layer has a thickness $\lambda/4n$, where n is the refractive index of the dielectric. We will use $n_{\text{AlAs}} = 3.0$ for the refractive index of AlAs and $n_{\text{GaAs}} = 3.5$ for that of GaAs.

In part (a) of the exercise, we are asked to calculate and plot optical reflectivity in the wavelength range $900\,\text{nm} < \lambda < 1100\,\text{nm}$. In part (a) of the figure below results of performing the calculation are shown. The mirror has a reflectivity of close to unity over a wavelength band width $\Delta\lambda = 120\,\text{nm}$ centered at $\lambda = 980\,\text{nm}$.

For the VCSEL design in part (b) there is a 25 layer-pair mirror above and a 25 layer-pair mirror below a central GaAs layer which is one wavelength wide. The two mirrors form a high-Q resonant optical cavity. Again, the reflectivity is close to unity over a similar wavelength band width, with the addition of an optical band pass with near unity transmission and a FWHM of 13 pm centered at wavelength $\lambda = 980\,\text{nm}$.

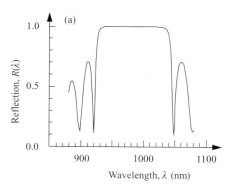
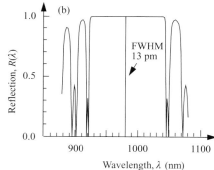

Listing of MATLAB program for Exercise 4.5

```
%Chapt4Exercise5.m
%VCSEL mirror design
clear
clf;

% '1'labels the GaAs layer and '2'labels the AlAs layer

wavelength = 980e-9;        %center wavelength (m)
d1=69.5e-9;                 %thickness of dielectric with refractive n1 (GaAs layer)
d2=82.3e-9;                 %thickness of dielectric with refractive n2 (AlAs layer)
d3=277.9e-9;                %thickness of dielectric with refractive n1 in center of pass-band filter
n1=wavelength/d1/4;         %3.5 is refractive index of GaAs
n2=wavelength/d2/4;         %3.0 is refractive index of AlAs
istep=0;
resolution=0.00005;                          %resolution (nm)
range=[-100:1:-2 -1:resolution:1 1:1:100]*1e-9+wavelength;   %wavelength scanning range

for wavelength=range                         %loop to increment wavelength
istep=istep+1;
k0=2*pi/wavelength;                          %input from air
k1=2*pi*n1/wavelength;
k2=2*pi*n2/wavelength;

%in EM, exp(-ikz) is considered positive propagation +z direction
%Assuming light coming in perpendicularly from the lower side (-z) to upper size (+z)
%A*exp(-ik1*z)+B*exp(ik1*z)=C*exp(-ik2*z)+D*exp(-ik2*z)
%D0,D1,D2 are interface matrices
%[D0]*[input from air]=[D1]*[output to n1],
%[D1]*[input from n1]=[D2]*[output to n2]

D0=[1 1
   k0 -k0];
D1=[1 1
   k1 -k1];
D2=[1 1
   k2 -k2];

%P1, P2, P3 are propagation matrices
%input from the below, output above material, [input from below]=[P]*[output above],
P1=[exp(i*k1*d1) 0
   0 exp(-i*k1*d1)];
P2=[exp(i*k2*d2) 0
   0 exp(-i*k2*d2)];
P3=[exp(i*k1*d3) 0                  %for dielectric with refractive index n1 in center of
   0 exp(-i*k1*d3)];                %passband filter

%Solution =
%[Input A;B]=inv(D0)*{D1*P1*inv(D1)*D2*P2*inv(D2)}*{2nd period}*{3rd}
%*..*{defect,wider n1}*...*{D1*P1*inv(D1)*D2*P2*inv(D2)}*D0*[output C;D]
```

```
G1=D1*P1*inv(D1)*D2*P2*inv(D2);        %for period of lower mirror
G2=D2*P2*inv(D2)*D1*P1*inv(D1);        %for period of upper mirror
lower=inv(D0)*(G1^25)*D1;              %propagation through lower 25 layer-pairs
upper=inv(D1)*(G2^25)*D0;             %propagation through upper 25 layer-pairs
Total=lower*P3*upper;

r(istep)=Total(2,1)/Total(1,1);        %reflected electromagnetic field
R(istep)=r(istep)*conj(r(istep));      %total reflection intensity is r × r*

lower_r(istep)=lower(2,1)/lower(1,1);
lower_R(istep)=lower_r(istep)*conj(lower_r(istep));   %total reflection

upper_r(istep)=upper(2,1)/upper(1,1);
upper_R(istep)=upper_r(istep)*conj(upper_r(istep));   %total reflection
                                                      %loop to next value of istep
end

figure(1);
plot(range,lower_R);
FWHM=size(find(R<max(R)/2),2)*resolution*1e-9   %accurate only when one peak in the range
figure(2);
plot(range,R);
grid;
ylabel('Reflection');
xlabel('Wavelength (m)');
title(['Chapt4Exercise5']);
temp=['FWHM = 'num2str(FWHM) 'm'];
text(980e-9, 0.5, temp);
```

Solution 4.6

To show that the Bloch wave function $\psi_k(x) = U_k(x)e^{ikx}$ is not an eigenfunction of the momentum operator $\hat{p} = -i\hbar \cdot \partial/\partial k$, we operate on the wave function, as follows:

$$\hat{p}\psi_k(x) = -i\hbar\frac{\partial}{\partial x}U_k(x)e^{ikx} = \hbar k\psi_k(x) - i\hbar e^{ikx}\frac{\partial}{\partial x}U_k(x)$$

Clearly, the wave function is not an eigenfunction of the operator $\hat{p}$, because the right-hand side of the equation is not a real number multiplied by the wave function $\psi_k(x)$.

Solution 4.7

(a) We are asked to repeat the calculation in Exercise 4.2, but in the presence of a uniform electric field that falls across the double-barrier and single-well structure only, so that the right-hand edge of the 3-nm-thick barrier is at a potential -0.63 eV below the left-hand edge of the 4-nm-thick barrier.

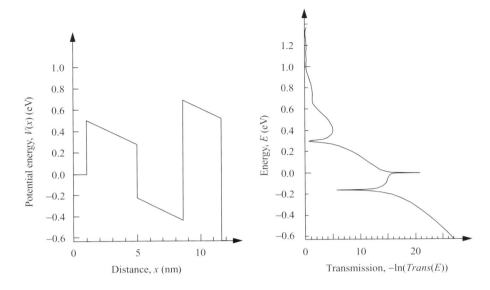

It is clear from the results shown in the above figure that a low-energy resonance exists at an energy below zero. This is possible because the applied electric field pulls the potential well below zero potential energy.

There is a reduction in transmission near zero energy as the velocity of a particle incident from the left tends to zero in the region before the first potential barrier. This causes velocity mismatch to become large giving rise to large reflection and low transmission. Such a situation can be avoided if one does not allow the incident particle to have a kinetic energy that approaches zero.

(b) We now consider transmission of the particle as a function of potential drop caused by the application of an electric field across the structure. Initial particle energy is $E = 0.025$ eV, and the particle is incident on the structure from the left-hand side. This value of particle energy ensures that initial particle velocity is high enough that we don't incur the problem with velocity mismatch that showed up in part (a).

Note that none of the resonances in the figure below approach unity transmission. By varying barrier energy, thickness, and well width, it is possible to design a resonant tunnel diode with a unity transmission resonance. However, such conditions can only be achieved over a small range of applied voltage.

To apply the calculation to a situation involving the motion of electrons passing current I through a resonant tunnel diode, one would anticipate including a number of improvements such as: (i) self-consistently including space-charging effects by solving Poisson's equation and modifying the potential according to charge density in the device, (ii) extending the calculation to include three dimensions, and (iii) including the electron distribution function and electron scattering when calculating current flow.

Of course, some of these improvements to the calculation are a little ambitious and may require some original research.

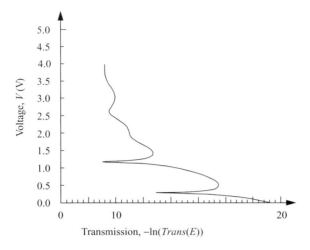

Solution 4.8

We would like to use the method outlined in Section 3.4 to numerically solve the one-dimensional Schrödinger wave equation for an electron of mass m_0 in a symmetrical double-well structure with finite barrier potential energy.

The main computer program deals with input parameters such as the length L, the electron mass, the number of discretization points N, and the plotting routine. Because we use a nontransmitting boundary condition, it is important to choose L large enough so that the wave function is approximately zero at the boundaries $x_0 = 0$ and $x_N = L$. Also, a large enough value of N should be chosen so that the wave function does not vary significantly between adjacent discretization points. This ensures that the three-point finite-difference approximation used in Eqn (3.40) is accurate.

The main computer program calls solve_schM, which was used in solution of Exercise 3.7. It solves the discretized Hamiltonian matrix (Eqn (3.44)).

In this exercise, the first two energy eigenvalues are $E_0 = 0.3226$ eV and $E_1 = 0.4053$ eV. The separation in energy is $\Delta E = 0.0827$ eV. The eigenfunctions generated by the program and plotted in the figure below are not normalized.

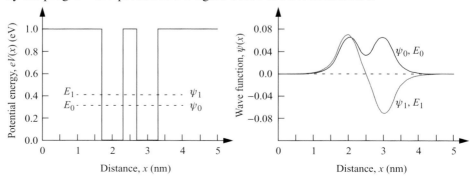

Listing of Matlab program for Exercise 4.8

```
%Chapt4Exercise8.m
%eigenstates of resonant tunnel barrier
%
clear
clf;
length = 5;                              %length of well (nm)
N=1000;                                 %number of sample points
x=0:length/N:length;                    %position of sample points in potential
mass=1.00;                              %effective electron mass
num_sol=2;                              %number of solutions sought
v0=1;                                   %potential scale (eV)

v=v0.*ones(1,N+1);

%first well
for i=341:460                           %well width 120*length/N = 0.6 nm
    v(i)=0;
end
                                        %barrier width 81*length/N = 0.4 nm
%second well
for i=540:659                           %well width 120*length/N = 0.6 nm
    v(i)=0;
end

[energy,phi]=solve_schM(length,N,v,mass,num_sol);        %call function solve_schM

    for i=1:num_sol
sprintf(['eigenfunction (',num2str(i),') = ',num2str(energy(i)),'eV'])    %energy eigenvalues
end

figure(1);
plot(x,v,'b');xlabel('Distance (nm)'),ylabel('Potential energy, (eV)');
ttl=['Chapt4Exercise8, m* = ',num2str(mass),'m0, Length = ',num2str(length),'nm'];
title(ttl);

s=char('y','k','r','g','b','m','c');                     %plot curves in different colors

figure(2);
for i=1:num_sol
    j=1+mod(i,7);                       %select color for plot
    plot(x,phi(:,i),s(j));              %plot eigenfunctions
        hold on;
end
xlabel('Distance (nm)'),ylabel('Wave function');
title(ttl);
hold off;
```

Solution 4.9

We follow the previous exercise and write a computer program to solve the Schrödinger wave equation for the first three eigenvalues and eigenstates of an electron with effective mass $m_e^* = 0.07 \times m_0$ confined to the periodic potential consisting of the eight quantum wells of width 6.25 nm, each separated by a potential barrier of thickness 3.75 nm and with barrier potential energy 1 eV. The total width of the multiple quantum well structure is 76.25 nm.

The energy eigenvalues for the first three eigenstates are $E_0 = 0.0885$ eV, $E_1 = 0.0886$ eV, and $E_2 = 0.0887$ eV. The corresponding eigenstates, ψ_0, ψ_1, and ψ_2, shown in the figure below, are the first of eight states that form the lowest energy allowed band in the periodic potential.

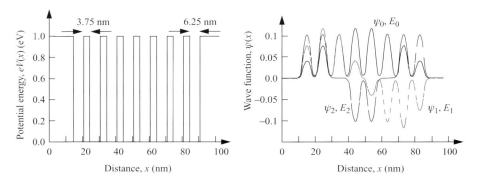

According to Bloch's theorem, states in a periodic potential are of the form $\psi_k(x) = U_k(x)e^{ikx}$, where $U_k(x)$ has the same periodicity as the potential and the term e^{ikx} carries phase information via the Bloch wave vector k. It is clear from the results shown in the above figure that $U_k(x)$ may be approximated as a symmetric function (let us say Gaussian) centered in each quantum well with some spatial overlap into adjacent quantum wells. The wave function $\psi_k(x)$ is then formed by modulating $U_k(x)$ with the sinusoidal envelope function e^{ikx}. The Bloch wave vector $k_n = 2\pi/\lambda_n$ takes on the values $\pi(n + 1)/L_{\text{eff}}$, where $n = 0, 1, 2, \ldots$ and $L_{\text{eff}} \sim 85$ nm is an effective size for the multiple quantum well structure, which takes into account the penetration of the wave functions into the outermost potential barriers. In this case, the envelope functions are similar to the standing waves we have calculated previously for states in a rectangular potential with infinite barrier energy. The figure below illustrates this for the first three eigenfunctions.

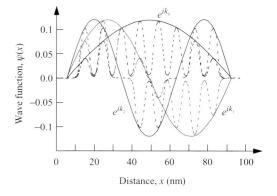

The eight quantum wells have eight states that form the lowest-energy allowed band. These eight states consist of a cell periodic function $U_k(x)$ modulated by the envelope function e^{ikx}, the Bloch wave vector k of which can have one of eight values ($k_n = n\pi/L_{\text{eff}}$, where $n = 1, 2, \ldots, 8$).

The next allowed band is formed from eight states, each of which consists of a new cell periodic function (this time antisymmetric) modulated by the same envelope function e^{ikx}.

5 Eigenstates and operators

5.1 Introduction

Quantum mechanics is a very successful description of atomic-scale systems. The simplicity of the mathematical description in terms of noncommuting linear operators is truly remarkable. The elegant, embedded symmetries are, in themselves, aesthetically pleasing and have been a source of inspiration for some studying this subject. Of course, this mathematical description uses postulates to provide a logical framework with which to make contact with the results of experimental measurements.

5.1.1 The postulates of quantum mechanics

From the material developed in the previous chapters, we may write down four assumptions or postulates for quantum mechanics.

5.1.1.1 Postulate 1

Associated with every physical observable is a corresponding operator $\hat{\mathbf{A}}$ from which results of measurement of the observable may be deduced.

We assume that each operator is linear and satisfies an eigenvalue equation of the form $\hat{\mathbf{A}}\psi_n = a_n\psi_n$, in which the eigenvalues a_n are real numbers and the eigenfunctions ψ_n form a complete orthogonal set in state-function space. The eigenvalues, which may take on discrete values or exist for a continuous range of values, are guaranteed to be real (and hence measurable) if the corresponding operator is Hermitian. We also note that, in general, the eigenfunctions themselves are complex and hence not directly measurable.

5.1.1.2 Postulate 2

The only possible result of a measurement on a single system of a physical observable associated with the operator $\hat{\mathbf{A}}$ is an eigenvalue of the operator $\hat{\mathbf{A}}$.

In this way, the result of measurement is related to the eigenvalue of the mathematical operator $\hat{\mathbf{A}}$. The act of measurement on the system gives an eigenvalue a_n, which is a real number. The eigenfunction associated with this eigenvalue is stationary. As a consequence, after the measurement has been performed, the system remains in the measured eigenstate unless acted upon by some external force.

5.1.1.3 Postulate 3

For every system there always exists a state function Ψ *that contains all of the information that is known about the system.*

The state function Ψ contains all of the information on all observables in the system. It may be used to find the relative probability of obtaining eigenvalue a_n associated with operator $\hat{\mathbf{A}}$ for a particular system at a given time.

5.1.1.4 Postulate 4

The time evolution of Ψ *is determined by* $i\hbar \partial \Psi / \partial t = H \Psi$*, where H is the Hamiltonian operator for the system.*

We recognize the time evolution of the state function as Schrödinger's equation (Eqn (2.40)).

The postulates of quantum mechanics are the underlying assumptions on which the theory is built. They may only be justified to the extent that results of physical experiments do not contradict them. The postulates, which are a connection between mathematics and the physical aspects of the model, contain the strangeness of quantum mechanics.

The probabilistic interpretation of measurement and the associated collapse of the state function to an eigenfunction are *physical* aspects of the model that in no way detract from the beauty of the mathematics. The correspondence principle, in which as $\hbar \to 0$ one obtains the result known from classical mechanics, is also a constraint imposed by physics. These and other aspects of the physical model are what introduce inconsistencies. When one talks about the weirdness of quantum mechanics, one is usually struggling to come to terms with the physical aspects of the model and implicitly asking for a new, more complete theory.

In this chapter, the idea is to introduce some of the mathematics used in our description of quantum phenomena. Our approach to the mathematics is going to be pragmatic. We are only going to introduce a concept if it is useful. We will then illustrate the idea with an example.

5.2 One-particle wave-function space

The probabilistic interpretation of the wave function means that $|\psi(\mathbf{r}, t)|^2 d^3 r$ represents the probability of finding the particle at time t in volume $d^3 r$ about the point $\mathbf{r}$ in space. Physical experience suggests that it is reasonable to assume that the total probability of finding the particle somewhere in space is unity, so that

$$\int_{-\infty}^{\infty} |\psi(\mathbf{r}, t)|^2 d^3 r = 1 \tag{5.1}$$

We require that the wave functions $\psi(\mathbf{r}, t)$ be defined, continuous, and differentiable. Also, the wave functions exist in a wave-function space $\mathcal{J}$ which is linear. The integrands for which this equation converges are square integrable functions. This is a set called L^2 by mathematicians and it has the structure of Hilbert space.

There are analogies between an ordinary N-dimensional vector space consisting of N orthonormal unit vectors and the eigenfunction space in quantum mechanics. They are, for example, both linear spaces. However, an important difference becomes apparent when one considers scalar products.

If the vector

$$\mathbf{A} = \sum_{j}^{N} a_j \mathbf{a}_j \tag{5.2}$$

where a_j is the j-th coefficient and $\mathbf{a}_j$ is the j-th orthonormal unit vector, and similarly the vector

$$\mathbf{B} = \sum_{j}^{N} b_j \mathbf{b}_j \tag{5.3}$$

then the scalar product of the two vectors $\mathbf{A}$ and $\mathbf{B}$ is just

$$\mathbf{A} \cdot \mathbf{B} = \sum_{j}^{N} a_j b_j \tag{5.4}$$

In quantum mechanics there are wave functions such as $\psi_A(\mathbf{r})$ and $\psi_B(\mathbf{r})$. In this case, the scalar product is an integral $\int \psi_A^*(\mathbf{r}) \psi_B(\mathbf{r}) d^3 r$. The integral is needed because wave-function space is continuously infinite dimensional. This is one of the characteristics of Hilbert space that distinguish it from an ordinary N-dimensional vector space.

5.3 Properties of linear operators

A linear operator $\hat{\mathbf{A}}$ associates with every function $\psi(\mathbf{r}) \in \mathcal{I}$ another function $\phi(\mathbf{r})$ in a linear way. This associativity may be expressed mathematically as

$$\phi(\mathbf{r}) = \hat{\mathbf{A}}\psi(\mathbf{r}) \tag{5.5}$$

If we let the wave function $\psi(\mathbf{r})$ be a linear combination $\psi(\mathbf{r}) = \lambda_1\psi_1(\mathbf{r}) + \lambda_2\psi_2(\mathbf{r})$, where λ_1 and λ_2 are numbers that weight the contribution of $\psi_1(\mathbf{r})$ and $\psi_2(\mathbf{r})$, then

$$\hat{\mathbf{A}}(\lambda_1\psi_1(\mathbf{r}) + \lambda_2\psi_2(\mathbf{r})) = \lambda_1\hat{\mathbf{A}}\psi_1(\mathbf{r}) + \lambda_2\hat{\mathbf{A}}\psi_2(\mathbf{r}) \tag{5.6}$$

As an example, consider the momentum operator for a particle moving in one dimension. The operator is $\hat{\mathbf{A}} = \hat{p}_x = -i\hbar \cdot \partial/\partial x$. If we let the wave function $\psi(x) = \lambda_1\psi_1(x) + \lambda_2\psi_2(x)$, then the expression for $\phi(x)$ in Eqn (5.5) becomes

$$\phi(x) = -i\hbar\frac{\partial}{\partial x}(\lambda_1\psi_1(x) + \lambda_2\psi_2(x)) = -\lambda_1 i\hbar\frac{\partial}{\partial x}\psi_1(x) - \lambda_2 i\hbar\frac{\partial}{\partial x}\psi_2(x) \tag{5.7}$$

5.3.1   Product of operators

If $\hat{\mathbf{A}}$ and $\hat{\mathbf{B}}$ are linear operators, then the product of operators acting upon the function $\psi(\mathbf{r})$ is

$$(\hat{\mathbf{A}}\hat{\mathbf{B}})\psi(\mathbf{r}) = \hat{\mathbf{A}}(\hat{\mathbf{B}}\psi(\mathbf{r})) \tag{5.8}$$

Equation (5.8) indicates that operator $\hat{\mathbf{B}}$ acts first upon $\psi(\mathbf{r})$ to give $\phi(\mathbf{r}) = \hat{\mathbf{B}}\psi(\mathbf{r})$. Operator $\hat{\mathbf{A}}$ then acts upon the new function $\phi(\mathbf{r})$. The order in which operators act upon a function is critical because, in general, $\hat{\mathbf{A}}\hat{\mathbf{B}} \neq \hat{\mathbf{B}}\hat{\mathbf{A}}$.

To illustrate this important property, consider the one-dimensional momentum operator in real space $\hat{\mathbf{A}} = \hat{p}_x = -i\hbar \cdot \partial/\partial x$ and the one-dimensional position operator $\hat{\mathbf{B}} = x$. Let the wave function $\psi = \psi(x)$. Then

$$\hat{\mathbf{A}}\hat{\mathbf{B}}\psi(x) = -i\hbar\frac{\partial}{\partial x}(x\psi(x)) = -i\hbar\psi(x) - i\hbar x\frac{\partial}{\partial x}\psi(x) \tag{5.9}$$

and

$$\hat{\mathbf{B}}\hat{\mathbf{A}}\psi(x) = -i\hbar x\frac{\partial}{\partial x}\psi(x) \tag{5.10}$$

Comparing Eqn (5.9) and Eqn (5.10), we may conclude that $\hat{\mathbf{A}}\hat{\mathbf{B}} \neq \hat{\mathbf{B}}\hat{\mathbf{A}}$. Clearly, the order in which operators are applied is something that needs to be handled with care. One way to determine the sensitivity of pairs of operators to the order in which a product is applied is to evaluate the *commutator*.

5.3.2   The commutator for operator pairs

The commutator for the pair of operators $\hat{A}$ and $\hat{B}$ is defined as

$$[\hat{A}, \hat{B}] = \hat{A}\hat{B} - \hat{B}\hat{A}$$ (5.11)

Mathematically, one may think of quantum mechanics as the description of physical systems with noncommuting operators. As with matrix algebra, in general $\hat{A}\hat{B} \neq \hat{B}\hat{A}$.

By way of an example, we consider the one-dimensional momentum operator in real space $\hat{A} = \hat{p}_x = -i\hbar \cdot \partial/\partial x$ and the one-dimensional position operator $\hat{B} = x$. These are the same operators used in Eqn (5.9) and Eqn (5.10), which, when substituted into Eqn (5.11) gives the commutator

$$[\hat{p}_x, x] = -i\hbar$$ (5.12)

The fact that the right-hand side of Eqn (5.12) is nonzero means that the pair of linear operators we used in this particular example are noncommuting. Of course, if the order in which the operators appear in Eqn (5.12) is interchanged, then the sign of the commutator is reversed. In our example this gives $[x, \hat{p}_x] = -i\hbar$.

5.3.3   Properties of Hermitian operators

The results of physical measurements are real numbers. This means that a physical model of reality is restricted to prediction of real numbers. Hermitian operators play a special role in quantum mechanics, because these operators guarantee real eigenvalues. Hence, a physical system described using a Hermitian operator will provide information on measurable quantities.

$\hat{A}$ is a *Hermitian operator* if the expectation value is such that

$$\int (\phi_n^*(\mathbf{r})\hat{A}\psi_m(\mathbf{r}))^* d^3r = \int \psi_m^*(\mathbf{r})\hat{A}\phi_n(\mathbf{r})d^3r = \int (\hat{A}\psi_m(\mathbf{r}))^*\phi_n(\mathbf{r})d^3r$$

$$\int (\phi_n^*(\mathbf{r})\hat{A}\psi_m(\mathbf{r}))^* d^3r = \int (\hat{A}\phi_n(\mathbf{r})\psi_m(\mathbf{r}))^* d^3r$$

or, equivalently, in matrix notation

$$A_{nm}^* = A_{mn}$$ (5.13)

where the matrix elements $A_{nm}^* = \int (\phi_n^*(\mathbf{r})\hat{A}\psi_m(\mathbf{r}))^* d^3r$ and $A_{mn} = \int \psi_m^*(\mathbf{r})\hat{A}\phi_n(\mathbf{r})d^3r$.

If the linear operator $\hat{A}$ is not Hermitian, it is always possible to define a *Hermitian adjoint operator* $\hat{A}^\dagger$ in such a way that

$$\int \phi^*(\mathbf{r})\hat{A}^\dagger \psi(\mathbf{r})d^3r = \int (\hat{A}\phi(\mathbf{r}))^* \psi(\mathbf{r})d^3r$$ (5.14)

It follows from the definition of Hermitian operators that an operator is Hermitian when it is its own Hermitian adjoint – i.e., $\hat{A}^{\dagger} = \hat{A}$.

To show that the eigenvalues of a Hermitian operator are real and that the associated eigenfunctions are orthogonal, we start by considering the operator $\hat{A}$ such that $\hat{A}\phi_m = a_m\phi_m$, where ϕ_m is an eigenfunction of $\hat{A}$ and a_m is the corresponding eigenvalue. We can always write

$$\hat{A}\phi_n = a_n\phi_n \tag{5.15}$$

If we multiply both sides of Eqn (5.15) by ϕ_m^* and integrate over all space we obtain

$$\int \phi_m^* \hat{A}\phi_n d^3r = a_n \int \phi_m^* \phi_n d^3r \tag{5.16}$$

Similarly, interchanging the subscripts m and n, we have

$$\int \phi_n^* \hat{A}\phi_m d^3r = a_m \int \phi_n^* \phi_m d^3r \tag{5.17}$$

which can be rewritten as

$$\int (\hat{A}\phi_n)^* \phi_m d^3r = a_m \int \phi_n^* \phi_m d^3r \tag{5.18}$$

If now one takes the complex conjugate, this gives

$$\int \phi_m^* \hat{A}\phi_n d^3r = a_m^* \int \phi_m^* \phi_n d^3r \tag{5.19}$$

Subtracting Eqn (5.19) from Eqn (5.16) gives

$$0 = (a_n - a_m^*) \int \phi_m^* \phi_n d^3r \tag{5.20}$$

For the case when $n = m$, we have

$$0 = (a_n - a_n^*) \int \phi_n^* \phi_n d^3r \tag{5.21}$$

Since $|\phi_n|^2$ is finite, $a_n = a_n^*$, and we conclude that *eigenvalues of Hermitian operators are real numbers*. This is useful in quantum mechanics, because it guarantees that the eigenvalue of a Hermitian operator results in a real measurable quantity.

For the case in which $n \neq m$, then the integral is zero provided $a_n \neq a_m$. Hence the nondegenerate eigenfunctions of Hermitian operators are *orthogonal* to each other, and we may write

$$\boxed{0 = \int \phi_m^* \phi_n d^3r} \tag{5.22}$$

for $n \neq m$.

5.3.4   Normalization of eigenfunctions

Because eigenvalue equations involve linear operators, we may specify eigenfunctions to within an arbitrary constant. It is convention that the constant is chosen in such a way that the integral over all space is unity. This means that the eigenfunctions are normalized to unity. Eigenfunctions that are orthogonal and normalized are called *orthonormal*. The orthonormal properties of Hermitian operator eigenfunctions can be expressed as

$$\int \phi_n^* \phi_m d^3 r = \delta_{nm} \tag{5.23}$$

where the Kronecker delta $\delta_{nm} = 0$ if $n \neq m$ and $\delta_{nm} = 1$ if $n = m$.[1]

5.3.5   Completeness of eigenfunctions

The eigenfunctions of a Hermitian operator can be used to expand an *arbitrary function* $\psi(\mathbf{r})$. This means that

$$\psi(\mathbf{r}) = \sum_n a_n \phi_n(\mathbf{r}) \tag{5.24}$$

where $\hat{\mathbf{A}}\phi_n = a_n \phi_n$. The expansion coefficient a_m is obtained by multiplying both sides of the equation by $\phi_m^*(\mathbf{r})$ and integrating

$$\int \phi_m^* \psi(\mathbf{r}) d^3 r = \sum_n a_n \int \phi_m^* \phi_n d^3 r \tag{5.25}$$

Using the fact that $\int \phi_m^* \phi_n d^3 r = \delta_{mn}$, one obtains

$$\int \phi_m^* \psi(\mathbf{r}) d^3 r = a_m \tag{5.26}$$

so that a_m is the projection of $\psi(\mathbf{r})$ on $\phi_m(\mathbf{r})$. The wave function $\psi(\mathbf{r})$ is an arbitrary function, and $\sum_n a_n \phi_n(\mathbf{r})$ is the expansion of that function in terms of the unit eigenfunctions $\phi_n(\mathbf{r})$.

5.4 Dirac notation

This is a particularly compact and efficient way to represent eigenfunctions and expectation values in quantum mechanics. This notation represents individual state functions by a *ket* or *bra*:

$$\phi \rightarrow |\phi\rangle \quad \text{ket} \tag{5.27}$$

$$\phi^* \rightarrow \langle\phi| \quad \text{bra} \tag{5.28}$$

---

[1] The Kronecker delta is not to be confused with the Dirac delta function described in Appendix C-4.

Often, a wave function $\phi_n(\mathbf{r})$ with quantum number n is represented as $|n\rangle$ and its complex conjugate as $\langle n|$.

In Dirac notation, the integral over all space is defined as

$$\int \phi^*(\mathbf{r})\psi(\mathbf{r})d^3r \equiv \langle \phi|\psi \rangle \tag{5.29}$$

where $\langle \quad \rangle$ is a braket (bra-ket).

The act of operating on wave function ϕ with operator $\hat{\mathbf{A}}$ is

$$\hat{\mathbf{A}}\phi \rightarrow \hat{\mathbf{A}}|\phi\rangle \tag{5.30}$$

and

$$\int \phi^*(\mathbf{r})\hat{\mathbf{A}}\psi(\mathbf{r})d^3r = \langle \phi|\hat{\mathbf{A}}|\psi\rangle \tag{5.31}$$

is the expectation value of the operator $\hat{\mathbf{A}}$.

The orthonormal condition is expressed as

$$\int \phi_n^*\phi_m d^3r = \langle \phi_n|\phi_m\rangle = \langle n|m\rangle = \delta_{nm} \tag{5.32}$$

The projection of $\psi(\mathbf{r})$ on $\phi_m(\mathbf{r})$ is expressed as

$$a_m = \langle \phi_m|\psi\rangle \tag{5.33}$$

and the expansion of an arbitrary state function $|\psi\rangle$ is

$$|\psi\rangle = \sum_n b_n|n\rangle \tag{5.34}$$

$$|\psi\rangle = \sum_n (\langle n|\psi\rangle)|n\rangle = \sum_n |n\rangle\langle n|\psi\rangle \tag{5.35}$$

Hence, $\sum_n |n\rangle\langle n| = \hat{\mathbf{I}}$, where $\hat{\mathbf{I}}$ is the identity operator.

The Schrödinger equation

$$\left(\frac{-\hbar^2\nabla^2}{2m} + V(\mathbf{r},t)\right)\psi_n(\mathbf{r},t) = H\psi_n(\mathbf{r},t) = i\hbar\frac{\partial}{\partial t}\psi_n(\mathbf{r},t) \tag{5.36}$$

can be written

$$H|\psi\rangle = i\hbar\frac{\partial}{\partial t}|\psi\rangle = i\hbar\left|\frac{\partial}{\partial t}\psi\right\rangle \tag{5.37}$$

5.5 Measurement of real numbers

In quantum mechanics, each type of physical observable is associated with a Hermitian operator. As mentioned in Section 5.3.3, Hermitian operators ensure that any eigenvalue is a real quantity. In this way, the result of a measurement is a real number that corresponds to one of the set of continuous or discrete eigenvalues for the system:

$$\hat{A}|n\rangle = a_n|n\rangle \tag{5.38}$$

$\hat{A}$ is a Hermitian operator, $|n\rangle$ is an eigenfunction, and a_n is its eigenvalue.

If there are two different physical observables with eigenvalues a_n and b_n, respectively, then there are two different associated operators $\hat{A}$ and $\hat{B}$. For a given system, measurement of $\hat{A}$ followed by measurement of $\hat{B}$ is denoted by $\hat{B}\hat{A}$, and the result may be different for $\hat{A}\hat{B}$. If the measurements interfere with each other, then the commutator

$$[\hat{A}, \hat{B}] = \hat{A}\hat{B} - \hat{B}\hat{A} \neq 0 \tag{5.39}$$

Measurements of position and momentum are good examples of measurements that interfere with each other. The commutation relation for the position operator $\hat{x}$ and the momentum operator $\hat{p}_x = -i\hbar \cdot \partial/\partial x$ for a particle moving in one dimension is

$$[\hat{x}, \hat{p}_x] = i\hbar \tag{5.40}$$

The momentum and position operators do not commute. A measurement on one observable influences the value of the other. The coupling between the two observables through the commutation relation has the physical consequence that the observable quantities cannot be measured simultaneously with arbitrary accuracy.

So far, we have considered an example of measurements that interfere with each other. The other possibility is that the measurements do not interfere with each other. In this case, the operators corresponding to the measurement commute. If two operators commute, then they possess common eigenfunctions. Since, in this case, $\hat{B}\hat{A} = \hat{A}\hat{B}$, we can write $\hat{A}\hat{B}\phi_B = \hat{B}\hat{A}\phi_B = \hat{A}b\phi_B = b\hat{A}\phi_B$. The function $\hat{A}\phi_B$ is thus an eigenfunction of $\hat{B}$ with eigenvalue b. If there is only one eigenfunction of $\hat{B}$ associated with eigenvalue b, then $\hat{A}\phi_B = c\phi_B$, where c is a constant, so that ϕ_B is an eigenfunction of $\hat{A}$.

5.5.1   Expectation value of an operator

Previously, we identified $\psi^*(\mathbf{r})\psi(\mathbf{r})d^3r$ as the probability of finding a particle in volume element d^3r at position $\mathbf{r}$. The fact that we characterize the position of a particle using probability means that we will be concerned with its statistical properties whether tunneling or moving in free space. Fortunately, we know enough about finding solutions to wave functions that we can explore this now.

If $\psi^*(\mathbf{r})\psi(\mathbf{r})d^3r$ is the probability of finding the particle in volume element d^3r at position $\mathbf{r}$, then, because it must be somewhere in space with certainty, the integral over all space is unity. This property of normalization requires $\langle\psi|\psi\rangle = \int \psi^*(\mathbf{r})\psi(\mathbf{r})d^3r = 1$, so the expectation of finding the particle somewhere is unity. Other *expectation values* can be found. Consider the Schrödinger equation

$$-\frac{\hbar^2}{2m}\nabla^2\psi(\mathbf{r}) + V(\mathbf{r})\psi(\mathbf{r}) = E\psi(\mathbf{r}) \tag{5.41}$$

Multiplying by $\psi^*(\mathbf{r})$ and integrating over all space gives

$$-\frac{\hbar^2}{2m}\int \psi^*(\mathbf{r})\nabla^2\psi(\mathbf{r})d^3r + \int \psi^*(\mathbf{r})V(\mathbf{r})\psi(\mathbf{r})d^3r = E\int \psi^*(\mathbf{r})\psi(\mathbf{r})d^3r \tag{5.42}$$

The first term in the integrand on the left-hand side is the local kinetic energy probability at position $\mathbf{r}$. The second term in the integrand on the left-hand side is the local potential energy at position $\mathbf{r}$. We are weighting the kinetic energy operator and potential operator at position $\mathbf{r}$ with the probability that the particle is at position $\mathbf{r}$. We then integrate over all space to get the average value or *expectation value*. The expectation values for kinetic energy $\langle T \rangle$, potential energy $\langle V \rangle$, and position $\langle \mathbf{r} \rangle$, are

$$\langle T \rangle = \langle \psi | \hat{T} | \psi \rangle = \frac{-\hbar^2}{2m}\int \psi^*(\mathbf{r})\nabla^2\psi(\mathbf{r})d^3r \tag{5.43}$$

$$\langle V \rangle = \langle \psi | \hat{V} | \psi \rangle = \int \psi^*(\mathbf{r})V(\mathbf{r})\psi(\mathbf{r})d^3r \tag{5.44}$$

$$\langle \mathbf{r} \rangle = \langle \psi | \hat{\mathbf{r}} | \psi \rangle = \int \psi^*(\mathbf{r})\mathbf{r}\psi(\mathbf{r})d^3r \tag{5.45}$$

Given that we have defined an average value for the result of a measurement, it is natural to consider the time evolution of the expectation value as well as the spread or deviation from the average value when a measurement is performed many times. The time dependence of an expectation value is considered in Section 5.5.2. The deviation from the mean result of a measurement performed many times is called the uncertainty in expectation value, and this is discussed in Section 5.5.3.

5.5.2   Time dependence of expectation value

To find the time dependence of an expectation value, we start by writing down the expectation value of the operator $\hat{\mathbf{A}}$:

$$\langle \hat{\mathbf{A}} \rangle = \langle \psi | \hat{\mathbf{A}} | \psi \rangle \tag{5.46}$$

The time dependence of this equation can be expressed in terms of the Schrödinger equation (Eqn (5.37)):

$$\frac{-i}{\hbar}H|\psi\rangle = \left|\frac{\partial\psi}{\partial t}\right\rangle \tag{5.47}$$

Taking the complex conjugate of both sides gives

$$\frac{i}{\hbar}\langle\psi|H = \left\langle\frac{\partial\psi}{\partial t}\right| \tag{5.48}$$

We now find the time derivative of Eqn (5.46) using the chain rule for differentiation and substituting in Eqn (5.47) and Eqn (5.48):

$$\frac{d}{dt}\langle \hat{\mathbf{A}} \rangle = \left\langle \frac{\partial \psi}{\partial t} \middle| \hat{\mathbf{A}} \middle| \psi \right\rangle + \left\langle \psi \middle| \frac{\partial}{\partial t}\hat{\mathbf{A}} \middle| \psi \right\rangle + \left\langle \psi \middle| \hat{\mathbf{A}} \middle| \frac{\partial \psi}{\partial t} \right\rangle \tag{5.49}$$

$$\frac{d}{dt}\langle \hat{\mathbf{A}} \rangle = \frac{i}{\hbar}\langle \psi | H\hat{\mathbf{A}} | \psi \rangle - \frac{i}{\hbar}\langle \psi | \hat{\mathbf{A}} H | \psi \rangle + \left\langle \psi \middle| \frac{\partial}{\partial t}\hat{\mathbf{A}} \middle| \psi \right\rangle \tag{5.50}$$

$$= \frac{i}{\hbar}\langle \psi | H\hat{\mathbf{A}} - \hat{\mathbf{A}}H | \psi \rangle + \left\langle \psi \middle| \frac{\partial}{\partial t}\hat{\mathbf{A}} \middle| \psi \right\rangle$$

$$\boxed{\frac{d}{dt}\langle \hat{\mathbf{A}} \rangle = \frac{i}{\hbar}\langle [H, \hat{\mathbf{A}}] \rangle + \left\langle \frac{\partial}{\partial t}\hat{\mathbf{A}} \right\rangle} \tag{5.51}$$

If the operator $\hat{\mathbf{A}}$ has no explicit time dependence, then

$$\left\langle \frac{\partial}{\partial t}\hat{\mathbf{A}} \right\rangle = 0$$

and

$$\frac{d}{dt}\langle \hat{\mathbf{A}} \rangle = \frac{i}{\hbar}\langle [H, \hat{\mathbf{A}}] \rangle \tag{5.52}$$

5.5.2.1 Time dependence of position operator of particle moving in free space

To check this result, consider a particle of mass m moving in free space in such a way that the Hamiltonian describing motion in the x direction is

$$H = -\frac{\hbar^2}{2m}\frac{d^2}{dx^2} \tag{5.53}$$

To evaluate the time dependence of the expectation value of the position operator $\hat{x}$, we need to find

$$\frac{d}{dt}\langle \hat{x} \rangle = \frac{i}{\hbar}\langle [H, \hat{x}] \rangle \tag{5.54}$$

The commutator operating on the wave function $\psi(x, t)$ that describes the particle gives

$$\frac{i}{\hbar}[H, \hat{x}]\psi = -\frac{\hbar^2}{2m}\frac{i}{\hbar}\left(\frac{d}{dx}\left(\frac{d}{dx} \cdot x\psi \right) - x\frac{d}{dx}\left(\frac{d}{dx}\psi \right) \right)$$

$$= \frac{-i\hbar}{2m}\left(\frac{d}{dx}\psi + \frac{d}{dx}\left(x\frac{d}{dx}\psi \right) - x\frac{d}{dx}\left(\frac{d}{dx}\psi \right) \right) \tag{5.55}$$

$$\frac{i}{\hbar}[H, \hat{x}]\psi = \frac{-i\hbar}{2m}\left(\frac{d}{dx}\psi + \frac{d}{dx}\psi + x\frac{d}{dx}\left(\frac{d}{dx}\psi \right) - x\frac{d}{dx}\left(\frac{d}{dx}\psi \right) \right) = \frac{-i\hbar}{m}\frac{d}{dx}\psi \tag{5.56}$$

Using the fact that the wave function of a free particle moving in the x direction is of the form $\psi = e^{i(k_x x - \omega t)}$, we may conclude that

$$\frac{d}{dt}\langle \hat{x} \rangle = \frac{\hbar k_x}{m} \qquad (5.57)$$

As expected, this is just the x component of momentum divided by the mass or, equivalently, the speed of the particle in the x direction.

5.5.3   Uncertainty of expectation value

Here we are interested in establishing a measure of the deviation of the result of a measurement from the mean value. Let $\hat{A}$ be an operator corresponding to an observable when the system is in state $\psi(\mathbf{r})$. The mean (expectation) value of the observable A is

$$\langle \hat{A} \rangle = \int \psi^*(\mathbf{r})\hat{A}\psi(\mathbf{r})d^3r \qquad (5.58)$$

However, we are interested in obtaining a measure of the spread in values of the observable A. The deviations in the observable A can be defined in terms of the mean of squares of the deviations

$$(\Delta A)^2 = \langle (\hat{A} - \langle \hat{A} \rangle)^2 \rangle = \langle \hat{A}^2 + \langle \hat{A} \rangle^2 - 2\hat{A}\langle \hat{A} \rangle \rangle$$
$$= \langle \hat{A}^2 \rangle + \langle \hat{A} \rangle^2 - 2\langle \hat{A} \rangle \langle \hat{A} \rangle \qquad (5.59)$$

where $\langle \hat{A} \rangle$ is the mean of the measured value A and $(\Delta A)^2$ is the square of the deviations. It follows that

$$\boxed{\Delta A^2 = \langle \hat{A}^2 \rangle - \langle \hat{A} \rangle^2} \qquad (5.60)$$

or

$$\Delta A = (\langle \hat{A}^2 \rangle - \langle \hat{A} \rangle^2)^{1/2} \qquad (5.61)$$

We can also express this in integral form:

$$\Delta A^2 = \int \psi^*(\mathbf{r})\hat{A}^2\psi(\mathbf{r})d^3r - \left(\int \psi^*(\mathbf{r})\hat{A}\psi(\mathbf{r})d^3r \right)^2 \qquad (5.62)$$

The physical meaning of this is that $\langle \hat{A} \rangle$ is the average value of many observations on the system, and ΔA is a measure of the *root-mean-square* (rms) deviations or spread in the values of the measurement. Of course, there are other ways to measure a spread in the values of a measurement. However, we chose the above approach based on rms deviations because it is the most commonly used.

5.5.3.1 Uncertainty in expectation value of a particle confined by a one-dimensional, infinite, rectangular potential

As usual, we start out by defining the potential in which the particle moves. Figure 5.1(a) is a sketch of the one-dimensional potential. To simplify our expression for the eigenfunctions, we chose the position $x = 0$ to be the left-hand boundary of the potential, so that

$$V(x) = 0 \qquad 0 < x < L \tag{5.63}$$

and

$$V(x) = \infty \qquad \text{elsewhere} \tag{5.64}$$

We wish to find the expectation value of the particle position and the uncertainty in the position when the particle is in the n-th energy state. To solve this problem one starts by writing down the time-independent Schrödinger equation:

$$\left(-\frac{\hbar^2}{2m} \nabla^2 + V(\mathbf{r}) \right) \psi_n(\mathbf{r}) = E_n \psi_n(\mathbf{r}) \tag{5.65}$$

The boundary conditions are $\psi_n(x) = 0$ at $x = 0, x = L$. The solutions to the wave function are

$$\psi_n = A_n \sin(k_n x) \tag{5.66}$$

where $k_n = n\pi/L$ and n is a nonzero positive integer $n = 1, 2, 3, \ldots$.

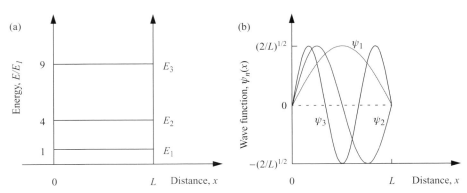

Fig. 5.1. (a) Sketch of a one-dimensional, rectangular potential well with infinite barrier energy showing the energy eigenvalues E_1, E_2, and E_3. (b) Sketch of the eigenfunctions ψ_1, ψ_2, and ψ_3 for the potential shown in (a).

The normalization constant A_n is found from the normalization condition

$$\int\limits_{x=0}^{x=L} \psi_n^*(x)\psi_n(x)dx = A_n^2 \int\limits_{x=0}^{x=L} \sin^2(k_n x)dx = 1 \tag{5.67}$$

$$\frac{1}{A_n^2} = \int\limits_{x=0}^{x=L} \left(\frac{1}{2} - \frac{1}{2}\cos(2k_n x)\right)dx = \left[\frac{x}{2} + \frac{1}{4k_n}\sin(2k_n x)\right]_0^L = \frac{L}{2} + 0 \tag{5.68}$$

where we used the relation $2\sin(x)\sin(y) = \cos(x-y) - \cos(x+y)$. Hence, $A_n = \sqrt{2/L}$, and we may write the wave function as

$$\boxed{\psi_n(x) = \sqrt{\frac{2}{L}}\sin\left(\frac{n\pi x}{L}\right)} \tag{5.69}$$

To find the expectation value of x one must solve the integral

$$\langle x_n \rangle = \int \psi_n^*(x)x\psi_n(x)dx = A_n^2 \int x\sin^2(k_n x)dx$$

$$= A_n^2 \int x\left(\frac{1}{2} - \frac{1}{2}\cos(2k_n x)\right)dx \tag{5.70}$$

Written in the form shown on the right-hand side, the integral may be found by inspection of odd and even functions to give

$$\langle x_n \rangle = A^2\left[\frac{x^2}{4}\right]_0^L \tag{5.71}$$

Alternatively, working a little harder, we may solve Eqn (5.70) by integrating by parts using $\int UV'dx = UV - \int U'Vdx$.

In this case, $U = x$ and $V' = (\frac{1}{2} - \frac{1}{2}\cos(2k_n x))$, so that

$$V = \frac{x}{2} - \frac{1}{4k_n}\sin(2k_n x)$$

and

$$\langle x_n \rangle = A_n^2\left(\left[\frac{x^2}{2} - \frac{x}{4k_n}\sin(2k_n x)\right]_0^L - \int\left(\frac{x}{2} - \frac{1}{4k_n}\sin(2k_n x)\right)dx\right) \tag{5.72}$$

$$\langle x_n \rangle = A_n^2\left[\frac{x^2}{2} - \frac{x}{4k_n}\sin(2k_n x) - \frac{x^2}{4} - \frac{1}{8k_n^2}\cos(2k_n x)\right]_0^L \tag{5.73}$$

But $k_n = n\pi/L$ and $n = 1, 2, 3, \ldots$, so that

$$\langle x_n \rangle = A_n^2\left(\frac{L^2}{2} + 0 - \frac{L^2}{4} + 0\right) = A_n^2\frac{L^2}{4} \tag{5.74}$$

And, since $A_n^2 = 2/L$, we finally have

$$\boxed{\langle x_n \rangle = \frac{L}{2}}$$
(5.75)

To check this quantum mechanical result for the average value of the position of the particle, it makes sense to compare it with the predictions of classical mechanics. In classical mechanics, the particle in the potential well moves at constant velocity, v, and traverses the well in time $\tau = L/v$. The average position x is given by

$$\langle x \rangle = \int\limits_{t=0}^{t=\tau} \frac{vt\,dt}{\tau} = \frac{1}{2}v\frac{\tau^2}{\tau} = \frac{1}{2}v\tau = \frac{1}{2}v\frac{L}{v}$$
(5.76)

Hence, $\langle x \rangle = L/2$, which is quite satisfying, since it is the same as the quantum result.

To find the expectation value of x^2 in the quantum mechanical case, we must solve

$$\langle x_n^2 \rangle = A_n^2 \int x^2 \sin^2(k_n x)dx = A_n^2 \int \left(\frac{x^2}{2} - \frac{x^2}{2}\cos(2k_n x) \right) dx$$
(5.77)

$$\langle x_n^2 \rangle = A_n^2 \left(\left[\frac{x^3}{6} - \frac{x^2}{2}\frac{1}{2k_n}\sin(2k_n x) \right]_0^L + \int \frac{x}{2k_n}\sin(2k_n x)dx \right)$$
(5.78)

$$\langle x_n^2 \rangle = A_n^2 \left(\left[\frac{x^3}{6} - \frac{x^2}{2}\frac{1}{2k_n}\sin(2k_n x) + \frac{x}{2k_n}\left(-\frac{1}{2k_n} \right)\cos(2k_n x) \right]_0^L \right.$$
$$\left. + \int \frac{1}{4k_n^2}\cos(2k_n x)dx \right)$$
(5.79)

$$\langle x_n^2 \rangle = A_n^2 \left[\frac{x^3}{6} - \frac{x^2}{4k_n}\sin(2k_n x) - \frac{x}{4k_n^2}\cos(2k_n x) + \frac{1}{8k_n^3}\sin(2k_n x) \right]_0^L$$
(5.80)

where the second and fourth terms contribute zero, since $k_n = n\pi/L$ and $n = 1, 2, 3, \ldots$. Hence,

$$\langle x_n^2 \rangle = A_n^2 \left(\frac{L^3}{6} - \frac{L}{4k_n^2} \right) = A_n^2 \left(\frac{L^3}{6} - \frac{L}{4}\cdot\frac{L^2}{n^2\pi^2} \right) = A_n^2 \left(\frac{L^3}{6} - \frac{L^3}{4n^2\pi^2} \right)$$
(5.81)

but $A_n^2 = 2/L$, so

$$\boxed{\langle x_n^2 \rangle = \frac{L^2}{3} - \frac{L^2}{2n^2\pi^2}}$$
(5.82)

The uncertainty in the position of the particle in the n-th state is given by the rms deviation $\Delta x_n = (\langle x_n^2 \rangle - \langle x_n \rangle^2)^{1/2}$, which we calculate using

$$\Delta x_n^2 = \langle x_n^2 \rangle - \langle x_n \rangle^2 = \frac{L^2}{3} - \frac{L^2}{2n^2\pi^2} - \frac{L^2}{4} = \frac{L^2}{12}\left(4 - \frac{6}{n^2\pi^2} - 3\right) \tag{5.83}$$

$$\boxed{\Delta x_n^2 = \frac{L^2}{12}\left(1 - \frac{6}{n^2\pi^2}\right)} \tag{5.84}$$

This result is interesting, because in the limit of very high-energy eigenvalues ($n \to \infty$) the rms deviation in particle position approaches the classical result $\Delta x_{\text{Classical}} = L/\sqrt{12}$. It is always a good idea to compare the predictions of quantum mechanics with the classical result, as this helps us to appreciate and develop an intuitive feel for quantum mechanics.

5.5.4   The generalized uncertainty relation

In Section 5.5.3 we considered the spread in results of measurement about some average value. We went on to consider the specific example of finding the expectation value and uncertainty in particle position in a one-dimensional, rectangular potential well with infinite barrier energy.

There is another important concept in quantum mechanics that links the uncertainty in results of measurement between a given pair of associated noncommuting operators. The spread in results of one set of measurements associated with one operator is related to the spread in measured values of the associated noncommuting operator. This is the uncertainty relation, which we now discuss by considering a pair of noncommuting operators $\hat{\mathbf{A}}$ and $\hat{\mathbf{B}}$.

Consider an operator $\hat{\mathbf{A}}$:

$$\langle \hat{\mathbf{A}}\hat{\mathbf{A}}^\dagger \rangle \geq 0 \tag{5.85}$$

because

$$\langle \hat{\mathbf{A}}\hat{\mathbf{A}}^\dagger \rangle = \langle \psi | \hat{\mathbf{A}}^\dagger \hat{\mathbf{A}} | \psi \rangle = \langle \hat{\mathbf{A}}\psi | \hat{\mathbf{A}}\psi \rangle \geq 0 \tag{5.86}$$

from the definition of Hermitian conjugate. Or, in terms of integrals,

$$\langle \hat{\mathbf{A}}\hat{\mathbf{A}}^\dagger \rangle = \int \psi^* (\hat{\mathbf{A}}^\dagger \hat{\mathbf{A}}\psi) = \int (\hat{\mathbf{A}}\psi)^* (\hat{\mathbf{A}}\psi) = \int (\hat{\mathbf{A}}\psi)^2 \geq 0 \tag{5.87}$$

Noting that

$$\int \phi^* i\hat{\mathbf{A}}\phi = \int (\hat{\mathbf{A}}^\dagger\phi)^* i\phi = -i \int (\hat{\mathbf{A}}^\dagger\phi)^*\phi = -\int (i\hat{\mathbf{A}}^\dagger\phi)^*\phi \tag{5.88}$$

we can create a linear combination $\hat{\mathbf{A}} + i\hat{\mathbf{B}}$, so that

$$\langle \hat{\mathbf{A}} + i\hat{\mathbf{B}} \rangle = \langle \hat{\mathbf{A}} \rangle + i \langle \hat{\mathbf{B}} \rangle \tag{5.89}$$

and

$$(\hat{\mathbf{A}} + i\hat{\mathbf{B}})^{\dagger} = \hat{\mathbf{A}}^{\dagger} - i\hat{\mathbf{B}}^{\dagger} \tag{5.90}$$

If $\hat{\mathbf{A}}$ and $\hat{\mathbf{B}}$ are Hermitian, then $(\hat{\mathbf{A}} + i\hat{\mathbf{B}})^{\dagger} = \hat{\mathbf{A}} - i\hat{\mathbf{B}}$. If one now considers an operator $(\hat{\mathbf{A}} + i\lambda\hat{\mathbf{B}})$, where λ is real and $\hat{\mathbf{A}}$ and $\hat{\mathbf{B}}$ are Hermitian operators, then one may write

$$\langle (\hat{\mathbf{A}} + i\lambda\hat{\mathbf{B}})(\hat{\mathbf{A}} + i\lambda\hat{\mathbf{B}})^{\dagger} \rangle = \langle (\hat{\mathbf{A}} + i\lambda\hat{\mathbf{B}})(\hat{\mathbf{A}}^{\dagger} - i\lambda\hat{\mathbf{B}}^{\dagger}) \rangle \geq 0 \tag{5.91}$$

$$\langle \hat{\mathbf{A}}^2 \rangle + \lambda^2 \langle \hat{\mathbf{B}}^2 \rangle - i\lambda \langle \hat{\mathbf{A}}\hat{\mathbf{B}} - \hat{\mathbf{B}}\hat{\mathbf{A}} \rangle \geq 0 \tag{5.92}$$

Therefore, the last term on the left-hand side $\langle \hat{\mathbf{A}}\hat{\mathbf{B}} - \hat{\mathbf{B}}\hat{\mathbf{A}} \rangle = [\langle \hat{\mathbf{A}}, \hat{\mathbf{B}} \rangle]$ must be zero or pure imaginary. The minimum value of λ is found by taking the derivative with respect to λ in such a way that

$$0 = \frac{d}{d\lambda}(\langle \hat{\mathbf{A}}^2 \rangle + \lambda^2 \langle \hat{\mathbf{B}}^2 \rangle - i\lambda \langle \hat{\mathbf{A}}\hat{\mathbf{B}} - \hat{\mathbf{B}}\hat{\mathbf{A}} \rangle) \tag{5.93}$$

$$0 = 2\lambda_{\min}\langle \hat{\mathbf{B}}^2 \rangle - i\langle \hat{\mathbf{A}}\hat{\mathbf{B}} - \hat{\mathbf{B}}\hat{\mathbf{A}} \rangle = 2\lambda_{\min}\langle \hat{\mathbf{B}}^2 \rangle - i[\langle \hat{\mathbf{A}}, \hat{\mathbf{B}} \rangle] \tag{5.94}$$

$$\lambda_{\min} = \frac{i}{2} \frac{[\langle \hat{\mathbf{A}}, \hat{\mathbf{B}} \rangle]}{\langle \hat{\mathbf{B}}^2 \rangle} \tag{5.95}$$

Substituting the minimum value $\lambda_{\min}$ into Eqn (5.92) gives

$$\langle \hat{\mathbf{A}}^2 \rangle - \frac{[\langle \hat{\mathbf{A}}, \hat{\mathbf{B}} \rangle]^2 \langle \hat{\mathbf{B}}^2 \rangle}{4\langle \hat{\mathbf{B}}^2 \rangle^2} + \frac{[\langle \hat{\mathbf{A}}, \hat{\mathbf{B}} \rangle]^2}{2\langle \hat{\mathbf{B}}^2 \rangle} \geq 0 \tag{5.96}$$

so that

$$\boxed{\langle \hat{\mathbf{A}}^2 \rangle \langle \hat{\mathbf{B}}^2 \rangle \geq -\frac{[\langle \hat{\mathbf{A}}, \hat{\mathbf{B}} \rangle]^2}{4}} \tag{5.97}$$

The product of the expectation value of the square of a Hermitian operator with the expectaion value of the square of another Hermitian operator has a minimum value that is proportional to the square of the commutator of the two operators. To show that this applies to the root-mean-square (rms) value we create a new set of operators in such a way that

$$\hat{\mathbf{A}} \to \hat{\mathbf{A}} - \langle \hat{\mathbf{A}} \rangle \equiv \delta\hat{\mathbf{A}} \tag{5.98}$$

$$\hat{\mathbf{B}} \to \hat{\mathbf{B}} - \langle \hat{\mathbf{B}} \rangle \equiv \delta\hat{\mathbf{B}} \tag{5.99}$$

so that

$$\langle (\delta\hat{\mathbf{A}})^2 \rangle = \langle (\hat{\mathbf{A}} - \langle \hat{\mathbf{A}} \rangle)^2 \rangle = \langle \hat{\mathbf{A}}^2 \rangle - \langle 2\hat{\mathbf{A}}\langle \hat{\mathbf{A}} \rangle \rangle + \langle \hat{\mathbf{A}} \rangle^2$$
$$= \langle \hat{\mathbf{A}}^2 \rangle - \langle \hat{\mathbf{A}} \rangle^2 = \Delta\hat{\mathbf{A}}^2 \tag{5.100}$$

which is the rms deviation. We can relate $[\delta\hat{A}, \delta\hat{B}]$ to operators $\hat{A}$ and $\hat{B}$:

$$[\delta\hat{A}, \delta\hat{B}] = \hat{A}\hat{B} - \hat{A}\langle\hat{B}\rangle - \langle\hat{A}\rangle\hat{B} + \langle\hat{A}\rangle\langle\hat{B}\rangle$$

$$- \hat{B}\hat{A} + \langle\hat{B}\rangle\hat{A} + \hat{B}\langle\hat{A}\rangle - \langle\hat{B}\rangle\langle\hat{A}\rangle \tag{5.101}$$

$$[\delta\hat{A}, \delta\hat{B}] = \hat{A}\hat{B} - \hat{B}\hat{A} = [\hat{A}, \hat{B}] \tag{5.102}$$

Substituting $\delta\hat{A}$ and $\delta\hat{B}$ into our previous expression $\langle\hat{A}^2\rangle\langle\hat{B}^2\rangle \geq -[\langle\hat{A}, \hat{B}\rangle]^2/4$, we obtain

$$\langle\delta\hat{A}^2\rangle\langle\delta\hat{B}^2\rangle \geq -[\langle\delta\hat{A}, \delta\hat{B}\rangle]^2/4 \tag{5.103}$$

Using $\langle\delta\hat{A}^2\rangle = \Delta A^2$ from Eqn (5.100) and the relation given by Eqn (5.102) allows us to rewrite Eqn (5.103) as

$$\Delta A^2 \Delta B^2 \geq -[\langle\hat{A}, \hat{B}\rangle]^2/4 \tag{5.104}$$

or

$$\boxed{\Delta A \Delta B \geq \frac{i}{2}[\langle\hat{A}, \hat{B}\rangle]} \tag{5.105}$$

which is the generalized uncertainty relation. This relationship between a conjugate pair of noncommuting linear operators may be considered a consequence of the mathematics that is built into our description of quantum phenomena.

As a specific example of the uncertainty relation, consider a particle moving in one dimension. To find the uncertainty in position and momentum we let the operator $\hat{A} = \hat{p}_x = -i\hbar \cdot \partial/\partial x$, which is the x component of the momentum operator, and the operator $\hat{B} = \hat{x}$, which is the x-position operator. Then, from the commutation relation,

$$[\langle\hat{p}_x, \hat{x}\rangle] \equiv [\hat{p}_x, \hat{x}] = -i\hbar \tag{5.106}$$

and the uncertainty relation for position and momentum operators can be found from Eqn (5.105):

$$\Delta p_x \Delta x \geq \frac{i}{2}[\langle\hat{p}_x, \hat{x}\rangle] = \frac{-i}{2} \cdot i\hbar \tag{5.107}$$

$$\Delta p_x \Delta x \geq \frac{\hbar}{2} \tag{5.108}$$

Suppose the particle is an electron confined to some region of space. If we perform a measurement to determine electron position once and then repeat the measurement in a large number of identically prepared systems containing an electron, we might obtain a Gaussian distribution of position results with spread $\Delta x = 1$ nm. The uncertainty relation given by Eqn (5.108) means that in this case we cannot know the momentum of the electron to an accuracy better than $\Delta p = \hbar/2\Delta x = 5.27 \times 10^{-26}$ kg m s^{-1}. The spread in momentum has a corresponding spread in velocity, which is $\Delta v = \Delta p / m_0 = 5.7 \times 10^4$ m s^{-1}. It is interesting to note that this value of velocity would cause a classical particle to traverse a distance of 1 nm in just 1.7×10^{-14} s.

5.6 Density of states

So far in this chapter we have discussed some of the one-particle properties of wave-function space. This included linearity, completeness, Hermitian operators, expectation values, and the measurement of noncommuting conjugate pairs of operators. In this section we will introduce the idea of a density of states in quantum mechanics. This concept is important because we will be interested in controlling and changing the occupation probability of certain states in a system.

5.6.1   Density of electron states

Suppose an electron is known to occupy a particular eigenstate up to a certain moment in time. The probability of occupation may be changed at some later time by changing the potential seen by the electron. This change in potential causes the electron to occupy other eigenstates, which, in general, have different energy eigenvalues. One might expect the number of distinct states available in a given energy range to have an influence on the probability of the electron changing its state. This is one reason why we are interested in the number of electron states per unit energy interval. Such a *density of states* is a very useful quantity when calculating many different properties of materials. As an application, we will show how the density of states leads to quantization of electron conduction. Of course, the density of electron states is important in many other applications. In a semiconductor laser diode, it plays a key role in determining device behavior.

So far, we have introduced the concept of energy eigenvalues and eigenfunctions as solutions to the time-independent Schrödinger equation. Our calculations showed that the number of states varies per unit energy interval. This fact makes it a little difficult to calculate the density of states (the number of states per unit energy interval). However, instead of looking for solutions to Schrödinger's equation in *real* space, we can consider solutions in k space. When we do this, it is often trivial to calculate the density of electron states.

As an example of how to calculate a density of states, we consider a particle free to move in space. The particle is described by a wave function $\psi(\mathbf{r}, t)$ of the form $e^{i\mathbf{k}\cdot\mathbf{r}}e^{-i\omega t}$ and a nonlinear dispersion relation $\omega = \hbar k^2 / 2m$. The energy of the particle is $E = \hbar^2 k^2 / 2m$. To make further progress, consider a large volume of space defined by a cube of side L along the Cartesian coordinates x, y, and z so that volume $V = L_x L_y L_z = L^3$. We now apply periodic boundary conditions to the wave function in

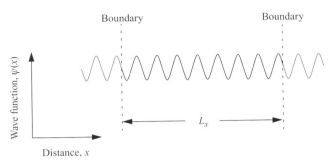

Fig. 5.2. Illustration showing periodic boundary conditions applied to wave function $\psi(x)$ in such a way that $\psi(x) = \psi(x + L_x)$.

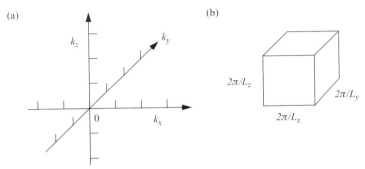

Fig. 5.3. (a) Counting particle states in k space is often easier than counting states in energy space. In this example, the particle can move in the positive direction with velocity $\hbar k/m$ or in the negative direction with velocity $-\hbar k/m$. (b) The volume of k space occupied by one k state is $(2\pi)^3/L^3$.

such a way that $\psi(x) = \psi(x + L_x)$, $\psi(y) = \psi(y + L_y)$, and $\psi(z) = \psi(z + L_z)$. The application of periodic boundary conditions to the wave function $\psi(x)$ is illustrated in Fig. 5.2. Periodic boundary conditions discretize the wave vector components in such a way that $k_x = 2n_x\pi/L_x$, $k_y = 2n_y\pi/L_y$, and $k_z = 2n_z\pi/L_z$, where n_x, n_y, and n_z are integers. Because each wave vector component is linear in the integer n, we see in Fig. 5.3(a) that quantum states are *equally spaced* in k *space*. For this reason, it is often easier to count particle states in k space than to count states in energy space. As illustrated in Fig. 5.3(b), each k state takes up a volume $(2\pi)^3/L^3$ or $(2\pi)^3$ if we normalize to unit length.

The density of states in k space is the *number of states* between k and $k + dk$ per *unit volume*. For a large number of equally spaced states in k space in three dimensions, this gives

$$D_3(k)dk = \frac{1}{V} \cdot \frac{L^3}{(2\pi)^3} 4\pi k^2 dk = \frac{4\pi k^2}{(2\pi)^3} dk \tag{5.109}$$

where $D_3(k)$ is the density of states and the subscript 3 indicates we are referring to three dimensions. In Eqn (5.109), we calculated the volume of a shell of radius k and thickness dk, divided by the volume occupied by each k state to obtain the number of states in the shell, and divided by the volume V to obtain a density of states.

If we are considering a particle with spin, this would be included as a multiplicative term in Eqn (5.109). For example, an electron of spin quantum number $s = \pm 1/2$ (or eigenvalue $\pm \hbar/2$) multiplies the density of states by a factor 2 because there are two possible spin states the electron could be in.

To convert the three-dimensional density of states in k space to a density of states in energy, we note that for a particle of mass m, energy $E = \hbar^2 k^2/2m$ and $dE = (\hbar^2 k/m)dk$. Hence,

$$D_3(E)dE = D_3(k)\frac{dk}{dE} \cdot dE = \frac{4\pi k^2}{(2\pi)^3}\frac{m}{\hbar^2 k} \cdot dE = \frac{1}{2\pi^2} \cdot \frac{km}{\hbar^2} \cdot dE \tag{5.110}$$

$$D_3(E)dE = \frac{1}{2 \cdot 2\pi^2}\left(\frac{2m}{\hbar^2}\right)^{3/2}\left(\frac{\hbar^2 k^2}{2m}\right)^{1/2} dE \tag{5.111}$$

so that the density of states in three dimensions is

$$D_3(E) = \frac{1}{4\pi^2}\left(\frac{2m}{\hbar^2}\right)^{3/2} E^{1/2} \tag{5.112}$$

It is straightforward to show that in two dimensions

$$D_2(E) = \frac{m}{2\pi \hbar^2} \tag{5.113}$$

and that for one dimension

$$D_1(E) = \frac{1}{4\pi}\left(\frac{2m}{\hbar^2}\right)^{1/2} E^{-1/2} \tag{5.114}$$

One may summarize the results by plotting the density of states for different dimensions. This is done in Fig. 5.4.

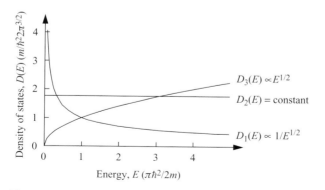

Fig. 5.4. Density of states as a function of energy plotted for one, two, and three dimensions.

For a particle of mass m, the densities of states in three and one dimensions are equal when the particle energy has a value $E = \pi \hbar^2 / 2m$. At this value of particle energy, the density of states has a value $m / \hbar^2 \pi^{3/2}$. The densities of states in three and two dimensions are equal with value $m / 2\pi \hbar^2$ when the particle energy is $E = \pi^2 \hbar^2 / 2m$.

Typically, a *quantum well* potential formed from a semiconductor heterostructure such as epitaxially grown thin layers of GaAs and AlGaAs has a two-dimensional density of electron states for low-energy electrons. Quantum wells formed from heterostructures are of practical importance in many semiconductor devices. For example, such quantum wells are often used as the active region of a semiconductor laser diode. The reason for this is that the small volume of the quantum well reduces the current needed to achieve lasing and, over some range of emission wavelengths, differential optical gain can increase compared with bulk values.

The atomic precision with which quantum wells can be fabricated is well illustrated in Fig. 5.5. The figure shows a transmission electron micrograph of an InGaAs quantum well that is just three monolayers thick sandwiched between InP barrier layers. The spots in the image represent tunnels between pairs of atoms. The minimum separation between tunnels in InP is 0.34 nm.

A laterally patterned quantum well can be made to form a *quantum wire* that has a one-dimensional density of electronic states. Structures of this type can be designed to exhibit quantized electrical conductance. Quantized conductance is not predicted classically and is another example of an effect that may play an important role in future very small (scaled) electronic structures such as transistors.

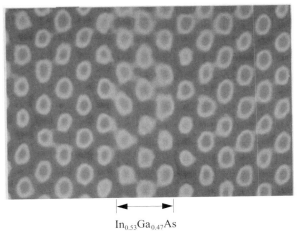

$In_{0.53}Ga_{0.47}As$

Fig. 5.5. Transmission electron micrograph showing an InGaAs quantum well in cross-section that is three monolayers thick and is sandwiched between InP barrier layers. The spots in the image represent tunnels between pairs of atoms. The minimum separation between tunnels in InP is 0.34 nm. Image courtesy of M. Gibson, Argonne National Laboratory.

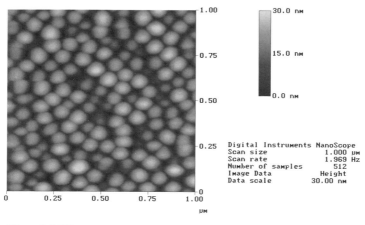

RMS rough=5.976nm
e3718a.м00

Fig. 5.6. Area view of InP self-assembled quantum dots grown using low-pressure MOCVD on an InAlP matrix layer lattice-matched to a GaAs substrate. As measured from AFM images, areal density of quantum dots is 1.5×10^{10} cm^{-2} and dominant size is in the range of 15–20 nm for a 15-monolayer "planar-growth-equivalent" deposition time at a growth temperature of 650 °C. Dominant sizes are controllable by changing the deposition time. Image courtesy of R. Dupuis, University of Texas at Austin.

One may also extend the density of states idea to "zero dimensions". In this case, the states are confined by a potential in all three dimensions similar to an isolated atom. In semiconductor devices, the structures that give rise to such a potential are called *quantum dots*. An example of InP quantum dots imaged by an atomic force microscope (AFM) is shown in Fig. 5.6. In this case most of the quantum dots have a diameter in the range 15–20 nm.

Quantum wells, wires, and dots formed in semiconductor structures do not have potentials with infinite barrier energies. However, there are still bound states that exist in potential minima formed by potentials with finite barrier energy.

As an application of density of states we will now show that in one dimension electron velocity and density of states exactly cancel to give quantized conductance.

5.6.1.1 Quantum conductance

To illustrate the quantization of electrical conductance, we consider the situation in which current flows through a region where electrons are confined by a potential to motion in one dimension. This is shown schematically in Fig. 5.7. The one-dimensional conduction region created by a confining potential is attached to electrodes placed on the left and right. We will be considering an electron moving from left to right. The transverse electron wave number for an electron moving in the x direction is quantized by the confining potential. Each quantized level defines a channel for electron

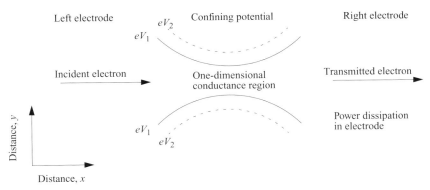

Fig. 5.7. Diagram showing the top view of left- and right-hand electrodes connected via a one-dimensional conductance region defined by a confining potential. Contours of constant potential energy eV_1 and eV_2, where $eV_1 < eV_2$, are shown. Transverse electron wave number is quantized by the confining potential.

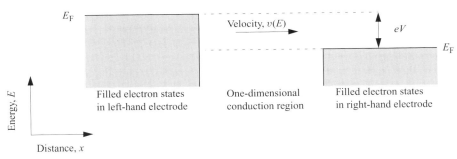

Fig. 5.8. At low temperatures, electrons in the left- and right-hand electrodes occupy states up to the Fermi energy, E_F. A potential energy eV applied between the left- and right-hand electrodes allows occupied electron states in the left-hand electrode to traverse the one-dimensional region and enter unoccupied electron states in the right-hand electrode.

transmission. In the following, we will assume that each channel is independent and hence uncorrelated.

At low temperatures, electrons in the electrodes occupy states up to the Fermi energy, E_F. If a voltage, V, is applied between the electrodes, we expect current to flow. As shown in Fig. 5.8, potential energy eV applied between the left- and right-hand electrodes allows occupied electron states in the left-hand electrode to traverse the one-dimensional region and enter unoccupied electron states in the right-hand electrode. At first sight, one might anticipate that current is proportional to the applied voltage, electron velocity, v, the one-dimensional transmission coefficient, T, and the one-dimensional density of electron states, D_1.

We start by evaluating v, T, and D_1, which depend on electron energy E. We will assume that the velocity of an electron characterized by wave vector k and mass m moving from left to right in the one-dimensional region is simply $v(E) = \hbar k/m$. The density of electron states in the one-dimensional region is just $D_1(E)dE = 2 \cdot 2dk/2\pi = 2mdE/\pi\hbar^2 k$, where electron energy $E = \hbar^2 k^2/2m$, so that $dk = mdE/\hbar^2 k$. The first factor of 2 in $D_1(E)$ is because there are two possible spins an electron can have per k state. The second factor of 2 is because the k state can represent an electron moving to the left or to the right. Since we only consider electrons moving from left to right, we must remember to divide $D_1(E)$ by 2 when evaluating the current.

The current at bias V is given by the integral

$$I = e \int_{E_F}^{E_F + eV} v(E)T(E)\frac{D_1(E)}{2}dE = e \int_{E_F}^{E_F + eV} \frac{1}{\pi\hbar}T(E)dE \qquad (5.115)$$

The key point is that terms in k in the expressions for v and D_1 cancel. Simplifying further by only considering small voltage bias V, the transmission coefficient $T(E) \to T(E_F)$, so that this may be taken out of the integral, and current becomes $I = e^2 T(E_F)V/\pi\hbar$. In this situation, the conductance $G_{cond} = I/V$ is

$$G_n = \frac{e^2}{\pi\hbar}T(E_F) \qquad (5.116)$$

This is sometimes called the Landauer formula. Conductance G_n has a subscript n because conductance is quantized. To see this, all one need do is consider the situation in which the transmission coefficient has its maximum value, $T(E_F) = 1$. In this case, the maximum conductance per electron *per spin* is $e^2/2\pi\hbar = 1/R_k = 25.8 \text{ k}\Omega^{-1}.^2$

The only way to increase conduction is to increase the number of parallel paths an electron can take from left to right through the region between the electrodes. Electrical conduction will then increase in a step-wise fashion to a value proportional to the number of parallel electron paths available between the electrodes. As illustrated in Fig. 5.9, one way to increase the number of parallel paths is to increase the width of the one-dimensional potential channel, thereby fitting more transverse *electron waveguide modes* through. One talks of electron waveguide modes in the confining one-dimensional potential because the wave nature of the electron suggests an analogy with classical electromagnetic waveguides.

Conduction is not limited by electron scattering or dissipation; rather it is limited by the quantum mechanical wavy nature of the electron. In our simple model system, no electron scattering takes place in the one-dimensional conduction region. The one-dimensional potential acts as a loss-less electron waveguide. Electron scattering and power dissipation take place in the electrodes.

If one creates electronic devices such as transistors in which electrons are constrained to move through regions comparable to the electron wavelength, it is necessary to consider quantum conductance. The value of quantum conductance per electron per

---

[2] The factor 2 appears because we only consider one spin state. The value of R_K is known as the von Klitzing constant (see Appendix A).

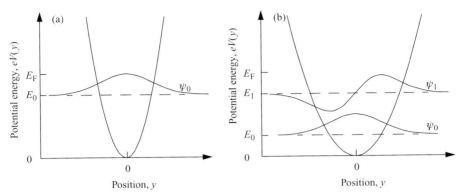

Fig. 5.9. At low temperatures, electrons occupy states up to the Fermi energy, E_F. For the strong one-dimensional confining potential shown in (a) electrons occupy the lowest-energy transverse state ψ_0, which has energy $E_0 < E_F$. This limits maximum conductance per electron per spin to $25.8 \text{ k}\Omega^{-1}$. Conductance may be increased by increasing the number of parallel paths an electron can access in traversing the confining potential. As shown in (b), one way to achieve this is to maintain the Fermi energy while increasing the width of the one-dimensional potential channel, thereby fitting more transverse electron waveguide modes through.

spin of only $25.8 \text{ k}\Omega^{-1}$ limits the current drive performance of small devices and hence the speed at which these devices can operate.

From a practical point of view, it might be more productive to consider devices that do not operate in the linear, near-equilibrium regime. In this case, one needs to adopt a somewhat more complex description of electrical conductivity.

5.6.2   Density of photon states

Photons, like electrons, may be characterized by a wavelength, λ, or k state in which $k = 2\pi/\lambda$. The three-dimensional density of states in k-space, $D_3^{\text{opt}}(k)$ follows directly and is given by Eqn (5.109). This density of states may be expressed in terms of angular frequency ω if we know the relationship between ω and k. Since the dispersion of polarized light propagating in three-dimensional free space is $\omega = c_k$ (where c is the speed of light), it follows that

$$D_3^{\text{opt}}(\omega)d\omega = \frac{4\pi}{(2\pi)^3}\frac{\omega^2}{c^2}\frac{1}{c}d\omega = \frac{\omega^2}{2\pi^2c^3}d\omega \tag{5.117}$$

In general, since a photon has a spin quantum number of $s = \pm 1$ corresponding to angular momentum eigenvalue $\pm\hbar$ (there are two orthogonal polarizations of light in free space), this density of states should be multiplied by a factor of 2. Hence, the density of photon states (or field modes) in three-dimensional free space in the frequency range ω to $\omega + d\omega$ is

$$\boxed{D_3^{\text{opt}}(\omega) = \frac{\omega^2}{\pi^2c^3}} \tag{5.118}$$

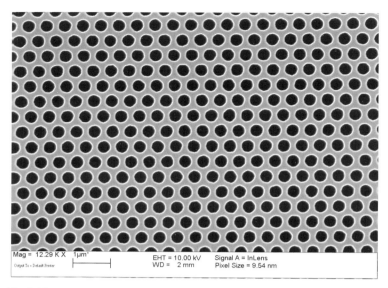

Fig. 5.10. Scanning electron microscope image of two-dimensional photonic crystal in plan view. The photonic crystal is a triangular lattice of 500-nm period with 350-nm holes etched into a 0.4-μm-thick silicon layer bonded to a 2-μm-thick silica layer.

In an isotropic loss-less dielectric medium characterized by a refractive index n_r, the dispersion relation is modified to $\omega = ck/n_r$, and our expression for $D_3^{\text{opt}}(\omega)$ becomes

$$D_3^{\text{opt}}(\omega) = \frac{\omega^2 n_r^3}{\pi^2 c^3} \tag{5.119}$$

In direct analogy with the electron density of states discussed in Section 5.6.1, there is interest in understanding the behavior of photons in situations in which large changes in the density of states and highly nonlinear dispersion relations exist in loss-less dielectric and active semiconductor nanostructures. It is straightforward to show that the photon density of states is modified by using dielectric structures that vary with a half-wavelength period in space. Such structures are called photonic crystals[3] and belong to a larger class of meta-materials.[4] For typical infrared laser light at a wavelength near 1500 nm and an effective refractive index near 1.5, this implies periods of approximately 500 nm and features with sizes less than this. Such nanoscale dielectrics are easily fabricated in two dimensions using existing semiconductor fabrication techniques. Figure 5.10 shows a scanning electron microscope image of a two-dimensional

---

[3] The concept of dispersion in photonic crystals was first introduced by K. Ohtaka, *Phys. Rev.* **B19**, 5057 (1979). For an introduction see J. D. Joannopoulos, R. D. Meade, and J. N. Winn, *Photonic Crystals*, Princeton University Press, Princeton, 1995 (ISBN 0 691 03744 2).

[4] Meta-materials are purely artificial structures which sometimes exhibit remarkable properties not usually found in nature.

photonic crystal with a triangular lattice created by etching 350-nm-diameter holes into 0.4-μm-thick single-crystal silicon. The thin silicon layer is bonded to a 2-μm-thick layer of silica grown on a silicon substrate. Light of wavelength near 1500 nm that is waveguided in the plane sees large periodic changes in refractive index from silicon, with $n_r = 3.47$, to air, with $n_r = 1$. The periodicity in refractive index can sometimes result in dramatic changes in photon density of states, including ranges of photon energy, called photonic band gaps, where no photons can propagate.

While the description of light propagation in periodic dielectrics is simplified by the existence of spatial symmetry, analysis of the photon density of states in nonperiodic nanoscale dielectrics is a significantly more challenging task.

5.7 Example exercises

Exercise 5.1
Show that, if wave functions $\psi_1(\mathbf{r})$ and $\psi_2(\mathbf{r})$ belong to the space of linear square-integrable wave functions, the linear combination $\psi(\mathbf{r}) = \lambda_1 \psi_1(\mathbf{r}) + \lambda_2 \psi_2(\mathbf{r})$, where λ_1 and λ_2 are complex numbers, is also square-integrable.

Exercise 5.2
Show that the density of states for a particle of mass m confined to a two-dimensional square potential well with infinite energy barriers is

$$D_2(E) = \frac{m}{2\pi\hbar^2}$$

and that for a similar one-dimensional square potential well is

$$D_1(E) = \frac{1}{4\pi}\left(\frac{2m}{\hbar^2}\right)^{1/2} E^{-1/2}$$

Exercise 5.3
An electron in an infinite, one-dimensional, rectangular potential well of width L is in the simple superposition state consisting of the ground and first excited state so that

$$\psi(x, t) = \frac{1}{\sqrt{2}}(\psi_1(x, t) + \psi_2(x, t))$$

Find expressions for:
(a) the probability density, $|\psi(x, t)|^2$;
(b) the average particle position, $\langle x(t)\rangle$;
(c) the momentum probability density, $|\psi(p_x, t)|^2$;
(d) the average momentum, $\langle p_x(t)\rangle$;
(e) the current flux, $\mathbf{J}(x, t)$.

Exercise 5.4

In Section 5.5.3.1 we found $\langle x \rangle$, $\langle x^2 \rangle$, and Δx^2 for a particle confined by the potential $V(x) = 0$ for $0 < x < L$ and $V(x) = \infty$ elsewhere. Repeat the calculation and show that as the state number $n \to \infty$ the average values approach those obtained from classical mechanics. Calculate the average particle momentum $\langle p_x \rangle$, $\langle p_x^2 \rangle$, and Δp_x^2 as a function of state n. How does $\Delta x \Delta p$ depend upon n?

Exercise 5.5

Find $\psi(k)$ for a particle with state function $\psi(x) = 1/\sqrt{2L}$ for $|x| < L$ and $\psi(x) = 0$ for $|x| > L$. Show that the uncertainty (rms deviation) in its momentum Δp_x is infinite, and plot $|\psi(k)|^2$ and $|\psi(x)|^2$. Calculate the uncertainty in position, Δx.

Exercise 5.6

A hydrogen atom in its ground state has electron wave function $\psi_0(r) = Ae^{-r/r_0}$. The electron is subject to a radially symmetric coulomb potential given by $V(r) = -e^2/4\pi\varepsilon_0\varepsilon_r r$. Find the normalization constant A. Find the minimized energy expectation value $\langle E_0 \rangle$, and show that $\langle E_{\text{kinetic}} \rangle = -\langle E_{\text{potential}} \rangle/2$ (which is a result predicted by the virial theorem). What is the value of r_0, and to what does it physically correspond? Show that the expectation value of momentum $\langle p \rangle = 0$.

Exercise 5.7

Prove that the expectation value of the (Hermitian) momentum operator in Cartesian coordinates is real. Show that $-i\hbar\partial/\partial r$ is not a Hermitian operator in radial coordinates. Show that the radial momentum operator

$$\hat{p}_r = -i\hbar \frac{1}{r} \frac{\partial}{\partial r} r$$

is Hermitian.

Exercise 5.8

Consider a particle of mass m in a finite, one-dimensional, rectangular potential well for which $V(x) = 0$ for $-L < x < L$ and $V(x) = V_0$ elsewhere. The value of V_0 is a finite positive constant. Calculate the average kinetic energy of the particle ground state, and show that the contribution from the region outside the quantum well is negative.

Exercise 5.9

What can be said about the time dependence of the expectation value an operator $\hat{\mathbf{A}}$ that commutes with a Hamiltonian used to describe a physical system?

Suppose a Hamiltonian with eigenfunctions ϕ_1 and ϕ_2 and corresponding eigenvalues E_1 and E_2 does not commute with an operator $\hat{\mathbf{A}}$. The operator $\hat{\mathbf{A}}$ has eigenfunctions $u_1 = (\phi_1 + \phi_2)/\sqrt{2}$ and $u_2 = (\phi_1 - \phi_2)/\sqrt{2}$ and corresponding eigenvalues a_1 and a_2.

At time $t = 0$, the system is in state $\psi = u_1$. Show that at time t the state of the system is $\psi(t) = (\phi_1 e^{-iE_1 t/\hbar} + \phi_2 e^{iE_2 t/\hbar})/\sqrt{2}$, and determine how the expectation value of the operator $\hat{A}$ varies with time.

Exercise 5.10
Discuss the similarities and differences between classical electrodynamics and quantum mechanics.

SOLUTIONS

Solution 5.1
To illustrate the L^2 nature of space $\mathcal{J}$ consider wave functions $\psi_1(\mathbf{r})$ and $\psi_2(\mathbf{r})$ that belong to $\mathcal{J}$. In a linear space one may form the linear combination

$$\psi(\mathbf{r}) = \lambda_1 \psi_1(\mathbf{r}) + \lambda_2 \psi_2(\mathbf{r})$$

where λ_1 and λ_2 are complex numbers. To show that $\psi(\mathbf{r})$ is also square-integrable, we expand:

$$|\psi(\mathbf{r})|^2 = |\lambda_1|^2 |\psi_1(\mathbf{r})|^2 + |\lambda_2|^2 |\psi_2(\mathbf{r})|^2 + \lambda_1^* \lambda_2 \psi_1^*(\mathbf{r}) \psi_2(\mathbf{r}) + \lambda_1 \lambda_2^* \psi_1(\mathbf{r}) \psi_2^*(\mathbf{r})$$

The last two terms have the same modulus with an upper limit:

$$|\lambda_1||\lambda_2|(|\psi_1(\mathbf{r})|^2 + |\psi_2(\mathbf{r})|^2)$$

$|\psi(\mathbf{r})|^2$ is therefore smaller than a function the integral of which converges, since $\psi_1(\mathbf{r})$ and $\psi_2(\mathbf{r})$ are square-integrable.

Solution 5.2
In this exercise, we are asked to show that the density of states for a particle of mass m confined to a two-dimensional square potential well with infinite energy barriers is

$$D_2(E) = \frac{m}{2\pi \hbar^2}$$

and that for a similar one-dimensional square potential well is

$$D_1(E) = \frac{1}{4\pi} \left(\frac{2m}{\hbar^2} \right)^{1/2} E^{-1/2}$$

Starting with the expression for the one-dimensional density of k states, we have

$$D_1(k)dk = \frac{dk}{(2\pi)}$$

$$D_1(E)dE = \frac{dk}{dE} = \frac{m}{\hbar^2 k} \frac{1}{(2\pi)} dE = \frac{m}{\hbar^2(2\pi)} \frac{\hbar\, dE}{(2mE)^{1/2}} = \frac{1}{4\pi} \left(\frac{2m}{\hbar^2} \right)^{1/2} E^{-1/2} dE$$

The density of states in two dimensions is just

$$D_2(E)dE = D_2(k)\frac{dk}{dE}dE = \frac{2\pi k}{(2\pi)^2}\frac{m}{\hbar^2 k}dE$$

$$D_2(E)dE = \frac{m}{2\pi\hbar^2}dE$$

where we have used the fact that wave number $k = \sqrt{2mE/\hbar^2}$, energy $E = \hbar^2 k^2/2m$, and the energy increment $dE = \hbar^2 k\,dk/m$.

Solution 5.3
An electron is in an infinite, one-dimensional, rectangular potential well of width L. The electron is in the simple superposition state consisting of the ground and first excited state so that

$$\psi(x,t) = \frac{1}{\sqrt{2}}(\psi_1(x,t) + \psi_2(x,t))$$

(a) To find the probability density $|\psi(x,t)|^2$, we must first find expressions for the wave functions $\psi_1(x,t)$ and $\psi_2(x,t)$. The first two lowest-energy wave functions for a particle of mass m confined to an infinite potential well of width L centered at $x = 0$ are

$$\psi_1(x,t) = \left(\frac{2}{L}\right)^{1/2}\cos\left(\frac{\pi x}{L}\right)\cdot e^{-i\omega_1 t}$$

and

$$\psi_2(x,t) = \left(\frac{2}{L}\right)^{1/2}\sin\left(\frac{2\pi x}{L}\right)\cdot e^{-i\omega_2 t}$$

where $E_n = \hbar\omega_n = (\hbar^2/2m)\cdot(n^2\pi^2/L^2)$ and n is a positive nonzero integer. The expression for probability density

$$|\psi|^2 = \frac{1}{2}\left(\psi_1\psi_1^* + \psi_1\psi_2^* + \psi_2\psi_1^* + \psi_2\psi_2^*\right)$$

$$|\psi|^2 = \frac{1}{2}\left(|\psi_1|^2 + |\psi_2|^2 + |\psi_1||\psi_2|e^{-i(\omega_1-\omega_2)t} + |\psi_1||\psi_2|e^{i(\omega_1-\omega_2)t}\right)$$

$$|\psi(x,t)|^2 = \frac{1}{2}\left(|\psi_1(x)|^2 + |\psi_2(x)|^2 + 2|\psi_1(x)||\psi_2(x)|\cos((\omega_1-\omega_2)t)\right)$$

shows an oscillatory solution in which the average position of the particle moves from one side of the well to the other. The sinusoidal oscillation frequency is $(\omega_2 - \omega_1) = 3\hbar\pi^2/2mL^2$.

(b) The average position of the particle is

$$\langle x(t) \rangle = \int_{-\frac{L}{2}}^{\frac{L}{2}} \psi^* x \psi \, dx = \frac{1}{2} \int_{-\frac{L}{2}}^{\frac{L}{2}} \left(|\psi_1|^2 + |\psi_2|^2 + 2|\psi_1||\psi_2| \cos((\omega_1 - \omega_2)t) \right) x \, dx$$

$$\langle x(t) \rangle = \frac{1}{2} \int_{-\frac{L}{2}}^{\frac{L}{2}} \left(2|\psi_1||\psi_2| \cos((\omega_1 - \omega_2)t) \right) x \, dx$$

$$\langle x(t) \rangle = (\cos((\omega_1 - \omega_2)t)) \left(\frac{2}{L} \right) \int_{\frac{-L}{2}}^{\frac{L}{2}} x \cos\left(\frac{\pi x}{L} \right) \sin\left(\frac{2\pi x}{L} \right) dx$$

$$\langle x(t) \rangle = (\cos((\omega_1 - \omega_2)t)) \left(\frac{1}{L} \right) \left(2 - \frac{2}{9} \right) \frac{L^2}{\pi^2}$$

$$\langle x(t) \rangle = (16L/9\pi^2) \cos((\omega_2 - \omega_1)t)$$

(c) The momentum probability density $|\psi(p_x, t)|^2$ can be found from the Fourier transform

$$\psi(p_x, t) = \frac{1}{\sqrt{2\pi\hbar}} \int_{-\infty}^{\infty} \psi(x, t) e^{-ip_x x} dx = \frac{1}{\sqrt{4\pi\hbar}} \left(\int_{-\infty}^{\infty} \psi_1 e^{-ip_x x} dx + \int_{-\infty}^{\infty} \psi_2 e^{-ip_x x} dx \right)$$

$$\psi(p_x, t) = \frac{1}{\sqrt{4\pi\hbar}} \sqrt{\frac{2}{L}} \left(e^{-i\omega_1 t} \int_{-\infty}^{\infty} \cos\left(\pi \frac{x}{L} \right) e^{-ip_x x} dx + e^{-i\omega_2 t} \int_{-\infty}^{\infty} \sin\left(2\pi \frac{x}{L} \right) e^{-ip_x x} dx \right)$$

$$|\psi(p_x, t)|^2 = \frac{1}{2\pi\hbar L} \left(\frac{1}{\frac{\pi^2}{L^2} - p_x^2} \cos\left(p_x \frac{L}{2} \right) \left(2\frac{\pi}{L} e^{-i\omega_2 t} - 2ip_x e^{-i\omega_1 t} \right) \right)$$

$$|\psi(p_x, t)|^2 = \frac{2}{\pi\hbar L} \times \frac{\cos\left(\frac{p_x L}{2} \right)}{\left(\frac{\pi^2}{L^2} - p_x^2 \right)^2} \left(\frac{\pi^2}{L^2} + p_x^2 \right)$$

(d) The average momentum of the particle is

$$\langle p_x(t) \rangle = -(8\hbar/(3L)) \sin((\omega_2 - \omega_1)t)$$

(e) The current flux is

$$\mathbf{J}(x, t) = -\frac{2e\pi\hbar}{mL^2} \left(\cos\left(\frac{\pi x}{L} \right) \cos\left(\frac{2\pi x}{L} \right) + \frac{1}{2} \sin\left(\frac{\pi x}{L} \right) \sin\left(\frac{2\pi x}{L} \right) \right) \sin((\omega_2 - \omega_1)t)$$

Solution 5.4

The values of $\langle x \rangle$ and $\langle x^2 \rangle$ for a particle confined by the potential $V(x) = 0$ for $0 < x < L$ and $V(x) = \infty$ elsewhere are

$$\langle x \rangle = \frac{L}{2}$$

and

$$\langle x^2 \rangle = \frac{L^2}{3} - \frac{L^2}{2n^2\pi^2}$$

We can now compare these results with those for a classical particle. The classical particle moves at constant velocity v and traverses the well in time $\tau = L/v$. Hence, classically,

$$\langle x \rangle = \int\limits_{t=0}^{t=\tau} \frac{vt}{\tau} dt = \frac{1}{2}v\tau = \frac{L}{2}$$

and

$$\langle x^2 \rangle = \int\limits_{t=0}^{t=\tau} \frac{(vt)^2}{\tau} dt = \frac{1}{3}v^2\tau^2 = \frac{L^2}{3}$$

Thus, as $n \to \infty$ the quantum results approach the classical solution.

The average values of $\langle p_x \rangle$ and $\langle p_x^2 \rangle$ are

$$\langle p_x \rangle = 0$$
$$\langle p_x^2 \rangle = 2m E_n$$

and since

$$E_n = \frac{\hbar^2 k_n^2}{2m} = \frac{\hbar^2 n^2 \pi^2}{2m L^2}$$

we have

$$\langle p_x^2 \rangle = \frac{\hbar^2 n^2 \pi^2}{L^2}$$

Hence,

$$\Delta x^2 = \frac{L^2}{12}\left(1 - \frac{6}{n^2\pi^2}\right)$$

$$\Delta p_x^2 = \frac{\hbar^2 \pi^2 n^2}{L^2}$$

$$\Delta x \Delta p_x = \frac{\hbar}{\sqrt{12}}(n^2\pi^2 - 6)^{1/2}$$

Solution 5.5

To find $\psi(k)$ for a particle the state function of which is $\psi(x) = 1/\sqrt{2L}$ for $|x| < L$ and $\psi(x) = 0$ for $|x| > L$, we take the Fourier transform:

$$\psi(k) = \frac{1}{\sqrt{2\pi}} \int_{-L}^{L} \frac{1}{\sqrt{2L}} e^{-ikx} dx = \left[\frac{1}{2\sqrt{\pi L}} \frac{-1}{ik} e^{-ikx} \right]_{-L}^{L} = \frac{1}{\sqrt{\pi L}} \frac{1}{k} \sin(kL)$$

$$|\psi(k)|^2 = \frac{\sin^2(kL)}{\pi L k^2}$$

The wave functions $\psi(x)$ and $\psi(k)$ are plotted in the following figures.

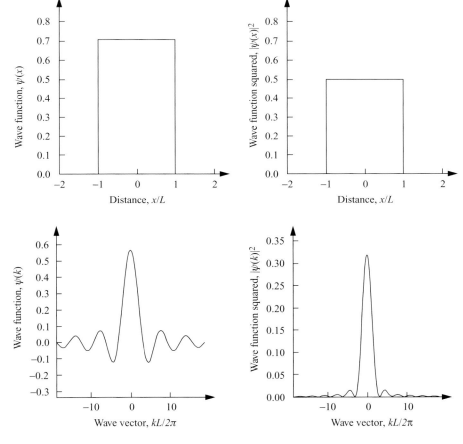

To show that the uncertainty (rms deviation) in particle momentum Δp is infinite it is necessary to calculate $\langle p_x \rangle$ and $\langle p_x^2 \rangle$:

$$\langle p_x \rangle = \int_{-\infty}^{\infty} \psi^*(k)\hbar k \psi(k) dk = \frac{\hbar}{L\pi} \int_{-\infty}^{\infty} \frac{1}{k} \sin^2(kL) dk = 0$$

by symmetry, and

$$\langle p_x^2 \rangle = \int_{-\infty}^{\infty} \psi^*(k)\hbar^2 k^2 \, \psi(k)dk = \frac{\hbar^2}{L\pi} \int_{-\infty}^{\infty} \sin^2(kL)dk = \frac{\hbar^2}{L\pi} \int_{-\infty}^{\infty} \left(\frac{1}{2} - \frac{1}{2}\cos(2kL) \right)dk$$

$$\langle p_x^2 \rangle = \frac{\hbar^2}{L\pi} \int_{-\infty}^{\infty} \left(\frac{1}{2} - \frac{1}{2}\cos(2kL) \right)dk = \frac{\hbar^2}{L\pi} \left[\frac{k}{2} - \frac{1}{4L}\sin(2kL) \right]_{-\infty}^{\infty}$$

$$\langle p_x^2 \rangle = \frac{\hbar^2}{L\pi} \left(\frac{\infty}{2} - \frac{-\infty}{2} - 0 + 0 \right) = \infty$$

$$\Delta p_x^2 = \langle p_x^2 \rangle - \langle p_x \rangle^2 = \infty$$

One understands this as $|\psi(k)|^2$ not decreasing to zero fast enough in the limit $k \to \infty$. The average value of position is

$$\langle x \rangle = \int_{-L}^{L} \psi(x)x\psi(x)dx = \left(\frac{1}{\sqrt{2L}} \right)^2 \frac{1}{2} \left[x^2 \right]_{-L}^{L} = 0$$

and the value of $\langle x^2 \rangle$ is

$$\langle x^2 \rangle = \int \psi(x)x^2\psi(x)dx = \frac{1}{2L}\frac{1}{3} \left[x^3 \right]_{-L}^{L} = \frac{1}{3L}L^3 = \frac{L^2}{3}$$

giving a measure of the spread in measured values of

$$\Delta x^2 = \langle x^2 \rangle - \langle x \rangle^2 = \frac{L^2}{3} - 0$$

It follows that

$$\Delta x = L/\sqrt{3}$$

i.e. finite nonzero, so that

$$\Delta x \Delta p_x = \infty$$

Solution 5.6

A hydrogen atom in its ground state has electron wave function $\psi_0(r) = Ae^{-r/r_0}$. The electron is subject to a radially symmetric coulomb potential given by $V(r) = -e^2/4\pi\varepsilon_0\varepsilon_r r$. We wish to find the normalization constant A, and the minimized energy expectation value $\langle E_0 \rangle$, and we wish to show that

$$\langle E_{\text{kinetic}} \rangle = -\langle E_{\text{potential}} \rangle/2$$

First, we normalize the wave function. This requires

$$\int \psi_0^*(r)\psi_0(r)dr = \int\limits_{r=0}^{r=\infty} A^2 4\pi r^2 \, e^{-2r/r_0}dr = 1$$

Integrating by parts using $\int UV' = UV - \int U'V$, where $U = r^2, U' = 2r, V' = e^{-2r/r_0}$, and

$$V = \frac{-r_0}{2}e^{-2r/r_0}$$

gives

$$\int r^2 e^{-2r/r_0}dr = \left[\frac{-r^2}{2}e^{-2r/r_0}\right]_0^\infty + \int\limits_0^\infty r r_0 e^{-2r/r_0}dr$$

Integrating by parts again with $U = r, U' = 1, V' = r_0 e^{-2r/r_0}, V = (-r_0^2/2)e^{-2r/r_0}$ gives

$$\int r r_0 e^{-2r/r_0}dr = \left[-\frac{r r_0^2}{2}e^{-2r/r_0}\right]_0^\infty + \int\limits_0^\infty \frac{r_0^2}{2}e^{-2r/r_0}dr = \left[-\frac{r_0^3}{4}e^{-2r/r_0}\right]_0^\infty = \frac{r_0^3}{4}$$

$$A^2 4\pi \frac{r_0^3}{4} = 1$$

$$A = \left(\frac{1}{\pi r_0^3}\right)^{1/2} \quad \text{and} \quad \psi_0(r) = \left(\frac{1}{\pi r_0^3}\right)^{1/2}e^{-r/r_0}$$

$Energy \ \langle E_0\rangle = \int \psi_0^* H \psi_0 d^3r$

$$H\psi_0 = \frac{1}{r^2}\frac{-\hbar^2}{2m}\frac{\partial}{\partial r}r^2\frac{\partial}{\partial r}\psi_0 - \frac{e^2}{4\pi\varepsilon_0 r}\psi_0$$

Since

$$\nabla^2 = \frac{1}{r^2}\frac{\partial}{\partial r}\left(r^2\frac{\partial}{\partial r}\right)$$

in spherical coordinates,

$$\langle p^2\rangle = -\hbar^2 \int \psi_0^* \nabla^2 \psi_0 d^3r = \frac{-\hbar^2}{r_0^2}\int \psi_0^*\left(\frac{-2r_0}{r}\psi'(r) + \psi(r)\right)d^3r$$

$$\langle p^2\rangle = \frac{-\hbar^2}{r_0^2}\int \psi_0^*\left(\frac{-2r_0}{r}\psi'(r) + \psi(r)\right)d^3r = \frac{-\hbar^2}{r_0^2}\int \psi_0^*\psi_0 d^3r = \frac{-\hbar^2}{r_0^2}$$

To calculate $\langle V \rangle$, we note that we are using spherical coordinates:

$$\langle V(r) \rangle = \int\limits_0^\infty A^2 e^{-2r/r_0} dr \cdot \frac{-e^2}{4\pi\varepsilon_0 r} \cdot 4\pi r^2 dr = \frac{-4\pi e^2 A^2}{4\pi\varepsilon_0} \int\limits_0^\infty r e^{-2r/r_0} dr$$

Integrating by parts $U = r, U' = 1, V' = e^{-2r/r_0}, V = (-r_0/2)e^{-2r/r_0}, \int U V' = UV - \int U'V$

$$\langle V(r) \rangle = \frac{-4\pi e^2 A^2}{4\pi\varepsilon_0} \left(\left[\frac{-r r_0}{2} e^{-2r/r_0} \right]_0^\infty + \int\limits_0^\infty \frac{r_0}{2} e^{-2r/r_0} dr \right)$$

$$\langle V(r) \rangle = \frac{+4\pi e^2 A^2}{4\pi\varepsilon_0} \left[\frac{r_0^2}{4} e^{-2r/r_0} \right]_0^\infty = \frac{-4\pi e^2 A^2}{4\pi\varepsilon_0} \cdot \frac{r_0^2}{4}$$

$$\langle V \rangle = \frac{e^2}{4\pi\varepsilon_0 r_0}$$

since $A^2 = 1/\pi r_0^3$. Hence,

$$\langle H \rangle = \frac{-p^2}{2m} - \frac{e^2}{4\pi\varepsilon_0 r_0} = \frac{\hbar^2}{2m r_0^2} - \frac{e^2}{4\pi\varepsilon_0 r_0}$$

Because we are at a local minimum we have

$$\frac{\partial}{\partial r_0} \langle E_0 \rangle = \frac{-\hbar^2}{m r_0^3} + \frac{e^2}{4\pi\varepsilon_0 r_0^2} = 0$$

Hence,

$$\frac{\hbar^2}{m r_0} = \frac{e^2}{4\pi\varepsilon_0}$$

and

$$r_0 = \frac{\hbar^2 4\pi\varepsilon_0}{m e^2}$$

the Bohr radius, which is $r_0 = 0.0529177$ nm, and

$$\langle E_0 \rangle = \frac{m e^4}{2\hbar^2 (4\pi)^2 \varepsilon_0^2} - \frac{m e^4}{\hbar^2 (4\pi)^2 \varepsilon_0^2}$$

where we identify the first term on the right-hand side with kinetic energy and the second term with potential energy. Hence,

$$\langle E_{\text{kinectic}} \rangle = \frac{1}{2} \langle E_{\text{potential}} \rangle$$

which is the result predicted by the *virial theorem*. Also,

$$\langle E_0 \rangle = -\frac{m e^4}{2\hbar^2 (4\pi)^2 \varepsilon_0^2} = -\frac{\hbar^2}{2m r_0^2} = -13.6058 \text{ eV}$$

which is the Rydberg constant.

To show that $\langle p \rangle = 0$, we apply the momentum operator $p = -i\hbar \nabla$. In radial coordinates,

$$\hat{p}_r = -i\hbar \frac{1}{r} \frac{\partial}{\partial r} r$$

$$\langle p \rangle = \int \psi^*(r)(-i\hbar \nabla)\psi(r)d^3r$$

$$r = (x^2 + y^2 + z^2)^{1/2}$$

$$\langle p \rangle = \sum_i \int_{-\infty}^{\infty} -i\hbar \psi_0^*(x_i)\frac{d}{dx_i}\psi_0(x_i)dx_i \hat{x}_i = \sum_i -i\hbar \hat{x}_i \int_{-\infty}^{\infty} \psi_0^*(x_i)\frac{d}{dx_i}\psi_0(x_i)dx_i$$

Notice that the integral has limits $\pm\infty$. Because the ground-state wave function is of even parity, its spatial derivative is of odd parity. We note that an even function times an odd function is an odd function, which, for convenience, we define as $\Gamma(x_i)$. It follows that the integral must be zero, giving

$$\langle p \rangle = \sum_i -i\hbar \hat{x}_i [\Gamma(x_i)]_{-\infty}^{\infty} = 0$$

Solution 5.7

The momentum operator is Hermitian – i.e., it must satisfy $A_{ij} = A_{ji}^*$ or equivalently $(\int \phi^* \hat{p}\psi dr)^* = \int \psi^* \hat{p}\phi dr = (\int \hat{p}\phi^*\psi dr)^*$. For simplicity, consider $\hat{p}_x = -i\hbar \cdot \partial/\partial x$, the x component of $\hat{p}$. We now have

$$\int \phi^* \hat{p}_x^* \psi dx = \int \hat{p}_x^* \phi^* \psi dx \text{ or } \langle \phi | \hat{p}_x \psi \rangle = (\langle \hat{p}_x \phi | \psi \rangle)^*$$

The operator $\hat{p}_x$ can be seen to be Hermitian if we integrate by parts, $\int UV'dx = UV - \int U'Vdx$, so

$$\int \phi^* \hat{p}_x \psi dx = -i\hbar \int \phi^* \frac{\partial}{\partial x}\psi dx = \left[-i\hbar \phi^* \psi\right]_{-\infty}^{\infty} + i\hbar \int_{-\infty}^{\infty} \frac{\partial \phi^*}{\partial x}\psi dx$$

the term $[-i\hbar \phi^* \psi]_{-\infty}^{\infty} = 0$ and

$$i\hbar \int_{-\infty}^{\infty} \frac{\partial \phi^*}{\partial x}\psi dx = \int \hat{p}_x^* \phi^* \psi dx$$

We have thus shown that $\langle \phi | \hat{p}\psi \rangle = (\langle \hat{p}\phi | \psi \rangle)^*$, provided that the wave function $\phi \to 0$ at $x \pm \infty$. To show that $\langle \hat{p} \rangle$ is real is now trivial. If $\psi = \phi$, then $\langle \hat{p} \rangle = \int \phi^* \hat{p}\phi dr = \int \hat{p}^*\phi^*\phi = \langle \hat{p} \rangle^*$, which can only be true if $\langle \hat{p} \rangle$ is real. Hence, the expectation value of the momentum operator is real.

To show that $-i\hbar \cdot \partial/\partial r$ is not Hermitian in radial coordinates, we find the expectation value by integrating by parts in radial coordinates:

$$\int_{r=0}^{r=\infty} 4\pi r^2 \psi_m^* \left(-i\hbar \frac{\partial}{\partial r} \right) \psi_n dr = \left[4\pi r^2 \psi_m^* \left(-i\hbar \frac{\partial}{\partial r} \psi_n \right) \right]_{r=0}^{r=\infty}$$

$$- \int_{r=0}^{r=\infty} -i\hbar \psi_n \cdot \frac{\partial}{\partial r} (4\pi r^2 \psi_m^*) dr$$

The first term on the right-hand side is zero, assuming that the eigenfunction vanishes at $r = \infty$, so that

$$\int_{r=0}^{r=\infty} 4\pi r^2 \psi_m^* \left(-i\hbar \frac{\partial}{\partial r} \right) \psi_n dr = \int i\hbar 4\pi \psi_n \cdot \left(r^2 \frac{\partial}{\partial r} \psi_m^* - 2r \psi_m^* \right) dr$$

$$= \int 4\pi r^2 \psi_n \cdot \left(i\hbar \left(2r - \frac{\partial}{\partial r} \right) \psi_m \right)^* dr$$

Obviously, this does not satisfy $A_{ij}^* = A_{ji}$ for a Hermitian operator.

Solution 5.8

A particle mass m is in a finite, one-dimensional rectangular potential well for which $V(x) = 0$ for $-L < x < L$ and $V(x) = V_0$ elsewhere, and for which the value of V_0 is a finite positive constant. The expectation value of kinetic energy is

$$\langle \hat{T} \rangle = \int_{-\infty}^{\infty} \psi^*(x) \cdot \frac{\hat{p}^2}{2m} \cdot \psi(x) dx = \frac{-\hbar^2}{2m} \int_{-\infty}^{\infty} \psi^*(x) \frac{d^2}{dx^2} \psi(x) dx$$

For even-parity *bound-state* solutions, including the ground state, the spatial wave functions in the well are of the form

$$\psi_n(x) = A_n \cos(k_n x)$$

and in the barrier they are of the form

$$\psi_n(x) = C_n e^{-\kappa_n x}$$

where the index n is an odd positive integer that labels the bound-state eigenvalue. For eigenenergy E_n,

$$k_n = \sqrt{2m E_n}/\hbar$$

and

$$\kappa_n = \sqrt{2m(V_0 - E)}/\hbar$$

For the ground state, $n = 1$ and we have an expectation value for kinetic energy that is

$$\langle \hat{T} \rangle = \frac{\hbar^2 k^2 A^2}{m} \cdot \int\limits_{x=0}^{x=L} \cos^2(k_1 x') dx' - \frac{\hbar^2 k^2 C^2}{m} \cdot \int\limits_{x=L}^{x=\infty} e^{-2\kappa_1 x'} dx'$$

This shows that the contribution to kinetic energy from the barrier region is negative.

Solution 5.9

The time dependence of the expectation value of the operator $\hat{A}$ is found from

$$\frac{d}{dt} \langle \hat{A} \rangle = \left\langle \frac{\partial \psi}{\partial t} \middle| \hat{A} \middle| \psi \right\rangle + \left\langle \psi \middle| \frac{\partial}{\partial t} \hat{A} \middle| \psi \right\rangle + \left\langle \psi \middle| \hat{A} \middle| \frac{\partial \psi}{\partial t} \right\rangle$$

which may be rewritten as

$$\frac{d}{dt} \langle \hat{A} \rangle = \frac{i}{\hbar} \langle \psi | H \hat{A} | \psi \rangle - \frac{i}{\hbar} \langle \psi | \hat{A} H | \psi \rangle + \left\langle \psi \middle| \frac{\partial}{\partial t} \hat{A} \middle| \psi \right\rangle$$

$$\frac{d}{dt} \langle \hat{A} \rangle = \frac{i}{\hbar} \langle \psi | H \hat{A} - \hat{A} H | \psi \rangle + \left\langle \psi \middle| \frac{\partial}{\partial t} \hat{A} \middle| \psi \right\rangle$$

where we used the fact that the Schrödinger equation is

$$\frac{-i}{\hbar} H | \psi \rangle = \left| \frac{\partial \psi}{\partial t} \right\rangle$$

Hence,

$$\frac{d}{dt} \langle \hat{A} \rangle = \frac{i}{\hbar} \langle [H, \hat{A}] \rangle + \left\langle \frac{\partial}{\partial t} \hat{A} \right\rangle$$

If the Hamiltonian commutes with the operator $\hat{A}$, then $[H, \hat{A}] = 0$, and we may conclude that

$$\frac{d}{dt} \hat{A} = \left\langle \frac{\partial}{\partial t} \hat{A} \right\rangle$$

If a Hamiltonian with eigenfunctions ϕ_1 and ϕ_2 and corresponding eigenvalues E_1 and E_2 does not commute with an operator $\hat{A}$, then $[H, \hat{A}] \neq 0$. Each orthonormal eigenfunction with quantum number n must satisfy the Schrödinger equation, so that

$$E_n \phi_n = i\hbar \frac{\partial \phi_n}{\partial t}$$

$$\phi_n(t) = \phi_n(0) e^{-i E_n t / \hbar}$$

and

$$\langle n | m \rangle = \delta_{nm}$$

If at time $t = 0$ the system is in state $\psi(0) = (\phi_1 + \phi_2)/\sqrt{2}$, then at time t the state is

$$\psi(t) = \left(\phi_1 e^{-iE_1 t/\hbar} + \phi_2 e^{-iE_2 t/\hbar}\right)/\sqrt{2}$$

The expectation value of operator $\hat{\mathbf{A}}$ is $\langle \hat{\mathbf{A}} \rangle = \langle \psi(t)|\hat{\mathbf{A}}|\psi(t)\rangle$, or

$$\langle \hat{\mathbf{A}} \rangle = \frac{1}{2} \int \left(\phi_1^* e^{iE_1 t/\hbar} + \phi_2^* e^{iE_2 t/\hbar}\right)\hat{\mathbf{A}}\left(\phi_1 e^{-iE_1 t/\hbar} + \phi_2 e^{-iE_2 t/\hbar}\right)d^3 r$$

Because we are told that the operator $\hat{\mathbf{A}}$ has eigenfunction

$$\psi(t) = \left(\phi_1 e^{-iE_1 t/\hbar} + \phi_2 e^{-iE_2 t/\hbar}\right)/\sqrt{2}$$

and corresponding eigenvalue a_1, it must satisfy

$$\hat{\mathbf{A}}\psi(t) = a_1 \psi(t)$$

It follows that the expecation value may be writen

$$\langle \hat{\mathbf{A}} \rangle = \frac{a_1}{2} \int \left(\phi_1^* e^{iE_1 t/\hbar} + \phi_2^* e^{iE_2 t/\hbar}\right)\left(\phi_1 e^{-iE_1 t/\hbar} + \phi_2 e^{-iE_2 t/\hbar}\right)d^3 r$$

$$\langle \hat{\mathbf{A}} \rangle = \frac{a_1}{2}(\langle 1|1\rangle + \langle 2|2\rangle + \langle 1|2\rangle + \langle 2|1\rangle) = a_1$$

where we have adopted the notation $\phi_n = |n\rangle$ and made use of the fact that the eigenfunctions ϕ_n are orthonormal in such a way that $\langle n|m\rangle = \delta_{nm}$. We may conclude that in this particular system the expectation value of $\hat{\mathbf{A}}$ does not vary with time.

Solution 5.10

In Chapter 1 we showed that for a source-free, linear, frequency-independent, lossless dielectric with $\varepsilon(\mathbf{r}) = \varepsilon_0(\mathbf{r})\varepsilon_r(\mathbf{r})$, and with relative magnetic permeability $\mu_r = 1$, Maxwell's equations lead directly to a wave equation for electromagnetic fields. The linearity allows us to separate out the time and space dependences into a set of harmonic solutions of the form $\mathbf{E}(\mathbf{r}, t) = \mathbf{E}_0 e^{i\mathbf{k}\cdot\mathbf{r}} e^{i\omega t}$ and $\mathbf{H}(\mathbf{r}, t) = \mathbf{H}_0 e^{i\mathbf{k}\cdot\mathbf{r}} e^{i\omega t}$, where we have used complex numbers for mathematical convenience (always remembering to take the real part to obtain the *physical fields*). The $\mathbf{H}$ and $\mathbf{E}$ fields in Maxwell's equations are *real vector fields*. Because the dielectric is source-free, the divergence $\nabla \cdot \mathbf{D} = 0$ and $\nabla \cdot \mathbf{H} = 0$. There are no sources or sinks of displacement ($\mathbf{D}$) or magnetic fields ($\mathbf{H}$). Field configurations are built up out of plane waves that are transverse in such a way that $\mathbf{E}_0 \cdot \mathbf{k} = 0$ and $\mathbf{H}_0 \cdot \mathbf{k} = 0$.

Since the $\mathbf{H}$ and $\mathbf{E}$ fields are related through

$$\nabla \times \mathbf{E} = -\mu_0 \frac{\partial \mathbf{H}}{\partial t}$$

$$\nabla \times \mathbf{H} = \varepsilon(\mathbf{r}) \frac{\partial \mathbf{E}}{\partial t}$$

dividing the equation for $\nabla \times \mathbf{H}$ by $\varepsilon(\mathbf{r})$ and taking the curl gives

$$\nabla \times \left(\frac{1}{\varepsilon(\mathbf{r})} \nabla \times \mathbf{H}(\mathbf{r}) \right) = \nabla \times \frac{\partial \mathbf{E}(\mathbf{r})}{\partial t} = -\mu_0 \frac{\partial^2 \mathbf{H}(\mathbf{r})}{\partial t^2}$$

And since $\mathbf{H}(\mathbf{r}, t) = \mathbf{H}(\mathbf{r})e^{i\omega t}$, one may write

$$\nabla \times \left(\frac{1}{\varepsilon_r(\mathbf{r})} \nabla \times \mathbf{H}(\mathbf{r}) \right) = \omega^2 \mu_0 \varepsilon_0 \frac{\partial^2 \mathbf{H}(\mathbf{r})}{\partial t^2} = \left(\frac{\omega}{c} \right)^2 \frac{\partial^2 \mathbf{H}(\mathbf{r})}{\partial t^2}$$

After solving this wave equation for $\mathbf{H}(\mathbf{r})$, one may obtain the transverse electric field via

$$\mathbf{E}(\mathbf{r}) = \frac{1}{i\omega\varepsilon(\mathbf{r})} \nabla \times \mathbf{H}(\mathbf{r})$$

Recognizing the wave equation for $\mathbf{H}(\mathbf{r})$ as an eigenvalue problem in which the differential operator $\tilde{H} = \nabla \times (1/\varepsilon_r(\mathbf{r}))\nabla$ is analogous to the Hamiltonian used in the Schrödinger equation gives

$$\tilde{H}\mathbf{H}(\mathbf{r}) = \left(\frac{\omega}{c} \right)^2 \mathbf{H}(\mathbf{r})$$

The eigenvectors $\mathbf{H}(\mathbf{r})$ are the field patterns of the harmonic modes oscillating at frequency ω and the eigenvalues are $(\omega/c)^2$.

As with the Hamiltonian used in quantum mechanics, $\tilde{H}$ is a linear Hermitian operator with real eigenvalues. In fact, comparing the electrodynamics discussed above and quantum mechanics we find a number of similarities, but also important differences. We should expect this because, as discussed in Section 1.3.2, classical electrodynamics is just the macroscopic, incoherent, large-photon-number limit of the quantum treatment. In electrodynamics $\mathbf{H}(\mathbf{r})$ is a *real vector field* with a simple time dependence $e^{i\omega t}$ that is used only as a mathematical convenience. When we want to find a measurable value of $\mathbf{H}(\mathbf{r})$, we take the real part. In quantum mechanics, Schrödinger's wave function $\psi(\mathbf{r})$ is a *complex scalar field*. When we want to find the value of a measured quantity, we do not find the real part of the wave function; rather we evaluate the expectation value of the operator $\hat{\mathbf{A}}$:

$$\langle \hat{\mathbf{A}} \rangle = \int \psi(\mathbf{r}, t) \hat{\mathbf{A}} \psi(\mathbf{r}, t) d^3 r$$

The eigenvalues of the Schrödinger wave equation are related to the oscillation frequency through the energy $E = \hbar\omega$. In electrodynamics, eigenvalues are proportional to ω^2. In quantum mechanics, the Hamiltonian is separable if the potential is separable (e.g., $V(\mathbf{r}) = V(x)V(y)V(z)$). In electrodynamics, there is no such simplification as $\tilde{H}$ couples the different directions even if $\varepsilon(\mathbf{r})$ is separable. In electrodynamics there is no *absolute* scale. Hence, in electrodynamics we may scale our solutions from radio waves to visible light and beyond based simply upon geometry and material parameters such as $\varepsilon(\mathbf{r})$. In quantum mechanics, there *is* an absolute scale, because $\hbar \neq 0$. Planck's constant $\hbar$ sets the scale. The corresponding length scale in atomic systems is set by the Bohr radius $a_B = 4\pi\varepsilon_0\hbar^2/m_0e^2 = 0.529177 \times 10^{-10}$ m (or an *effective* Bohr radius in materials with $\varepsilon \neq \varepsilon_0$ and $m_e \neq m_0$). It is on this length scale (and corresponding energy and time scales) that quantum effects ($\hbar \neq 0$) are important.

Classical electromagnetic theory uses real magnetic and electric fields coupled via Maxwell's equations. The magnetic and electric fields each have physical meaning. Both fields are needed to describe both the instantaneous state and time evolution of the system.

Quantum mechanics uses one complex wavefunction to describe both the instantaneous state and time evolution of the system. It is also possible to describe quantum mechanics using two coupled real wave functions corresponding to the real and imaginary parts of the complex wave function. However, such an approach is more complicated. In addition, the real and imaginary parts of the wave function have no special physical meaning.

Time-reversal symmetry can occur in both quantum mechanics and classical electrodynamics. This time-reversal symmetry exists when the system under consideration is conservative.

6 The harmonic oscillator

6.1 The harmonic oscillator potential

We know from our experience with classical mechanics that a particle of mass m subject to a linear restoring force $F(x) = -\kappa x$, where κ is the force constant, results in one-dimensional simple harmonic motion with an oscillation frequency $\omega = \sqrt{\kappa/m}$. The potential the particle moves in is quadratic $V(x) = \kappa x^2/2$, and so in this case the potential has a minimum at position $x = 0$. The idea that a quadratic potential may be used to describe a local minimum in an otherwise more complex potential turns out to be a very useful concept in both classical and quantum mechanics. An underlying reason why it is of practical importance is that a local potential minimum often describes a point of stability in a system. For example, the positions of atoms that form a crystal are stabilized by the presence of a potential that has a local minimum at the location of each atom. If we wish to understand how the vibrational motion of atoms in a crystal determines properties such as the speed of sound and heat transfer, then we need to develop a model that describes the oscillatory motion of an atom about a local potential minimum. The same is true if we wish to understand the vibrational behavior of atoms in molecules.

As a starting point of our investigation of the vibrational properties of atomic systems, let's assume a static potential and then expand the potential function in a power series about the classically stable equilibrium position x_0 of one particular atom. In one dimension,

$$V(x) = \sum_{n=0}^{\infty} = \frac{1}{n!} \frac{d^n}{dx^n} V(x) \bigg|_{x=x_0} (x - x_0)^n \tag{6.1}$$

Assuming that higher-order terms in the polynomial expansion are of decreasing importance, we need only keep the first few terms:

$$V(x) = V(x_0) + \frac{d}{dx} V(x) \bigg|_{x=x_0} (x - x_0) + \frac{1}{2} \frac{d^2}{dx^2} V(x) \bigg|_{x=x_0} (x - x_0)^2 + \cdots \tag{6.2}$$

Because the atom position is stabilized by the potential, we know that the potential is at a local minimum, so the term in the first derivative in our series expansion about the equilibrium position x_0 can be set to zero. This leaves us with

$$V(x) = V(x_0) + \frac{1}{2}\frac{d^2}{dx^2}V(x)\bigg|_{x=x_0}(x - x_0)^2 + \cdots \tag{6.3}$$

The first term on the right-hand side of the equation, $V(x_0)$, is a constant, and so it has no impact on the particle dynamics. The second term is just the quadratic potential of a one-dimensional harmonic oscillator for which the force constant is easily identified as a measure of the curvature of the potential about the equilibrium point:

$$\kappa = \frac{d^2}{dx^2}V(x)\bigg|_{x=x_0} \tag{6.4}$$

We now see the importance of the harmonic oscillator in describing the dynamics of a particle in a local potential minimum. Very often a local minimum in potential energy can be approximated by the quadratic function of a harmonic oscillator.

While it is often convenient to visualize the harmonic oscillator in classical terms as illustrated in Fig. 6.1, if we are dealing with atomic-scale particles then we will have to solve for the particle motion using quantum mechanics. As a starting point, let's consider the time-independent Schrödinger equation for a particle of mass m subject

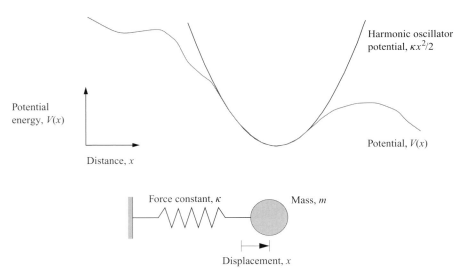

Fig. 6.1. Illustration of a one-dimensional potential with a local minimum that may be approximated by the parabolic potential of a harmonic oscillator. Also shown is a representation of a physical system that has a harmonic potential for small displacement from equilibrium. The classical system consists of a particle of mass m attached to a light spring with force constant κ. The one-dimensional displacement of the particle from its equilibrium position is x.

to a restoring force $F(x) = -\kappa x$ in one dimension. The equation is

$$\left(\frac{-\hbar^2}{2m}\frac{d^2}{dx^2} + \frac{\kappa}{2}x^2\right)\psi_n(x) = E_n\psi_n(x) \tag{6.5}$$

where the first term in the brackets is the kinetic energy and the second term is the potential energy

$$eV(x) = -\int\limits_{x'=0}^{x'=x} F(x')dx' = \int\limits_{x'=0}^{x'=x} \kappa x'dx' = \frac{\kappa}{2}x^2 \tag{6.6}$$

Note that we have used our definition of a scalar potential that relates force to the potential via $\mathbf{F}(\mathbf{r}) = -\nabla V(\mathbf{r})$.

Our next step is to solve Eqn (6.5). However, before finding the quantized eigenstates and eigenvalues of the harmonic oscillator, we can predict the form of the results using our previous experience developed in Chapter 3. We start by noting that the potential $V(x) = \kappa x^2/2$ has inversion symmetry in such a way that $V(x) = V(-x)$. A consequence of this fact is that the wave functions that describe the bound states of the harmonic oscillator must have definite parity. In addition, we can state that the lowest-energy state of the system (the ground state) will have even parity. With these basic facts in mind, we now turn our effort to finding the quantum mechanical solution for the harmonic oscillator.

6.2 Creation and annihilation operators

In classical mechanics, a particle of mass m moving in the potential $V(x) = \kappa x^2/2$ oscillates at frequency $\omega = \sqrt{\kappa/m}$, where κ is the force constant. The Hamiltonian for this one-dimensional harmonic oscillator consists of kinetic energy and potential energy terms such that

$$H = T + V = \frac{p_x^2}{2m} + \frac{m\omega^2}{2}x^2 \tag{6.7}$$

where the x-directed particle momentum p_x is $m \cdot dx/dt$.

In quantum mechanics, the classical momentum p_x is replaced by the operator $\hat{p} = -i\hbar \cdot \partial/\partial x$, so that

$$H = \frac{\hat{p}_x^2}{2m} + \frac{m\omega^2}{2}\hat{x}^2 \tag{6.8}$$

Mathematically, this equation is nicely symmetric, since the two operators $\hat{p}$ and $\hat{x}$ only appear as simple squares. This immediately suggests that the equation can be factored into two operators that are linear in $\hat{p}$ and $\hat{x}$:

$$H = \frac{m\omega^2}{2}\left(\frac{\hat{p}_x^2}{m^2\omega^2} + \hat{x}^2\right) = \frac{m\omega^2}{2}\left(\hat{x} + \frac{i\hat{p}_x}{m\omega}\right)\left(\hat{x} - \frac{i\hat{p}_x}{m\omega}\right) \tag{6.9}$$

Defining new operators,

$$\hat{b} = \left(\frac{m\omega}{2\hbar}\right)^{1/2} \left(\hat{x} + \frac{i\hat{p}_x}{m\omega}\right) \tag{6.10}$$

$$\hat{b}^\dagger = \left(\frac{m\omega}{2\hbar}\right)^{1/2} \left(\hat{x} - \frac{i\hat{p}_x}{m\omega}\right) \tag{6.11}$$

so that

$$\hat{x} = \left(\frac{\hbar}{2m\omega}\right)^{1/2} \left(\hat{b} + \hat{b}^\dagger\right) \tag{6.12}$$

and

$$\hat{p}_x = i\left(\frac{\hbar m\omega}{2}\right)^{1/2} \left(\hat{b}^\dagger - \hat{b}\right) \tag{6.13}$$

The Hamiltonian expressed in terms of the new operators is

$$\boxed{H = \frac{\hbar\omega}{2}\left(\hat{b}\hat{b}^\dagger + \hat{b}^\dagger\hat{b}\right)} \tag{6.14}$$

The symmetry of this equation will both help in simplifying problem solving and provide new insight into the quantum mechanical nature of the harmonic oscillator.

The commutation relations for the operators $\hat{b}^\dagger$ and $\hat{b}$ can be found by writing out the differential form and operating on a dummy wave function. For example, to find the commutation relation

$$[\hat{b}, \hat{b}^\dagger] = \hat{b}\hat{b}^\dagger - \hat{b}^\dagger\hat{b} \tag{6.15}$$

we reexpress each term on the right-hand side using differential operators. This gives

$$\left(\hat{b}\hat{b}^\dagger\right)\psi = \left(\frac{m\omega}{2\hbar}\right)\left(x + \frac{\hbar}{m\omega}\frac{\partial}{\partial x}\right)\left(x - \frac{\hbar}{m\omega}\frac{\partial}{\partial x}\right)\psi \tag{6.16}$$

$$\left(\hat{b}\hat{b}^\dagger\right)\psi = \left(\frac{m\omega}{2\hbar}\right)\left(x^2 + \frac{\hbar}{m\omega} + \frac{\hbar}{m\omega}x\frac{\partial}{\partial x} - \frac{\hbar}{m\omega}x\frac{\partial}{\partial x} - \frac{\hbar^2}{m^2\omega^2}\frac{\partial^2}{\partial x^2}\right)\psi \tag{6.17}$$

and

$$\left(\hat{b}^\dagger\hat{b}\right)\psi = \left(\frac{m\omega}{2\hbar}\right)\left(\left(x - \frac{\hbar}{m\omega}\frac{\partial}{\partial x}\right)\left(x + \frac{\hbar}{m\omega}\frac{\partial}{\partial x}\right)\right)\psi \tag{6.18}$$

$$\left(\hat{b}^\dagger\hat{b}\right)\psi = \left(\frac{m\omega}{2\hbar}\right)\left(x^2 - \frac{\hbar}{m\omega} - \frac{\hbar}{m\omega}x\frac{\partial}{\partial x} + \frac{\hbar}{m\omega}x\frac{\partial}{\partial x} - \frac{\hbar^2}{m^2\omega^2}\frac{\partial^2}{\partial x^2}\right)\psi \tag{6.19}$$

so that the commutation relation simplifies to just

$$\left(\hat{b}\hat{b}^\dagger - \hat{b}^\dagger\hat{b}\right)\psi = \left(\frac{m\omega}{2\hbar}\right)\left(\frac{2\hbar}{m\omega} - \frac{2\hbar}{m\omega}x\frac{\partial}{\partial x} + \frac{2\hbar}{m\omega}x\frac{\partial}{\partial x}\right)\psi = \psi \tag{6.20}$$

or, in even more compact form,

$$[\hat{b}, \hat{b}^\dagger] = \hat{b}\hat{b}^\dagger - \hat{b}^\dagger\hat{b} = 1 \tag{6.21}$$

One can now go through the same process and obtain all the commutation relations for the operators $\hat{b}$ and $\hat{b}^\dagger$. The results are

$$\boxed{[\hat{b}, \hat{b}^\dagger] = \hat{b}\hat{b}^\dagger - \hat{b}^\dagger\hat{b} = 1} \tag{6.22}$$

$$\boxed{[\hat{b}^\dagger, \hat{b}] = \hat{b}^\dagger\hat{b} - \hat{b}\hat{b}^\dagger = -1} \tag{6.23}$$

$$\boxed{[\hat{b}, \hat{b}] = \hat{b}\hat{b} - \hat{b}\hat{b} = 0} \tag{6.24}$$

$$\boxed{[\hat{b}^\dagger, \hat{b}^\dagger] = \hat{b}^\dagger\hat{b}^\dagger - \hat{b}^\dagger\hat{b}^\dagger = 0} \tag{6.25}$$

Thus, the Hamiltonian given by Eqn (6.14) may be rewritten as

$$H = \frac{\hbar\omega}{2}\left(\hat{b}\hat{b}^\dagger + \hat{b}^\dagger\hat{b}\right) = \frac{\hbar\omega}{2}\left(\hat{b}\hat{b}^\dagger - \hat{b}^\dagger\hat{b} + 2\hat{b}^\dagger\hat{b}\right) = \frac{\hbar\omega}{2}\left(1 + 2\hat{b}^\dagger\hat{b}\right) \tag{6.26}$$

Notice that we made use of the fact that $[\hat{b}, \hat{b}^\dagger] = \hat{b}\hat{b}^\dagger - \hat{b}^\dagger\hat{b} = 1$. Hence, the Hamiltonian is

$$\boxed{H = \hbar\omega\left(\hat{b}^\dagger\hat{b} + \frac{1}{2}\right)} \tag{6.27}$$

The commutation relations, the Hamiltonian, and the constraint that a lowest energy (ground state) exists *completely* specify the harmonic oscillator in terms of operators. What remains, of course, is to find the condition that expresses the fact that a ground state exists.

6.2.1 The ground state of the harmonic oscillator

To find the ground state wave function and energy of the one-dimensional harmonic oscillator we start with the Schrödinger equation:

$$H\psi_n = \hbar\omega\left(\hat{b}^\dagger\hat{b} + \frac{1}{2}\right)\psi_n = E_n\psi_n \tag{6.28}$$

Now we multiply from the *left* by $\hat{b}$ to give

$$\hbar\omega\left(\hat{b}\hat{b}^\dagger\hat{b} + \frac{\hat{b}}{2}\right)\psi_n = E_n\hat{b}\psi_n \tag{6.29}$$

But $[\hat{b}, \hat{b}^\dagger] = \hat{b}\hat{b}^\dagger - \hat{b}^\dagger\hat{b} = 1$, so that $\hat{b}\hat{b}^\dagger = 1 + \hat{b}^\dagger\hat{b}$. Hence, Eqn (6.29) may be written

$$\hbar\omega\left((1 + \hat{b}^\dagger\hat{b})\hat{b} + \frac{\hat{b}}{2}\right)\psi_n = E_n\hat{b}\psi_n \tag{6.30}$$

Factoring out the term $\hat{b}\psi_n$ on the left-hand side of Eqn (6.30) gives

$$\hbar\omega\left(\left(1 + \hat{b}^\dagger\hat{b}\right) + \frac{1}{2}\right)\left(\hat{b}\psi_n\right) = E_n\left(\hat{b}\psi_n\right) \tag{6.31}$$

Subtracting the term $\hbar\omega(\hat{b}\psi_n)$ from both sides allows one to write

$$\hbar\omega\left(\hat{b}^\dagger\hat{b} + \frac{1}{2}\right)\left(\hat{b}\psi_n\right) = (E_n - \hbar\omega)(\hat{b}\psi_n) \tag{6.32}$$

$$\hbar\omega\left(\hat{b}^\dagger\hat{b} + \frac{1}{2}\right)\psi_{n-1} = E_{n-1}\psi_{n-1} \tag{6.33}$$

This shows that $\psi_{n-1} = (\hat{b}\psi_n)$ is a new eigenfunction with energy eigenvalue $(E_n - \hbar\omega)$. From this we can conclude that the operator $\hat{b}$ acting on the Schrödinger equation creates a new eigenfunction with eigenenergy $(E_n - \hbar\omega)$. In a similar way, it can be shown that the operator $\hat{b}^\dagger$ creates a new eigenfunction ψ_{n+1} with eigenenergy $(E_n + \hbar\omega)$.

We now know enough to define the ground state. Clearly, the operator $\hat{b}$ can only be used to reduce the energy eigenvalue of any eigenstate except the ground state. Notice that we *assume the existence* of a ground state. Because there are, by definition, no energy eigenstates with energy less than the ground state, the ground state must be defined by

$$\boxed{\hat{b}\psi_0 = 0} \tag{6.34}$$

This, when combined with Eqn (6.22)–Eqn (6.25) and Eqn (6.27), completes our definition of the harmonic oscillator in terms of the operators $\hat{b}^\dagger$ and $\hat{b}$.

It is now possible to use our definition of the ground state to find the ground-state wave function. Since

$$\hat{b} = \left(\frac{m\omega}{2\hbar}\right)^{1/2}\left(\hat{x} + \frac{i\hat{p}_x}{m\omega}\right) = \left(\frac{m\omega}{2\hbar}\right)^{1/2}\left(x + \frac{\hbar}{m\omega}\frac{\partial}{\partial x}\right) \tag{6.35}$$

our definition $\hat{b}\psi_0 = 0$ requires

$$\left(\frac{m\omega}{2\hbar}\right)^{1/2}\left(x + \frac{\hbar}{m\omega}\frac{\partial}{\partial x}\right)\psi_0 = 0 \tag{6.36}$$

The solution for the wave function is of Gaussian form

$$\boxed{\psi_0 = A_0 e^{-x^2 m\omega/2\hbar}} \tag{6.37}$$

where the normalization constant A_0 is found in the usual way from the requirement that $\int \psi_0^* \psi_0 dx = 1$. This gives

$$A_0 = \left(\frac{m\omega}{\pi\hbar}\right)^{1/4} \tag{6.38}$$

Notice that the ground-state wave function ψ_0 for the harmonic oscillator has the even parity we predicted earlier based solely on symmetry arguments.

To find the eigenenergy of the ground state ψ_0 one substitutes Eqn (6.37) into the Schrödinger equation for the one-dimensional harmonic oscillator:

$$\left(\frac{-\hbar^2}{2m} \frac{d^2}{dx^2} + \frac{m\omega^2}{2} x^2 \right) \psi_0 = \left(\frac{-\hbar^2}{2m} \left(-\frac{2m\omega}{2\hbar} + 4x^2 \left(\frac{m\omega}{2\hbar} \right)^2 \right) + \frac{m\omega^2}{2} x^2 \right) \psi_0 = E_0 \psi_0$$

(6.39)

$$\left(\frac{\hbar\omega}{2} - \frac{m\omega^2}{2} x^2 + \frac{m\omega^2}{2} x^2 \right) \psi_0 = E_0 \psi_0 \tag{6.40}$$

so that the value of the ground-state energy, E_0, is

$$\boxed{E_0 = \frac{\hbar\omega}{2}} \tag{6.41}$$

6.2.1.1 Uncertainty in position and momentum for the harmonic oscillator in the ground state

The ground-state wave function ψ_0 given by Eqn (6.37) is of even parity. This symmetry will be helpful when we evaluate integrals that give us the expectation values for position and momentum.

We start by considering uncertainty in position $\triangle x = (\langle x^2 \rangle - \langle x \rangle^2)^{1/2}$. To evaluate $\triangle x$, we will need to calculate the expectation values of x and of x^2. This is done by expressing the position operator and the position operator squared in terms of $\hat{b}^\dagger$ and $\hat{b}$:

$$\hat{x} = \left(\frac{\hbar}{2m\omega} \right)^{1/2} \left(\hat{b} + \hat{b}^\dagger \right) \tag{6.42}$$

$$\hat{x}^2 = \left(\frac{\hbar}{2m\omega} \right) \left(\hat{b} + \hat{b}^\dagger \right)^2 = \left(\frac{\hbar}{2m\omega} \right) \left(\hat{b}\hat{b} + \hat{b}^\dagger\hat{b}^\dagger + \hat{b}\hat{b}^\dagger + \hat{b}^\dagger\hat{b} \right) \tag{6.43}$$

The expectation values of x and x^2 in the ground state are now easy to evaluate.

$$\langle x \rangle = \int \psi_0^* \hat{x} \psi_0 dx = 0 \tag{6.44}$$

The fact that $\langle x \rangle = 0$ follows directly from the observation that ψ_0 is an even function and x is an odd function, so the integral must, by symmetry, be zero.

The result for $\langle x^2 \rangle$ is almost as straightforward to evaluate. We start by writing down the expectation value in integral form:

$$\langle x^2 \rangle = \int \psi_0^* \hat{x}^2 \psi_0 dx = \frac{\hbar}{2m\omega} \int \psi_0^* \left(\hat{b}\hat{b} + \hat{b}^\dagger\hat{b}^\dagger + \hat{b}\hat{b}^\dagger + \hat{b}^\dagger\hat{b} \right) \psi_0 dx \tag{6.45}$$

The terms involving $\hat{b}\psi_0$ must, by definition of the ground state, be zero. The term $\hat{b}^\dagger\hat{b}^\dagger \psi_0$ creates a state ψ_2 that is orthogonal to ψ_0^* and so must contribute zero to the integral. This leaves the term $\hat{b}\hat{b}^\dagger\psi_0 = \psi_0$, which means that

$$\int \psi_0^* \left(\hat{b}\hat{b}^\dagger\right)\psi_0 dx = \int \psi_0^*\psi_0 dx = 1$$

Hence

$$\langle x^2 \rangle = \frac{\hbar}{2m\omega} \tag{6.46}$$

The same approach may be used to evaluate the uncertainty in momentum $\triangle p_x = (\langle p_x^2 \rangle - \langle p_x \rangle^2)^{1/2}$. As before, we express the momentum operator and the momentum operator squared in terms of $\hat{b}^\dagger$ and $\hat{b}$. The momentum operator can be written

$$p_x = i\left(\frac{\hbar m\omega}{2}\right)^{1/2}\left(\hat{b}^\dagger - \hat{b}\right) \tag{6.47}$$

so that

$$p_x^2 = \left(\frac{\hbar m\omega}{2}\right)\left(-\hat{b}\hat{b} - \hat{b}^\dagger\hat{b}^\dagger + \hat{b}\hat{b}^\dagger + \hat{b}^\dagger\hat{b}\right) \tag{6.48}$$

It follows that

$$\langle p_x \rangle = 0 \tag{6.49}$$

and

$$\langle p_x^2 \rangle = \frac{\hbar m\omega}{2} \tag{6.50}$$

We now have expressions for $\triangle x^2 = \langle x^2 \rangle - \langle x \rangle^2$ and $\triangle p_x^2 = \langle p_x^2 \rangle - \langle p_x \rangle^2$, which, because $\langle x \rangle^2 = 0$ and $\langle p_x \rangle^2 = 0$, give an uncertainty product:

$$\triangle x^2 \triangle p_x^2 = \langle x^2 \rangle\langle p_x^2 \rangle = \frac{\hbar^2 m\omega}{4m\omega} = \frac{\hbar^2}{4} \tag{6.51}$$

Taking the square root of both sides gives the uncertainty product of position and momentum

$$\triangle x \triangle p_x = \frac{\hbar}{2} \tag{6.52}$$

which satisfies the uncertainty relation $\triangle p \triangle x \geq \hbar/2$.

6.2.1.2 Using the uncertainty relation to obtain the ground state energy

In this approach, we use the uncertainty relation $\triangle x \triangle p_x \geq \hbar/2$ to calculate the minimum energy of a harmonic oscillator. One starts by expressing the total ground-state energy as the sum of the potential and kinetic energy terms involving displacement $\triangle x$ and momentum $\triangle p_x$:

$$E = \frac{1}{2}\left(\kappa(\triangle x)^2 + \frac{(\triangle p)^2}{m}\right) \tag{6.53}$$

Using the relationship $\triangle p \triangle x \geq \hbar/2$ to eliminate $\triangle x$, we will assume *minimum uncertainty* so that $\triangle p \triangle x = \hbar/2$. This gives

$$E = \frac{1}{2}\left(\frac{\kappa \hbar^2}{4(\triangle p_x)^2} + \frac{(\triangle p_x)^2}{m}\right) \tag{6.54}$$

Minimizing with respect to $\triangle p$,

$$\frac{d}{d\triangle p_x}\left(\frac{\kappa \hbar^2}{4(\triangle p_x)^2} + \frac{(\triangle p_x)^2}{m}\right) = -\frac{\kappa \hbar^2}{2(\triangle p_x)^3} + \frac{2\triangle p_x}{m} = 0 \tag{6.55}$$

$$\frac{m\kappa \hbar^2}{4} = \triangle p_x^4 \tag{6.56}$$

And since $\kappa = m\omega^2$,

$$\frac{m^2\omega^2\hbar^2}{4} = \triangle p_x^4 \tag{6.57}$$

Hence,

$$E_{min} = \frac{1}{2}\left(\frac{m\omega^2\hbar^2}{4 \cdot m\omega\hbar/2} + \frac{m\omega\hbar}{2m}\right) = \frac{1}{2}\left(\frac{\hbar\omega}{2} + \frac{\hbar\omega}{2}\right) = \frac{1}{2}\hbar\omega \tag{6.58}$$

Thus, we may conclude that the ground-state energy of the harmonic oscillator is just $E_0 = \hbar\omega/2$. The important physical interpretation of this result is that, according to the uncertainty relation, this ground-state energy represents a minimum uncertainty in the product of position and momentum.

In contrast, the lowest energy of a *classical* harmonic oscillator is *zero*. In the classical case, the minimum energy of a particle in the harmonic potential $V(x) = \kappa x^2/2$ corresponds to both momentum and position simultaneously being zero. In quantum mechanics, this is impossible, since $\triangle x \triangle p_x \geq \hbar/2$. As we have seen, the uncertainty product between position and momentum that minimizes total energy gives the ground-state energy $E_0 = \hbar\omega/2$.

6.2.2 Excited states of the harmonic oscillator

What we now need to know is how to use the operators $\hat{b}^\dagger$ and $\hat{b}$ to find the eigenstates and eigenenergies of all the other states of the system. These nonground states are called *excited states*.

Fortunately, it turns out that if we know ψ_0 we can generate all other ψ_n using the creation (or raising) operator $\hat{b}^\dagger$. To see that this is the case, we multiply the ground state of the harmonic oscillator by $\hat{b}^\dagger$ in the Schrödinger equation:

$$\hat{b}^\dagger \hbar\omega \left(\hat{b}^\dagger \hat{b} + \frac{1}{2} \right) \psi_n = \hat{b}^\dagger E_n \psi_n = \hat{b}^\dagger H \psi_n \tag{6.59}$$

$$\hbar\omega \left(\hat{b}^\dagger \hat{b}^\dagger \hat{b} + \frac{\hat{b}^\dagger}{2} \right) \psi_n = E_n \hat{b}^\dagger \psi_n = \hat{b}^\dagger H \psi_n \tag{6.60}$$

Now, using the commutation relation $\hat{b}\hat{b}^\dagger - \hat{b}^\dagger\hat{b} = 1$ and substituting for $\hat{b}^\dagger\hat{b} = \hat{b}\hat{b}^\dagger - 1$,

$$\hbar\omega \left(\hat{b}^\dagger \left(\hat{b}\hat{b}^\dagger - 1 \right) + \frac{\hat{b}^\dagger}{2} \right) \psi_n = \hbar\omega \left(\left(\hat{b}^\dagger \hat{b}\hat{b}^\dagger - \hat{b}^\dagger \right) + \frac{\hat{b}^\dagger}{2} \right) \psi_n = E_n \hat{b}^\dagger \psi_n \tag{6.61}$$

$$\hbar\omega \left(\left(\hat{b}^\dagger\hat{b} - 1 \right) \hat{b}^\dagger + \frac{\hat{b}^\dagger}{2} \right) \psi_n = \hbar\omega \left(\left(\hat{b}^\dagger\hat{b} - 1 \right) + \frac{1}{2} \right) \hat{b}^\dagger \psi_n = E_n \hat{b}^\dagger \psi_n \tag{6.62}$$

$$\hbar\omega \left(\hat{b}^\dagger\hat{b} + \frac{1}{2} \right) \left(\hat{b}^\dagger \psi_n \right) = \left(E_n + \hbar\omega \right) \left(\hat{b}^\dagger \psi_n \right) \tag{6.63}$$

This shows that the operator $\hat{b}^\dagger$, acting on the eigenstate ψ_n, generates a new eigenstate $(\hat{b}^\dagger \psi_n)$ with eigenvalue energy $(E_n + \hbar\omega)$.

It is now clear that $\hat{b}^\dagger$, operating on ψ_n, increases the eigenenergy by an amount $\hbar\omega$, so that the eigenenergy for the n-th state is

$$\boxed{E_n = \hbar\omega \left(n + \frac{1}{2} \right)} \tag{6.64}$$

where n is a positive integer $n = 0, 1, 2, \ldots$.

Summarizing what we know so far, we may think of $\hat{b}^\dagger$ and $\hat{b}$ as creation (or raising) and annihilation (or lowering) operators, respectively, that act upon the state ψ_n in such a way that

$$\hat{b}^\dagger \psi_n = A_{n+1} \psi_{n+1} \tag{6.65}$$

and

$$\hat{b} \psi_n = A_{n-1} \psi_{n-1} \tag{6.66}$$

where A_{n+1} and A_{n-1} are normalization constants, which we will now find. The way we do this is to start by assuming that the n-th state is correctly normalized and then find the relationship between the normalization of the n-th state and the $(n + 1)$-th state. Rather than write ψ_n, we use the notation $|n\rangle$, and for ψ_n^* we use $\langle n|$. Since we have assumed that the n-th state is normalized, we may write

$$\int \psi_n^* \psi_n dx = \langle n|n \rangle = 1 \tag{6.67}$$

Because the state $\hat{b}^{\dagger}\psi_n = |n+1\rangle$ is also required to be normalized, we may write

$$|A_{n+1}|^2 \langle \hat{b}^{\dagger}n | \hat{b}^{\dagger}n \rangle = 1 \tag{6.68}$$

where A_{n+1} is the normalization constant we wish to find. Because $\hat{b}^{\dagger}$ is a Hermitian operator, Eqn (6.68) may be written

$$|A_{n+1}|^2 \langle n | \hat{b}\hat{b}^{\dagger}n \rangle = |A_{n+1}|^2 \langle n | \left(\hat{b}\hat{b}^{\dagger} - 1\right)n \rangle = |A_{n+1}|^2 (n+1) \langle n|n \rangle = 1 \tag{6.69}$$

where we used the commutation relation $\hat{b}^{\dagger}\hat{b} = \hat{b}\hat{b}^{\dagger} - 1$ (Eqn (6.23)) and the fact that $|n\rangle$ is an eigenfunction of the Hamiltonian operator $\hat{b}^{\dagger}\hat{b}$. Hence,

$$|A_{n+1}|^2 = \frac{1}{(n+1)} \tag{6.70}$$

Choosing A_{n+1} to be real, then

$$A_{n+1} = \frac{1}{(n+1)^{1/2}} \tag{6.71}$$

or

$$|n+1\rangle = \frac{1}{(n+1)^{1/2}} |\hat{b}^{\dagger}n\rangle \tag{6.72}$$

Using this approach we may conclude that

$$\boxed{|\hat{b}^{\dagger}n\rangle = (n+1)^{1/2} \, | \, n+1 \, \rangle} \tag{6.73}$$

and

$$\boxed{| \, \hat{b}n \, \rangle = n^{1/2} \, | \, n-1 \, \rangle} \tag{6.74}$$

Suppose we normalize the ground state in such a way that $\langle 0|0 \rangle = 1$. We then can write a generating function for the state $|n\rangle$:

$$\boxed{|n\rangle = \frac{\left(\hat{b}^{\dagger}\right)^n}{(n!)^{1/2}} \cdot |0\rangle} \tag{6.75}$$

and

$$\hat{b}^{\dagger}|n\rangle = \frac{\left(\hat{b}^{\dagger}\right)^{n+1}}{(n!)^{1/2}} \cdot |0\rangle = \frac{(n+1)^{1/2}\left(\hat{b}^{\dagger}\right)^{n+1}}{((n+1)!)^{1/2}} \cdot |0\rangle \tag{6.76}$$

$$\hat{b}^{\dagger}|n\rangle = (n+1)^{1/2} \cdot |n+1\rangle \tag{6.77}$$

6.2.2.1 Matrix elements

The eigenstates $\psi_n = |n\rangle$ of the harmonic oscillator are orthonormal, so that

$$\langle m|n \rangle = \delta_{m=n} \tag{6.78}$$

In our notation, $\langle n|\hat{b}^{\dagger}|m\rangle = \int \psi_n^* \hat{b}^{\dagger} \psi_m dx$ is a matrix element. It can be shown that the matrix elements involving $\hat{b}^{\dagger}$ and $\hat{b}$ only exist between adjacent states, so that

$$\langle n|\hat{b}^{\dagger}|m\rangle = (m+1)^{1/2}\delta_{m=n-1} \tag{6.79}$$

$$\langle n|\hat{b}|m\rangle = m^{1/2}\delta_{m=n+1} \tag{6.80}$$

6.2.2.2 The operator $\hat{n}$

Sometimes it is convenient to define a *number* operator $\hat{n} = \hat{b}^{\dagger}\hat{b}$. The eigenvalue of this operator applied to an eigenstate labeled by quantum number n is just n:

$$\hat{b}^{\dagger}\hat{b}|n\rangle = \hat{b}^{\dagger}n^{1/2}|n-1\rangle = n^{1/2}\hat{b}^{\dagger}|n-1\rangle = n^{1/2}(n-1+1)^{1/2}|n\rangle = n|n\rangle \tag{6.81}$$

This operator commutes with $\hat{b}$ and $\hat{b}^{\dagger}$ in the following way:

$$\left[\hat{n}, \hat{b}\right] = \left[\hat{b}^{\dagger}\hat{b}, b\right] = \hat{b}^{\dagger}\left[\hat{b}, \hat{b}\right] + \left[\hat{b}^{\dagger}, \hat{b}\right]\hat{b} \tag{6.82}$$

$$\left[\hat{n}, \hat{b}^{\dagger}\right] = \left[\hat{b}^{\dagger}\hat{b}, \hat{b}^{\dagger}\right] = \hat{b}^{\dagger}\left[\hat{b}, \hat{b}^{\dagger}\right] + \left[\hat{b}^{\dagger}, \hat{b}^{\dagger}\right]\hat{b} \tag{6.83}$$

However, we know from our previous work that $[\hat{b}, \hat{b}] = 0$, $[\hat{b}, \hat{b}^{\dagger}] = 1$, and $[\hat{b}^{\dagger}, b] = -1$, so that

$$\left[\hat{n}, \hat{b}\right] = -\hat{b} \tag{6.84}$$

and

$$\left[\hat{n}, \hat{b}^{\dagger}\right] = \hat{b}^{\dagger} \tag{6.85}$$

Obviously, since the Hamiltonian operator for the harmonic oscillator is $H = \hbar\omega(\hat{n} + 1/2)$, the eigenfunctions of the Hamiltonian H are also eigenfunctions of the number operator $\hat{n}$.

We can summarize pictorially the results obtained so far in this chapter. In Fig. 6.2, the ground-state energy level and excited-state energy levels near the n-th state of the one-dimensional harmonic oscillator are shown schematically. Transition between eigenstates of neighboring energy is achieved by applying the operators $\hat{b}^{\dagger}$ or $\hat{b}$ to a given eigenstate. The energy of the n-th eigenstate is $\hbar\omega(n + 1/2)$, and the value of n is found by applying the operator $\hat{b}^{\dagger}\hat{b} = \hat{n}$ to the eigenstate. The ground state ψ_0 is defined by $\hat{b}\psi_0 = 0$.

Classical simple harmonic oscillation occurs in a single mode of frequency ω. The vibrational energy can be changed continuously by varying the oscillation amplitude. The quantum mechanical oscillator also has a single oscillatory mode characterized by frequency ω but the vibrational energy is quantized in such a way that

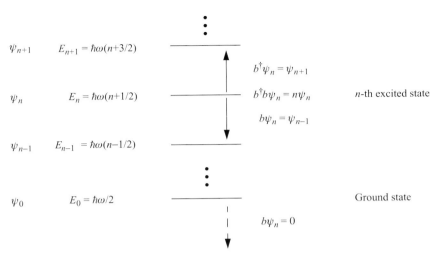

Fig. 6.2. Diagram showing the equally spaced energy levels of the one-dimensional harmonic oscillator. The raising or creation operator $b^{\dagger}$ acts upon eigenstate ψ_n with eigenenergy E_n to form a new eigenstate ψ_{n+1} with eigenenergy E_{n+1}. In a similar way, the annihilation operator b acts upon eigenstate ψ_n with eigenenergy E_n to form a new eigenstate ψ_{n-1} with eigenenergy E_{n-1}. Energy levels are equally spaced in energy by $\hbar w$. The ground state ψ_0 of the harmonic oscillator is the single state for which $b\psi_{n+1} = 0$. The ground state energy is $\hbar\omega/2$.

$E_n = \hbar\omega(n + 1/2)$. If we associate a particle with each quanta $\hbar\omega$, then there can be n particles in a given mode. Each n-particle state of the system is associated with a different wave function ψ_n.

The manipulation of operators is similar to ordinary algebra, with the obvious exception that the order of operators must be accurately maintained. There is another important rule. One must not divide by an operator $\hat{b}$. To show this, consider the state formed by

$$\psi = \hat{b}\psi_0 \tag{6.86}$$

Now, if we divide both sides by $\hat{b}$, then

$$\frac{1}{b}\psi = \psi_0 \tag{6.87}$$

To show that Eqn (6.86) and Eqn (6.87) are inconsistent with each other, consider the situation in which ψ_0 is the ground state. In this case, $\psi = \hat{b}\psi_0 = 0$ by our definition of a ground state (Eqn (6.34)). However, Eqn (6.87) states $(1/\hat{b})\psi = \psi_0 \neq 0$. While one cannot multiply by $1/\hat{b}$, it is possible to multiply by an operator of the form $1/(\alpha + \hat{b})$ since this may be expanded as a power series in $\hat{b}$.

6.3 The harmonic oscillator wave functions

Previously, we derived expressions for the creation or raising operator $\hat{b}^\dagger$ and the ground-state wave function ψ_0 for the one-dimensional harmonic oscillator so that

$$\hat{b}^\dagger = \left(\frac{m\omega}{2\hbar}\right)^{1/2}\left(x - \frac{\hbar}{m\omega}\frac{\partial}{\partial x}\right) \tag{6.88}$$

and

$$\psi_0(x) = A_0 e^{-x^2 \cdot \frac{m\omega}{2\hbar}} \tag{6.89}$$

where the normalization constant for the Gaussian wave function is given by

$$A_0 = \left(\frac{m\omega}{\pi\hbar}\right)^{1/4} \tag{6.90}$$

To simplify the notation, it is convenient to introduce a new spatial variable

$$\xi = \left(\frac{m\omega}{\hbar}\right)^{1/2} \cdot x \tag{6.91}$$

Eqn (6.88) may now be written as

$$\hat{b}^\dagger = \frac{1}{\sqrt{2}}\left(\left(\frac{m\omega}{\hbar}\right)^{1/2} \cdot x - \left(\frac{\hbar}{m\omega}\right)^{1/2}\frac{\partial}{\partial x}\right) = \frac{1}{\sqrt{2}}\left(\xi - \frac{\partial}{\partial\xi}\right) \tag{6.92}$$

and Eqn (6.89) for the ground-state wave function becomes

$$\psi_0(\xi) = A_0 e^{-\xi^2/2} \tag{6.93}$$

where the normalization constant is simply

$$A_0 = \left(\frac{1}{\pi}\right)^{1/4} \tag{6.94}$$

We can now generate the other higher-order states by using the operator $\hat{b}^\dagger$. Starting with the ground state and using Eqn (6.75) to ensure correct normalization, a natural sequence of wave functions is created:

$$\psi_0 \tag{6.95}$$

$$\psi_1 = \hat{b}^\dagger \psi_0 \tag{6.96}$$

$$\psi_2 = \frac{1}{\sqrt{2}}\hat{b}^\dagger \psi_1 = \frac{1}{\sqrt{2}}\left(\hat{b}^\dagger\right)^2 \psi_0 \tag{6.97}$$

$$\psi_3 = \frac{1}{\sqrt{3}}\hat{b}^\dagger \psi_2 = \frac{1}{\sqrt{2}\sqrt{3}}\left(\hat{b}^\dagger\right)^2 \psi_1 = \frac{1}{\sqrt{3!}}\left(\hat{b}^\dagger\right)^3 \psi_0 \tag{6.98}$$

$$\psi_n = \frac{1}{\sqrt{n!}}\left(\hat{b}^\dagger\right)^n \psi_0 \tag{6.99}$$

Because we know the ground-state wave function (Eqn (6.93)), it is now possible to generate all the other excited states of the system The first few states of the system are

$$\psi_0 = A_0 e^{-\xi^2/2} \tag{6.100}$$

$$\psi_1 = \hat{b}^\dagger \psi_0 = \frac{1}{\sqrt{1!}} \frac{1}{\sqrt{2}} \left(\xi - \frac{\partial}{\partial \xi} \right) A_0 e^{-\xi^2/2} = \frac{1}{\sqrt{2}} 2\xi A_0 e^{-\xi^2/2} = \frac{1}{\sqrt{2}} 2\xi \psi_0 \tag{6.101}$$

$$\psi_2 = \hat{b}^\dagger \psi_1 = \frac{1}{\sqrt{2!}} \left(\hat{a}^\dagger \right)^2 \psi_0 = \frac{1}{\sqrt{2!}} \frac{1}{\sqrt{2}} \left(\xi - \frac{\partial}{\partial \xi} \right) \frac{1}{\sqrt{2}} 2\xi A_0 e^{-\xi^2/2}$$

$$= \frac{1}{\sqrt{2}} \frac{1}{\sqrt{4}} \left(4\xi^2 - 2 \right) \psi_0 \tag{6.102}$$

$$\psi_3 = \hat{b}^\dagger \psi_2 = \frac{1}{\sqrt{3!}} \left(\hat{a}^\dagger \right)^3 \psi_0 = \frac{1}{\sqrt{3!}} \frac{1}{\sqrt{2}} \left(\xi - \frac{\partial}{\partial \xi} \right) \frac{1}{2} \left(4\xi^2 - 2 \right) \psi_0$$

$$= \frac{1}{\sqrt{6}} \frac{1}{\sqrt{8}} \left(8\xi^3 - 12\xi \right) \psi_0 \tag{6.103}$$

$$\psi_4 = \hat{b}^\dagger \psi_3 = \frac{1}{\sqrt{4!}} \left(\hat{a}^\dagger \right)^4 \psi_0 = \frac{1}{\sqrt{24}} \frac{1}{\sqrt{16}} \left(16\xi^4 - 48\xi^2 + 12 \right) \psi_0 \tag{6.104}$$

$$\psi_5 = \hat{b}^\dagger \psi_4 = \frac{1}{\sqrt{5!}} \left(\hat{a}^\dagger \right)^5 \psi_0 = \frac{1}{\sqrt{120}} \frac{1}{\sqrt{32}} \left(32\xi^5 - 160\xi^3 + 120\xi \right) \psi_0 \tag{6.105}$$

Notice that the wave functions are alternately even and odd functions.

It is clear from Eqn (6.100)–Eqn (6.105) that there is a relationship between the wave functions that can be expressed as a Hermite polynomial $H_n(\xi)$ so that

$$\psi_n(\xi) = \hat{b}^\dagger \psi_{n-1}(\xi) = \frac{1}{\sqrt{2^n n!}} \cdot H_n(\xi) \psi_0(\xi) \tag{6.106}$$

Learning more about $H_n(\xi)$ one finds that the n-th polynomial is related to the $n - 1$ and $n - 2$ polynomials via

$$H_n(\xi) = 2\xi H_{n-1}(\xi) - 2(n - 1) H_{n-2}(\xi) \tag{6.107}$$

The Hermite polynomials themselves may be obtained from the generating function

$$e^{-t^2 + 2t\xi} = \sum_{n=0}^{\infty} \frac{H_n(\xi)}{n!} t^n \tag{6.108}$$

or

$$H_n(\xi) = \left(\frac{d^n}{dt^n} e^{-t^2 + 2t\xi} \right)_{t=0} = (-1)^n e^{\xi^2} \frac{d^n}{d\xi^n} e^{-\xi^2} \tag{6.109}$$

The first few Hermite polynomials are

$$H_0(\xi) = 1 \tag{6.110}$$

$$H_1(\xi) = 2\xi \tag{6.111}$$

$$H_2(\xi) = 4\xi^2 - 2 \tag{6.112}$$

$$H_3(\xi) = 8\xi^3 - 12\xi \tag{6.113}$$

$$H_4(\xi) = 16\xi^4 - 48\xi^2 + 12 \tag{6.114}$$

$$H_5(\xi) = 32\xi^5 - 160\xi^3 + 120\xi \tag{6.115}$$

The Schrödinger equation for the one-dimensional harmonic oscillator can be written in terms of the variable ξ to give

$$\left(\frac{d^2}{d\xi^2} + \left(\frac{2E}{\hbar\omega} - \xi^2 \right) \right) \psi_n(\xi) = 0 \tag{6.116}$$

The solutions are the Hermite–Gaussian functions

$$\psi_n(\xi) = \left(\frac{1}{\sqrt{\pi}\, 2^n n!} \right)^{1/2} H_n(\xi) e^{-\xi^2/2} = \frac{1}{\sqrt{2^n n!}} \cdot H_n(\xi) \psi_0(\xi) \tag{6.117}$$

where $H_n(\xi)$ are Hermite polynomials. These satisfy the differential equation

$$\left(\frac{d^2}{d\xi^2} - 2\xi \frac{d}{d\xi} + 2n \right) H_n(\xi) = 0 \tag{6.118}$$

and n is related to the energy E_n by

$$E_n = \left(n + \frac{1}{2} \right) \hbar\omega \tag{6.119}$$

where $n = 0, 1, 2, \ldots$. Alternatively, if we know the two starting functions ψ_0 and ψ_1 then the n-th wave function can be generated by using

$$\psi_n(\xi) = \sqrt{\frac{2}{n}} \left(\xi \psi_{n-1}(\xi) - \sqrt{\frac{n-1}{2}} \psi_{n-2}(\xi) \right) \tag{6.120}$$

In Fig. 6.3, the wave function and probability function for the three lowest-energy states of the one-dimensional harmonic oscillator are plotted.

6.3.1   The classical turning point of the harmonic oscillator

Consider a one-dimensional classical harmonic oscillator consisting of a particle of mass m subject to a restoring force $-\kappa x$. The frequency of oscillation is $\omega = \sqrt{\kappa/m}$, and the total energy is $E_{\text{total}} = m\omega^2 A^2/2 = \kappa A^2/2$, where A is the classical amplitude of oscillation. If we equate the total energy of the classical harmonic oscillator with the energy of a one-dimensional quantum mechanical oscillator in the n-th state,

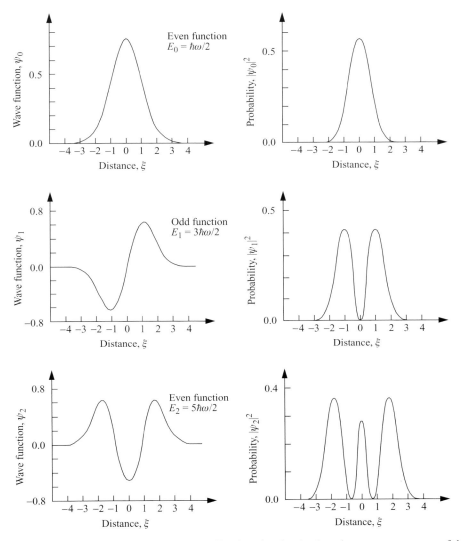

Fig. 6.3. Plot of wave function and probability function for the three lowest-energy states of the one-dimensional harmonic oscillator. Distance is measured in normalized units of $\xi = x(m\omega/\hbar)^{1/2}$.

we have

$$E_{\text{total}} = \kappa A_n^2/2 = \hbar\omega\left(n + \frac{1}{2}\right) \tag{6.121}$$

The classical turning point for the harmonic oscillator occurs at a distance x_n, corresponding to the classical amplitude A_n. This value is just

$$x_n = (\hbar\omega/\kappa)^{1/2}(2n+1)^{1/2} = \left(\frac{\hbar}{m\omega}\right)^{1/2}(2n+1)^{1/2} \tag{6.122}$$

The eigenfunctions of the quantum mechanical harmonic oscillator extend beyond the classical turning point. A portion of each wave function tunnels into a region of the potential that is not accessible classically. This means that there is a finite probability of finding the particle outside the region bounded by the potential.

Using Eqn (6.122) we see that the classical turning points for the ground state and the first two excited states of the harmonic oscillator are

$$x_0 = \left(\frac{\hbar}{m\omega}\right)^{1/2} \tag{6.123}$$

$$x_1 = \sqrt{3}\left(\frac{\hbar}{m\omega}\right)^{1/2} \tag{6.124}$$

$$x_0 = \sqrt{5}\left(\frac{\hbar}{m\omega}\right)^{1/2} \tag{6.125}$$

or, in terms of the parameter $\xi = (m\omega/\hbar)^{1/2} \cdot x$,

$$\xi_0 = 1 \tag{6.126}$$
$$\xi_1 = \sqrt{3} \tag{6.127}$$
$$\xi_2 = \sqrt{5} \tag{6.128}$$
$$\xi_n = (2n + 1)^{1/2} \tag{6.129}$$

Figure 6.4 illustrates the classical turning point $\pm x_n$ for the ground state ψ_0 and the first two excited states ψ_1 and ψ_2 of the one-dimensional harmonic oscillator. In the figure the potential $V(x) = \kappa x^2/2$, the energy levels E_n and the position of x_n are indicated.

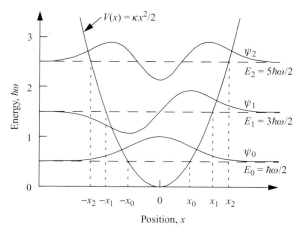

Fig. 6.4. Diagram showing the first three lowest-energy eigenfunctions of the one-dimensional harmonic oscillator. The wave functions penetrate into the regions of the potential that are not accessible according to classical mechanics. The classical turning points are x_0, x_1, and x_2 for the ground state and first two excited states of the harmonic oscillator respectively.

The portion of an eigenstate that is outside the classically allowed region can be used to obtain the probability of finding a particle in that region. If the particle is in a particular eigenstate, then all that needs to be done is integrate the square of the wave function in the classically inaccessible region.

As an example, consider a particle that is in the ground state ψ_0 with eigenenergy $E_0 = \hbar\omega/2$. The region where $V(x) > \hbar\omega/2$ is not accessible classically. Rewriting

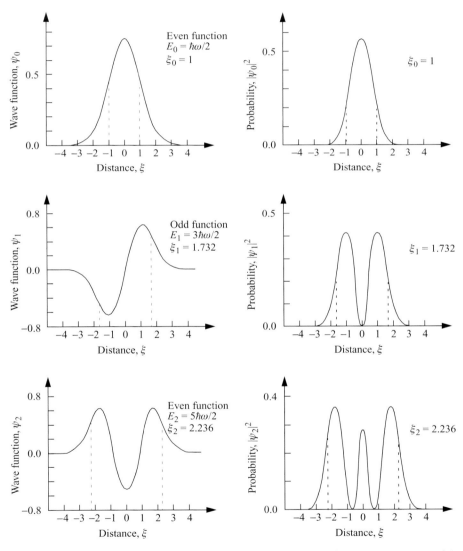

Fig. 6.5. Plot of wave function and probability function for the three lowest-energy states of the one-dimensional harmonic oscillator. Distance is measured in normalized units of $\xi = x(m\omega/\hbar)^{1/2}$. The classical turning points are ξ_0, ξ_1, and ξ_2 for the ground state and first two excited states of the harmonic oscillator, respectively.

this condition,

$$\frac{1}{2}m\omega^2 x^2 > \frac{1}{2}\hbar\omega \tag{6.130}$$

$$\left| x\sqrt{\frac{m\omega}{\hbar}} \right| > 1 \tag{6.131}$$

$$|\xi| > 1 \tag{6.132}$$

The probability of finding the particle in this nonclassical region is given by

$$\int_{\left|x\sqrt{\frac{m\omega}{\hbar}}\right|>1} \psi_0^*(x)\psi_0(x)dx = \int_{|\xi|>1} \psi_0^*(\xi)\psi_0(\xi)d\xi = \frac{1}{\sqrt{\pi}} \int_{|\xi|>1} e^{-\xi^2}d\xi = 0.157 \tag{6.133}$$

where we have used the fact that the ground-state wave function written as a function of the variable ξ (Eqn (6.91)) is (Eqn (6.93)):

$$\psi_0(\xi) = \left(\frac{1}{\pi}\right)^{1/4} e^{-\xi^2/2} \tag{6.134}$$

The numerical value of the integral in Eqn (6.133) is found from tables of values for the error function.

The excited states of the harmonic oscillator have a reduced probability of finding the particle in this nonclassical region. This probability decreases slowly as the energy eigenstate increases.

Figure 6.5 illustrates the classical turning points for the wave function and probability function of the one-dimensional harmonic oscillator. Distance is measured in normalized units of $\xi = (m\omega/\hbar)^{1/2} \cdot x$, so that the classical turning points $\xi_n = (2n+1)^{1/2}$ for the ground state and first two excited states are $\xi_0 = 1$, $\xi_1 = \sqrt{3}$, and $\xi_2 = \sqrt{5}$, respectively.

6.4 Time dependence

We know from experience gained in Chapter 3 that the probability distribution of bound-state eigenfunctions is time-independent and cannot carry flux or current. If the particle is in an eigenstate labeled by the positive integer quantum number n, then $\psi_n(x,t) = \psi_n(x)e^{-i\omega_n t}$, and no current flows because $|\psi_n(x,t)|^2$ is time-independent. However, if the particle is in a linear superposition of eigenstates,

$$\psi(x,t) = \sum_n a_n \psi_n(x)e^{-i\omega_n t} \tag{6.135}$$

where a_n is a coefficient that weights each eigenfunction, then we expect there to be a time dependence to the spatial probability distribution of the particle.

To investigate this more, consider a linear superposition of the ground-state and first excited-state eigenfunctions, $\psi_0(x, t)$ and $\psi_1(x, t)$. In this case, the total wave function describing the particle is

$$\psi(x, t) = a_0\psi_0(x)e^{-i\omega t/2} + a_1\psi_1(x)e^{-i3\omega t/2} \tag{6.136}$$

The energy of the n-th eigenstate is $E_n = \hbar\omega(n + 1/2)$, where $\hbar\omega$ is the energy difference between adjacent eigenstates. The coefficients a_0 and a_1 could be real or complex. Complex coefficients can be viewed as adding an initial phase to the eigenstate.

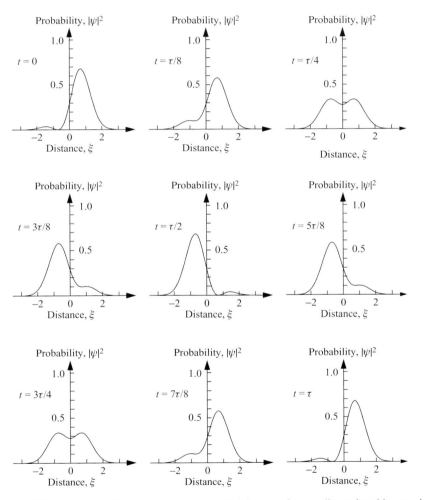

Fig. 6.6. Probability distribution at the indicated times t of a one-dimensional harmonic oscillator superposition state. In this particular case, the superposition state consists of the ground state and the first excited state, with equal weights and same initial phase. The probability distribution oscillates with a period $\tau = 2\pi/\omega$. The oscillation frequency is $\omega = (\kappa/m)^{1/2}$, where the force constant $\kappa = m\omega^2$. Distance is normalized to units of $\xi = x(m\omega/\hbar)^{1/2}$.

The probability density distribution for the superposition state $\psi(x, t)$ given by Eqn (6.136) is

$$|\psi(x, t)|^2 = a_0^2 |\psi_0(x)|^2 + a_1^2 |\psi_1(x)|^2 + 2a_0 a_1 \psi_0(x)\psi_1(x) \cos(\omega t) \qquad (6.137)$$

where a_0 and a_1 have been assumed real and $\hbar\omega = E_1 - E_0$. For the special case in which $a_0 = a_1$, we have a particularly simple expression for the probability density distribution:

$$|\psi(x, t)|^2 = a_0^2 \left(|\psi_0(x)|^2 + |\psi_1(x)|^2 + 2\psi_0(x)\psi_1(x)\cos(\omega t)\right) \qquad (6.138)$$

The probability distribution oscillates with frequency ω, where $\hbar\omega$ is the difference in energy between the two eigenstates. There is a coherent superposition of ψ_0 and ψ_1, with equal weight in each state giving a total wave function probability distribution and an expectation value of position that oscillates at frequency ω. The state of the system evolves according to a superposition of eigenfunctions. In general, unless the system is in a pure eigenstate, the probability distribution will be a function of time.

The time evolution of Eqn (6.138) is illustrated in Fig. 6.6. The superposition state $\psi(x, t)$ consists of the ground state and the first excited state, with equal weights and the same initial phase. In this case, the probability distribution $|\psi(x, t)|^2$ for the superposition state oscillates at frequency ω. This oscillation frequency is characterized by a period $\tau = 2\pi/\omega$, where $\omega = \sqrt{\kappa/m}$ for the force constant κ. In the figure, distance is normalized to units of $\xi = (m\omega/\hbar)^{1/2} \cdot x$.

It is apparent from Eqn (6.138) and Fig. 6.6 that a superposition of harmonic oscillator eigenstates can be used to create a spatial oscillation in the probability distribution function. If the probability distribution function describes a charged particle, such as an electron, then such an oscillation may give rise to a dipole moment and hence to a source of dipole radiation.

6.4.1   The superposition operator

It is possible to form an operator that creates a superposition state. This is best shown by example. Suppose we wish to create a superposition state consisting of the ground state of the harmonic oscillator and the first excited state so that the total wave function is

$$\psi = a_0 \psi_0 + a_1 \psi_1 \qquad (6.139)$$

where a_0 and a_1 are weights on each eigenfunction. This equation may always be written as

$$\psi = \left(a_0 + a_1 \hat{b}^\dagger\right)\psi_0 = a_1 \left(\frac{a_0}{a_1} + \hat{b}^\dagger\right)\psi_0 = a_1\left(\alpha + \hat{b}^\dagger\right)\psi_0 \qquad (6.140)$$

This shows that by adding a number α to the operator $\hat{b}^\dagger$ we create a new operator of the form $(\alpha + \hat{b}^\dagger)$, which acts to create a superposition state.

6.4.2   Measurement of a superposition state

In the previous section we showed that a coherent superposition state of the one-dimensional harmonic oscillator can dramatically alter the time dependence of probability density distributions and, by inference, the time dependence of expectation values. If the particle described by the superposition state is charged, then measurable effects such as dipole radiation may result.

However, one cannot *directly measure* a superposition state. When an energy measurement is performed on the system, the only result possible is a single eigenenergy corresponding to a single eigenfunction. These are the only measurable long-lived (stationary) energy states of the system that can be measured. In the case of a superposition state consisting of equal weights of the ground state ψ_0 and first excited state ψ_1 of the one-dimensional harmonic oscillator, we will obtain a result with energy E_0, with probability $1/2$ and energy E_1, also with probability $1/2$. This occurs because the superposition state has equal weight in the states ψ_0 and ψ_1. By probability, we mean that if the measurement is performed once on each of many separately prepared systems with the same initial state, the cumulative results are proportional to probabilities. After the measurement has been performed, the state of the system remains in the measured eigenstate. This phenomenon is called *the collapse of the wave function*. This mysterious behavior has to do with the nature of measurement, something that quantum mechanics does not describe very well. Energy is quantized and we can only measure discrete energy values. Energy measurement disturbs the system, forcing an initially coherent superposition of eigenstates into a definite stationary eigenstate. If we measure the probability distribution $\langle\psi|\psi\rangle$ or an expectation value such as the expectation value of position $\langle\psi|\hat{x}|\psi\rangle$, we find that the probability distribution is created from a number of discrete events. Only when the measurement is performed a large number of times, and each time on an identically prepared system, do the cumulative results asymptotically approach the predictions of the probability distribution or expectation value.

6.4.3   Time dependence of creation and annihilation operators

In quantum mechanics, we aim to solve the Schrödinger equation

$$H\psi(t) = i\hbar \frac{\partial}{\partial t}\psi(t) \tag{6.141}$$

So far, we have considered the case in which operators are time-independent and the wave functions are time-dependent. There is an alternative approach in which *operators are time-dependent and the wave functions are time-independent.* This is called the *Heisenberg representation.* To understand where this alternative view comes from, we start by noting that Eqn (6.141) can always be written

$$\psi(t) = e^{-iHt/\hbar} \psi(0) \tag{6.142}$$

where $\psi(0)$ is the initial wave function at time $t = 0$. While Eqn (6.142) does not formally change anything mathematically, it does suggest a new viewpoint. Because in quantum mechanics we want to predict the outcome of experiments, we will be interested in calculating expectation values of a time-independent operator $\hat{A}$ where

$$\langle \hat{A} \rangle = \langle \psi(t) | \hat{A} | \psi(t) \rangle \tag{6.143}$$

Using Eqn (6.142) we can now write the expectation value as

$$\langle \hat{A} \rangle = \langle \psi(0) | e^{iHt/\hbar} \hat{A} e^{-iHt/\hbar} | \psi(0) \rangle = \langle \psi(0) | \tilde{A}(t) | \psi(0) \rangle \tag{6.144}$$

where

$$\tilde{A}(t) = e^{iHt/\hbar} \hat{A} e^{-iHt/\hbar} \tag{6.145}$$

is a new time-dependent operator that acts on the initial wave function $\psi(0)$. These, the new operators in the Heisenberg representation, are time-dependent, and the wave functions they operate on are time-independent initial states.

Differentiating Eqn (6.145) with respect to time using the chain rule gives

$$\frac{d}{dt} \tilde{A}(t) = \frac{i}{\hbar} H \tilde{A}(t) - \frac{i}{\hbar} \tilde{A}(t) H + e^{iHt/\hbar} \frac{\partial}{\partial t} \hat{A} e^{-iHt/\hbar} \tag{6.146}$$

and since $(\partial/\partial t)\hat{A} = 0$, we may write

$$\frac{d}{dt} \tilde{A}(t) = \frac{i}{\hbar} [H, \tilde{A}] \tag{6.147}$$

Sometimes it is useful to know the time dependence of the operators $\hat{b}$ and $\hat{b}^\dagger$ for the harmonic oscillator. In the Heisenberg representation, the time development is given by Eqn (6.145), where $\hat{A}$ is an operator such as $\hat{b}$ or $\hat{b}^\dagger$. For the lowering or destruction operator $\hat{b}$, the time dependence becomes

$$\frac{d}{dt} \tilde{b}(t) = \frac{i}{\hbar} [H, \hat{b}] = i\omega(\hat{b}^\dagger \hat{b} \hat{b} - \hat{b} \hat{b}^\dagger \hat{b}) = i\omega[\hat{b}^\dagger, \hat{b}]\hat{b} = -i\omega \tilde{b}(t) \tag{6.148}$$

where we made use of the fact that $[\hat{b}, \hat{b}^\dagger] = -[\hat{b}^\dagger, \hat{b}] = 1$. The solution to the equation

$$\frac{d}{dt} \tilde{b}(t) = -i\omega \tilde{b}(t) \tag{6.149}$$

is simply

$$\tilde{b}(t) = e^{-i\omega t} \hat{b} \tag{6.150}$$

The Hermitian conjugate of this expression is

$$\tilde{b}^\dagger(t) = e^{i\omega t}\hat{b}^\dagger \tag{6.151}$$

Thus, the time development of the position operator

$$\hat{x} = \left(\frac{\hbar}{2m\omega}\right)^{1/2}(\hat{b} + \hat{b}^\dagger) \tag{6.152}$$

becomes

$$\hat{x}(t) = \left(\frac{\hbar}{2m\omega}\right)^{1/2}(\hat{b}e^{-i\omega t} + \hat{b}^\dagger e^{i\omega t}) \tag{6.153}$$

6.4.3.1 Charged particle in a harmonic potential subject to a constant electric field **E**

We now wish to apply what we know about the time dependence of raising and lowering operators to find out what happens when a particle of mass m and charge e in an initially one-dimensional harmonic potential is subject to a constant applied electric field of strength $|\mathbf{E}|$ in the positive x direction.

We begin by writing down the Hamiltonian for the system. This must contain contributions from both the oscillator and the electric field, as follows:

$$H = \frac{p^2}{2m} + \frac{\kappa}{2}x^2 + e|\mathbf{E}|x \tag{6.154}$$

Equation (6.154) may be rewritten in terms of the operators $\hat{b}^\dagger$ and $\hat{b}$:

$$H = \hbar\omega\left(\hat{b}^\dagger\hat{b} + \frac{1}{2}\right) + \hbar\lambda\left(\hat{b} + \hat{b}^\dagger\right) \tag{6.155}$$

where $\lambda = e|\mathbf{E}|\,(1/2m\hbar\omega)^{1/2}$. We may now define a new set of operators

$$\hat{B} = \hat{b} + \frac{\lambda}{\omega} \tag{6.156}$$

$$\hat{B}^\dagger = \hat{b}^\dagger + \frac{\lambda}{\omega} \tag{6.157}$$

They obey

$$\frac{\partial}{\partial t}\hat{B}(t) = -i\omega\hat{B}(t) \tag{6.158}$$

and so they have time dependence

$$\hat{B}(t) = e^{-i\omega t}\hat{B} \tag{6.159}$$

$$\hat{B}^\dagger(t) = e^{i\omega t}\hat{B}^\dagger \tag{6.160}$$

The operators $\hat{B}^\dagger$ and $\hat{B}$ obey the usual commutation relations

$$\left[\hat{B}, \hat{B}^\dagger\right] = \left[\hat{b} + \frac{\lambda}{\omega}, \hat{b}^\dagger + \frac{\lambda}{\omega}\right] = 1 \tag{6.161}$$

$$\left[\hat{B}^\dagger, \hat{B}\right] = -1 \tag{6.162}$$

$$\left[\hat{B}^\dagger, \hat{B}^\dagger\right] = 0 \tag{6.163}$$

$$\left[\hat{B}, \hat{B}\right] = 0 \tag{6.164}$$

The Hamiltonian for a particle of mass m and charge e in a one-dimensional harmonic potential subject to an applied electric field in the x direction may now be written in terms of the operators $\hat{B}^\dagger$ and $\hat{B}$:

$$H = \hbar\omega\left(\left(\hat{B}^\dagger - \frac{\lambda}{\omega}\right)\left(\hat{B} - \frac{\lambda}{\omega}\right) + \frac{1}{2}\right) + \hbar\lambda\left(\hat{B} + \hat{B}^\dagger - \frac{2\lambda}{\omega}\right) \tag{6.165}$$

$$H = \hbar\omega\left(\hat{B}^\dagger\hat{B} + \frac{1}{2}\right) - \frac{\hbar\lambda^2}{\omega} \tag{6.166}$$

The Schrödinger equation for this Hamiltonian is

$$H|n\rangle = \left(\hbar\omega\left(n + \frac{1}{2}\right) - \frac{\hbar\lambda^2}{\omega}\right)|n\rangle \tag{6.167}$$

where the state $|n\rangle$ is related to the ground state $|0\rangle$ by

$$|n\rangle = \frac{1}{(n!)^{1/2}}\left(\hat{B}^\dagger\right)^n|0\rangle \tag{6.168}$$

The operator for the position of the particle is

$$x(t) = \left(\frac{\hbar}{2m\omega}\right)^{1/2}\left(\hat{B}e^{-i\omega t} + \hat{B}^\dagger e^{i\omega t} - \frac{2\lambda}{\omega}\right) \tag{6.169}$$

Hence, the physics of the Hamiltonian describing the charged particle in a harmonic potential subject to an electric field is relatively straightforward. The particle oscillates at the same frequency, ω, established by the harmonic potential, but it is displaced by a distance

$$x_0 = \left(\frac{\hbar}{2m\omega}\right)^{1/2}\frac{2\lambda}{\omega} = \frac{e|\mathbf{E}|}{\kappa} \tag{6.170}$$

from the original position. The applied electric field has strength $|\mathbf{E}|$ in the positive x direction. If the sign of the electric field is changed to the negative x direction, then so is the sign of the displacement x_0. The change in equilibrium position due to application of the electric field causes the energy levels of the system to be changed by an amount

$$\Delta E = \frac{-\hbar\lambda^2}{\omega} = -\frac{e^2|\mathbf{E}|^2}{2m\omega^2} = -\frac{e^2|\mathbf{E}|^2}{2\kappa} \tag{6.171}$$

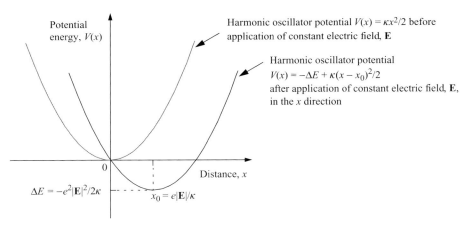

Fig. 6.7. Illustration of the potential of a particle of mass m and charge e in a one-dimensional harmonic potential before and after being subject to an applied constant electric field of strength $|\mathbf{E}|$ in the x direction. The effect of the electric field is to shift the equilibrium position of the oscillator to x_0 and lower the potential energy by an amount ΔE.

The energy levels are always shifted by $-\lambda^2/\omega$ independently of the sign of the electric field. This is so because the extra energy stored in the potential due to a displacement x_0 in equilibrium position enters as x_0^2 and so is positive, independent of the direction of x_0. The extra energy stored in the potential effectively removes a fixed amount of energy from the oscillator portion of the system. Figure 6.7 illustrates these ideas in graphical form.

6.4.3.2 Comparison with the classical result for a charged particle in a harmonic potential subject to a constant electric field **E**

In the classical case, the new equilibrium position of the particle will be shifted by Δx with $e|\mathbf{E}| = \kappa \Delta x$, where $\mathbf{E}$ is the electric field in the x direction and κ is the force constant. The total energy will be

$$E_{\text{total}} = \frac{1}{2}\kappa(x + \Delta x)^2 - e|\mathbf{E}|(x + \Delta x)$$

$$= \frac{1}{2}\kappa x^2 + \kappa x \Delta x + \frac{1}{2}\kappa \Delta x^2 - e|\mathbf{E}|x - e|\mathbf{E}|\Delta x \tag{6.172}$$

Now, since $e|\mathbf{E}| = \kappa \Delta x$, we can write

$$E_{\text{total}} = \frac{1}{2}\kappa x^2 + \frac{1}{2}\kappa \Delta x^2 - e|\mathbf{E}|\Delta x = \frac{1}{2}\kappa x^2 + \frac{1}{2}\kappa \frac{e^2|\mathbf{E}|^2}{\kappa^2} - \frac{e^2|\mathbf{E}|^2}{\kappa} \tag{6.173}$$

$$E_{\text{total}} = \frac{1}{2}\kappa x^2 + \frac{1}{2}\frac{e^2|\mathbf{E}|^2}{\kappa} - \frac{e^2|\mathbf{E}|^2}{\kappa} = \frac{1}{2}m\omega^2 x^2 - \frac{e^2|\mathbf{E}|^2}{2m\omega^2} \tag{6.174}$$

where we used the fact that $\kappa = m\omega^2$. With no electric field, the total energy is just

$$E_{total}^{(0)} = \frac{1}{2}m\omega^2 x^2 \tag{6.175}$$

and with an applied electric field the total energy is

$$E_{total} = E_{total}^{(0)} - \frac{e^2|\mathbf{E}|^2}{2m\omega^2} \tag{6.176}$$

From this we may conclude that the classical result is the same as the quantum one – the energy is reduced by $e^2|\mathbf{E}|^2/2m\omega^2$.

6.5 Quantization of electromagnetic fields

In Chapter 1 we introduced the idea that in free space the total electromagnetic energy density at position $\mathbf{r}$ and time t could be written

$$U = \frac{1}{2}(\mathbf{E} \cdot \mathbf{D} + \mathbf{B} \cdot \mathbf{H}) = \frac{\varepsilon_0}{2}|\mathbf{E}(\mathbf{r}, t)|^2 + \frac{1}{2\mu_0}|\mathbf{B}(\mathbf{r}, t)|^2 \tag{6.177}$$

In free space there is no charge density, so that $\nabla V(\mathbf{r}, t) = 0$ and the electric and magnetic fields can be written in terms of the vector potential $\mathbf{A}(\mathbf{r}, t)$ in such a way that $\mathbf{E}(\mathbf{r}, t) = -(d/dt)(\mathbf{A}(\mathbf{r}, t))$ and $\mathbf{B}(\mathbf{r}, t) = \nabla \times \mathbf{A}(\mathbf{r}, t)$. To make connection with the harmonic oscillator, we express vector potential in terms of its Fourier components:

$$\mathbf{A}(\mathbf{r}, t) = \frac{1}{(2\pi)^{3/2}} \int \mathbf{A}(\mathbf{k}, t)e^{i\mathbf{k} \cdot \mathbf{r}}d\mathbf{k} \tag{6.178}$$

Substituting into our expression for energy density gives

$$U = \frac{\varepsilon_0}{2}\left|-\frac{d}{dt}(\mathbf{A}(\mathbf{k}, t))\right|^2 + \frac{k^2}{2\mu_0}|\mathbf{A}(\mathbf{k}, t)|^2 \tag{6.179}$$

Comparing this with the energy of a particle of mass m in a harmonic oscillator potential

$$E = \frac{\hat{p}^2}{2m} + \frac{\kappa}{2}\hat{x}^2 \tag{6.180}$$

which oscillates at frequency $\omega = \sqrt{\kappa/m}$, it is clear that the electromagnetic field may be viewed as a number of harmonic oscillators of amplitude $\mathbf{A}(\mathbf{k}, t)$ and frequency $\omega = \sqrt{k^2/\mu_0\varepsilon_0} = kc$, where $c = 1/\sqrt{\mu_0\varepsilon_0}$ is the speed of light.

We are now able to draw a rather surprising conclusion. Because the electromagnetic field must be quantized in the same way as a harmonic oscillator, the energy density of the electromagnetic field in vacuum is never zero and through random processes can fluctuate. This result is a direct consequence of the fact that each electromagnetic mode of wave vector $\mathbf{k}$ and amplitude $\mathbf{A}(\mathbf{k}, t)$ has nonzero ground-state energy. Fluctuations in vacuum electromagnetic field density cause spontaneous transitions from excited electronic states of a system to lower energy states via emission of electromagnic energy.

The quantum of the electromagnetic field is the *photon*. Each photon particle carries energy $\hbar\omega$ and integer spin quantum number $s = \pm 1$ carrying angular momentum $\pm\hbar$. Photon spin corresponds to left- or right-circular polarization of plane waves in classical electromagnetism. All integer spin particles such as photons are classified as *bosons*.

6.5.1   Laser light

We have already shown that electromagnetic waves in free space can be described as the oscillation of the vector potential $\mathbf{A}\,(\mathbf{k}, t)$ at frequency $\omega = \sqrt{k^2/\mu_0\varepsilon_0} = kc$. Laser light emission is usually dominated by a single frequency of oscillation, say ω. Each component of the electromagnetic field obeys

$$\frac{\partial^2\phi}{\partial t^2} = -\omega^2\phi \tag{6.181}$$

This means that ϕ describes simple harmonic motion. Quantization of the electromagnetic field follows naturally. The energy levels for one spatial component of the field will be quantized so that each photon has energy $\hbar\omega$. The total energy due to all photons in the laser beam is $E = \hbar\omega(n + 1/2)$, where n is the number of photons in the electromagnetic field. Every time we add an additional photon to the system we add additional energy, $\hbar\omega$. Applying the operator $\hat{b}^\dagger$ to a state function ψ, which describes coherent laser light, increases the energy by $\hbar\omega$. Thus, we can think of $\hat{b}^\dagger$ as creating a photon. In the same way, $\hat{b}$ annihilates a photon. The operators $\hat{b}^\dagger$ and $\hat{b}$ are creation and annihilation operators for photon particles in the photon field. We may think of ψ_n as a multi-particle state function because it contains n photons.

The requirement that $\hat{b}\psi_0 = 0$ describes the ground state, ψ_0, corresponds to the idea that the vacuum contains no photons that $\hat{b}$ can annihilate. The ground-state energy $\hbar\omega/2$ is the energy in the electromagnetic field before the laser has been turned on. This is the so-called vacuum energy. In this picture, the energy level $E_n = \hbar\omega(n + 1/2)$ describes the situation when n photons have been added to the vacuum.

6.5.2   Quantization of an electrical resonator

The LC circuit of Exercise 1.12 has a resonant frequency such that $\omega = 1/\sqrt{LC}$, where L is an inductor and C is a capacitor placed in a series. Current flow in the circuit oscillates in time as $I(t) = I_0 e^{i\omega t}$, and electromagnetic energy is stored in the capacitor and inductor. There are, of course, many practical applications in which such a circuit may be used, including the production of electromagnetic radiation at the resonant frequency ω. The circuit behaves as a harmonic oscillator with electromagnetic energy quantized as photons in such a way that $E_n = \hbar\omega(n + 1/2)$, where $\hbar\omega = \hbar/\sqrt{LC}$.

The electromagnetic energy stored in the LC circuit is the sum of the energy in the inductor $LI^2/2$ and the energy in the capacitor $CV^2/2$. Since magnetic flux $\phi_B = LI$, charge $Q = CV$, and, on resonance, $C = 1/L\omega^2$, we may write the stored electromagnetic energy as

$$E = \frac{\phi_B^2}{2L} + \frac{L\omega^2 Q^2}{2} \tag{6.182}$$

Comparing this with energy of a particle of mass m in a harmonic oscillator potential (Eqn (6.180)) the "mass" of the resonator is L, the "spring constant" is $L\omega^2$, and we identify the coordinates $\hat{p} = \hat{\phi}_B$ and $\hat{x} = \hat{Q}$. From our previous experience, we expect operators $\hat{\phi}_B$ and $\hat{Q}$ to form a conjugate pair so that $\hat{\phi}_B = -i\hbar.\partial/\partial\hat{Q}$ and $[\hat{\phi}_B, \hat{Q}] = -i\hbar$. Quantum mechanics predicts that charge and magnetic flux obey the uncertainty relation $\Delta Q \Delta \phi_B \geq \hbar/2$ and so cannot be measured simultaneously to arbitrary accuracy. Quantum mechanics also predicts that the minimum electromagnetic energy in the resonant circuit is $E_0 = \hbar\omega/2 = \hbar/\sqrt{4LC}$.

6.6 Quantization of lattice vibrations

The classical Hamiltonian of a linear monatomic chain in the harmonic nearest-neighbor interaction approximation is given by Eqn (1.16) and was discussed in Section 1.2.3. The chain consists of N particles, each of mass m and equilibrium nearest-neighbor spacing L. The displacement from equilibrium position of the j-th particle is x_j. A vibrational normal mode of the chain is characterized by frequency $\omega(q)$ and wave vector $q = 2\pi/\lambda$. We can write the Hamiltonian for the linear chain in quantum form by first substituting the momentum operator $\hat{p}$ and the displacement operator $\hat{x}$ into the classical Hamiltonian Eqn (1.16). This gives

$$H = \sum_{j}^{N} \frac{\hat{p}_j^2}{2m} + \frac{\kappa}{2} \sum_{j}^{N} \left(2\hat{x}_j^2 - \hat{x}_j\hat{x}_{j+1} - \hat{x}_j\hat{x}_{j-1}\right) \tag{6.183}$$

where the sum is over all N particles in the chain. To transform Eqn (6.183) into a more convenient diagonal form, it is necessary to perform a canonical transformation. To do this, one defines new operators in terms of a linear combination of displacements and the momenta of each particle so that

$$\hat{b}_q = \frac{1}{\sqrt{N}} \sum_{j}^{N} \left(\frac{m\omega_q}{2\hbar}\right)^{1/2} \cdot e^{-iqjL} \left(\hat{x}_j + \frac{i\hat{p}_j}{m\omega_q}\right) \tag{6.184}$$

$$\hat{b}_q^\dagger = \frac{1}{\sqrt{N}} \sum_{j}^{N} \left(\frac{m\omega_q}{2\hbar}\right)^{1/2} \cdot e^{iqjL} \left(\hat{x}_j - \frac{i\hat{p}_j}{m\omega_q}\right) \tag{6.185}$$

where e^{iqjL} is a Bloch phase factor and ω_q will be chosen to diagonalize the Hamiltonian. The new operators, $\hat{b}_q^\dagger$ and $\hat{b}_q$, which obey the usual commutation relations

(Eqn (6.22)–Eqn (6.25)), may be used in linear combination to give

$$\hat{x}_j = \frac{1}{\sqrt{N}} \sum_q^N \left(\frac{\hbar}{2m\omega_q}\right)^{1/2} \cdot e^{iqjL} \left(\hat{b}_q + \hat{b}_{-q}^\dagger\right) \tag{6.186}$$

$$\hat{p}_j = \frac{-i}{\sqrt{N}} \sum_q^N \left(\frac{m\hbar\omega_q}{2}\right)^{1/2} \cdot e^{iqjL} \left(\hat{b}_q - \hat{b}_{-q}^\dagger\right) \tag{6.187}$$

Substitution of these new expressions for $\hat{x}_j$ and $\hat{p}_j$ into Eqn (6.183) results in a Hamiltonian

$$H = -\frac{1}{4} \sum_q \hbar\omega_q \left(\hat{b}_q - \hat{b}_{-q}^\dagger\right)\left(\hat{b}_{-q} - \hat{b}_q^\dagger\right)$$

$$+ \frac{1}{4} \sum_q \frac{\hbar}{\omega_q} \frac{\kappa}{m} \left(\hat{b}_q + \hat{b}_{-q}^\dagger\right)\left(\hat{b}_{-q} + \hat{b}_q^\dagger\right)\left(2 - e^{iqL} - e^{-iqL}\right) \tag{6.188}$$

Recognizing that if we choose

$$\omega_q^2 = \frac{\kappa}{m}\left(2 - e^{iqL} - e^{-iqL}\right) \tag{6.189}$$

the Hamiltonian takes on the familiar diagonal form (Eqn (6.14) and Eqn (6.27)), as follows:

$$H = \sum_q \frac{\hbar\omega_q}{2}\left(\hat{b}_q\hat{b}_q^\dagger + \hat{b}_q^\dagger\hat{b}_q\right) = \sum_q \hbar\omega_q \left(\hat{b}_q^\dagger\hat{b}_q + \frac{1}{2}\right) = \sum_q \hbar\omega_q \left(n_q + \frac{1}{2}\right) \tag{6.190}$$

This is the sum of independent linear oscillators of frequency $\omega(q)$. Modes of vibrational frequency ω_q and wave vector q are described by the dispersion relation $\omega = \omega(q)$, which, for the case we are considering, is the same as the classical result (Eqn (1.20)). Each quantized vibrational mode of the linear chain is made up of N individual particles of mass m oscillating about their equilibrium positions coupled via the interaction potential. A lattice vibration of wave vector $q = 2\pi/\lambda$ is quantized with energy $E_n = \hbar\omega(q)(n + 1/2)$ and contains $n_q = \hat{b}_q^\dagger\hat{b}_q$ phonons. Phonons have zero integer spin and, like photons, are bosons.

6.7 Quantization of mechanical vibration

In this chapter we have shown that a harmonic oscillator is quantized in energy so that in one dimension $E_n = \hbar\omega(n + 1/2)$. Lattice vibrations in a crystal are quantized and so are oscillations of small mechanical structures. As an example of a small mechanical structure, consider the cantilever beam shown schematically in Fig. 6.8.

In Exercise 1.10 it was stated that the lowest-frequency vibrational mode of a long, thin cantilever beam is

$$\omega = 3.25 \frac{d}{l^2}\sqrt{\frac{E_{\text{Young}}}{12\rho}} \tag{6.191}$$

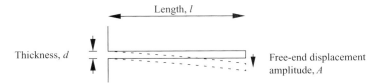

Fig. 6.8. Cross-section of a cantilever beam of length l and thickness d with a free-end displacement amplitude A.

where l is the length, d is the thickness, ρ is the density of the beam, and E_{Young} is Young's modulus. Classical mechanics predicts that the vibrational energy of a cantilever with width w and free-end displacement amplitude A is

$$E_{\text{classical}} = \frac{wd^3 A^2 E_{\text{Young}}}{6l^3} \tag{6.192}$$

If one equates $E_{\text{classical}}$ with the characteristic energy of a quantized one-dimensional harmonic oscillator $\hbar\omega$, then one may estimate the amplitude of oscillation of the free end to be

$$A = \left(\frac{21.12 \times \hbar \times l}{w \times d^2 \sqrt{12 \times \rho \times E_{\text{Young}}}} \right)^{1/2} \tag{6.193}$$

Suppose the cantilever is made of silicon by a microelectromechanical systems (MEMS) process. In this case, we might use the bulk values for density $\rho = 2.328 \times 10^3 \, \text{kg m}^{-3}$ and Young's modulus $E_{\text{Young}} = 1.96 \times 10^{11} \, \text{N m}^{-2}$ to estimate oscillation frequency and amplitude. Equations (6.191) and (6.193) predict that a small cantilever structure with dimensions $l = 10 \, \mu\text{m}$, $w = 50 \, \text{nm}$, and $d = 10 \, \text{mn}$ has a free-end oscillation frequency $\nu = 148 \, \text{kHz}$ and an amplitude $A = 0.0078 \, \text{mn}$. To make sure that low-energy vibrational motion dominates, we require temperature $T \leq \hbar\omega/k_B = 7 \, \mu\text{K}$.

This combination of very small vibrational amplitude and low temperature makes direct measurement of quantized mechanical motion in a MEMS structure quite challenging. Nevertheless, it is a prediction of our quantum theory that the motion of small mechanical structures is quantized in such a way that vibrational energy is $E_n = \hbar\omega(n + 1/2)$.

6.8 Example exercises

Exercise 6.1

A one-dimensional harmonic oscillator is in the $n = 1$ state, for which

$$\psi_{n=1}(x) = 2(2\pi^{1/2}x_0)^{-1/2}(x/x_0)e^{-0.5(x/x_0)^2}$$

with $x_0 = (\hbar/m\omega)^{1/2}$. Calculate the probability of finding the particle in the interval x to $x + dx$. Show that, according to classical mechanics, the probability is

$$P_{\text{classical}}(x)dx = \frac{1}{\pi}(A_0^2 - x^2)^{-1/2}dx$$

for $-A_0 < x < A_0$ and zero elsewhere (A_0 is the classical amplitude). With the aid of sketches compare this probability distribution with the quantum mechanical one. Locate the maxima for the probability distribution for the $n = 1$ quantum state relative to the classical turning point. Assume that the classical amplitude, A_0, is such that the total energy is identical in both cases.

Exercise 6.2
Show that $\langle \hat{b}^\dagger n | \hat{b}^\dagger n \rangle = \langle n | \hat{b}\hat{b}^\dagger n \rangle$, where $\hat{b}^\dagger$ is the creation operator for the harmonic oscillator.

Exercise 6.3
A two-dimensional potential for a particle of mass m is of the form

$$V(x, y) = m\omega^2(x^2 + xy + y^2)$$

Write the potential as a 2×2 matrix and find new coordinates u and v that diagonalize the matrix. Find the energy levels of the particle.

Exercise 6.4
The purpose of this exercise is to show that a coherent superposition of high-quantum-number energy eigenstates of a harmonic oscillator with a spread in energies that is small compared with their mean energy behaves as a classical oscillator. Consider a one-dimensional harmonic oscillator characterized by mass m and angular frequency ω. The eigenstate of the Hamiltonian corresponding to the quantum number n is $|n\rangle$. The time-dependent state $\psi(t)$ of the oscillator at time $t = 0$ is

$$\psi(t = 0) = \frac{1}{\sqrt{2\Delta N}} \sum_{n=N-\Delta N}^{n=N+\Delta N} |n\rangle$$

where $1 \ll \Delta N \ll N$.

(a) Find the expectation value of position as a function of time.

(b) Compare your result in (a) with the predictions of a classical harmonic oscillator.

(c) Write a computer program that plots the n-th eigenstate $|n\rangle$ and probability $\langle n|n \rangle$ of the one-dimensional harmonic oscillator. Use Eqn (6.120) to generate the wave function, and plot the $|n = 18\rangle$ eigenstate and its probability function.

(d) In general, a superposition wave function of the one-dimensional harmonic oscillator can be formed so that $\psi(t) = \sum a_n |n\rangle e^{-i\omega_n t}$, where a_n is a weighting factor that contains amplitude and phase information for each eigenstate $|n\rangle$ and $\omega_n = \omega(n + 1/2)$.

However, if one assumes equal weights and a contiguous sum, then at time $t = 0$ the superposition wave function

$$\psi(t = 0) = \frac{1}{\sqrt{2\Delta N}} \sum_{n=N-\Delta N}^{n=N+\Delta N} |n\rangle$$

represents a particle at an extreme of its motion. Use (c) and write a computer program that plots the superposition wave function and particle probability function for the specific case in which $N = 18$ and $\Delta N = 2$. Compare the peak in probability with the classical turning point for the $|n = 18\rangle$ eigenstate.

Exercise 6.5

A particle of mass m in a one-dimensional harmonic potential $V(x) = m\omega^2 x^2/2$ is in an eigenstate $\psi_n(\xi)$ with eigenvalue n and eigenenergy $E_n = \hbar\omega(n + 1/2)$. The probability of the particle being found in the nonclassical region of the harmonic oscillator is given by

$$P_n^{\text{nonclassical}} = \int\limits_{|\xi|>1} \psi_n^*(\xi)\psi_n(\xi)d\xi$$

where $\xi = (m\omega/\hbar)^{1/2} \cdot x$ and the classical turning point is $\xi_n = (2n + 1)^{1/2}$.

Find the values of $P_n^{\text{nonclassical}}$ for the first excited state $n = 1$ and for the second excited state $n = 2$.

Exercise 6.6

In Section 6.2 it was shown that the creation operator $\hat{b}^\dagger$ and the annihilation operator $\hat{b}$ for the one-dimensional harmonic oscillator with Hamiltonian H are related to each other through the commutation relation $[\hat{b}, \hat{b}^\dagger] = (\hat{b}\hat{b}^\dagger - \hat{b}^\dagger\hat{b}) = 1$.

Verify that the expressions for the commutation relations $[\hat{b}, \hat{b}^\dagger]$, $[\hat{b}, \hat{b}]$, and $[\hat{b}^\dagger, \hat{b}^\dagger]$ given in Section 6.2 are correct.

If we define a new operator $\hat{n} = \hat{b}^\dagger\hat{b}$, show that

$$H = \hbar\omega\left(\hat{n} + \frac{1}{2}\right)$$

and derive expressions for the commutation relations $[\hat{n}, \hat{b}]$ and $[\hat{n}, \hat{b}^\dagger]$.

Exercise 6.7

What are the degeneracies of the three lowest levels of a symmetric three-dimensional harmonic oscillator? Find a general expression for the degeneracy of the n-th level. Derive and explain the meaning of the matrix elements $\langle n|\hat{b}^\dagger|m\rangle = (m + 1)^{1/2}\,\delta_{m=n-1}$, $\langle n|\hat{b}|m\rangle = m^{1/2}\delta_{m=n+1}$, and $\langle m|n\rangle = \delta_{mn}$ between states m and n of a harmonic oscillator.

Exercise 6.8

The ground-state wave function of a particle of mass m in a harmonic potential is $\psi_0 = A_0 e^{-x^2/4\alpha^2}$, where $\alpha^2 = \hbar/2m\omega$. Derive the uncertainties in position and momentum, and show that they satisfy the uncertainty relation.

Exercise 6.9

The ground state and the *second* excited state of a charged particle of mass m in a one-dimensional harmonic oscillator potential are both occupied. What is the expectation value of the particle position x as a function of time? What happens to the expectation value if the potential is subject to a constant electric field $\mathbf{E}$ in the x direction?

Exercise 6.10

Using the method outlined in Section 3.4, write a computer program to solve the Schrödinger wave equation for the first four eigenvalues and eigenstates of an electron with effective mass $m_e^* = 0.07 \times m_0$ confined to a parabolic potential well in such a way that $V(x) = ((x - L/2)^2/(L/2)^2)$ eV and $L = 100$ nm.

SOLUTIONS

Solution 6.1

Starting with a one-dimensional harmonic oscillator in the $n = 1$ state for which

$$\psi_{n=1}(x) = 2(2\pi^{1/2}x_0)^{-1/2}(x/x_0)e^{-0.5(x/x_0)^2}$$

with $x_0 = (\hbar/m\omega)^{1/2}$, we wish to calculate the probability of finding the particle in the interval x to $x + dx$. The quantum mechanical probability is

$$P_{\text{quantum}}(x)dx = |\psi_1|^2 dx = \frac{2}{\pi^{1/2}x_0}(x/x_0)^2 e^{-(x/x_0)^2}dx$$

The maxima in this distribution occur when

$$0 = \frac{d}{dx}P_{\text{quantum}}(x)$$

$$0 = \frac{2}{\pi^{1/2}x_0^3}2xe^{-(x/x_0)^2} - \frac{2}{\pi^{1/2}x_0^3}2x(x/x_0)^2 e^{-(x/x_0)^2}$$

$$0 = 1 - \frac{x^2}{x_0^2}$$

so that $x_{\text{max}} = \pm x_0$ is the peak in the quantum mechanical probability distribution.

Classically, $x(t) = A_0 \sin \omega t$, where A_0 is the classical amplitude of oscillation. The energy of the classical harmonic oscillator is found from the solution to the equation of motion

$$\kappa x + m\frac{d^2x}{dt^2} = 0$$

or the Hamiltonian

$$H = T + V = \frac{1}{2}m\left(\frac{dx}{dt}\right)^2 + \frac{1}{2}\kappa x^2$$

where κ is the force constant and m is the particle mass. The potential energy is

$$V = \frac{1}{2}\kappa A^2 \cos^2(\omega_0 t + \phi)$$

and the kinetic energy is

$$T = \frac{1}{2}m\omega_0^2 A^2 \sin^2(\omega_0 t + \phi)$$

where ϕ is an arbitrary phase factor. Hence, the total energy

$$H = \frac{1}{2}m\omega_0^2 A^2 = \frac{1}{2}\kappa A^2$$

since $\sin^2(\theta) + \cos^2(\theta) = 1$ and $\kappa = m\omega_0^2$.

Equating the total energy for the classical and quantum mechanical cases gives

$$E_{\text{total}} = \frac{1}{2}m\omega^2 A_0^2 = \frac{3}{2}\hbar\omega$$

Therefore,

$$A_0 = \sqrt{\frac{3\hbar\omega}{m\omega^2}} = \sqrt{\frac{\hbar}{m\omega}}\sqrt{3} = x_0\sqrt{3} = 1.73x_0$$

Classically, $x(t) = A_0 \sin \omega t$, where A_0 is the classical amplitude of oscillation and $\tau = 2\pi/\omega$ is the oscillation period. For any time interval during the period of oscillation for which $0 < t < \tau = 2\pi/\omega$, the oscillator will be between x and $x + dx$ at the time

$$t = \frac{1}{\omega}\sin^{-1}(x/A_0)$$

Hence, the probability of finding the classical particle in the interval x to $x + dx$ is

$$P_{\text{classical}}(x)dx = \frac{2dt}{\tau} = \frac{2dt}{2\pi/\omega} = \frac{\omega dt}{\pi}$$

where we note that the factor of 2 arises because the particle passes position x twice during one oscillation period. Substituting in our expression for dt, we have

$$dt = \frac{dt}{dx}dx = \frac{d}{dx}\left(\frac{1}{\omega}\sin^{-1}(x/A_0)\right)dx = \frac{1}{\omega}\cdot\frac{1/A_0}{\sqrt{1-(x/A_0)^2}}\cdot dx$$

$$dt = \frac{1}{\omega}\left(A_0^2 - x^2\right)^{-1/2}\cdot dx$$

so that

$$P_{\text{classical}}(x)dx = \frac{1}{\pi}\left(A_0^2 - x^2\right)^{-1/2}dx$$

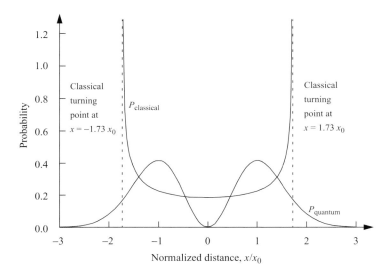

Solution 6.2

We wish to show that $\langle \hat{b}^\dagger n | \hat{b}^\dagger n \rangle = \langle n | \hat{b} \hat{b}^\dagger n \rangle$, where $\hat{b}^\dagger$ is the creation operator for the harmonic oscillator. The fact that $\langle \hat{b}^\dagger n | \hat{b}^\dagger n \rangle = \langle n | \hat{b} \hat{b}^\dagger n \rangle$ follows from the Hermitian property of the operator $\hat{b}^\dagger$. What we need to show is that $\langle \hat{b}^\dagger n | m \rangle = \langle n | \hat{b} m \rangle$, since $|\hat{b}^\dagger n\rangle = (n+1)^{1/2} |n+1\rangle$, which we can chose to define as a new state $|m\rangle$.

We start by writing down the definition of the creation and annihilation operator for the one-dimensional harmonic oscillator in terms of the parameter $\xi = x \cdot \sqrt{m\omega/\hbar}$:

$$\hat{b}^\dagger = \frac{1}{\sqrt{2}} \left(\left(\frac{m\omega}{\hbar}\right)^{1/2} \cdot x - \left(\frac{\hbar}{m\omega}\right)^{1/2} \frac{\partial}{\partial x} \right) = \frac{1}{\sqrt{2}} \left(\xi - \frac{\partial}{\partial x} \right)$$

$$\hat{b} = \frac{1}{\sqrt{2}} \left(\left(\frac{m\omega}{\hbar}\right)^{1/2} \cdot x + \left(\frac{\hbar}{m\omega}\right)^{1/2} \frac{\partial}{\partial x} \right) = \frac{1}{\sqrt{2}} \left(\xi + \frac{\partial}{\partial x} \right)$$

We will suppose that the eigenfunction $\psi_n(\xi)$ tends to zero at plus and minus infinity. This is in fact the case for the harmonic oscillator. Now, we write $\langle \hat{b}^\dagger n | m \rangle = \langle n | \hat{b} m \rangle$ in integral form

$$\int \left(\hat{b}^\dagger \psi_n^*(\xi) \right) \psi_m(\xi) d\xi = \frac{1}{\sqrt{2}} \int \left(\left(\xi - \frac{\partial}{\partial \xi} \right) \psi_n^*(\xi) \right) \psi_m(\xi) d\xi$$

$$\int \left(\hat{b}^\dagger \psi_n^*(\xi) \right) \psi_m(\xi) d\xi = \frac{1}{\sqrt{2}} \int \xi \psi_n^*(\xi) \psi_m(\xi) d\xi - \frac{1}{\sqrt{2}} \int \left(\frac{\partial}{\partial \xi} \psi_n^*(\xi) \right) \psi_m(\xi) d\xi$$

ξ in the first term on the right-hand side is just a multiplicative factor, and so it can be placed between the two wave functions without changing the integral. The second integral on the right-hand side can be done by parts: $\int U V' dx = U V - \int U' V dx$, where $U = \psi_m^*(\xi)$, and $V' = \partial\psi_n(\xi)/\partial\xi$. This gives

$$\int \left(\hat{b}^\dagger \psi_n^*(\xi)\right)\psi_m(\xi)d\xi = \frac{1}{\sqrt{2}} \int \psi_n^*(\xi)\xi\psi_m(\xi)d\xi$$

$$- \left(\frac{1}{\sqrt{2}}\psi_n^*(\xi)\psi_m(\xi)\right)\Bigg|_{\xi\to-\infty}^{\xi\to\infty} + \frac{1}{\sqrt{2}} \int \psi_n^*(\xi)\frac{\partial}{\partial\xi}\psi_m(\xi)d\xi$$

Since we have assumed that $\psi_n(\xi) \to 0$ as $\xi \to \pm\infty$, the second term on the right-hand side is zero, leaving

$$\int \left(\hat{b}^\dagger \psi_n^*(\xi)\right)\psi_n(\xi)d\xi = \frac{1}{\sqrt{2}} \int \psi_n^*(\xi)\left(\left(\xi + \frac{\partial}{\partial\xi}\right)\psi_m(\xi)\right)d\xi$$

$$\int \left(\hat{b}^\dagger \psi_n^*(\xi)\right)\psi_m(\xi)d\xi = \frac{1}{\sqrt{2}} \int \psi_n^*(\xi)(\hat{b}\psi_m(\xi))d\xi$$

or

$$\langle\hat{b}^\dagger n|m\rangle = \langle n|\hat{b}m\rangle$$

As a check of the algebraic formalism, we note that

$$\langle\hat{b}^\dagger n|\hat{b}^\dagger n\rangle = \langle\hat{b}^\dagger n|(n+1)^{1/2}|n+1\rangle = (n+1)\langle n+1|n+1\rangle = n+1$$

and, noting that $\hat{b}\hat{b}^\dagger = 1 + \hat{b}^\dagger\hat{b}$, we can write

$$\langle n|\hat{b}\hat{b}^\dagger n\rangle = \langle n|(1 + \hat{b}^\dagger\hat{b})n\rangle = 1 + \langle n|\hat{b}^\dagger\hat{b}n\rangle = 1 + n^{1/2}\langle n|\hat{b}^\dagger(n-1)\rangle = 1 + n\langle n|n\rangle$$

$$\langle n|\hat{b}\hat{b}^\dagger n\rangle = 1 + n$$

We may conclude that $\langle\hat{b}^\dagger n|\hat{b}^\dagger n\rangle = \langle n|\hat{b}\hat{b}^\dagger n\rangle$, which agrees with the previous result using integration of wave functions.

Solution 6.3
The symmetric potential $V(x, y) = m\omega^2(x^2 + xy + y^2)$ can be written as a 2×2 matrix

$$V(x, y) = \frac{m}{2}\omega^2[x \quad y]\begin{bmatrix} 2 & 1 \\ 1 & 2 \end{bmatrix}\begin{bmatrix} x \\ y \end{bmatrix}$$

Because the eigenvalues of the 2×2 matrix are 3 and 1, there must be coordinates u and v that diagonalize the matrix to give

$$V(u, v) = \frac{m}{2}\omega^2[u \quad v]\begin{bmatrix} 3 & 0 \\ 0 & 1 \end{bmatrix}\begin{bmatrix} u \\ v \end{bmatrix}$$

Notice that the symmetric 2×2 matrix has been diagonalized by an orthogonal change of variables

$$u = \frac{1}{\sqrt{2}}(x + y) \quad \text{and} \quad v = \frac{1}{\sqrt{2}}(x - y)$$

Multiplying out the matrix, we see that the potential is now given by

$$V(u, v) = \frac{m}{2}\omega^2 \left(3u^2 + v^2\right)$$

and so the eigenenergy values of the particle are

$$E_{n_u, n_v} = \sqrt{3} \cdot \hbar\omega \left(n_u + \frac{1}{2}\right) + \hbar\omega \left(n_v + \frac{1}{2}\right)$$

where the quantum numbers n_u and n_v are positive integers $0, 1, 2, \ldots$.

Solution 6.4

(a) The state function at time $t = 0$ of the one-dimensional harmonic oscillator is given as

$$\psi(0) = \frac{1}{\sqrt{2\Delta N}} \sum_{n=N-\Delta N}^{n=N+\Delta N} |n\rangle$$

where $1 \ll \Delta N \ll N$. This is a coherent superposition of high-quantum-number eigenfunctions the relative phase of which is specified at time $t = 0$. It follows that the wave function evolves in time as

$$\psi(t) = \frac{1}{\sqrt{2\Delta N}} \sum_{n=N-\Delta N}^{n=N+\Delta N} |n\rangle e^{-iE_n(t/\hbar)}$$

$$\psi(t) = \frac{1}{\sqrt{2\Delta N}} \sum_{n=N-\Delta N}^{n=N+\Delta N} |n\rangle e^{-i\left(n+\frac{1}{2}\right)\omega t}$$

since the energy of the n-th eigenfunction is $E_n = (n + \frac{1}{2})\hbar\omega$. The position operator

$$x = \left(\frac{\hbar}{2m\omega}\right)^{1/2} \left(\hat{b} + \hat{b}^\dagger\right)$$

has an expectation value given by

$$\langle x \rangle = \left(\frac{\hbar}{2m\omega}\right)^{1/2} \langle \psi(t)|\hat{b} + \hat{b}^\dagger|\psi(t)\rangle$$

$$\langle x \rangle = \left(\frac{\hbar}{2m\omega}\right)^{1/2} \sum_{nn'} \langle n'|\hat{b} + \hat{b}^\dagger|n\rangle \exp(i(n' - n)\omega t)$$

The relations $b|n\rangle = \sqrt{n}|n-1\rangle$ and $\hat{b}^\dagger|n\rangle = \sqrt{(n+1)}|n+1\rangle$ show that the matrix element is zero unless $n' = n \pm 1$. Hence,

$$\langle x \rangle = \left(\frac{\hbar}{2m\omega}\right)^{1/2} \frac{1}{2\Delta N} \sum_{n=N-\Delta N}^{n=N+\Delta N} \left(\sqrt{n}e^{-i\omega t} + \sqrt{n+1}e^{i\omega t}\right)$$

The summation over n is from $n = N - \Delta N$ to $n = N + \Delta N$. However, since $1 \ll \Delta N \ll N$, one may approximate $\sqrt{n} \cong \sqrt{n+1} \cong \sqrt{N}$. Since the sum is from

$n = N - \Delta N$ to $n = N + \Delta N$, there are $2\Delta N + 1 \cong 2\Delta N$ equal terms in the summation, and thus

$$\sum_n \left(\sqrt{n} \cdot e^{-i\omega t} + \sqrt{n+1} \cdot e^{i\omega t} \right) \approx 2 \cdot 2\Delta N \sqrt{N} \cos(\omega t)$$

giving an expectation value for position

$$\langle x \rangle = \left(\frac{2\hbar N}{m\omega} \right)^{1/2} \cos(\omega t)$$

(b) Classically, $m\left(d^2x/dt^2\right) + \kappa x = 0$, where $\kappa = m\omega^2$. This has solution for the position coordinate $x = A_0 \cos \omega t$ and total energy $E = \kappa x^2/2 + m(dx/dt)^2/2 = m\omega^2 A_0^2/2$. In the quantum calculation, the energy is $E \approx (N+1/2)\hbar\omega \approx N\hbar\omega$.

Therefore, the quantum amplitude $(2\hbar N/m\omega)^{1/2} \approx \left(2E/m\omega^2\right)^{1/2} = A_0$. Hence, the coherent combination of a large number of energy eigenstates with a spread in energies that is small compared with their mean energy behaves as a classical oscillator.

(c) The figure below plots the wave function $\psi\xi$ and the wave function squared $|\psi(\xi)|^2$ for the $n = 18$ state of the one-dimensional harmonic oscillator. The normalized spatial coordinate $\xi = (m\omega/\hbar) \cdot x$, where m is the particle mass and ω is the oscillation frequency. The eigenenergy of the state $\psi_{18}(\xi)$ is $E_{n=18} = (n + 1/2)\hbar\omega = 18.5 \times \hbar\omega$, and the classical turning point occurs at $\xi_{n=18} = \pm(2n + 1)^{1/2} = \pm 6.083$.

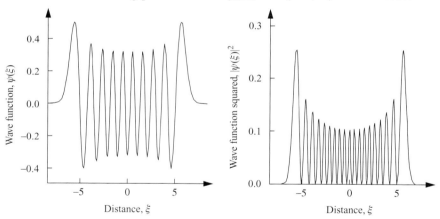

Listing of MATLAB program for Exercse 6.4(c)

```
%Chapt6Exercise4c.m
%simple harmonic oscillator wave function psi_n(xi)and wave function squared |psi_n(xi)|^2
%xi=(m*w/hbar)*x
%using relation psi_n=((2/n)^0.5)*((xi*psi_n-1)-((((n-1)/2)^0.5)*psi_n-2))
clear;
clf;
npoints=500;                        %(number of data points in plot) - 1
nlim=100;                           %arbitrary limit to value of n that can be plotted
```

n=input('Input quantum index n = ');   %Input from keyboard value of quantum index n
                                                        %of wave function to be plotted (n=0,1,2,3,...)
if n < 0;   error('minimum value of n must be greater or equal to 0');   end;
if n > nlim; error('maximum value of n must be less than or equal to 100'); end;

ximax=sqrt((2*n)+1);   %classical turning point
xiplot=(3/((n+1)^(1/3)))+(1.2*ximax);   %plot range of x-axis
deltaxi=2*xiplot/npoints;   %increment in xi
Ao=(1/pi)^(1/4);   %normalization amplitude for n=0
An=1.1*Ao/((n+1)^0.1);   %fix vertical scale
xi(1)=-xiplot;   %first value of xi
for j=2:1:npoints+1
   xi(j)=xi(j-1)+deltaxi;
   psi1(j)=Ao*exp((-xi(j)^2)/2);   %known n=0 ground state wave function
   psi2(j)=(sqrt(2))*xi(j)*psi1(j);   %known n=1 first excited-state wave function
   psi(j)=psi2(j);
end
if n < 1; psi=psi1; end;

if n>=2
     for ni=2:1:n
for j=2:1:npoints+1
   xi(j)=xi(j-1)+deltaxi;   %increment to new value of xi
   psi(j)=(sqrt(2/ni))*((xi(j)*psi2(j))-((sqrt((ni-1)/2))*psi1(j)));
   psi1(j)=psi2(j);   %update new value of psi_(n-2)
   psi2(j)=psi(j);   %update new value of psi_(n-1)
end
     end
end

figure(1);
subplot(1,2,1),plot(xi,psi);
axis([-xiplot,xiplot,-An,An]),xlabel('Distance, xi (m)'),ylabel('Wave function, psi(xi)');
ttl = sprintf('xi=(m*w/hbar)*x, n=%3.0f, E=%3.1 f hbar*w',n,n+0.5);
title (ttl);
subplot(1,2,2),plot(xi,abs(psi.^2));
axis([-xiplot,xiplot,0,An^2]),xlabel('Distance, xi (m)'),ylabel('Wave function squared,
|psi(xi)|^2');
ttl2=sprintf('SHO classical turning point = +/- %5.3f',ximax);
title (ttl2);

(d) The figure below plots the superposition wave function

$$\psi(t=0) = \frac{1}{\sqrt{2\Delta N}} \sum_{n=N-\Delta N}^{n=N+\Delta N} |n\rangle$$

and the wave function squared for the specific situation when $N = 18$ and $\Delta N = 2$. The total wave function is a coherent sum of five states centered on the $|n = 18\rangle$ state. In this case, for which time $t = 0$, the particle may be viewed as at an extreme of its motion with the peak probability occurring at position $\xi = 5.7$, which is almost the same as the classical turning point $\xi_{n=18} = 6.1$ for the $|n = 18\rangle$ eigenstate.

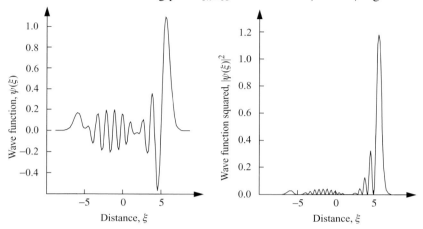

Solution 6.5

$$P_n^{\text{nonclassical}} = \int_{|\xi| > \xi_n} \psi_n^*(\xi)\xi\,\psi_n(\xi)\,d\xi$$

where $\xi = (m\omega/\hbar)^{1/2} \cdot x$ and the classical turning point is $\xi_n = (2n+1)^{1/2}$. For the first excited state and the second excited state we have

$$\psi_1(\xi) = \left(\frac{4}{\pi}\right)^{1/4} \xi e^{-\xi^2/2}$$

$$\psi_2(\xi) = \left(\frac{1}{64\pi}\right)^{1/4} (4\xi^2 - 2)e^{-\xi^2/2}$$

The probability of finding the particle in the nonclassical region is

$$P_1^{\text{nonclassical}} = \int_{|\xi| > \sqrt{3}} \psi_1^*(\xi)\psi_1(\xi)\,d\xi = 0.112$$

$$P_2^{\text{nonclassical}} = \int_{|\xi| > \sqrt{5}} \psi_2^*(\xi)\psi_2(\xi)\,d\xi = 0.095$$

With increasing quantum number n, the probability of finding the particle in the nonclassical region decreases.

Solution 6.6

To find $[\hat{n}, \hat{b}]$ and $[\hat{n}, \hat{b}^\dagger]$, we expand the expressions using $\hat{n} = \hat{b}^\dagger \hat{b}$ and rearrange in terms of commutator relations we know. Hence,

$$[\hat{n}, \hat{b}] = [\hat{b}^\dagger\hat{b}, b] = \hat{b}^\dagger\hat{b}\hat{b} + (-\hat{b}^\dagger\hat{b}\hat{b} + \hat{b}^\dagger\hat{b}\hat{b}) - \hat{b}\hat{b}^\dagger\hat{b} = \hat{b}^\dagger[\hat{b}, \hat{b}] + [\hat{b}^\dagger, \hat{b}]\hat{b} = -\hat{b}$$

and

$$[\hat{n}, \hat{b}^\dagger] = [\hat{b}^\dagger\hat{b}, \hat{b}^\dagger] = \hat{b}^\dagger\hat{b}\hat{b}^\dagger + (-\hat{b}^\dagger\hat{b}^\dagger\hat{b} + \hat{b}^\dagger\hat{b}^\dagger\hat{b}) - \hat{b}^\dagger\hat{b}^\dagger\hat{b} = \hat{b}^\dagger[\hat{b}, \hat{b}^\dagger] + [\hat{b}^\dagger, \hat{b}^\dagger]\hat{b} = \hat{b}^\dagger$$

where we made use of the fact that $[\hat{b}, \hat{b}] = 0$, $[\hat{b}^\dagger, \hat{b}] = -1$, $[\hat{b}^\dagger, \hat{b}^\dagger] = -0$, and $[\hat{b}, \hat{b}^\dagger] = 1$.

Solution 6.7

The n-th energy level of the symmetric three-dimensional harmonic oscillator has energy $E_n = (n + 3/2)\hbar\omega$ for which $n = n_x + n_y + n_z$ or $n - n_x = n_y + n_z$. To find the degeneracy of the n-th state, consider n_x fixed. There are now $(n - n_x + 1)$ possibilities for the pair $\{n_y, n_z\}$, which are

$$\{0, n - n_x\}, \{1, n - n_x - 1\}, \ldots, \{n - n_x, 0\}$$

The degeneracy of the n-th energy level is therefore

$$g_n = \sum_{n_x=0}^{n} (n - n_x + 1) = (n + 1)\sum_{n_x=0}^{n} 1 - \sum_{n_x=0}^{n} n_x = \frac{(n + 1)(n + 2)}{2}$$

or

$$g_n = \frac{(n + 2)}{n!2!} = \frac{(n + 2)(n + 1)n!}{2n!} = \frac{1}{2}(n + 2)(n + 1)$$

Solution 6.8

For the operator $\hat{\mathbf{A}}$

$$(\Delta A)^2 = \langle(\hat{\mathbf{A}} - \langle\hat{\mathbf{A}}\rangle)^2\rangle = \langle\hat{\mathbf{A}}^2 + \langle\hat{\mathbf{A}}\rangle^2 - 2\hat{\mathbf{A}}\langle\hat{\mathbf{A}}\rangle\rangle = \langle\hat{\mathbf{A}}\rangle^2 + \langle\hat{\mathbf{A}}\rangle^2 - 2\langle\hat{\mathbf{A}}\rangle\langle\hat{\mathbf{A}}\rangle$$

Hence,

$$(\Delta A)^2 = \langle\hat{\mathbf{A}}^2\rangle - \langle\hat{\mathbf{A}}\rangle^2$$

or

$$\Delta A = \left(\langle\hat{\mathbf{A}}^2\rangle - \langle\hat{\mathbf{A}}\rangle\right)^{1/2}$$

There are two possible approaches to solving the problem: (i) one could calculate the matrix elements by performing the integrals directly, or (ii) one can use the properties of the raising and lowering operators of the harmonic oscillator to obtain the solution.

First approach to a solution:

$\Delta\hat{x}^2 = \langle\hat{x}^2\rangle - \langle\hat{x}\rangle^2$. We note that $\langle\hat{x}\rangle = 0$ from symmetry, so we need only find $\langle\hat{x}^2\rangle$. To do so, we need to solve the integral

$$\langle\hat{x}^2\rangle = \frac{\int\limits_{-\infty}^{\infty} x^2 e^{-x^2/2\alpha^2} dx}{\int\limits_{-\infty}^{\infty} |\psi_0|^2 dx}$$

Using the standard integral

$$\int\limits_0^\infty x^2 e^{-\beta x^2} dx = \frac{1}{4\beta}\sqrt{\frac{\pi}{\beta}}$$

and

$$\int\limits_0^\infty e^{-\beta x^2} dx = \frac{1}{2}\sqrt{\frac{\pi}{\beta}}$$

we obtain

$$\langle \hat{x}^2 \rangle = \Delta x^2 = \frac{2\alpha^2}{2}\frac{\alpha\sqrt{2\pi}}{\alpha\sqrt{2\pi}} = \alpha^2 = \frac{\hbar}{2m\omega}$$

and we see that

$$A_0 = (2\pi)^{-1/4}\alpha^{-1/2} = \left(\frac{m\omega}{\pi\hbar}\right)^{1/4}$$

The uncertainty in momentum is

$$\Delta p_x^2 = \langle \hat{p}_x^2 \rangle - \langle \hat{p}_x \rangle^2$$

where $\hat{p} = -i\hbar\partial/\partial x$

$$\langle \hat{p}_x \rangle = A_0^2 \int\limits_{-\infty}^\infty e^{-x^2/4\alpha^2}\frac{\partial}{\partial x}e^{-x^2/4\alpha^2}dx = A_0^2 \int\limits_{-\infty}^\infty e^{-x^2/2\alpha^2}\frac{-2x}{4\alpha^2}dx = 0$$

from symmetry. Hence,

$$\Delta p_x^2 = \langle \hat{p}_x^2 \rangle = \frac{-\hbar^2}{\alpha\sqrt{2\pi}} \int\limits_{-\infty}^\infty e^{-x^2/4\alpha^2}\frac{\partial^2}{\partial x^2}e^{-x^2/4\alpha^2}dx$$

We note that

$$\frac{\partial^2}{\partial x^2}e^{-x^2/4\alpha^2} = -\frac{2}{4\alpha^2}e^{-x^2/2\alpha^2} + \frac{4x^2}{16\alpha^4}e^{-x^2/2\alpha^2}$$

so

$$\Delta p_x^2 = \frac{-\hbar^2}{\alpha\sqrt{2\pi}}\left(\frac{-1}{\alpha}\int\limits_{-\infty}^\infty e^{-x^2/2\alpha^2}dx + \frac{1}{2\alpha^4}\int\limits_0^\infty x^2 e^{-x^2/2\alpha^2}dx\right)$$

$$\Delta p_x^2 = \frac{-\hbar^2}{\alpha\sqrt{2\pi}}\left(\frac{-1}{2\alpha^2}\alpha\sqrt{2\pi} + \frac{1}{2\alpha^4}\frac{\alpha^2}{2}\alpha\sqrt{2\pi}\right)$$

$$\Delta p_x^2 = \frac{-\hbar^2}{\alpha\sqrt{2\pi}}\left(\frac{-1}{4\alpha^2}\alpha\sqrt{2\pi}\right)$$

This gives

$$\Delta p_x^2 = \frac{\hbar^2}{4\alpha^2} = \frac{\hbar\omega m}{2}$$

and so

$$\Delta p_x^2 \Delta x^2 = \frac{\hbar^2}{4\alpha^2}\alpha^2 = \frac{\hbar^2}{4}$$

or

$$\Delta p_x \Delta x = \frac{\hbar}{2}$$

which is the uncertainty relation.

Second approach to a solution:

We begin by noting that the ground state of the harmonic oscillator is defined by $\hat{b}|0\rangle = 0$ and that $\langle j|k\rangle = \delta_{jk}$.

Since we have

$$\hat{x} = \left(\frac{\hbar}{2m\omega}\right)^{1/2}(\hat{b} + \hat{b}^\dagger)$$

it follows that

$$\langle\hat{x}\rangle = \left(\frac{\hbar}{2m\omega}\right)^{1/2}\langle 0|(\hat{b} + \hat{b}^\dagger)|0\rangle = \left(\frac{\hbar}{2m\omega}\right)^{1/2}(\langle 0|\hat{b}|0\rangle + \langle 0|1\rangle) = 0$$

$$\langle\hat{x}^2\rangle = \left(\frac{\hbar}{2m\omega}\right)\langle 0|(\hat{b}\hat{b} + \hat{b}^\dagger\hat{b}^\dagger + \hat{b}\hat{b}^\dagger + \hat{b}^\dagger\hat{b})|0\rangle$$

$$\langle\hat{x}^2\rangle = \left(\frac{\hbar}{2m\omega}\right)^{1/2}\langle 0|(\hat{b}\hat{b} + \hat{b}^\dagger\hat{b}^\dagger + 1 + \hat{b}^\dagger\hat{b})|0\rangle$$

$$\langle\hat{x}^2\rangle = \left(\frac{\hbar}{2m\omega}\right)$$

and

$$\langle\hat{x}^2\rangle = \left(\frac{\hbar}{2m\omega}\right)(1 + 2n)$$

for the general state $|n\rangle$.

Similarly one finds that $\langle\hat{x}\rangle = 0$, $\langle\hat{p}_x^2\rangle = (\hbar m\omega/2)$, and $\langle\hat{p}_x^2\rangle = (\hbar m\omega/2)(1 + 2n)$ for the general state $|n\rangle$. Hence, we have

$$\Delta x^2 \Delta p_x^2 = \langle\hat{x}^2\rangle\langle\hat{p}_x^2\rangle = \hbar^2 m\omega/4m\omega = \hbar^2/4$$

for the ground state, or

$$\Delta x^2 \Delta p_x^2 = \langle\hat{x}^2\rangle\langle\hat{p}_x^2\rangle = \frac{\hbar^2}{4}(1 + 2n)^2$$

for the general state $|n\rangle$

$$\Delta x \Delta p_x = \hbar/2$$

for the ground state and is the minimum value given by the uncertainty relation. For the general state $|n\rangle$, we have $\Delta x \Delta p_x = \hbar(1 + 2n)/2$.

Solution 6.9

The ground state and the *second* excited state of a charged particle of mass m in a one-dimensional harmonic oscillator potential are both occupied. We wish to find the expectation value of the particle position x as a function of time. We start by noting that $|0\rangle$ and $|2\rangle$ have no overlap when operated on by the position operator $\hat{x}$. Hence,

$$\langle \hat{x}(t) \rangle = \frac{\hbar}{2m\omega} \langle a_0 \langle 0| + a_2 \langle 2| \, | \hat{b} + \hat{b}^\dagger | \, a_0|0\rangle + a_2|2\rangle \rangle$$

$$\langle \hat{x}(t) \rangle = |a_0|^2 \langle 0|0\rangle + |a_2|^2 \langle 2|2\rangle = \langle x \rangle$$

and we conclude that there is no time dependence for the position expectation operator.

We note that $|0\rangle = \phi_0 e^{-i\hbar\omega t/2}$ and $|2\rangle = \phi_2 e^{-i5\hbar\omega t/2}$, where ϕ_0 and ϕ_2 are the spatial wave functions for the ground state and the second excited state respectively.

In the presence of a constant electric field $\mathbf{E}$ in the x direction, a particle of charge e in the harmonic potential experiences a constant x-directed force $e|\mathbf{E}|$, which shifts the equilibrium position of the oscillator by Δx.

The new Hamiltonian is $H = \hat{p}^2/2m + \kappa \hat{x}'^2/2 + e|\mathbf{E}|x'$, and the amount of the shift is given by $e|\mathbf{E}| = \kappa \Delta x$, where $\kappa = m\omega^2$ is the oscillator force constant and $x' = x + \Delta x$. The total energy is

$$E_{\text{total}} = \frac{1}{2}\kappa(x + \Delta x)^2 - e|\mathbf{E}|(x + \Delta x) = \frac{1}{2}\kappa x^2 + \kappa x \Delta x + \frac{1}{2}\kappa \Delta x^2 - e|\mathbf{E}|x - e|\mathbf{E}|\Delta x$$

$$E_{\text{total}} = \frac{1}{2}\kappa x^2 + \frac{1}{2}\kappa \Delta x^2 - e|\mathbf{E}|\Delta x$$

since $e|\mathbf{E}| = \kappa \Delta x$ and

$$E_{\text{total}} = \frac{1}{2}\kappa x^2 + \frac{1}{2}\kappa \frac{e^2|\mathbf{E}|^2}{\kappa^2} - \frac{e^2|\mathbf{E}|^2}{\kappa} = \frac{1}{2}\kappa x^2 + \frac{1}{2}\frac{e^2|\mathbf{E}|^2}{\kappa} - \frac{e^2|\mathbf{E}|^2}{\kappa}$$

$$E_{\text{total}} = \frac{1}{2}m\omega^2 x^2 - \frac{1}{2}\frac{e^2|\mathbf{E}|^2}{\kappa}$$

since $\kappa = m\omega^2$.

The oscillator frequency remains the same, but all of the energy levels are uniformly reduced by a fixed amount

$$\frac{-e^2|\mathbf{E}|}{2m\omega^2}$$

Solution 6.10

We would like to use the method outlined in Section 3.4 to numerically solve the one-dimensional Schrödinger wave equation for the first four eigenvalues and eigenstates of an electron with effective mass $m_e^* = 0.07 \times m_0$ confined to a parabolic potential well in such a way that $V(x) = ((x - L/2)^2/(L/2)^2)$ eV and $L = 100$ nm.

Because we use a nontransmitting boundary condition, the value of L must be large enough to ensure that the wave function is approximately zero at the boundaries $x_0 = 0$ and $x_N = L$. The number of sample points N should also have a large enough value to ensure that the wave function does not vary significantly between adjacent discretization points.

The main computer program calls solve_schM, which was used in the solution of Exercise 3.7.

In this exercise, the first four energy eigenvalues are $E_0 = 0.0147$ eV, $E_1 = 0.0443$ eV, $E_2 = 0.0738$ eV, and $E_3 = 0.1032$ eV. As expected for a harmonic oscillator, the separation in energy between adjacent states is independent of eigenvalue, in this case $\hbar\omega = 0.0295$ eV. The eigenfunctions generated by the program and plotted in the figure below are not normalized.

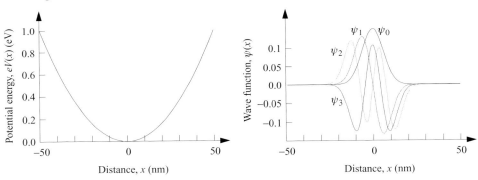

Listing of MATLAB program for Exercise 6.10

```
%Chapt6Exercise10.m
%simple harmonic oscillator
clear;
clf;
length = 100;                              %length of well (nm)
n=400;                                     %number of sample points
x=-length/2:length/n:length/2;            %position of sample points in potential
mass=0.07;                                 %effective electron mass
num_sol=4;                                 %number of solutions sought
v0=1;                                      %potential scale (eV)

for i=1:n+1                                %potential (eV)
    v(i)=v0*(x(i)^2)/((length/2)^2);
end
```

```
[energy,phi]=solve_schM(length,n,v,mass,num_sol);

  for i=1:num_sol
  sprintf(['eigenfunction (',num2str(i),') = ',num2str(energy(i)),'eV'])    %energy eigenvalues
  end

figure(1);
plot(x,v,'b');xlabel('Distance (nm)'),ylabel('Potential energy, (eV)');
title(['m* = ',num2str(mass),'m0, Length = ',num2str(length),'nm']);

s=char('y','k','r','g','b','m','c');                  %plot curves in different colors

figure(2);
for i=1:num_sol
  j=1+mod(i,7);
  plot(x,phi(:,i),s(j));                              %plot eigenfunctions
    hold on;
  end
  xlabel('Distance (nm)'),ylabel('Wave function');
  title(['m* = ',num2str(mass),'m0, Length = ',num2str(length),'nm']);
  hold off;
```

7 Fermions and bosons

7.1 Introduction

For most of this book we have considered a single particle moving in a potential. In this chapter we will briefly examine the behavior of many identical particles. Our main focus will be to appreciate the statistical distribution function for large numbers of particles in thermal equilibrium.

The Hamiltonian for N particles subject to mutual two-body interactions is

$$H = \sum_n^N \frac{p_n^2}{2m_n} + \sum_n^N V_n(x_n) + \sum_{n,m}^{n>m} V_{n,m}(x_n - x_m) \tag{7.1}$$

where $n > m$ in the sum avoids double counting. The corresponding multi-particle wave function obeys the Schrödinger equation

$$H\psi(x_1, x_2, x_3, \ldots, x_N, t) = i\hbar \frac{\partial}{\partial t} \psi(x_1, x_2, x_3, \ldots, x_N, t) \tag{7.2}$$

where $|\psi(x_1, x_2, x_3, \ldots, x_N, t)|^2 dx_1 dx_2 dx_3 \ldots dx_N$ is the probability of finding particle 1 in the interval x_1 to $x_1 + dx_1$, particle 2 in the interval x_2 to $x_2 + dx_2$, and so on. The key idea is that there is a single multi-particle wave function that describes the state of the N-particle system.

As it stands, this is a complex multi-particle, or *many-body*, problem that is difficult to solve. However, if we remove the mutual two-body interactions in Eqn (7.1) then the Hamiltonian takes on a much simpler form:

$$H = \sum_n^N \left(\frac{p_n^2}{2m_n} + V_n(x_n) \right) = \sum_n^N H_n \tag{7.3}$$

If the potential is time-independent, then the multi-particle wave function is a product

$$\psi(x_1, x_2, x_3, \ldots, x_N) = \psi_1(x_1)\psi_2(x_2)\psi_3(x_3) \cdots \psi_N(x_N) = \prod_n^N \psi_n(x_n) \tag{7.4}$$

that satisfies the time-independent Schrödinger equation

$$H_n \psi_n(x_n) = E_n \psi_n(x_n) \tag{7.5}$$

We may now describe the system in terms of a product of N one-dimensional solutions.

7.1.1   The symmetry of indistinguishable particles

The next key idea we need to introduce is the concept of indistinguishable particles. Unlike in classical mechanics, where one may assign labels to distinguish mechanically identical objects, in quantum mechanics elementary particles are indistinguishable. This fact introduces a symmetry to a multi-particle wave function describing a system containing many identical particles. One consequence is that the statistically most likely energy distribution of identical indistinguishable particles in thermal equilibrium falls into one of two classes. Identical integer-spin particles behave as *bosons* and half-odd-integer-spin particles behave as *fermions*.

The Bose–Einstein energy distribution applies to *boson* particles such as photons and phonons. Photons have spin quantum number ± 1, and phonons have spin zero.

The Fermi–Dirac energy distribution applies to identical indistinguishable half-odd-integer-spin particles, an example of which are electrons that have spin quantum number $\pm 1/2$. Such *fermion* particles obey the Pauli exclusion principle, which states that no identical indistinguishable half-odd-integer-spin particles may occupy the same state. An almost equivalent statement is that in a system of N fermion particles the total eigenfunction must be antisymmetric (must change sign) upon the permutation of any two particles.

To illustrate this, we start by considering just two noninteracting identical particles which, for our convenience only, are labelled (1) and (2). The particles satisfy their respective Hamiltonians so that

$$H(1)\psi(1) = E_1 \psi(1) \tag{7.6}$$
$$H(2)\psi(2) = E_2 \psi(2) \tag{7.7}$$

where E_1 and E_2 are energy eigenvalues. The total Hamiltonian is just the sum of individual Hamiltonians

$$H = H(1) + H(2) \tag{7.8}$$

with the solution characterized by

$$H\psi = (E_1 + E_2)\psi \tag{7.9}$$

where $E_1 + E_2$ is the total energy in the system and ψ is the multi-particle wave function.

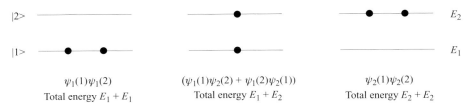

$\psi_1(1)\psi_1(2)$ $(\psi_1(1)\psi_2(2) + \psi_1(2)\psi_2(1))$ $\psi_2(1)\psi_2(2)$

Total energy $E_1 + E_1$ Total energy $E_1 + E_2$ Total energy $E_2 + E_2$

Fig. 7.1. Ways of arranging two identical indistinguishable particles of integer spin between two eigenfunctions and their associated energy eigenvalues.

Because the particles are identical, we should be able to interchange (or permute) them without affecting the total energy. If we introduce a *permutation operator* $\mathcal{P}_{12}$ that interchanges the identical particles (1) and (2), it follows that $\mathcal{P}_{12}$ and H commute so that $\mathcal{P}_{12}H = H\mathcal{P}_{12}$, or

$$[\mathcal{P}_{12}, H] = 0 \tag{7.10}$$

It is therefore possible to choose common eigenvalues for the two operators H and $\mathcal{P}_{12}$. For eigenvalue λ such that

$$\mathcal{P}_{12}\psi = \lambda\psi \tag{7.11}$$

$$\mathcal{P}_{12}\mathcal{P}_{12}\psi = \lambda^2\psi = \psi \tag{7.12}$$

and so $\lambda^2 = 1$ or $\lambda = \pm 1$. This means that the wave function ψ can only be symmetric or antisymmetric under particle exchange. This result is due to symmetry built into the system, namely that the particles are identical and noninteracting. The symmetric wave functions for the two-particle system under consideration are

$$\psi_s = \psi_1(1)\psi_1(2) \tag{7.13}$$

$$\psi_s = \psi_2(1)\psi_2(2) \tag{7.14}$$

$$\psi_s = \frac{1}{\sqrt{2}}(\psi_1(1)\psi_2(2) + \psi_1(2)\psi_2(1)) \tag{7.15}$$

The subscript labels the eigenstate. The particle is labeled by the number in the parentheses. Figure 7.1 illustrates the different ways of maintaining a symmetric multi-particle wave function while distributing two identical indistinguishable particles between two eigenfunctions $|1\rangle$ and $|2\rangle$ with eigenenergies E_1 and E_2, respectively. As may be seen, a symmetric multi-particle wave function allows particles to occupy the same state of integer spin between two energy levels.

The antisymmetric wave function ψ_a may be found from the determinant of the 2×2 matrix that describes state and particle occupation.

$$\psi_a = \frac{1}{\sqrt{2}} \begin{bmatrix} \psi_1(1) & \psi_1(2) \\ \psi_2(1) & \psi_2(2) \end{bmatrix} \tag{7.16}$$

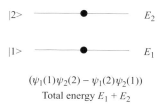

$(\psi_1(1)\psi_2(2) - \psi_1(2)\psi_2(1))$
Total energy $E_1 + E_2$

Fig. 7.2. The way of arranging two identical indistinguishable half-odd-integer-spin particles between two eigenfunctions and their associated energy eigenvalues.

Here, rows label the state and columns label the particle. Equation (7.16) is an example of a *Slater determinant*, which, in this case, gives

$$\psi_a = \frac{1}{\sqrt{2}}(\psi_1(1)\psi_2(2) - \psi_1(2)\psi_2(1)) \tag{7.17}$$

Figure 7.2 illustrates the only way of arranging two identical half-odd-integer-spin particles between two eigenfunctions and their associated energy eigenvalues.

For situations in which we wish to describe more than two identical noninteracting fermion particles, the antisymmetric state can be found using the Slater determinant of a larger matrix. For N particles, this gives

$$\psi_a(1, 2, \ldots, N) = \frac{1}{\sqrt{N!}} \begin{vmatrix} \psi_1(1) & \psi_1(2) & \cdots & \psi_1(N) \\ \psi_2(1) & \psi_2(2) & \cdots & \psi_2(N) \\ \vdots & & & \vdots \\ \psi_N(1) & & \cdots & \psi_N(N) \end{vmatrix} \tag{7.18}$$

The interchange of any two particles causes the sign of the multi-particle wave function ψ_a to change, since it involves the interchange of two columns. Expansion of the determinant has $N!$ terms, which take into account all possible permutations of the particles among N states. If any two single-particle eigenfunctions are the same, then those two particles are in the same state, and it follows that the multi-particle wave function $\psi_a = 0$, since the determinant will vanish. This fact is known as the Pauli exclusion principle. The Slater determinant ensures that no two noninteracting identical fermion particles can possess the same quantum numbers.

7.1.1.1 Ferminon creation and annihilation operators

In Chapter 6 it was shown that the creation and annihilation operators $\hat{b}^\dagger$ and $\hat{b}$ could be used to change the energy of a harmonic oscillator by one quantum of energy, $\hbar\omega$. The same operators were able to generate the corresponding harmonic oscillator wave functions. The number of boson particles in a single mode could be increased to n by applying the operator $(\hat{b}^\dagger)^n$ to the ground state. The number n might correspond to the number of photons in coherent laser light emission or the number of phonons in a vibrational mode.

The concept of creation and annihilation operators can also be applied to multi-particle fermion systems. It is, however, necessary to take into account the Pauli exclusion principle and the antisymmetry of the wave functions. Unlike bosons, where the number of particles in a single mode of frequency ω can be increased to an arbitrary value n, the number of fermion particles in a given state is limited to unity by the Pauli exclusion principle.

Suppose the occupation of a fermion state $|\mu\rangle$ is zero, so that $|\mu = 0\rangle$. Application of the fermion creation operator $\hat{c}_\mu^\dagger$ will create one fermion in that state, so that $|\mu = 1\rangle$. However, creation of an additional fermion in the same state is not allowed because of the Pauli exclusion principle. Hence, we require

$$\left(\hat{c}_\mu^\dagger\right)^2 |\psi\rangle = 0 \tag{7.19}$$

where $|\psi\rangle$ is any state of the multi-particle system. The same must be true of the fermion annihilation operator $\hat{c}_\mu$, so that

$$(\hat{c}_\mu)^2 |\psi\rangle = 0 \tag{7.20}$$

Consider the sequence of operations $(\hat{c}_\mu^\dagger \hat{c}_\mu + \hat{c}_\mu \hat{c}^\dagger)$. If the state $|\mu\rangle$ is empty, the first term is zero because $\hat{c}_\mu |\mu = 0\rangle = 0$. The second term creates a fermion and then annihilates it, so that $\hat{c}_\mu \hat{c}^\dagger |\mu = 0\rangle = |\mu = 0\rangle$, and from this we may conclude that $\hat{c}_\mu \hat{c}^\dagger = 1$ (this result also follows from Eqn (6.73) and Eqn (6.74)). If the state $|\mu\rangle$ is occupied, then the first term is $\hat{c}_\mu^\dagger \hat{c}_\mu = 1$ and the second term $\hat{c}_\mu \hat{c}^\dagger = 0$ because of the Pauli exclusion principle. These results suggest the existence of *anticommutation* relations between fermion creation and annihilation operators. The anticommutation relations are

$$\{\hat{c}_\mu, \hat{c}_\mu^\dagger\} = \hat{c}_\mu \hat{c}_\mu^\dagger + \hat{c}_\mu^\dagger \hat{c}_\mu = 1 \tag{7.21}$$
$$\{\hat{c}_\mu^\dagger, \hat{c}_\mu\} = \hat{c}_\mu^\dagger \hat{c}_\mu + \hat{c}_\mu \hat{c}_\mu^\dagger = 1 \tag{7.22}$$
$$\{\hat{c}_\mu, \hat{c}_\mu\} = \hat{c}_\mu \hat{c}_\mu + \hat{c}_\mu \hat{c}_\mu = 0 \tag{7.23}$$
$$\{\hat{c}^\dagger, \hat{c}^\dagger\} = \hat{c}_\mu^\dagger \hat{c}_\mu^\dagger + \hat{c}_\mu^\dagger \hat{c}_\mu^\dagger = 0 \tag{7.24}$$

We use curly brackets to distinguish the anticommutation relations for fermions from the commutation relations for bosons. The anticommutation relations are identical to the commutation relations with the exception that we replace the minus signs in Eqn (6.22), Eqn (6.23), Eqn (6.24), and Eqn (6.25) with plus signs. This small change in the equations has a dramatic effect on the quantum mechanical behavior of particles in the system. In particular, it forces the multi-particle wave functions to be antisymmetric.

Since we characterized the multi-particle wave function ψ_a by specifying the number of particles in each state, it seems natural to adopt a particle number representation with basis vectors

$$|n_1, n_2, n_3, \ldots, n_N\rangle \tag{7.25}$$

where n_μ is the number of particles in state ψ_μ. In mathematics, the particle number representation is said to exist in *Fock space*.

If there are no particles in any of the states, then we have $|0, 0, 0, \ldots, 0\rangle$. If there is one particle in one state, then any 0 may be replaced by a 1. Likewise, if two states are occupied, then any two 0s may be replaced with 1s, and so on.

For a total of N particles in the system, the antisymmetric wave function ψ_a is

$$\psi_a(x_1, x_2, x_3, \ldots, x_N) = \langle x_1, x_2, x_3, \ldots, x_N | n_1, n_2, n_3, \ldots, n_N \rangle \tag{7.26}$$

To make use of the particle number representation, we need to be able to create and annihilate particles in the state $|n_\mu\rangle$. We do this by applying either the fermion creation operator $\hat{c}_\mu^\dagger$ or the annihilation operator $\hat{c}_\mu$.

The existence of the vacuum state $|0\rangle = |n_1 = 0, n_2 = 0, n_3 = 0, \ldots, n_N = 0\rangle$ allows us to conclude that

$$\hat{c}_\mu | n_\mu = 0 \rangle = 0 \tag{7.27}$$

for all values of μ. It follows that the many-electron wave function can be written

$$|n_1, n_2, n_3, \ldots, n_N\rangle = \prod_\mu \left(\hat{c}_\mu^\dagger\right)^{n_\mu} |n_\mu = 0\rangle \tag{7.28}$$

where the values of n_μ can only be 1 or 0 because $(\hat{c}_\mu^\dagger)^2 = 0$ and $(\hat{c}_\mu^\dagger)^0 = 1$. The energy of the wave function is simply

$$E = \sum_\mu E_\mu n_\mu \tag{7.29}$$

and the total number of fermions is found by using the number operator in such a way that

$$N = \sum_\mu \hat{c}_\mu^\dagger \hat{c}_\mu \tag{7.30}$$

The annihilation operator $\hat{c}_\mu$, acting on the many-electron wave function, gives

$$\hat{c}_\mu | n_1, \ldots, n_\mu = 1, \ldots, n_N \rangle = (-1)^{\sum_{\lambda < \mu} n_\lambda} |n_1, \ldots, n_\mu = 0, \ldots, n_N\rangle \tag{7.31}$$

when $n_\mu = 1$ and

$$\hat{c}_\mu | n_1, \ldots, n_\mu = 0, \ldots, n_N \rangle = 0 \tag{7.32}$$

when $n_\mu = 0$. The term $(-1)^{\sum_{\lambda < \mu} n_\lambda}$ in Eqn (7.31) gives a factor of -1 for each occupied state to the left of μ. The origin of the term is best seen by considering removal of the ψ_μ row in the Slater determinant (Eqn (7.18)). The required interchange of rows introduces a factor $(-1)^{\mu-1}$ and the total number of electrons is reduced by one.

Likewise, the creation operator $\hat{c}_\mu^\dagger$, acting on the many-electron wave function, gives

$$\hat{c}_\mu^\dagger | n_1, \ldots, n_\mu = 1, \ldots, n_N \rangle = 0 \tag{7.33}$$

when $n_\mu = 1$ and

$$\hat{c}_\mu |n_1, \ldots, n_\mu = 0, \ldots, n_N\rangle = (-1)^{\sum\limits_{\lambda < \mu} n_\lambda} |n_1, \ldots, n_\mu = 1, \ldots, n_N\rangle \qquad (7.34)$$

when $n_\mu = 0$.

Just as we did when considering the harmonic oscillator, we may now proceed to express the Hamiltonian and other ordinary quantum mechanical operators in terms of Fermi creation and annihilation operators. This so-called *second quantization* method is quite a powerful way of dealing with many-particle systems and, importantly, is the common starting point for the quantum field theory description of solids.[1]

7.2 Fermi–Dirac distribution and chemical potential

Understanding the distribution function of particles such as electrons or photons is of great practical significance. For example, it plays a crucial role in determining the behavior of a semiconductor laser diode. The distribution in energy of electrons in the conduction band and holes in the valence band of a direct band-gap semiconductor such as GaAs determines the presence or absence of optical gain. Electrons, which have spin of one-half, obey Fermi–Dirac statistics, and so we introduce this function first.

The Fermi–Dirac probability distribution for half-odd-integer-spin particles of energy $E_{\mathbf{k}}$ in thermal equilibrium characterized by absolute temperature T is

$$\boxed{f_{\mathbf{k}}(E_{\mathbf{k}}) = \frac{1}{e^{(E_{\mathbf{k}} - \mu)/k_B T} + 1}} \qquad (7.35)$$

where k_B is the Boltzmann constant and μ is the chemical potential. The appearance of the chemical potential in Eqn (7.35) is due to the fact that particle number is conserved. The chemical potential is defined as the energy needed to place an extra particle in the system of N particles.

The fact that half-integer-spin particles are quantized according to Fermi–Dirac statistics can be justified using relativistic quantum field theory along with the assumption that the system has a lowest-energy state.[2] The same theory shows that integer-spin particles are quantized according to Bose–Einstein statistics. Unfortunately, a discussion of quantum field theory[3] is beyond the scope of this book.

The total number of spin one-half electrons in three-dimensional free space is just the integral over k states multiplied by the distribution function $f_{\mathbf{k}}(E_{\mathbf{k}})$. For electrons,

---

[1] For an introduction see H. Haken, *Quantum Field Theory of Solids*, North Holland, Amsterdam, 1988 (ISBN 0 444 86737 6).

[2] W. Pauli, *Phys. Rev.* **58**, 716 (1940).

[3] For an introduction see J. J. Sakurai *Advanced Quantum Mechanics*, Addison Wesley, Reading, Massachusetts, 1967 (ISBN 0 201 06710 2).

this gives an electron density

$$n = \int \frac{d^3k}{(2\pi)^3} \cdot 2 \cdot f_{\mathbf{k}}(E_{\mathbf{k}}) \qquad (7.36)$$

where the factor 2 appears in the integral because each electron may be in a state of either $+\hbar/2$ or $-\hbar/2$ spin (corresponding to spin quantum number $s = \pm 1/2$).

In the low-temperature limit $T \to 0$ K and we have $f(E_{\mathbf{k}}) = 1$ for $E_{\mathbf{k}} \le \mu$ and $f(E_{\mathbf{k}}) = 0$ for $E_{\mathbf{k}} > \mu$. This is an important limit, and so we define

$$\mu_{T=0} \equiv E_F \qquad (7.37)$$

as the Fermi energy, or

$$E_F = \frac{\hbar^2 k_F^2}{2m} \qquad (7.38)$$

where k_F is the Fermi wave vector for electrons of mass m. In a semiconductor the mass m may often be replaced by an effective electron mass m^*. For electron density in three dimensions of n, and taking the low-temperature limit, we may use Eqn (7.36) to find a simple relationship between n and the Fermi wave vector k_F:

$$n_{T=0} = \int \frac{2}{(2\pi)^3} \cdot d^3k = \frac{2}{(2\pi)^3} \int_0^{k_F} 4\pi k^2 \sin(\theta) d\theta dk = \frac{2}{(2\pi)^3} \cdot \frac{4\pi}{3} k_F^3 = \frac{k_F^3}{3\pi^2} \qquad (7.39)$$

Hence, the Fermi wave vector in three dimensions is

$$k_F = (3\pi^2 n)^{1/3} \qquad (7.40)$$

where $k_F = 2\pi/\lambda_F$. In this case λ_F is the de Broglie wavelength associated with an electron of energy E_F.

To get a feel for the numbers, consider an electron carrier density $n = 10^{18}$ cm^{-3} in the conduction band of GaAs with effective electron mass $m_e^* = 0.07 \times m_0$, a Fermi wave vector $k_F = 3.1 \times 10^6$ cm^{-1}, a de Broglie Fermi wavelength $\lambda_F = 20$ nm, and a Fermi energy $E_F = 52$ meV. These values set a scale for a number of physical effects, so it is worth making note of a few additional values. Table 7.1 lists values of Fermi energy for different carrier concentrations and two representative values of effective electron mass. The effective electron mass in the conduction band of GaAs is $m_e^* = 0.07 \times m_0$, and $m_{hh}^* = 0.50 \times m_0$ is the effective heavy-hole mass in the valence band of GaAs.

The three-dimensional density of states at the Fermi energy can also be calculated since

$$D_3(k_F) = 2 \cdot \frac{4\pi k_F^2}{(2\pi)^3} = \frac{k_F^2}{\pi^2} \qquad (7.41)$$

Table 7.1. *Fermi energy for different three-dimensional carrier concentrations*

Carrier concentration, n (cm^{-3})	Fermi wave vector, $k_F(\times 10^6$ cm^{-1})	Fermi wavelength, λ_F (nm)	Fermi energy, E_F (meV) ($m_e^* = 0.07 \times m_0$)	Fermi energy, E_F (meV) ($m_{hh}^* = 0.50 \times m_0$)
1×10^{19}	6.66	9.4	241.6	33.8
1×10^{18}	3.09	20.3	52.1	7.3
1×10^{17}	1.44	43.8	11.2	1.6
1×10^{16}	0.67	94.3	2.4	0.3

This is just Eqn (5.109) evaluated at k_F and multiplied by a factor 2 to account for electron spin. However, $E_F = \hbar^2 k_F^2 / 2m$, so that $dE_F = \hbar^2 k_F dk_F / m$, which means that the density of states at energy E_F is

$$D_3(E_F) = \frac{k_F^2}{\pi^2} \frac{m}{\hbar^2 k_F} = \frac{mk_F}{\hbar^2 \pi^2} \tag{7.42}$$

For finite temperatures whereby $k_B T \ll E_F$, we can perform a Taylor expansion of Eqn (7.36) about the energy E_F to give

$$n \sim \int_0^{E_F} D_3(E)dE + \left((\mu - E_F)D_3(E_F) + \frac{\pi^2}{6}(k_B T)^2 \frac{dD_3(E_F)}{dE_F} \right) \tag{7.43}$$

Since the first term on the right-hand side is the carrier density n at a temperature of absolute zero and n is assumed to be independent of temperature, we may write

$$0 \sim (\mu - E_F)D_3(E_F) + \frac{\pi^2}{6}(k_B T)^2 \frac{dD_3(E_F)}{dE_F} \tag{7.44}$$

Hence, the chemical potential is

$$\mu \sim E_F - \frac{\pi^2}{6} \frac{(k_B T)^2}{D_3(E_F)} \frac{dD_3(E_F)}{dE_F} \tag{7.45}$$

In three dimensions, the chemical potential may be approximated to second order in temperature as

$$\mu \sim E_F - \frac{1}{3} \frac{(\pi k_B T)^2}{4E_F} \tag{7.46}$$

It may be shown (see Exercise 7.2) that to fourth order in the temperature the chemical potential in three dimensions is

$$\mu \sim E_F - \frac{\pi^2}{12} \frac{(k_B T)^2}{E_F} - \frac{7\pi^4}{960} \frac{(k_B T)^4}{E_F^3} \tag{7.47}$$

Of course, the chemical potential at finite temperature T may be calculated numerically without the limitations associated with the Taylor expansion, and we will discuss how to do that in the next section.

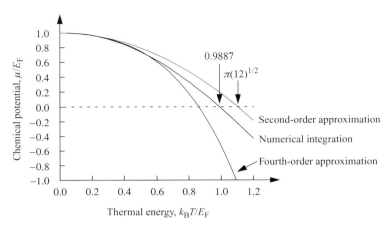

Fig. 7.3. Chemical potential for a three-dimensional gas of electrons as a function of thermal energy calculated using numerical integration and compared with results from second-order and fourth-order approximations. The chemical potential μ and thermal energy $k_B T$ are normalized to the Fermi energy, E_F.

Figure 7.3 plots the exact, second-order, and fourth-order approximations of chemical potential energy as a function of thermal energy. The axes are normalized to Fermi energy, E_F. Notice that both approximations become less accurate as $k_B T$ approaches E_F. The temperature at which the chemical potential is zero that was estimated using Eqn (7.46) is $T = \sqrt{12} E_F / \pi k_B$, while more accurate calculations (see Exercise 7.5) give a temperature $T = 0.9887 \times E_F / k_B$.

7.2.1   Writing a computer program to calculate the chemical potential

We wish to calculate the chemical potential for a three-dimensional electron gas of fixed density n and temperature T. The expression for carrier density in Eqn (7.36) is an integral over k space that we need to convert to an integral over energy. Using the three-dimensional density of states $D_3(E)$ (Eqn (5.112)) and remembering to multiply by 2 to take into account electron spin gives

$$n = \int_0^\infty D_3(E) \cdot 2 \cdot f(E) dE = \frac{1}{2\pi^2}\left(\frac{2m}{\hbar^2}\right)^{3/2} \int_{E_{min}=0}^{E_{max}=\infty} E^{1/2} \cdot \frac{1}{e^{(E-\mu)/k_B T}+1} \cdot dE$$

$$(7.48)$$

We will proceed by making a guess at the value of the chemical potential, use a computer to numerically integrate Eqn (7.48), and then iterate to a better value of chemical potential.

First, one needs to estimate an initial value of the chemical potential. We know that the maximum possible value is given by the Fermi energy

$$\mu_{\text{max}} = E_{\text{F}} = \frac{\hbar^2 k_{\text{F}}^2}{2m} \tag{7.49}$$

where $k_{\text{F}} = (3\pi^2 n)^{1/3}$ for a three-dimensional carrier density n (Eqn (7.40)). The minimum possible value of the chemical potential, μ_{min}, is given by the high-temperature limit $(T \rightarrow \infty)$ which, for fixed particle density n, is $\mu / k_{\text{B}} T \rightarrow -\infty$. In this limit, the Fermi–Dirac distribution function becomes

$$f(E)|_{T \rightarrow \infty} = \left. \frac{1}{e^{(E-\mu)/k_{\text{B}}T} + 1} \right|_{T \rightarrow \infty} = e^{(\mu - E)/k_{\text{B}}T} \tag{7.50}$$

which is the Boltzmann distribution. In the limit $T \rightarrow \infty$, occupation probability at energy $E = 0$ takes on the value $e^{\mu / k_{\text{B}} T}$.

One may use classical thermodynamics to show that a three-dimensional electron gas in this high-temperature limit has chemical potential[4]

$$\mu_{\text{min}} = k_{\text{B}} T \ln \left(\frac{n}{2} \left(\frac{2\pi \hbar^2}{m k_{\text{B}} T} \right)^{3/2} \right) \tag{7.51}$$

A computer program may now be used to calculate a carrier density n' for given temperature T by using an initial estimate for the chemical potential $\mu' = \mu_{\text{min}} + (\mu_{\text{max}} - \mu_{\text{min}})/2$ and numerically integrating Eqn (7.48). Notice that we have to choose a cut-off for E_{max} in Eqn (7.48). In practice, choosing a value $E_{\text{max}} = E_{\text{F}} + 15 k_{\text{B}} T$ works well for most cases of interest.

If the value of n' calculated using μ' is less than the actual value n, then the new best estimate for $\mu_{\text{min}} = \mu'$. If $n' \geq n$, then $\mu_{\text{max}} = \mu'$. A new value of μ' can now be calculated and the integration to calculate a new value of n' performed again. In this way, it is possible to iterate to the desired level of accuracy in μ. See Exercise 7.3.

To give us some experience with numerical values, Fig. 7.4(a) and (b) plots the temperature dependence of chemical potential $\mu(T)$ for the indicated carrier concentrations and effective electron masses. A number of features are worth pointing out. First, in the limit of low temperature $(T \rightarrow 0 \text{ K})$ the chemical potential approaches the Fermi energy, E_{F}. With increasing temperature, the chemical potential monotonically decreases in value, eventually taking on a negative value. Equation (7.36) can be used to find the value of the temperature at which the chemical potential is zero (Exercise 7.5).

7.2.2 Writing a computer program to plot the Fermi–Dirac distribution

Now that we know how to calculate the chemical potential for a given carrier concentration and temperature, we can plot the Fermi–Dirac distribution function. By plotting

[4] L. D. Landau and E. M. Lifshitz, *Statistical Physics*, Pergamon Press, Oxford, 1985 (ISBN 0 08 023039 3).

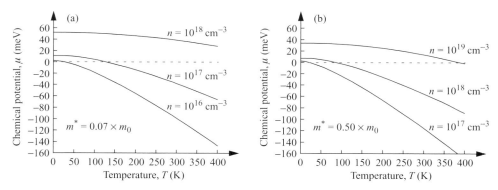

Fig. 7.4. Calculated chemical potential for electrons of carrier concentration n with (a) effective mass $m^* = 0.07 m_0$ and (b) effective mass $m^* = 0.50 m_0$ as a function of temperature T. The broken line corresponds to chemical potential $\mu = 0$ meV.

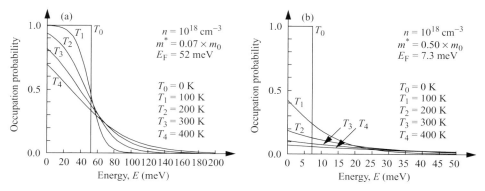

Fig. 7.5. The Fermi–Dirac distribution function as a function of electron energy for carrier concentration $n = 10^{18}$ cm^{-3} with (a) effective electron mass $m^* = 0.07 \times m_0$ and (b) effective electron mass $m^* = 0.5 \times m_0$. The distribution functions are calculated for the indicated temperatures.

the Fermi–Dirac function for electrons in the conduction band of GaAs for different temperatures one may clearly see its behavior.

At zero temperature with increasing electron energy there is a step-like distribution going from 1 to 0 at energy $E = E_F$. For the situation shown in Fig. 7.5, the Fermi energy is $E_F = 52$ meV above the conduction-band minimum. At finite temperatures, the step function is smeared out in energy. The broadening of the step transition is controlled by the value of the chemical potential μ and the temperature T in Eqn (7.35).

Notice that in the limit in which the chemical potential tends to a large negative value ($\mu \to -\infty$) the distribution function tends to a Boltzmann function. To learn more about this high-energy tail of the distribution, it is convenient to plot the occupation probability on a natural logarithmic scale. As may be seen in Fig. 7.6, when

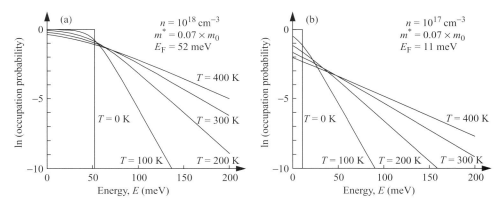

Fig. 7.6. Natural logarithm of Fermi–Dirac distribution as a function of electron energy for carrier concentration (a) $n = 10^{18}$ cm^{-3} and (b) $n = 10^{17}$ cm^{-3} for the indicated temperatures. The effective electron mass is $m^* = 0.07 \times m_0$.

the occupation probability is displayed in this way, at high electron energy, it is linear with increasing energy. This means that high-energy occupation probability scales as a Boltzmann factor $g_{MB}(t) = e^{-E/k_B T}$.

This may be confirmed by comparing Fig. 7.6(a) with Fig. 7.6(b). Here, one may see that while for given temperature T carrier density differs by a factor of 10 and Fermi energy by a factor of almost 5 the slope of the high-energy tail in the distribution is the same and has the value $-1/k_B T$.

7.2.3   Fermi–Dirac distribution function and thermal equilibrium statistics

There are a number of ways to explore the origin of the Fermi–Dirac statistical distributions. In the following, we will set out to find the statistically most likely arrangement of a large number of particle states when the particles are in equilibrium. To figure this out we will have to make use of a postulate that defines equilibrium:

In equilibrium, any two microscopically distinguishable arrangements of a system with the same total energy are equally likely.

This idea means that the probability of finding the system with n_1 particles in the energy range E to $E + \Delta E$ is proportional to the number of microscopically distinguishable arrangements that correspond to the same macroscopic arrangement.

To find the distribution function for identical indistinguishable particles such as electrons that have half-odd-integer spin, we start by considering an energy level labeled by E_j for which each energy level has a degeneracy n_j. We want to place a total of N electrons into the system using the following rules:

(i) Electrons are indistinguishable, and each state can only accommodate one electron (Pauli exclusion principle).

(ii) The total number of electrons is fixed by the sum rule. $N = \sum_j N_j$ is a constant, where N_j is the number of electrons in an energy level E_j.

(iii) The total energy of the system is fixed so that $E_{\text{total}} = \sum_j E_j N_j$.

At equilibrium there is a distribution function that describes the most probable arrangement of electrons as a function of energy. This is the Fermi–Dirac function by which the probability of finding the particle at energy E_j is

$$f(E_j) = \frac{N_j}{n_j} \tag{7.52}$$

To find this probability distribution we need to work out the number of ways in which N_j indistinguishable electrons in the energy level E_j can be placed in n_j states. Because of the Pauli exclusion principle, we must have $n_j \geq N_j$. Starting with the first particle, there are n_j states from which to choose. For the second particle there are $(n_j - 1)$ states available, and so on, giving

$$n_j(n_j - 1) \cdots (n_j - N_j + 1) \tag{7.53}$$

The number of possible permutations of N_j particles among themselves is $N_j!$. Because the particles are indistinguishable, these permutations do not lead to distinguishable arrangements, so the total number of distinct arrangements with N_j particles is

$$\frac{n_j(n_j - 1) \cdots (n_j - N_j + 1)}{N_j!} = \frac{n_j(n_j - 1) \cdots (n_j - N_j + 1)}{N_j!} \frac{(n_j - N_j)!}{(n_j - N_j)!} \tag{7.54}$$

$$\frac{n_j(n_j - 1) \cdots (n_j - N_j + 1)}{N_j!} = \frac{n_j!}{(n_j - N_j)! N_j!} \tag{7.55}$$

The total number of ways of arranging N electrons in the multi-level system is

$$\mathcal{P} = \prod_j \frac{n_j!}{(n_j - N_j)! N_j!} \tag{7.56}$$

To find the most probable set of values for N_j we must find an extreme value of $\mathcal{P}$ so that $d\mathcal{P} = 0$. Taking the logarithm of both sides,

$$\ln(\mathcal{P}) = \sum_j (\ln(n_j!) - \ln((n_j - N_j)!) - \ln(N_j!)) \tag{7.57}$$

For *large* values of N_j and n_j we can use Sterling's approximation so that $\ln(x!) = x \ln(x) - x$. This gives

$$\ln(\mathcal{P}) \sim \sum_j (n_j \ln(n_j) - n_j - (n_j - N_j) \ln(n_j - N_j)$$
$$+ (n_j - N_j) - N_j \ln(N_j) + N_j) \tag{7.58}$$

And, because $dn_j = 0$, the derivative is

$$d(\ln(\mathcal{P})) = \sum_j \frac{\partial}{\partial N_j} \ln(P) dN_j = \sum_j (\ln(n_j - N_j) + 1 - \ln(N_j) - 1) dN_j \tag{7.59}$$

$$d(\ln(\mathcal{P})) = \sum_j \left(\ln\left(\frac{n_j - N_j}{N_j} \right) \right) dN_j \tag{7.60}$$

Finding the extremes of this function under the constraint of total energy and particle conservation requires that $\sum_j dN_j = 0$ and $\sum_j E_j dN_j = 0$, so that

$$\sum_j \left(\ln\left(\frac{n_j - N_j}{N_j} \right) - \alpha - \beta E_j \right) dN_j = 0 \tag{7.61}$$

where α and β are Lagrange multipliers. The sum vanishes if

$$\ln\left(\frac{n_j - N_j}{N_j} \right) - \alpha - \beta E_j = 0 \tag{7.62}$$

for all j. To find the Fermi–Dirac distribution, we need to find $f(E_j) = N_j/n_j$. Rearranging Eqn (7.62) results in

$$\ln\left(\frac{n_j - N_j}{N_j} \right) = \alpha + \beta E_j \tag{7.63}$$

Taking the exponential of both sides gives $n_j/N_j = 1 + e^{(\alpha + \beta E_j)}$, so that we may write Eqn (7.52) as

$$\frac{N_j}{n_j} = f(E_j) = \frac{1}{e^{(\alpha + \beta E_j)} + 1} \tag{7.64}$$

For a continuum of energy levels $E_j \to E$ and

$$f_{\mathbf{k}}(E_{\mathbf{k}}) = \frac{1}{e^{(\alpha + \beta E_{\mathbf{k}})} + 1} \tag{7.65}$$

The coefficients α and β are found from classical thermodynamics to be[5] $\alpha = -\mu/k_B T$ and $\beta = 1/k_B T$, where μ is the chemical potential, k_B is the Boltzmann constant, and T is the absolute temperature. There is, therefore, agreement with the Fermi–Dirac distribution given by Eqn (7.35).

In the limit $T \to \infty$, the Fermi–Dirac distribution function $f_{\mathbf{k}}(E_{\mathbf{k}})|_{T \to \infty} = e^{-E_{\mathbf{k}}/k_B T}$, which is the classical Maxwell–Boltzmann ratio (see Exercise 7.4).

7.3 The Bose–Einstein distribution function

The Bose–Einstein probability distribution for indistinguishable integer-spin particles of energy $\hbar\omega$ in thermal equilibrium characterized by absolute temperature T is

$$g_{BE}(\hbar\omega) = \frac{1}{e^{(\hbar\omega - \mu)/k_B T} - 1} \tag{7.66}$$

---

[5] See any text on statistical physics, such as L. D. Landau and E. M. Lifshitz, *Statistical Physics*, Pergamon Press, Oxford, 1985 (ISBN 0 08 023039 3), or F. Reif, *Fundamentals of Statistical and Thermal Physics*, McGraw Hill, Boston, Massachusetts, 1965 (ISBN 07 051800 9).

This distribution function applies to quantum particles such as phonons and photons. Typically, phonon and photon numbers are not conserved, and in such circumstances the chemical potential $\mu = 0$, so that Eqn (7.66) reduces to

$$g(\hbar\omega) = \frac{1}{e^{\hbar\omega/k_B T} - 1} \tag{7.67}$$

As an example of the use of Eqn (7.67), recall from Section 6.5 that an electromagnetic field may be described in terms of a number of harmonic oscillators. Each photon contributing to the electromagnetic field is quantized in energy in such a way that $E_n = \hbar\omega(n + 1/2)$, where the factor $1/2$ comes from the contribution of zero-point energy. Assuming that the oscillators are excited thermally and are in equilibrium characterized by an absolute temperature T, the probability of excitation into the n-th state $P_{\text{prob}}(n)$ is given by a Boltzmann factor

$$P_{\text{prob}}(n) = \frac{g_{\text{MB}}(E_n)}{\sum_n g_{\text{MB}}(E_n)} = \frac{e^{-E_n/k_B T}}{\sum_n e^{-E_n/k_B T}} = \frac{e^{-n\hbar\omega/k_B T}}{\sum_n e^{-n\hbar\omega/k_B T}} \tag{7.68}$$

Notice that in the expression for energy E_n the zero-point energy term cancels out. The sum in the denominator may be written

$$\sum_{n=0}^{n=\infty} e^{-n\hbar\omega/k_B T} = 1 + e^{-\hbar\omega/k_B T} \sum_{n=0}^{n=\infty} e^{-n\hbar\omega/k_B T} = \frac{1}{1 - e^{-\hbar\omega/k_B T}} \tag{7.69}$$

The probability of excitation in the n-th state becomes

$$P_{\text{prob}}(n) = (1 - e^{-\hbar\omega/k_B T})e^{-n\hbar\omega/k_B T} \tag{7.70}$$

and the average number of photons excited in the n-th field mode at temperature T is

$$g(\hbar\omega) = \sum_{n=0}^{n=\infty} n P_{\text{prob}}(n) = \left(1 - e^{-\hbar\omega/k_B T}\right) \sum_{n=0}^{n=\infty} n e^{-n\hbar\omega/k_B T} \tag{7.71}$$

$$g(\hbar\omega) = \left(1 - e^{-\hbar\omega/k_B T}\right)e^{-\hbar\omega/k_B T} \frac{\partial}{\partial\left(e^{-\hbar\omega/k_B T}\right)} \sum_{n=0}^{n=\infty} e^{-n\hbar\omega/k_B T} = \frac{e^{-\hbar\omega/k_B T}}{1 - e^{-\hbar\omega/k_B T}} \tag{7.72}$$

Thus, finally,

$$g(\hbar\omega) = \frac{1}{e^{\hbar\omega/k_B T} - 1} \tag{7.73}$$

which is in agreement with Eqn (7.67).

It follows that the average energy in excess of the zero-point energy for the electromagnetic field in thermal equilibrium is just $\hbar\omega \times g(\hbar\omega)$. The radiative energy density is this average energy multiplied by the density of field modes in the frequency range ω

to $\omega + d\omega$. In three dimensions and allowing for the fact that the electromagnetic field can have one of two orthogonal polarizations, the radiative energy density is

$$S(\omega) = 2g(\hbar\omega)\hbar\omega D_3(\omega) = \frac{2\hbar\omega}{e^{-\hbar\omega/k_B T} - 1} \cdot \frac{\omega^2}{\pi^2 c^3} = \frac{\hbar\omega^3}{\pi^2 c^3} \frac{1}{e^{-\hbar\omega/k_B T} - 1} \tag{7.74}$$

This is Planck's radiative energy density spectrum for thermal light discussed in Section 2.1.2. $S(\omega)$ is measured in units of J s m^{-3}.

7.4 Example exercises

Exercise 7.1
Calculate k_F in two dimensions for a GaAs quantum well with electron density $n = 10^{12}$ cm^{-2}. What is the de Broglie wavelength for an electron at the Fermi energy?

Exercise 7.2
In this chapter we showed that for temperatures $k_B T \ll E_F$ the chemical potential for carriers in three dimensions may be approximated to second order in $k_B T$ as

$$\mu \sim E_F - \frac{\pi^2}{12} \frac{(k_B T)^2}{E_F}$$

Derive an expression for the chemical potential to fourth order in $k_B T$.

Exercise 7.3
Write a computer program to calculate the chemical potential for n electrons per unit volume at temperature, T. The electrons have effective mass of $m^* = 0.067 \times m_0$.

(a) Calculate the value of the chemical potential when $n = 10^{18}$ cm^{-3} and $T = 300$ K. Compare the results of your calculation with the expressions for μ derived from second-order and fourth-order expansions. How do your results compare with the value of chemical potential when $T = 0$ K?

(b) Repeat the calculations of (a) but now for the case in which $n = 10^{14}$ cm^{-3}. Explain why the results from the second-order and fourth-order expansions are inaccurate in this case.

Exercise 7.4
Use the method of Section 7.2.3 to determine the classical Maxwell–Botzmann distribution function $g_{MB}(E)$ and ratio for distinguishable identical particles.

Exercise 7.5
Find the temperature at which the chemical potential of a three-dimensional electron gas is zero.

SOLUTIONS

Solution 7.1

We wish to calculate k_F in two dimensions for a GaAs quantum well with n electrons per unit area. In two dimensions

$$n_{T=0} = \int \frac{2}{(2\pi)^2} d^2k = \frac{2}{(2\pi)^2} \int\limits^{k_F} 2\pi k dk = \frac{2}{(2\pi)^2} \cdot \pi k_F^2 = \frac{1}{2\pi} \cdot k_F^2$$

where the factor 2 accounts for electron spin. Hence, in two dimensions

$$k_F = (2\pi n)^{1/2}$$

In GaAs with $n = 10^{12}$ cm^{-2}, we find $k_F = 2\pi/\lambda_{F_e} = 2.5 \times 10^6$ cm^{-1} and $\lambda_{F_e} = 4$ nm.

Solution 7.2

We wish to derive an expression for the chemical potential similar to Eqn (7.47), but to fourth order in $k_B T$. We proceed by noting that Eqn (7.36) is of the form

$$n = \int\limits_{E=-\infty}^{E=\infty} h(E) f(E) dE$$

where $f(E)$ is the Fermi–Dirac function given by Eqn (7.25). Our approach is to solve this using a Sommerfeld expansion.[6] We let

$$\kappa(E) = \int\limits_{E=-\infty}^{E=\infty} h(E') dE'$$

so that

$$h(E) = \frac{d}{dE} \kappa(E)$$

Integration by parts gives

$$n = \int\limits_{E=-\infty}^{E=\infty} h(E) f(E) dE = f(E)\kappa(E)\Big|_{E=-\infty}^{E=\infty} - \int\limits_{E=-\infty}^{E=\infty} \kappa(E') \frac{\partial}{\partial E'} f(E') dE'$$

$$n = - \int\limits_{E=-\infty}^{E=\infty} \kappa(E') \frac{\partial}{\partial E'} f(E') dE'$$

[6] We follow the standard approach found in the book by N. Ashcroft and N. D. Mermin, *Solid State Physics*, Saunders College, Philadelphia, 1976 (ISBN 0 03 049346 3).

If $h(E)$ is not too rapidly varying in E around μ, then $\kappa(E)$ may be expanded in a Taylor series about $E \sim \mu$.

$$\kappa(E) = \kappa(\mu) + (E - \mu) \frac{d}{dE} \kappa(E) \Big|_{E=\mu} + \frac{(E - \mu)^2}{2!} \frac{d^2}{dE^2} \kappa(E) \Big|_{E=\mu} + \cdots$$

$$+ \frac{(E - \mu)^n}{n!} \frac{d^n}{dE^n} \kappa(E)$$

Hence,

$$-\int_{E=-\infty}^{E=\infty} \kappa(E') \frac{\partial}{\partial E'} f(E') dE' =$$

$$-\int_{E=-\infty}^{E=\infty} \left(\kappa(E) + \sum_{n=1}^{n=\infty} \frac{(E' - \mu)^n}{n!} \frac{d^n}{dE'^n} \kappa(E') \Big|_{E=\mu} \right) \frac{\partial}{\partial E'} f(E') dE'$$

$$-\int_{E=-\infty}^{E=\infty} \kappa(E') \frac{\partial}{\partial E'} f(E') dE' = -\int_{E=-\infty}^{E=\infty} \kappa(E') \frac{\partial}{\partial E'} f(E') dE'$$

$$-\int \sum_{n=1}^{n=\infty} \frac{(E' - \mu)^n}{n!} \frac{d^n}{dE^n} \kappa(E') \Big|_{E=\mu} \frac{\partial}{\partial E'} f(E') dE'$$

$$\int_{E=-\infty}^{E=\infty} h(E') f(E') dE' = \int_{E=-\infty}^{E=\mu} h(E') dE'$$

$$+ \sum_{n=1}^{n=\infty} \int_{E=-\infty}^{E=\infty} \frac{(E' - \mu)^{2n}}{(2n)!} \left(-\frac{\partial}{\partial E'} f(E') \right) \frac{d^{2n-1}}{dE'^{2n-1}} h(E')|_{E'=\mu} dE'$$

Let $(E - \mu)/k_B T = x$ so that $dE = k_B T dx$. We may now write

$$\int_{E=-\infty}^{E=\infty} h(E') f(E') dE' = \int_{E=-\infty}^{E=\mu} h(E') dE' + \sum_{n=1}^{n=\infty} a_n (k_B T)^{2n} \frac{d^{2n-1}}{dE'^{2n-1}} h(E')|_{E'=\mu}$$

where

$$a_n = \int_{E=-\infty}^{E=\infty} \frac{x^{2n}}{(2n)!} \left(-\frac{d}{dx} \cdot \frac{1}{e^x + 1} \right) dx = 2 \left(1 - \frac{1}{2^{2n}} + \frac{1}{3^{2n}} - \frac{1}{4^{2n}} + \frac{1}{5^{2n}} - \cdots \right)$$

$$a_n = \left(2 - \frac{1}{2^{2(n-1)}} \right) \xi(2n)$$

$$\xi(n) = 1 + \frac{1}{2^n} + \frac{1}{3^n} + \frac{1}{4^n} + \cdots$$

$$\xi(2n) = 2^{2n-1} \frac{\pi^{2n}}{(2n)!} B_n$$

The first few values of B_n are

$$B_1 = \frac{1}{6}, \ B_2 = \frac{1}{30}, \ B_3 = \frac{1}{42}, \ B_4 = \frac{1}{30}, \ B_5 = \frac{5}{66}$$

so that

$$\xi(2) = \frac{\pi^2}{6} \text{ and } \xi(4) = \frac{\pi^4}{90}$$

$$\int_{E=-\infty}^{E=\infty} h(E')f(E')dE' =$$

$$\int_{E=-\infty}^{E=\mu} h(E')dE' + \frac{\pi^2}{6}(k_{\mathrm{B}}T)^2 h'(\mu) + \frac{7\pi^4}{360}(k_{\mathrm{B}}T)^4 h'''(\mu) + 0(k_{\mathrm{B}}T)^6$$

$$\int_{E=-\infty}^{E=\mu} h(E')dE' = \int_0^{E_{\mathrm{F}}} h(E')dE' + (\mu - E_{\mathrm{F}})h(E_{\mathrm{F}})$$

$$n = \int_0^{E_{\mathrm{F}}} g(E')dE' + (\mu - E_{\mathrm{F}})g(E_{\mathrm{F}}) + \frac{\pi^2}{6}(k_{\mathrm{B}}T)^2 g'(\mu) + \frac{7\pi^4}{360}(k_{\mathrm{B}}T)^4 g'''(\mu) + 0(k_{\mathrm{B}}T)^6$$

$$(\mu - E_{\mathrm{F}})g(E_{\mathrm{F}}) = \frac{-\pi^2}{6}(k_{\mathrm{B}}T)^2 g'(\mu) - \left(\frac{7\pi^4}{360}\right)(k_{\mathrm{B}}T)^4 g'''(\mu) - 0(k_{\mathrm{B}}T)^6$$

$$g(E_{\mathrm{F}}) = \frac{mk_{\mathrm{F}}}{\hbar^2\pi^2} = \frac{m}{\hbar^3\pi^2}(2m)^{1/2}E_{\mathrm{F}}^{1/2}$$

$$g'(E_{\mathrm{F}}) = \frac{m}{\hbar^2\pi^2 k_{\mathrm{F}}^2}\left(\frac{2m}{\hbar^2}\right)^{3/2} E_{\mathrm{F}}^{1/2} = \frac{m}{\hbar^3\pi^2}\frac{(2m)^{1/2}}{2}E_{\mathrm{F}}^{-1/2}$$

$$g''(E_{\mathrm{F}}) = \frac{m}{2\hbar^2\pi^2 k_{\mathrm{F}}^2}\left(\frac{2m}{\hbar^2}\right)^{3/2} E_{\mathrm{F}}^{-1/2} = \frac{-m}{\hbar^3\pi^2}\frac{(2m)^{1/2}}{4}E_{\mathrm{F}}^{-3/2}$$

$$g'''(E_{\mathrm{F}}) = \frac{m}{-4\hbar^2\pi^2 k_{\mathrm{F}}^2}\left(\frac{2m}{\hbar^2}\right)^{3/2} E_{\mathrm{F}}^{-3/2} = \frac{m}{\hbar^3\pi^2}\frac{3(2m)^{1/2}}{8}E_{\mathrm{F}}^{-5/2}$$

$$\mu = E_{\mathrm{F}} - \frac{\pi^2}{6}(k_{\mathrm{B}}T)^2\frac{g'(E_{\mathrm{F}})}{g(E_{\mathrm{F}})} - \frac{7\pi^2}{360}(k_{\mathrm{B}}T)^4\frac{g'''E_{\mathrm{F}}}{g(E_{\mathrm{F}})} - 0\left(\frac{k_{\mathrm{B}}T}{E_{\mathrm{F}}}\right)^6$$

Hence,

$$\mu = E_{\mathrm{F}} - \frac{\pi^2}{12}(k_{\mathrm{B}}T)^2\frac{1}{E_{\mathrm{F}}} - \frac{7\pi^2}{120}(k_{\mathrm{B}}T)^4\frac{1}{8}\frac{1}{E_{\mathrm{F}}^3} - 0\left(\frac{k_{\mathrm{B}}T}{E_{\mathrm{F}}}\right)^6$$

and, finally,

$$\mu = E_{\mathrm{F}} - \frac{\pi^2}{12}\frac{(k_{\mathrm{B}}T)^2}{E_{\mathrm{F}}} - \frac{7\pi^4}{960}\frac{(k_{\mathrm{B}}T)^4}{E_{\mathrm{F}}^3} - 0\left(\frac{k_{\mathrm{B}}T}{E_{\mathrm{F}}}\right)^6$$

Solution 7.3

(a) We are asked to write a computer program to calculate the chemical potential for n electrons per unit volume at temperatures $T = 0$ K and $T = 300$ K. The electrons have effective mass $m^* = 0.067 \times m_0$ and carrier density $n = 10^{18}$ cm^{-3}.

We start by using our analytic expressions in the low-temperature limit. We do this because it gives us a feel for the magnitude of the numbers we should expect when we later consider the finite temperature case. For temperature $T = 0$ K, we write down the value of the Fermi wave number k_F for a carrier density n and then substitute into the expression for the Fermi energy E_F of electrons with effective electron mass $m^* = 0.067 \times m_0$. In three dimensions, $k_F = (3\pi n)^{1/3}$ and $E_F = \hbar^2 k_F^2 / 2m^*$. Putting in the numbers gives $E_F = 54$ meV. We will use this value to confirm that our computer program gives the correct numbers.

We now need to write a computer program to calculate the chemical potential. We wish to calculate the chemical potential for a three-dimensional electron gas of fixed density n and temperature T. The expression for carrier density in Eqn (7.36) is an integral over k space that we need to convert to an integral over energy. Using the three-dimensional density of states $D_3(E)$ (Eqn (5.112)) gives

$$n = \int_0^\infty D_3(E) \cdot 2 \cdot f(E) dE = \frac{1}{2\pi^2} \left(\frac{2m}{\hbar^2} \right)^{3/2} \int_{E_{min}=0}^{E_{max}=\infty} E^{1/2} \cdot \frac{1}{e^{(E-\mu)/k_B T} + 1} \cdot dE$$

We will proceed by making a guess at the value of the chemical potential, use a computer to numerically integrate Eqn (7.48), and then iterate to a better value of chemical potential.

First, one needs to estimate an initial value of the chemical potential. We know that the maximum possible value is given by the Fermi energy

$$\mu_{max} = E_F = \frac{\hbar^2 k_F^2}{2m}$$

where $k_F = (3\pi^2 n)^{1/3}$ for a three-dimensional carrier density n (Eqn (7.40)). The minimum possible value of the chemical potential, μ_{min}, is given by the high-temperature limit ($T \to \infty$), which, for fixed particle density n, is $\mu/k_B T \to -\infty$. In this limit, the Fermi–Dirac distribution function becomes

$$f(E)|_{T \to \infty} = \frac{1}{e^{(E-\mu)/k_B T} + 1} \bigg|_{T \to \infty} = e^{(\mu-E)/k_B T}$$

which is the Maxwell–Boltzmann distribution. One may use classical thermodynamics to show that for a three-dimensional electron gas in this limit

$$\mu_{min} = k_B T \ln \left(\frac{n}{2} \left(\frac{2\pi \hbar^2}{m k_B T} \right)^{3/2} \right)$$

The computer program can now calculate a carrier density n' for given temperature T using an initial estimate for the chemical potential $\mu' = \mu_{min} + (\mu_{max} - \mu_{min})/2$ by numerically integrating Eqn (7.48). Notice that we have to choose a cut-off for E_{max} in Eqn (7.48). In practice, choosing a value $E_{max} = E_F + 15k_BT$ works well in most situations.

If the value of n' calculated using μ' is less than the actual value n, then the new best estimate for $\mu_{min} = \mu'$. If $n' \geq n$, then $\mu_{max} = \mu'$. A new value of μ' can now be calculated and the integration to calculate a new value of n' can be performed again. In this way it is possible to iterate to a desired level of accuracy in μ.

For temperature $T = 300$ K, a computer program gives $\mu_{T=300\ K} = 41.9$ meV, which may be compared with the second-order approximation

$$\mu \sim E_F - \frac{\pi^2}{12} \frac{(k_B T)^2}{E_F}$$

which gives $\mu_{T=300\ K} = 44.3$ meV. The agreement is quite good because thermal energy at temperature $T = 300$ K is $k_B T = 25.8$ meV and this value is enough less than the Fermi energy, $E_F = 54$ meV, to make the expansion quite accurate. The fourth-order approximation,

$$\mu \sim E_F - \frac{\pi^2}{12} \frac{(k_B T)^2}{E_F} - \frac{7\pi^4}{960} \frac{(k_B T)^4}{E_F^3}$$

gives a slightly more accurate result $\mu_{T=300\ K} = 42.4$ meV.

(b) When $n = 10^{14}$ cm^{-3} the computer program gives $\mu_{T=300\ K} = -217$ meV. The results from the second-order and fourth-order expansions are inaccurate, because now $k_B T$ is much greater than the Fermi energy $E_F = 0.12$ meV.

Listing of MATLAB program for Exercise 7.3

```
% Chapt7Exercise3.m
% uses function fermi.m
% carrier density n (cm-3), temperature kelvin (K), relative error rerr
% returns chemical potential mu1 measured in units of meV
  n=1.e18;                                    %carrier density (cm-3)
  kelvin=300.0;                               %absolute temperature (K)
  rerr=1.e-3;                                 %relative error
  m0=9.10956;                                 %bare electron mass (kg x 10^31)
  m1=0.07;                                    %effective electron mass
echarge=1.60219;                              %electron charge (C x 10^19)
hbar=1.0545928;                               %Planck's constant (J s x 10^34)
  kB=8.617e-5;                                %Boltzmann's constant (eV K-1)

  kF1=(3.0*(pi^2.0)*n)^(1/3);                 %Fermi wave vector (cm-1)
  eF=1.e-11*((hbar*kF1)^2)/(2.0*m0*m1*echarge);  %Fermi energy (meV)
  kBT=1000.*kelvin*kB;                        %thermal energy (meV)
  beta=1./kBT;                                %inverse thermal energy (meV-1)
```

```
mumax=eF;                    %maximum possible value of chemical potential
x=((n*1.e6)/2.)*(((2.e-15*pi*beta*(hbar^2))/(echarge*m0*m1))^1.5);
mumin=(+1./beta)*log(x);

   emax=eF+(15./beta);    %maximum limit of integration
   de=emax/1000.;            %energy step
const=1.e16*((20.*echarge*m1*m0)^.5)*echarge*m1*m0/((pi^2)*(hbar^3));

for j=1:25
   mu1=mumin+((mumax-mumin)/2.);
energy=0.0;
ainter=0.0;

%calculate carrier density n'
for i=1:1000;
   energy=energy+de;
   ainter=ainter+(((sqrt(energy))*de)*fermi(beta,energy,mu1));
end;
nprime=const*ainter;
% delta is relative error in carrier density
   delta=(n-nprime)/n;
   if((abs(delta)) < rerr)
      break;
         elseif(delta < 0.)
      mumax=mu1;
         else
         mumin=mu1;
   end;
end;
%print output: chemical potential mu1(meV), Fermi energy eF(meV), temperature kelvin(K)
%kBT (meV), carrier density nprime(cm-3), number of iterations j
ttl1=['chemical potential = ',num2str(mu1),' meV'];
ttl2=['Fermi energy    = ',num2str(eF),' meV'];
ttl3=['Fermi wave vector = ',num2str(kF1),' cm-1'];
ttl4=['Fermi wavelength = ',num2str(lambdaF),' cm'];
ttl5=['Temperature    = ',num2str(kelvin),' K'];
ttl6=['kBT            = ',num2str(kBT),' meV'];
ttl7=['carrier density = ',num2str(nprime),' cm-3'];
ttl8=['number of iterations = ',num2str(j)];
%compare with second-order mu2(meV) and fourth-order mu4(meV) expansion
ttl9=['chemical potential (second-order) = ',num2str(eF-(((pi*kBT)^2)/(12.*eF))),' meV'];
ttl10=['chemical potential (fourth-order) = ',num2str(eF-(((pi*kBT)^2)/(12.*eF))-
(7*((pi*kBT)^4)/(960.*(eF^3)))),' meV'];
Solution =strvcat(ttl1,ttl2,ttl3,ttl4,ttl5,ttl6,ttl7,ttl8,ttl9,ttl10)
if j >= 25
   'check convergence!'
end;
```

Listing of fermi function for MATLAB program used in Exercise 7.3

```
function [fermi]=fermi(beta,energy,mu1)
% Fermi is the Fermi-Dirac function
%
x=(energy-mu1)*beta;
   if(x > 180.0)              %check overflow
      x=180.;
   end;
   if(x < -180.);             %check underflow
      x=-180.;
   end;
   fermi=1./((exp(x))+1.);
   return;
```

Solution 7.4

We seek to determine the number of microscopically distinguishable arrangements of n_1 distinguishable particles among a total of N particles. Clearly, the first particle can be chosen from a total of N particles, the second from $(N-1)$ and so on, so that the total number of choices is

$$N(N-1)(N-2)\cdots(N-n_1+1) = (N(N-1)(N-2)\cdots(N-n_1+1))$$
$$\cdot\frac{(N-n_1)!}{(N-n_1)!} = \frac{N!}{(N-n_1)!}$$

We must now remember to divide by the number of ways of arranging n_1 distinguishable particles among themselves. The result is the number of microscopic arrangements of n_1 particles in the energy range E to $E+\Delta E$. Hence,

$$\mathcal{P}_1 = \frac{N!}{n_1!(N-n_1)!}$$

$$\mathcal{P}_2 = \frac{(N-n_1)!}{n_2!(N-n_1-n_2)!}$$

$$\mathcal{P}_3 = \frac{(N-n_1-n_2)!}{n_3!(N-n_1-n_2-n_3)!}$$

and so on. The total number of arrangements is

$$\mathcal{P}(n_j) = \mathcal{P}_1\mathcal{P}_2\mathcal{P}_3\cdots\mathcal{P}_j$$
$$= \frac{N!}{n_1!(N-n_1)!}\cdot\frac{(N-n_1)!}{n_2!(N-n_1-n_2)!}\cdot\frac{(N-n_1-n_2)!}{n_3!(N-n_1-n_2-n_3)!}\cdots$$

$$\mathcal{P}(n_j) = N!\prod_{j=1}^{j=\infty}\frac{1}{n_j!}$$

Since, at equilibrium, all microscopically distinguishable distributions with a fixed number of particles and the same total energy are equally likely, it follows that the most probable macroscopic distribution is one in which the number of microscopically distinguishable arrangements $\mathcal{P}(n_1, n_2, \ldots, n_j)$ is a *maximum* subject to the *constraints* of particle conservation

$$f = \left(\sum_{j=1}^{j=\infty} n_j \right) - N = 0$$

and conservation of total energy

$$g = \left(\sum_{j=1}^{j=\infty} E_j n_j \right) - E_{\text{total}} = 0$$

Actually, it is easier to maximize a new function $G = \ln(\mathcal{P}(n_j))$ instead of $\mathcal{P}(n_j)$ itself. Maximization of a function subject to constraints has been worked on a great deal in the past, especially in the context of classical mechanics. One approach uses the Lagrange method of undetermined multipliers. We wish to maximize the function

$$F(n_1, n_2, \ldots \alpha, \beta) = G(n_1, n_2, \ldots) - \alpha f(n_1, n_2, \ldots) - \beta g(n_1, n_2, \ldots)$$

and we solve for $n_1, n_2, \ldots \alpha$ and β in such a way that G is a maximum. This is done by requiring $\partial F / \partial n_j = 0$ for all j, $\partial F / \partial \alpha = 0$, and $\partial F / \partial \beta = 0$.

Since

$$F = \ln(\mathcal{P}) - \alpha \left(\left(\sum_j n_j \right) - N \right) - \beta \left(\left(\sum_j E_j n_j \right) - E \right)$$

we note that

$$\ln(\mathcal{P}) = \ln \left(N! \prod_{j=1}^{j=\infty} \frac{1}{n_j} \right) = \ln(N!) + \sum_{j=1}^{j=\infty} (\ln(1) - \ln(n_j!)) = \ln(N!) + \sum_{j=1}^{j=\infty} \ln(n_j!)$$

because $\ln(1) = 0$. For *large* n_j one may use Sterling's formula $\ln(n!) \sim n \ln(n) - n$ and rewrite:

$$\ln(\mathcal{P}) = \ln(N!) + \sum_{j=1}^{j=\infty} (n_j \ln(n_j) - n_j)$$

We now fix j and take the derivative of F with respect to n_j:

$$\frac{\partial F}{\partial n_j} = \frac{\partial}{\partial n_j} \left(\ln(N!) + \sum_{j=1}^{j=\infty} (n_j \ln(n_j) - n_j) - \alpha \left(\left(\sum_j n_j \right) - N \right) \right.$$

$$\left. - \beta \left(\left(\sum_j E_j n_j \right) - E \right) \right)$$

$$0 = 0 - \ln(n_j) - n_j \frac{\partial}{\partial n_j} \ln(n_j) + 1 - \alpha - \beta E_j = \ln(n_j) - \frac{n_j}{n_j} + 1 - \alpha - \beta E_j$$

$$= -\ln(n_j) - \alpha - \beta E_j$$

We now rewrite:

$$-\ln(n_j) = \alpha + \beta E_j$$

$$\frac{1}{n_j} = e^{\alpha + \beta E_j}$$

$$n_j = \frac{1}{e^{\alpha + \beta E_j}}$$

which is the Maxwell–Boltzmann distribution function

$$g_{MB}(E) = \frac{1}{e^{(E-\mu)/k_B T}}$$

where $\alpha = \mu/k_B T$ and $\beta = 1/k_B T$ may be obtained from the classical theory of gases. This distribution specifies the probability that an available state of energy E is occupied under equilibrium conditions.

The Maxwell–Boltzmann ratio $n_j/n_k = 1/e^{(E_j - E_k)\beta}$ is simply $e^{-E/k_B T}$.

Solution 7.5

To find the temperature at which the chemical potential is zero in a three-dimensional electron gas, we start by writing down Eqn (7.36):

$$n = \frac{1}{2\pi^2}\left(\frac{2m}{\hbar^2}\right)^{3/2} \int\limits_{E_{min}=0}^{E_{max}=\infty} E^{1/2} \cdot \frac{1}{e^{(E-\mu)/k_B T} + 1} \cdot dE$$

Setting the chemical potential to zero, $\mu = 0$, and normalizing energy to the Fermi energy, E_F, gives

$$n = \frac{1}{2\pi^2}\left(\frac{2m}{\hbar^2}\right)^{3/2} (E_F)^{3/2} \int\limits_{E_{min}=0}^{E_{max}=\infty} \left(\frac{E}{E_F}\right)^{1/2} \cdot \frac{1}{e^{(E/E_F)(E_F/k_B T)} + 1} \cdot d\left(\frac{E}{E_F}\right)$$

Introducing the variable $x = E/E_F$ and the value we wish to find $r = E_F/k_B T$, we then make use of the fact that $E_F = \hbar^2 k_F^2/2m$ (Eqn (7.38)), where $k_F = (3\pi^2 n)^{1/3}$ (Eqn (7.40)), which allows us to write

$$n = \frac{1}{2\pi^2}\left(\frac{2m}{\hbar^2}\right)^{3/2}\left(\frac{\hbar^2}{2m}\right)^{3/2} 3\pi^2 n \int\limits_0^\infty x^{1/2} \cdot \frac{1}{e^{rx} + 1} \cdot dx$$

The carrier density n cancels, and after some rearrangement we are left with the expression

$$\frac{3}{2} = \int\limits_0^\infty \frac{x^{1/2}}{e^{rx} + 1} \cdot dx = \int\limits_0^\infty \frac{x^{p-1}}{e^{rx} - q} \cdot dx = \frac{1}{qr^p}\Gamma(p)\sum\limits_{k=1}^\infty \frac{q^k}{k^p}$$

Here, we have rewritten the integral in a familiar form[7] in which $\Gamma(p)$ is the gamma function[8], $p > 0$, $r > 0$, and $-1 < q < 1$. In this particular, case $p = 3/2$, $r = E_F/k_B T$, and $q = -1$, giving

$$\frac{k_B T}{E_F} = \left(\frac{-2}{3 \times \Gamma(1.5) \sum_{k=1}^{\infty} \frac{-1^k}{k^{3/2}}} \right)^{2/3}$$

Putting in the numbers,

$$\Gamma(1.5) = \pi^{1/2}/2 = 0.886227$$

and the sum

$$\sum_{k=1}^{\infty} \frac{-1^k}{k^{3/2}} = -0.765147$$

Hence, we may conclude that the chemical potential of a three-dimensional electron gas is always zero when temperature T is such that

$$\frac{k_B T}{E_F} = \left(\frac{2}{3 \times 0.886227 \times 0.765147} \right)^{2/3} = 0.9887$$

---

[7] I. S. Gradshteyn and I. M. Ryzhik, *Table of Integrals, Series, and Products*, Academic Press, San Diego, 1980 p. 326 (ISBN 0 12 294760 6).

[8] M. Abramowitz and I. A. Stegun, *Handbook of Mathematical Functions*, Dover, New York, 1974 pp. 267–273 (ISBN 0 486 61272 4).

Time-dependent perturbation

8.1 Introduction

Engineers who design transistors, lasers, and other semiconductor components want to understand and control the cause of resistance to current flow so that they may better optimize device performance. A detailed microscopic understanding of electron motion from one part of a semiconductor to another requires the explicit calculation of electron scattering probability. One would like to know how to predict electron scattering from one state to another by application of a time-dependent potential. In this chapter we will see how to do this using powerful quantum mechanical techniques.

In addition to understanding electron motion in a semiconductor we also want to understand how to make devices that emit or absorb light. In Chapter 6 it was shown that a superposition of two harmonic oscillator eigenstates could give rise to dipole radiation and emission of a photon. The creation of a photon was only possible if a superposition state existed between a correct pair of eigenstates. This leads directly to the concept of rules determining pairs of eigenstates which can give rise to photon emission. Such selection rules are a useful tool to help us understand the emission and absorption of light by matter. However, the real challenge is to use what we know to make practical devices which operate using emission and absorption of photons. This usually requires imposing some control over atomic-scale physical processes. We will, of course, use quantum mechanics to describe such atomic-scale processes.

Our study begins by considering electronic transitions due to an abrupt time-dependent change in potential. We will then go on to calculate excitation of a charged particle in a harmonic potential due to a transient electric field pulse. Following this, we will derive important results from first-order time-dependent perturbation theory, also known as Fermi's golden rule, which will allow us to consider the effect of more general time-varying potentials. As an example, we will use Fermi's golden rule to calculate the elastic scattering rate from ionized impurities for electrons in the conduction band of n-type GaAs. Such calculations are of practical importance for the design of high-performance transistors and laser diodes. Our study will result in a number of predictions, such as the temperature dependence of conductivity and the fact that we must take into account the response of many mobile electrons to the presence of a scattering site. We will also learn that, by controlling the position of scattering sites on an atomic scale, the probability of elastic scattering can be dramatically altered.

As a basic starting point, and by way of example, we would like to know how to cause an electronic transition to take place from, say, a ground state to an excited state in a quantum mechanical system. The key idea is application of a time-dependent potential to change the distribution of occupied states. In principle, the change in potential could take place smoothly or abruptly in time. To explore the influence of a time-varying potential in a quantum system, an abrupt change in potential is considered first.

8.1.1   An abrupt change in potential

Let's start with a familiar system. A particle of mass m is in a one-dimensional rectangular potential well in such a way that $V(x) = 0$ for $0 < x < L$ and $V(x) = \infty$ elsewhere. The energy eigenvalues are $E_n = \hbar^2 k_n^2/2m$, and the eigenfunctions are $\psi_n = \sqrt{2/L} \cdot \sin(k_n x)$, where $k_n = n\pi/L$ for $n = 1, 2, 3, \ldots$. The energy levels and wave functions are illustrated in Fig. 8.1.

We assume that the particle is initially prepared in the ground state ψ_1 with eigenenergy E_1. Then, at some time, say $t = 0$, the potential is very rapidly changed in such a way that the original wave function remains the same but $V(x) = 0$ for $0 < x < 2L$ and $V(x) = \infty$ elsewhere. This situation is illustrated in Fig. 8.2. One would like to know what effect such a change in potential has on the expectation value of particle energy and the probability that the particle is in an excited state of the system.

We start by finding the expectation value of particle energy after the potential well is abruptly increased in width. We know that the energy of the particle $\langle E \rangle = E_1$ for

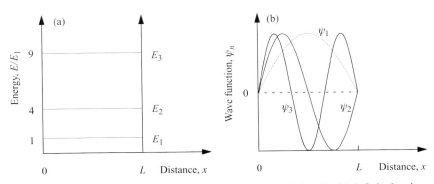

Fig. 8.1. (a) Sketch of a one-dimensional rectangular potential well with infinite barrier energy showing the energy eigenvalues E_1, E_2, and E_3. (b) Sketch of the eigenfunctions ψ_1, ψ_2, and ψ_3 for the potential shown in (a).

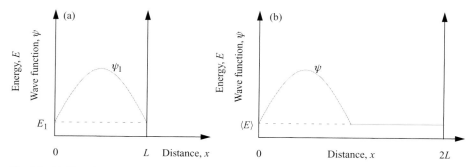

Fig. 8.2. (a) Sketch of a one-dimensional rectangular potential well with infinite barrier energy showing the lowest-energy eigenvalue E_1 and its associated ground-state wave function ψ_1. (b) The potential barrier at position L is suddenly moved to position $2L$, resulting in a new wave function ψ. The energy expectation value of the new state is $\langle E \rangle = E_1$.

$t < 0$. Since the wave function after the change in potential ψ is the same as the original wave function ψ_1 but with the addition of a constant zero value for $L < x < 2L$, one might anticipate that the expectation value in energy is $\langle E \rangle = E_1$ for time $t \geq 0$. It is important to check that the kink in the wave function at position $x = L$ does not contribute $\Delta \langle E \rangle = -\hbar^2 \psi(x = L)\Delta\psi/2m$ to the energy (see Section 3.1.1). Clearly, since $\psi(x = L) = 0$, the kink does not make a contribution, and so it is safe to conclude that $\langle E \rangle = E_1$ after the change in potential.

When time $t \geq 0$, the new state ψ is not an eigenfunction of the system. The new eigenfunctions of a rectangular potential well of width $2L$ with infinite barrier energy are $\psi_m = \sqrt{1/L} \sin(k_m x)$, where $k_m = m\pi/2L$ and the index $m = 1, 2, 3, \ldots$.

Since the state ψ is not an eigenfunction, it may be expressed as a sum of the new eigenfunctions, so that

$$\psi = \sum_m a_m \psi_m \tag{8.1}$$

The coefficients a_m are found by multiplying both sides by ψ_m^* and integrating over all space. This overlap integral gives the coefficients

$$a_m = \int \psi_m^* \psi \, dx \tag{8.2}$$

The effect of the overlap integral is to project out the components of the new eigenstates that contribute to the wave function ψ. The value of $|a_m|^2$ is the probability of finding the particle in the eigenstate ψ_m.

To illustrate how to find the contribution of the new eigenstates to the wave function ψ, we calculate the probability that the particle is in the new ground state $\psi_{m=1}$ when

$t \geq 0$. The probability is given by the square of the overlap integral, a_m:

$$a_1 = \langle \psi_{m=1} | \psi \rangle = \int\limits_{x=0}^{x=L} \sqrt{\frac{1}{L}} \sin\left(\frac{\pi x'}{2L}\right) \sqrt{\frac{2}{L}} \sin\left(\frac{\pi x'}{L}\right) dx' \tag{8.3}$$

$$a_1 = \frac{\sqrt{2}}{L} \int\limits_{x=0}^{x=L} \left(\frac{1}{2}\cos\left(\frac{\pi x'}{2L}\right) - \frac{1}{2}\cos\left(\frac{3\pi x'}{2L}\right)\right) dx' \tag{8.4}$$

$$a_1 = \frac{\sqrt{2}}{L}\left[\frac{1}{2}\cdot\frac{2L}{\pi}\sin\left(\left(\frac{\pi x'}{2L}\right) - \frac{1}{2}\cdot\frac{2L}{3\pi}\sin\left(\frac{3\pi x'}{2L}\right)\right)\right]_0^L \tag{8.5}$$

$$a_1 = \frac{\sqrt{2}}{L}\left(\frac{L}{\pi} + \frac{L}{3\pi}\right) = \frac{4\sqrt{2}}{3\pi} \tag{8.6}$$

Hence, the probability of finding the particle in the new ground state of the system is

$$|a_1|^2 = \frac{32}{9\pi^2} = 0.36025 \tag{8.7}$$

The idea that the potential can be modified so rapidly that the wave function does not change may seem a little extreme. However, it is possible to think of systems in which such an approach is possible. Figure 8.3 illustrates a system in which voltage applied to a gated semiconductor heterostructure potential well is used to control well width, L.

In this section, we have only considered a potential that changed instantaneously. In other words, the potential energy changed much faster than the particle's response time. Classically, this means that the potential wall moved faster than the velocity of a particle with energy E. If the potential wall moves at a velocity that is comparable to

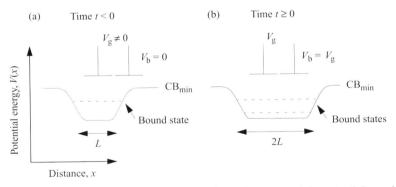

Fig. 8.3. Illustration showing use of a gate electrode to control the potential seen by an electron in the conduction band of a semiconductor heterostructure. The line labeled CB_{min} is the energy of the conduction-band minimum as a function of distance in the device. In (a) the potential well is of width L, and in (b) the width is increased to approximately $2L$ by application of a gate electrode potential $V_g = V_b$.

the velocity of a particle with energy E, then our previous approach is not suitable. A different method that goes beyond the abrupt or sudden approximation is needed.

8.1.2   Time-dependent change in potential

Consider a quantum mechanical system described by Hamiltonian $H_{(0)}$ and for which we know the solutions to the time-independent Schrödinger equation. That is,

$$H_{(0)}|n\rangle = E_n|n\rangle \tag{8.8}$$

are known. The time-independent eigenvalues are $E_n = \hbar\omega_n$, and the orthonormal eigenfunctions are $|n\rangle$. The eigenfunction $|n\rangle$ evolves in time according to

$$|n\rangle e^{-i\omega_n t} = \phi_n(x)e^{-i\omega_n t} \tag{8.9}$$

and satisfies

$$i\hbar\frac{\partial}{\partial t}|n\rangle e^{-i\omega_n t} = H_{(0)}|n\rangle e^{-i\omega_n t} \tag{8.10}$$

To introduce the basic idea, at time $t = 0$ we apply a time-dependent change in potential $W(t)$ the effect of which is to create a new Hamiltonian:

$$H = H_{(0)} + W(t) \tag{8.11}$$

and state $|\psi(t)\rangle$, which evolves in time according to

$$i\hbar\frac{\partial}{\partial t}|\psi(t)\rangle = \big(H_{(0)} + W(t)\big)|\psi(t)\rangle \tag{8.12}$$

The time-dependent change in potential energy $W(t)$ might, for example, be a step function or an oscillatory function.

We seek solutions to the time-dependent Schrödinger equation, which includes the change in potential (Eqn (8.12)), in the form of a sum over known eigenstates

$$|\psi(t)\rangle = \sum_n a_n(t)|n\rangle e^{-i\omega_n t} \tag{8.13}$$

where $a_n(t)$ are time-dependent coefficients.

Substituting Eqn (8.13) into Eqn (8.12) gives

$$i\hbar\frac{d}{dt}\sum_n a_n(t)|n\rangle e^{-i\omega_n t} = \big(H_{(0)} + W(t)\big)\sum_n a_n(t)|n\rangle e^{-i\omega_n t} \tag{8.14}$$

Using the product rule for differentiation $((fg)' = (f'g + fg'))$, one may rewrite the left-hand side as

$$i\hbar\sum_n\left(\left(\frac{\partial}{\partial t}a_n(t)\right)|n\rangle e^{-i\omega_n t} + a_n(t)\left(\frac{\partial}{\partial t}|n\rangle e^{-i\omega_n t}\right)\right)$$
$$= \big(H_{(0)} + W(t)\big)\sum_n a_n(t)|n\rangle e^{-i\omega_n t} \tag{8.15}$$

Making use of Eqn (8.10), we notice that the terms

$$i\hbar \sum_n a_n(t) \frac{\partial}{\partial t} |n\rangle e^{-i\omega_n t} = \sum_n a_n(t) H_{(0)} |n\rangle e^{-i\omega_n t} \qquad (8.16)$$

may be removed from Eqn (8.15) to leave

$$i\hbar \sum_n |n\rangle e^{-i\omega_n t} \frac{\partial}{\partial t} a_n(t) = \sum_n a_n(t) W(t) |n\rangle e^{-i\omega_n t} \qquad (8.17)$$

Multiplying both sides by $\langle m|$ and using the orthonormal relationship $\langle m|n\rangle = \delta_{mn}$ gives

$$i\hbar \frac{d}{dt} a_m(t) = \sum_n a_n(t) \langle m|W(t)|n\rangle e^{i\omega_{mn} t} \qquad (8.18)$$

However, since

$$\int \phi_m^*(x) e^{i\omega_m t} W \phi_n(x) e^{-i\omega_n t} dx = W_{mn} e^{i\omega_{mn} t} \qquad (8.19)$$

where $\hbar\omega_{mn} = E_m - E_n$ and W_{mn} is defined as the matrix element $\langle m|W|n\rangle = \int \phi_m^*(x) W \phi_n(x) dx$, we may write

$$\boxed{i\hbar \frac{d}{dt} a_m(t) = \sum_n a_n(t) W_{mn} e^{i\omega_{mn} t}} \qquad (8.20)$$

The intrinsic time dependence of each state can be factored out by introducing different time-dependent coefficients so that $c_n(t) = a_n(t) e^{-i E_n t/\hbar}$. This is called the *interaction picture* in contrast to the a-coefficient formalism that is the *Schrödinger picture*.

If the change in potential is turned off at time t, then the probability that the system can be found in a stationary state $|n\rangle$ is

$$P_n(t) = |a_n(t)|^2 = |c_n(t)|^2 \qquad (8.21)$$

It is important to recognize that Eqn (8.20) is an *exact* result. However, the right-hand side contains the time-dependent coefficients we want to find. At first sight, not a great deal of progress seems to have been made. To make a little more headway, it helps to be quite specific.

Suppose the system is initially in the ground state $|0\rangle$. Then, for times such that $t < 0$ one has

$$a_n(0) = c_n(0) = \delta_{n0} \qquad (8.22)$$

We now assume that there is a constant step change in potential $W(t)$ that is turned on at time $t = 0$ for duration τ so that $W(t) \neq 0$, $\partial W(t)/\partial t = 0$, and $W(t) = 0$ for $0 > t > \tau$. Figure 8.4 is an illustration of this time-dependent potential.

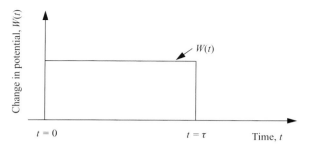

Fig. 8.4. Plot of the time-dependent change in potential $W(t)$ discussed in the text.

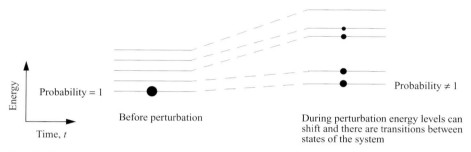

Fig. 8.5. Illustration of energy-level shifts and changes in population of energy levels during perturbation.

We are now in a position to make some qualitative statements about the time evolution of the system subject to the time-dependent change in potential, $W(t)$.

Before application of $W(t)$: Time-independent, stationary-state solutions satisfy $H_{(0)}|n\rangle = E_n|n\rangle$. The probability of finding the state, in this case $|0\rangle$, is assumed to be unity.

During application of $W(t)$: The state of the system evolves according to

$$|\psi(t)\rangle = \sum_n a_n(t)|n\rangle e^{-iE_n t/\hbar} \tag{8.23}$$

The time-dependent change in potential causes *transitions between eigenstates* of the initial system (*off-diagonal* matrix elements) and can *shift energy levels* of the initial eigenstates (*diagonal* matrix elements).

After application of $W(t)$: Stationary-state solutions satisfy $H_{(0)}|n\rangle = E_n|n\rangle$. Any time after $W(t)$ is turned off at time $t = \tau$ there is a probability of finding the state in any of the known initial time-independent stationary solutions. The final state of the system may be a *superposition* of these eigenstates

$$|\psi(t)\rangle = \sum_n a_n(t)|n\rangle e^{-iE_n t/\hbar} \tag{8.24}$$

For the above example, one may visualize the perturbation as changing the population of eigenstates and shifting energy eigenvalues. This is illustrated in Fig. 8.5.

8.2 Charged particle in a harmonic potential

Suppose a particle of mass m_0 and charge e is moving in a harmonic potential created by the electronic structure of a molecule. We want to use what we have learned so far to control the state of the electron in the molecule. The way we are going to proceed is to apply a macroscopic external pulse of electric field that has a Gaussian time dependence. While use of a macroscopic field to manipulate an atomic-scale entity may seem a little crude, it is in fact quite a powerful way to control single electrons.

We begin by assuming that the charged particle is initially in the ground state of the one-dimensional harmonic potential $V(x)$ of the molecule. For convenience, we assume that our initial condition applies at time $t = -\infty$. As already mentioned, we have decided to control the state of the charged particle by applying a pulse of electric field $\mathbf{E}(t) = |\mathbf{E}_0| e^{-t^2/\tau^2} \cdot \hat{\mathbf{x}}$, where $|\mathbf{E}_0|$ and τ are constants and $\hat{\mathbf{x}}$ is the unit vector in the x direction. $|\mathbf{E}_0|$ is the maximum strength of the applied electric field, and $2\tau \cdot \sqrt{\ln(2)}$ is the full-width-half-maximum (FWHM) of the electric field pulse. To demonstrate our ability to control the electron state, we are interested in finding the value of τ that gives the *maximum* probability of the system being in an excited state a long time after application of the pulse.

Starting from the exact result, Eqn (8.20),

$$i\hbar \frac{d}{dt} a_m(t) = \sum_n a_n(t) W_{mn} e^{i\omega_{mn}t} \tag{8.25}$$

we approximate $a_n(t)$ by its initial value $a_n(t = -\infty)$ (this approximation is called first-order time-dependent perturbation theory). So, if the system is initially in an eigenstate $|n\rangle$ of the Hamiltonian $H_{(0)}$, then $a_n(t = -\infty) = 1$ and $a_m(t = -\infty) = 0$ for $m \neq n$. There is now only one term on the right-hand side of the equation and we can write

$$i\hbar \frac{d}{dt} a_m(t) = a_n(t) W_{mn} e^{i\omega_{mn}t} \tag{8.26}$$

Integration gives

$$a_m(t) = \frac{1}{i\hbar} \int\limits_{t=-\infty}^{t=\infty} W_{mn} e^{i\omega_{mn}t'} dt' \tag{8.27}$$

The charged particle starts in the ground state $|0\rangle$ of the harmonic potential. The probability of the system being in an excited state after the electrical pulse has gone ($t \to \infty$) is given by the sum

$$P_{t\to\infty} = \sum_{m \neq n} |a_n(t = \infty)|^2 \tag{8.28}$$

The matrix element for transitions from the ground state of the harmonic oscillator in the presence of a uniform electric field is

$$W_{mn=0} = e|\mathbf{E}_0|e^{-t^2/\tau^2}\langle m|x|n = 0\rangle \tag{8.29}$$

The only matrix element that contributes to the sum for the probability P is that which couples the ground state to the first excited state. This is separated in energy by the energy spacing of the harmonic oscillator which is just $\hbar\omega = \hbar\sqrt{\kappa/m_0}$. The matrix element is

$$W_{10} = e|\mathbf{E}_0|e^{-t^2/\tau^2}\left(\frac{\hbar}{2m_0\omega}\right)^{1/2}\langle 1|\hat{b} + \hat{b}^\dagger|0\rangle = e|\mathbf{E}_0|e^{-t^2/\tau^2}\left(\frac{\hbar}{2m_0\omega}\right)^{1/2} \tag{8.30}$$

Hence, the probability after the electrical pulse has gone ($t \to \infty$) is

$$P_{t\to\infty} = \frac{1}{\hbar^2}\left|\int\limits_{t=-\infty}^{t=\infty} W_{10}e^{i\omega_{10}t'}dt'\right|^2 = \left(\frac{\hbar}{2m_0\omega}\right)\frac{e^2|\mathbf{E}_0|^2}{\hbar^2}\left|\int\limits_{t=-\infty}^{t=\infty} e^{-t'^2/\tau^2}e^{i\omega t'}dt'\right|^2 \tag{8.31}$$

where we note that the frequency $\omega_{mn} = \omega_{10} = \omega$. Completing the square in the exponent in such a way that $-(t'/\tau - i\omega\tau/2)^2 - \omega^2\tau^2/4$, one may write

$$P_{t\to\infty} = \left(\frac{e^2|\mathbf{E}_0|^2}{2m_0\hbar\omega}\right)\cdot e^{-\omega^2\tau^2/2}\left|\int\limits_{t=-\infty}^{t=\infty} e^{-(t'/\tau - i\omega t')^2}dt'\right|^2 \tag{8.32}$$

Fortunately, the integral is standard, with the solution

$$\int\limits_{t=-\infty}^{t=\infty} e^{-(t'/\tau + i\omega\tau/2)^2}dt' = \tau\sqrt{\pi} \tag{8.33}$$

Hence, Eqn (8.32) may be written

$$P_{t\to\infty} = \left(\frac{\pi e^2|\mathbf{E}_0|^2}{2m_0\hbar\omega}\right)\cdot\tau^2 e^{-\omega^2\tau^2/2} \tag{8.34}$$

The physics of how the transition is induced is illustrated in Fig. 8.6. The electric field pulse exerts a force on the charged particle through a change in the potential energy. We know from Chapter 6 that at any given instant that the electric field has the value $|\mathbf{E}_0|$, the parabolic harmonic potential energy is shifted in position by $x_0 = e|\mathbf{E}_0|/\kappa$ and reduced in energy by $\Delta E = -e^2|\mathbf{E}_0|^2/2\kappa$. Transitions from the ground state are induced by this change in potential energy.

The maximum transition probability occurs when the derivative of the probability function with respect to the variable τ is zero (one may check that this is a maximum by taking the second derivative):

$$0 = \frac{d}{d\tau}P_{t\to\infty} = 2\tau e^{-\omega^2\tau^2/2} - \frac{2\omega^2\tau}{2}\cdot\tau^2 e^{-\omega^2\tau^2/2} = 2\tau - \omega^2\tau^3 \tag{8.35}$$

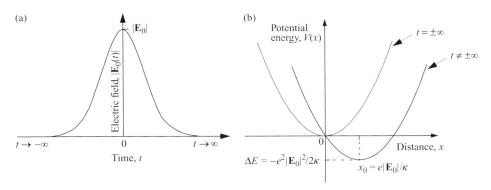

Fig. 8.6. (a) Electric field pulse is Gaussian in shape and centered at time $t = 0$. (b) The harmonic oscillator potential evolves in time.

Hence, the maximum transition probability occurs when the value of τ is

$$\tau = \frac{\sqrt{2}}{\omega} \tag{8.36}$$

Suppose $\omega = 10^{12}$ rad s^{-1}. Then, according to Eqn (8.36), the maximum transition probability occurs for $\tau = 1.4$ ps, corresponding to an electric field pulse with a FWHM of 2.3 ps.

The value of the maximum transition probability is found by substituting Eqn (8.36) into Eqn (8.34). This gives

$$P_{max} = \left(\frac{\pi e^2 |\mathbf{E}_0|^2}{m_0 \hbar \omega^3} \right) \cdot e^{-1} \tag{8.37}$$

There is an obvious difficulty with this result, since when $\omega \to 0$ or $|\mathbf{E}_0| \to \infty$ the maximum transition probability $P_{max} > 1$. According to Eqn (8.37) $P_{max} = 1$, when $\omega = (\pi e^2 |\mathbf{E}_0|^2 \cdot e^{-1} / m_0 \hbar)^{1/3}$.

The resolution of this inconsistency is that the perturbation theory we are using only applies when the time-dependent change in potential $W(t)$ is small. The assumption that the perturbation is weak means that the probability of scattering out of the initial state $|n\rangle$ is small, so that $|a_m|^2 \ll 1$. This condition is simply $P_{max} \ll 1$, which constrains the value of ω and $|\mathbf{E}_0|$ so that

$$\left(\frac{\pi e^2 |\mathbf{E}_0|^2 \hbar^2}{m_0} \right) \cdot e^{-1} \ll \hbar^3 \omega^3 \tag{8.38}$$

For the case we are considering, with $\omega = 10^{12}$ rad s^{-1}, this results in $|\mathbf{E}_0| \ll 5.7 \times 10^4$ V m^{-1}. Clearly, the lesson to be learned here is that, while perturbation theory can be used to calculate transition rates, it is always important to identify and understand the limitations of the calculation.

8.3 First-order time-dependent perturabtion

While Eqn (8.20) is an *exact* result, we often need to make approximations if we wish to calculate actual transition probabilities after a time-dependent change in potential $W(t)$ is turned on at time t > 0. One way to proceed is to approximate the values of $a_n(t > 0)$ by $a_n(t = 0)$ which is the same as first-order perturbation theory. Typically, this approach is valid when the time-dependent change in potential $W(t)$ is small and thus may be considered a perturbation to the initial system described by the Hamiltonian $H_{(0)}$.

Consider the case in which the system is in an eigenstate $|n\rangle$ of $H_{(0)}$ at $t \leq 0$, so $a_n(t = 0) = 1$ and $a_m(t = 0) = 0$ for $m \neq n$. There is now only one term on the right-hand side of Eqn (8.20) which, for $m \neq n$ and $t > 0$, becomes

$$i\hbar \frac{d}{dt}a_m(t) = W_{mn}e^{i\omega_{mn}t} \tag{8.39}$$

since all coefficients $a_m(t = 0) = 0$ except for $a_n(t = 0) = 1$. This means that the matrix element W_{mn} couples $|n\rangle$ to $|m\rangle$ and creates the coefficient $a_m(t)$ for times $t > 0$ and $m \neq n$.

To find how $a_m(t)$ evolves in time from $t = 0$, Eqn (8.39) is rewritten as an integral

$$a_m(t) = \frac{1}{i\hbar} \int_{t'=0}^{t'=t} W_{mn}e^{i\omega_{mn}t'}dt' \tag{8.40}$$

We assume that the state with eigenenergy E_n is not degenerate, so that the perturbed wave function can be expressed as a sum of unperturbed states weighted by coefficients $a_k(t)$

$$\psi(x, t) = \sum_k a_k(t)e^{-i\omega_k t}\phi_k(x) \tag{8.41}$$

If $|a_m|^2 \ll 1$, then we may assume that the coefficient $a_n(t) = 1$. In this case, Eqn (8.41) may be written as

$$\psi(x, t) = \phi_n(x)e^{-i\omega_n t} + \sum_{m \neq n} \frac{1}{i\hbar} \int_{t'=0}^{t'=t} W_{mn}e^{i\omega_{mn}t'}dt' \cdot e^{-i\omega_m t}\phi_m(x) \tag{8.42}$$

One should note that there is an obvious problem with normalization of the scattered state given by Eqn (8.42). Clearly, in a more complete theory correction for the normalization error must be performed self-consistently.

In a more pictorial way, depicted in Fig. 8.7, one may visualize the scattering event using arrows to indicate initial and final states. The lengths of the arrows are a measure

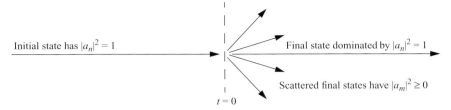

Fig. 8.7. Illustration of initial state with probability amplitude $|a_n|^2 = 1$ and final state at time $t = 0$ being either the same state or a scattered state with probability amplitude $|a_m|^2 \geq 0$. The arrows represent the k-state direction or the direction of motion of the particle. The length of the lines is related to the probability amplitude.

of $|a_m|^2$. The initial state is $|n\rangle$ with $|a_n|^2 = 1$ for time $t \leq 0$. After the scattering event, a number of states are excited in such a way that $|a_m|^2 \neq 0$ for $m \neq n$. However, because we have assumed that the perturbation is weak, the final states are dominated by the state $|n\rangle$ with $|a_n|^2 = 1$. The final states after scattering by a weak perturbation have probability $|a_m|^2 \geq 0$, which may be found using

$$a_m(t) = \frac{1}{i\hbar} \int_{t'=0}^{t'=t} W_{mn} e^{i\omega_{mn}t'} dt' \tag{8.43}$$

where t is the time the perturbation is applied and W_{mn} is the matrix element for scattering out of the initial state $|n\rangle$. The assumption that the perturbation is weak means that the probability of scattering out of the initial state is small. This means that $|a_m|^2 \ll 1$.

8.4 Fermi's golden rule

We already know from our previous work that a perturbing Hamiltonian brings about transitions between eigenstates of the unperturbed Hamiltonian. If the system is initially prepared in state $|n\rangle$ and $W(t) = 0$ for $t \leq 0$, then up to time $t = 0$ the system remains in state $|n\rangle$ with energy E_n. For times $t > 0$, we allow $W(t) \neq 0$. Hence, for times $t = t' > 0$ it is possible that the system is in a different state $|m\rangle$ with energy E_m.

In the following we will add the probabilities of a transition to each state $|m\rangle$, so that the total probability of a transition is proportional to time t. We will then find that the total transition probability is

$$\frac{dP}{dt} = \frac{2\pi}{\hbar} D(E_m) |W_{mn}|^2 \tag{8.44}$$

where $D(E_m) dE_m$ is the number of m states in the energy interval E_m to $E_m + dE_m$. The result assumes that the energy range is so small that both $D(E_m)$ and W_{mn} can be treated as constant.

To show how one obtains Eqn (8.44), we start with Eqn (8.20)

$$i\hbar \frac{d}{dt} a_m(t) = \sum_n a_n(t) W_{mn} e^{i\omega_{mn} t} \tag{8.45}$$

This may be integrated to give

$$a_m(t) = \frac{1}{i\hbar} \int\limits_{t'=0}^{t'=t} \sum_n a_n(t') W_{mn} e^{i\omega_{mn} t'} dt' \tag{8.46}$$

where

$$W_{mn} = \int \phi_m^*(x) W \phi_n(x) dx \tag{8.47}$$

If we assume that $a_n(t')$ and W_{mn} are *slowly varying functions* of time compared with the oscillatory term $e^{i(\omega_m - \omega_n) t'}$, then we can move them outside the integral. Under these circumstances, the equation becomes

$$a_m(t) = \frac{1}{i\hbar} \sum_n a_n(t=0) W_{mn} \int\limits_{t'=0}^{t'=t} e^{i(\omega_m - \omega_n) t'} dt' \tag{8.48}$$

Assuming an initial condition in such a way that $a_n(t \leq 0) = 1$ for only one eigenvalue and $a_m(t \leq 0) = 0$ for $m \neq n$, we may write

$$\boxed{a_m(t) = \frac{1}{i\hbar} W_{mn} \int\limits_{t'=0}^{t'=t} e^{i(\omega_m - \omega_n) t'} dt'} \tag{8.49}$$

Performing the integration gives

$$a_m(t) = \frac{-W_{mn}}{\hbar} \left[\frac{e^{i(\omega_m - \omega_n) t'}}{\omega_m - \omega_n} \right]_0^t = \frac{-W_{mn}}{\hbar} \left(\frac{e^{i(\omega_m - \omega_n) t} - 1}{\omega_m - \omega_n} \right) \tag{8.50}$$

$$a_m(t) = \frac{-W_{mn}}{\hbar} \cdot e^{i(\omega_m - \omega_n) t/2} \cdot \left(\frac{e^{i(\omega_m - \omega_n) t/2} - e^{-i(\omega_m - \omega_n) t/2}}{\omega_m - \omega_n} \right) \tag{8.51}$$

$$a_m(t) = \frac{-2W_{mn}}{\hbar} \cdot e^{i(\omega_m - \omega_n) t/2} \cdot \frac{i \sin((\omega_m - \omega_n) t/2)}{\omega_m - \omega_n} \tag{8.52}$$

so that the probability of a transition is

$$|a_m(t)|^2 = \frac{4}{\hbar^2} |W_{mn}|^2 \cdot \frac{\sin^2((\omega_m - \omega_n) t/2)}{(\omega_m - \omega_n)^2} \tag{8.53}$$

The probability of a transition out of state $|n\rangle$ into any state $|m\rangle$ is the *sum*

$$P_n(t) = \sum_m |a_m(t)|^2 \tag{8.54}$$

The total probability is a sum over the final states $|m\rangle$, because it is assumed that each state $|m\rangle$ is an independent parallel channel for a scattering process. If there are $D(E) = dN/dE$ states in the energy interval $dE = \hbar d\omega$, then the sum can be written as an integral

$$P_n(t) = \frac{4}{\hbar^2}|W_{mn}|^2 \cdot \frac{dN}{dE} \cdot \int \frac{\sin^2((\omega - \omega_n)t/2}{(\omega - \omega_n)^2} \cdot \hbar d\omega \tag{8.55}$$

To perform the integral, we change variables, so that $x = (\omega - \omega_n)/2$, and then we take the limit $t \to \infty$. This gives

$$\frac{\sin^2(tx)}{\pi t x^2}\bigg|_{t \to \infty} = \delta(x) \tag{8.56}$$

We now note that $dE/dx = 2\hbar$, since $E = \hbar 2x$. Hence, the integral can be written

$$\int \frac{\sin^2((\omega - \omega_n)t/2)}{(\omega - \omega_n)^2} \cdot \frac{dE}{dx} dx = 2\hbar \int \frac{\sin^2(tx) \cdot t\pi}{\pi t 4x^2} dx = \frac{\hbar \pi t}{2} \int \frac{\sin^2(tx)}{\pi t x^2} dx \tag{8.57}$$

so that in the limit $t \to \infty$

$$\int \frac{\sin^2((\omega - \omega_n)t/2)}{(\omega - \omega_n)^2} \cdot \frac{dE}{dx} dx = \frac{\hbar \pi t}{2} \int \delta(x) dx \tag{8.58}$$

One may now write the probability of a transition out of state $|n\rangle$ into any state $|m\rangle$ given by Eqn (8.55) as

$$P_n(t) = \frac{4}{\hbar^2}|W_{mn}|^2 \cdot D(E) \cdot \frac{\hbar \pi t}{2} \tag{8.59}$$

or

$$P_n(t) = \frac{2\pi}{\hbar}|W_{mn}|^2 \cdot D(E) \cdot t \tag{8.60}$$

We notice that the probability of a transition is *linearly* proportional to time. The reason for this is embedded in the approximations we have used to obtain this result.

The *transition rate* is the time derivative of the probability $P_n(t)$

$$\frac{d}{dt}P_n(t) = \frac{2\pi}{\hbar}|W_{mn}|^2 \cdot D(E) \tag{8.61}$$

Recognizing $dP_n(t)/dt$ as the inverse probability lifetime τ_n of the state $|n\rangle$, we can write Fermi's golden rule:

$$\boxed{\frac{1}{\tau_n} = \frac{2\pi}{\hbar}|W_{mn}|^2 \cdot D(E)} \tag{8.62}$$

in which the inverse lifetime $1/\tau_n$ of the initial state $|n\rangle$ only depends upon the matrix element squared coupling the initial state to any scattered state $|m\rangle$ multiplied by the final density of scattered states $D(E)$. This simple expression may be used for many calculations of practical importance.

The derivation of Fermi's golden rule involved a number of approximations that can limit its validity in some applications. For example, use of the $t \to \infty$ limit implies that the collision is completed. Hence, one should check to make sure that the perturbing potential is small, so that collisions do not overlap in space or time. It is also assumed that the probability of scattering out of the initial state $|n\rangle$ is so small that $|a_n|^2 = 1$ and conservation of the number of particles can be ignored. Also, *if*, as will often happen, we use a plane-wave initial state characterized by wave vector **k** and a final plane-wave state characterized by wave vector **k**′, then the actual *collision is localized in real space*, so that the use of Fourier components is justifiable. All of these assumptions can be violated in modern semiconductor devices in which scattering can be quite strong and nonlocal effects can become important. So, we need to proceed with caution.

In this section we used Fermi's golden rule to calculate the transition probability for an electron initially in the ground state of a one-dimensional harmonic potential of a molecule subject to an electric field pulse the strength of which has a Gaussian time dependence. We found that changing the electric-field pulse width changes the probability of transitions to excited states of the molecule.

In Section 8.5, we will use Fermi's golden rule to calculate the average distance (the mean free path) an electron travels before scattering in an n-type semiconductor with ionized substitutional impurity concentration n. The microscopic model we develop will allow us to estimate the electrical mobility and electrical conductivity of the material.

In Section 8.6, Fermi's golden rule will be used to calculate the probability of inducing stimulated optical transitions between electronic states. Stimulated emission of light is a key ingredient determining the operation of lasers.

8.5 Elastic scattering from ionized impurities

Establishing a method to control electrical conductivity in semiconductors is essential for many practical device applications. The performance of transistors and lasers depends critically upon flow of current through specific regions of a semiconductor. An important way to control the electrical conductivity of a semiconductor is by a technique called substitutional *doping*.

Substitutional doping involves introducing a small number of impurity atoms into the semiconductor crystal. Each impurity atom replaces an atom on a lattice site of the original semiconductor crystal. In the example, we will be considering a density n of Si donor impurity atoms that sit on Ga sites in a GaAs crystal. At each impurity site, three of the four chemically active Si electrons are used to replace Ga valence electrons. At low temperatures, the remaining Si electron is bound by the positive charge of the Si donor impurity ion. In a GaAs crystal, this extra electron has an effective electron mass of $m_e^* = 0.07m_0$ and its ground state is in a hydrogenic s-like electronic state (Table 2.6). The coulomb potential seen by the electron is screened by the presence of the semiconductor dielectric which is characterized by dielectric constant ε. The screened coulomb potential and the low effective electron mass in the conduction band of GaAs increase the Bohr radius (Eqn (2.70)) characterizing a hydrogenic state from

$$a_B = \frac{4\pi\varepsilon_0\varepsilon_{r0}\hbar^2}{m_0 e^2} = 0.0529\,\text{nm} \tag{8.63}$$

to an effective Bohr radius given by

$$a_B^* = \frac{4\pi\varepsilon_0\varepsilon_{r0}\hbar^2}{m_e^* e^2} = 10\,\text{nm} \tag{8.64}$$

where the use of the low-frequency relative dielectric (permittivity) constant, ε_{r0}, implies that we are considering low-frequency processes.

In addition, the hydrogenic binding energy is reduced from its value of a Rydberg in atomic hydrogen (see Section 2.2.3.2)

$$Ry = \frac{-m_0}{2}\frac{e^4}{(4\pi\varepsilon_0)^2\hbar^2} = -13.6058\,\text{eV} \tag{8.65}$$

to a new value (an effective Rydberg constant, Ry^*), which is given by

$$Ry^* = E_{\text{donor}} - E_{\text{CB}_{\text{min}}} = \frac{-m_e^*}{2}\frac{e^4}{(4\pi\varepsilon)^2\hbar^2} = Ry\left(\frac{m_e^*/m_0}{\varepsilon_{r0}^2}\right)$$

$$= -13.6\frac{0.07}{(13.2)^2} = -5.5\,\text{meV} \tag{8.66}$$

From Eqn (8.66) one concludes that the donor electron is only loosely bound to the donor ion in GaAs. The reasons for this are the small effective electron mass and the value of the dielectric constant ε_{r0} in the semiconductor.

At finite temperatures, lattice vibrations or interaction with freely moving electrons can easily excite the donor electron from its bound state into unbound states in the conduction band. This ionization process is shown schematically in Fig. 8.8(a).

For temperatures for which $k_B T > (E_{\text{CB}_{\text{min}}} - E_{\text{donor}})$ and low impurity concentrations, the loosely bound donor electron has a high probability of being excited into the conduction band, leaving behind a positive Si^+ ion core. For high impurity concentrations in which there is a significant overlap between donor wave functions, electrons can also move freely through the conduction band. In either case, the coulomb potential due to the positive ion core acts as a scattering potential for electrons moving in the conduction band. If we introduce a density n of Si impurities into the GaAs semiconductor to increase the number of electrons, we also increase the number of ionized impurity sites in the crystal that can scatter these electrons.

In the remainder of this section we are going to consider the interesting question of why it is possible to increase the conductivity of a semiconductor by increasing the number of electrons in the conduction band through substitutional impurity doping. It

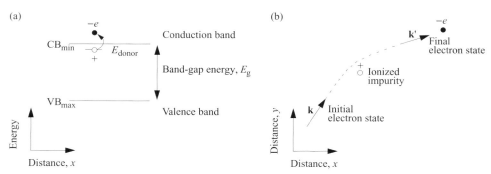

Fig. 8.8. (a) Conduction-band minimum (CB_{min}) and valence-band maximum (VB_{max}) of a semiconductor with band-gap energy E_g. The energy level of a donor impurity is shown. When an electron is excited from the donor level into the conduction band, a positively charged ion core is left at the donor site. This is called an ionized impurity. (b) An electron moving in the conduction band in initial state **k** can be elastically scattered into final state **k′** by the coulomb potential of an ionized impurity. The sketch represents the trajectory of the wave vector associated with the electron.

is, after all, not yet obvious that such a strategy would work, as we have to consider the role of electron scattering from the increased number of ionized impurity sites.

Anticipating our solution, we will assume that electrons in the conduction band can be described by well-defined $|\mathbf{k}\rangle$ states that scatter into final states $|\mathbf{k}'\rangle$, transferring momentum **q** in such a way that $\mathbf{k} = \mathbf{k}' + \mathbf{q}$. This is illustrated in Fig. 8.8(b), where an electron moving past an ionized impurity is scattered from an initial state $|\mathbf{k}\rangle$ to a final state $|\mathbf{k}'\rangle$. We expect to deal with a dilute number of impurity sites, weak scattering, and a plane-wave description for initial and final states. On average, the distance between scattering events is l_k. This distance, called the *mean free path*, is assumed to be longer than the electron wavelength. We also assume no energy transfer, so that we are in the elastic scattering limit.

A natural question concerns the justification for using time-dependent perturbation theory for an electron scattering from a static potential that has no explicit time dependence. The answer to this question is that an expansion of a state function belonging to one time-independent Hamiltonian in the eigenfunctions of another time-independent Hamiltonian has time-dependent expansion coefficients. In our case, the incident particle is described by a plane-wave state far away from the scattering center. The final state is also described by a plane-wave state far away from the scattering center. However, near the scattering center the particle definitely cannot be described as a plane wave.

Electrons scatter from a state $\psi_{\mathbf{k}} = Ae^{i(\mathbf{k}\cdot\mathbf{r})} = |\mathbf{k}\rangle$ of energy $E(\mathbf{k})$ to a final state $|\mathbf{k}'\rangle$ with the same energy. Fermi's golden rule requires that we evaluate the matrix element $\langle \mathbf{k}'|v(r)|\mathbf{k}\rangle$, where $v(r)$ is the coulomb potential. Simple substitution of initial and final plane-wave states into this matrix element reveals that $\langle \mathbf{k}'|v(r)|\mathbf{k}\rangle = v(\mathbf{q})$, where $v(\mathbf{q})$ is the Fourier transform of the coulomb potential in real space. It is clear then, that we would like to find an expression for the coulomb potential in momentum (wave vector) space.

8.5.1 The coulomb potential

The *bare* coulomb potential energy in real space is

$$v(r) = \frac{-e^2}{4\pi\varepsilon_0 r} \tag{8.67}$$

where $-e$ is electron charge. To obtain the potential energy in wave vector space, one may take the Fourier transform. The result for the bare coulomb potential (see Exercise 8.2) is

$$v(q) = \frac{-e^2}{\varepsilon_0 q^2} \tag{8.68}$$

In a uniform dielectric such as an isotropic semiconductor characterized by relative dielectric (permittivity) function ε_r, the term ε_0 in Eqn (8.68) is replaced with $\varepsilon = \varepsilon_0 \varepsilon_r$. In this case, ε is constant over real space, but the value of ε_r depends upon wave vector $\mathbf{q}$, so that $\varepsilon = \varepsilon(\mathbf{q}) = \varepsilon_0 \varepsilon_r(\mathbf{q})$. We expect such a dependence because many electrons can respond to the long-range electric field components of the real-space coulomb potential. This effect, called *screening*, reduces the coulomb potential for small values of $\mathbf{q}$. Because screening is embedded in ε one talks of a screened *dielectric response function*, $\varepsilon(\mathbf{q})$. In general, the dielectric response should also have a frequency dependence, so that $\varepsilon = \varepsilon(\mathbf{q}, \omega)$. However, we ignore this at present. In the following, we simply replace a constant value of ε with $\varepsilon(\mathbf{q})$. Later, we will find expressions for the functional form of $\varepsilon(\mathbf{q})$.

We now have a coulomb potential energy in wave vector space

$$v(\mathbf{q}) = \frac{-e^2}{\varepsilon(\mathbf{q}) q^2} \tag{8.69}$$

and in real space

$$v_{\mathbf{q}}(r) = \frac{-e^2}{4\pi\varepsilon_{\mathbf{q}} r} \tag{8.70}$$

where we note, the subscript $\mathbf{q}$ is used because $v(r)$ and ε depend upon the scattered wave vector.

The potential energy at position $\mathbf{r}$, due to ion charge at position $\mathbf{R}_j$, is

$$v_{\mathbf{q}}(\mathbf{r} - \mathbf{R}_j) = \frac{-e^2}{4\pi\varepsilon_{\mathbf{q}}|\mathbf{r} - \mathbf{R}_j|} \tag{8.71}$$

which depends only upon the separation of charges.

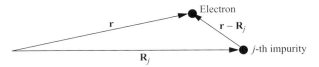

Fig. 8.9. Diagram illustrating the relative position $\mathbf{r} - \mathbf{R}_j$ of an electron at position $\mathbf{r}$ and the j-th ionized impurity at position $\mathbf{R}_j$.

To obtain $v(\mathbf{q})$, we take the Fourier transform

$$v(\mathbf{q}) = \int d^3r \, v(\mathbf{r} - \mathbf{R}_j) \, e^{-i\mathbf{q}\cdot(\mathbf{r}-\mathbf{R}_j)} \tag{8.72}$$

where the integral is over all space. Clearly, in a homogeneous medium, if we determine $v(\mathbf{q})$ at one position in space, then we have determined it for all space.

8.5.1.1 Elastic scattering of electrons by ionized impurities in GaAs

We wish to estimate the elastic scattering rate for electrons in GaAs doped to $n = 10^{18} \, \text{cm}^{-3}$ due to the presence of ionized impurities. Suppose $\mathbf{R}_j$ is the position of the j-th dopant atom in n-type GaAs and, as shown in Fig. 8.9, we are interested in an electron at position $\mathbf{r}$. The interaction potential in real space is the sum of the contributions from the n individual ions per cubic centimeter. Thus, the total potential is

$$V(\mathbf{r}) = \sum_{j=1}^{n} v(\mathbf{r} - \mathbf{R}_j) \tag{8.73}$$

In wave vector space we have a sum of Fourier transforms of $v(\mathbf{r} - \mathbf{R}_j)$

$$V(\mathbf{q}) = \sum_{j=1}^{n} \int d^3r \, v(\mathbf{r} - \mathbf{R}_j)e^{-i\mathbf{q}\cdot\mathbf{r}} = \sum_{j=1}^{n} \int d^3r \, v(\mathbf{r} - \mathbf{R}_j)e^{-i\mathbf{q}\cdot\mathbf{r}}e^{i\mathbf{q}\cdot\mathbf{R}_j}e^{-i\mathbf{q}\cdot\mathbf{R}_j} \tag{8.74}$$

$$V(\mathbf{q}) = \sum_{j=1}^{n} \int d^3r \, v(\mathbf{r} - \mathbf{R}_j)e^{-i\mathbf{q}\cdot(\mathbf{r}-\mathbf{R}_j)}e^{i\mathbf{q}\cdot\mathbf{R}_j} = v(\mathbf{q})\sum_{j=1}^{n} e^{-i\mathbf{q}\cdot\mathbf{R}_j} \tag{8.75}$$

Hence, the total potential seen by the electron in the presence of n ionized impurities per unit volume is

$$V(\mathbf{q}) = v(\mathbf{q})\sum_{j=1}^{n} e^{-i\mathbf{q}\cdot\mathbf{R}_j} \tag{8.76}$$

For elastic scattering, we consider transitions between a state $\psi_{\mathbf{k}} = Ae^{i(\mathbf{k}\cdot\mathbf{r})} = |\mathbf{k}\rangle$ of energy $E(\mathbf{k})$ and a final state $|\mathbf{k}'\rangle$ with the same energy. Fermi's golden rule (the first term in the Born series) involves evaluating the matrix element $\langle \mathbf{k}'|v(r)|\mathbf{k}\rangle$. Since $|\mathbf{k}\rangle$ and $\langle \mathbf{k}'|$ are plane-wave states of the form $e^{-i\mathbf{k}\cdot\mathbf{r}}$ (we have assumed that $kl_k \gg 1$, so the

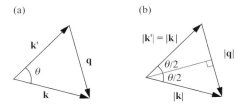

Fig. 8.10. (a) Diagram illustrating initial wave vector **k**, final wave vector **k′**, and transferred momentum **q**. (b) For elastic scattering, the scattered angle θ is related to q by $k \sin(\theta/2) = q/2$.

mean free path l_k is many electron wavelengths long), we write

$$\langle \mathbf{k'}|v(r)|\mathbf{k}\rangle = \int d^3r \; e^{i\mathbf{k'}\cdot\mathbf{r}} v(\mathbf{r}) e^{-i\mathbf{k}\cdot\mathbf{r}} = \int d^3r \; v(\mathbf{r}) e^{-i(\mathbf{k}-\mathbf{k'})\cdot\mathbf{r}}$$

$$= \int d^3r \; v(\mathbf{r}) e^{-i\mathbf{q}\cdot\mathbf{r}} = v(\mathbf{q}) \tag{8.77}$$

which is just the Fourier transform of the coulomb potential in real space. In this expression $\mathbf{q} = \mathbf{k} - \mathbf{k'}$, since momentum conservation requires $\mathbf{k} = \mathbf{k'} + \mathbf{q}$. As illustrated in Fig. 8.10, the scattering angle θ for elastic scattering (no energy loss) is such that $k \sin(\theta/2) = q/2$.

The probability of elastic scattering between the two states is

$$1/\tau_{\mathbf{kk'}} = \frac{2\pi}{\hbar} |v(\mathbf{q})|^2 \delta(E(\mathbf{k}) - E(\mathbf{k} - \mathbf{q})) \tag{8.78}$$

where the delta function ensures that no energy is exchanged. The total scattering rate is a sum over all transitions, so that for a *single impurity*

$$\boxed{\frac{1}{\tau_{\mathrm{el}}} = \frac{2\pi}{\hbar} \int \frac{d^3q}{(2\pi)^3} |v(\mathbf{q})|^2 \, \delta(E(\mathbf{k}) - E(\mathbf{k} - \mathbf{q}))} \tag{8.79}$$

Elastic scattering from n impurities can now be calculated using

$$|V(\mathbf{q})|^2 = |v(\mathbf{q})|^2 \left| \sum_{j=1}^{n} e^{-i\mathbf{q}\cdot\mathbf{R}_j} \right|^2 = |v(\mathbf{q})|^2 s(\mathbf{q}) \tag{8.80}$$

where

$$s(\mathbf{q}) = \left| \sum_{j=1}^{n} e^{-i\mathbf{q}\cdot\mathbf{R}_j} \right|^2 \tag{8.81}$$

$s(\mathbf{q})$ is a *structure factor* that contains phase information on the scattered wave from site $\mathbf{R}_j$. For n large and random $\mathbf{R}_j$, the sum over n random phases is $n^{1/2}$, and so the sum squared is n. It follows that if there are n spatially *uncorrelated* scattering sites corresponding to random impurity positions, we expect $s(\mathbf{q}) = n$. To show that this is so, see Exercise 8.3. For a large number of spatially random impurity positions, the matrix element squared given by Eqn (8.80) becomes

$$|V(\mathbf{q})|^2 = n|v(\mathbf{q})|^2 \tag{8.82}$$

and the *total elastic scattering rate* from n impurities per unit volume is

$$\frac{1}{\tau_{el}} = \frac{2\pi}{\hbar} \cdot n \int \frac{d^3q}{(2\pi)^3} \left| \frac{e^2}{\varepsilon(\mathbf{q})q^2} \right|^2 \delta(E(\mathbf{k}) - E(\mathbf{k} - \mathbf{q})) \tag{8.83}$$

where the integral over d^3q is the final density of states.

Physically, each impurity is viewed as contributing independently, so that the scattering rate is n times the scattering rate from a *single impurity atom*. We can also see why increasing the impurity concentration n does not necessarily result in a linear increase in scattering rate $1/\tau_{el}$. The integral contains a matrix element squared $|e^2/\varepsilon(\mathbf{q})q^2|^2$, which can influence scattering rate. The $1/q^2$ term reflects the fact that ionized impurity coulomb scattering is weighted toward final states with small q transfer. This means that electrons moving in a given direction are mainly scattered by small angles without too much deviation from the forward direction. The dielectric function $\varepsilon(\mathbf{q})$ also has an influence on scattering rate, in part because of its $\mathbf{q}$ dependence but also because the function depends upon carrier concentration, n.

8.5.1.2 Correlation effects due to spatial position of dopant atoms

Thus far, we have assumed that each substitutional dopant atom occupies a random crystal lattice site. However, the constraint that *substitutional* impurity atoms in a crystal *occupy crystal lattice sites* gives rise to a *correlation effect because double occupancy of a site is not allowed*. Suppose a fraction f of sites is occupied. In this case, we no longer have a truly random distribution, and, for small f, the scattering rate will be reduced by $s(\mathbf{q}) = n(1 - f)$. The factor $(1 - f)$ reflects the fact that not allowing double occupancy of a site is a correlation effect.

Other spatial correlation effects are possible and can, in principle, dramatically alter scattering rates.[1]

8.5.1.3 Calculating electron mean free path

We wish to calculate the mean free path of a conduction-band electron in an isotropic semiconductor that has been doped with n randomly positioned impurities. We start

---

[1] A. F. J. Levi, S. L. McCall, and P. M. Platzman, *Appl. Phys. Lett.* **54**, 940 (1989) and A. L. Efros, F. G. Pikus, and G. G. Samsonidze, *Phys. Rev.* **B41**, 8295 (1990).

from the expression for total elastic scattering rate of an electron mass m_e^* and charge e from a density of n random ionized impurities given by Eqn (8.83). Because the aim is to evaluate the total elastic scattering rate as a function of the incoming electron energy E, it is necessary to express the volume element d^3q in terms of energy $E = \hbar^2 k^2 / 2m_e^*$ and scattering angle θ. Since the material is isotropic, one may write $q = 2k \sin(\theta/2)$. It can be shown (Exercise 8.4) that

$$
\frac{1}{\tau_{el}} = \frac{2\pi m_e^*}{\hbar^3 k^3} \cdot n \cdot \left(\frac{e^2}{4\pi \varepsilon_0} \right)^2 \cdot \int_{\eta=0}^{\eta=1} \frac{d\eta}{(\varepsilon_r(q))^2 \eta^3}
\tag{8.84}
$$

or, as a function of energy,

$$
\frac{1}{\tau_{el}(E)} = \frac{\pi}{(2m_e^*)^{1/2}} \cdot n \cdot \left(\frac{e^2}{4\pi \varepsilon_0} \right)^2 \cdot E^{-3/2} \cdot \int_{\eta=0}^{\eta=1} \frac{d\eta}{(\varepsilon_r(2k\eta))^2 \eta^3}
\tag{8.85}
$$

where $\eta = \sin(\theta/2)$, scattered wave vector $q = 2k\eta$, and dielectric function $\varepsilon(q) = \varepsilon_0 \varepsilon_r(q)$.

Before estimating the value of the integral given by Eqn (8.84), we calculate the prefactor using parameters for GaAs with $n = 10^{18}$ cm^{-3}. In this case, the conduction-band electron has effective electron mass $m_e^* = 0.07 m_0$ and Fermi wave vector in three dimensions $k_F = (3\pi^2 n)^{1/3} = 3 \times 10^6$ cm^{-1}. The Fermi energy is $E_F = \hbar^2 k_F^2 / 2m_e^*$, and the wavelength associated with an electron at the Fermi energy is $\lambda_F = 2\pi / k_F = 20$ nm in this case.

Because we will be interested in relating the calculated elastic scattering rate to the measured low-temperature conductivity and mobility of the semiconductor, we need to estimate $1/\tau_{el}$ for an electron of energy $E = E_F$. This is because at low temperatures the motion of electrons with energy near the Fermi energy determines electrical conductivity.

The prefactor of the integral given by Eqn (8.84) is

$$
\frac{2\pi}{\hbar^3} \cdot n \cdot \frac{e^4 m_e^*}{(4\pi \varepsilon_0)^2 k_F^3} = \frac{2\pi n c (m_e^*/m_0)}{3\pi^2 n} \cdot \frac{m_0 e^2}{4\pi \varepsilon_0 \hbar^2} \cdot \frac{e^2}{4\pi \varepsilon_0 \hbar c} = \frac{2c(m_e^*/m_0)}{3\pi a_B \alpha^{-1}}
\tag{8.86}
$$

Putting in the numbers, we have in SI-MKS units

$$
\frac{2c(m_e^*/m_0)}{3\pi a_B \alpha^{-1}} = \frac{2 \times 3 \times 10^8 \times 0.07}{3\pi \times 0.53 \times 10^{-10} \times 137} = 6.14 \times 10^{14} \text{ s}^{-1}
\tag{8.87}
$$

Notice that when evaluating the prefactor we used known *physical values* and *dimensionless units* as much as possible. This helps us to avoid mistakes and confusion.

To estimate the integral in Eqn (8.84), we assume that $\varepsilon_r(q) \sim \varepsilon_{r0} \sim 10$, and we approximate the integral as $1/\varepsilon_{r0}^2 = 1/100$, so that

$$\frac{1}{\tau_{el}} \sim \frac{6.14 \times 10^{14}}{\varepsilon_{r0}^2} = 6.14 \times 10^{12} \text{ s}^{-1} \tag{8.88}$$

The mean free path is the characteristic length l_k between electron scattering events. For elastic scattering at the Fermi energy in n-type GaAs, we have $l_{k_F} = v_F \tau_{el}$, where Fermi velocity $v_F = \hbar k_F/m_e^* = 5 \times 10^7$ cm s^{-1}. Hence, $l_{k_F} = 5 \times 10^7/6 \times 10^{12} = 83$ nm. We can compare this length with the average spacing between impurity sites, which is only 10 nm for an impurity concentration $n = 10^{18}$ cm^{-3}. Obviously, this impurity concentration is *not* the *dilute* limit that we had previously assumed, since the electron wavelength $\lambda_F = 2\pi/k_F = 20$ nm is similar to the average spacing between impurities. However, $l_{k_F} > \lambda_F$, so that $k_F l_{k_F} \gg 1$, justifying our assumption of weak scattering. One may also compare the average spacing between impurity sites, which is 10 nm (many times the GaAs lattice constant $L = 0.56533$ nm), with the effective Bohr radius for a hydrogenic n-type impurity

$$a_B^* = \frac{4\pi\varepsilon_0\varepsilon_{r0}\hbar^2}{m_e^* e^2} \tag{8.89}$$

which, using a value $\varepsilon_{r0} = 13.2$ for the *low-frequency* dielectric constant, gives $a_B^* = 10$ nm. Because a_B^* is comparable to the average spacing between impurity sites, there should be a significant overlap between donor electron wave functions giving rise to metallic behavior. More formally, one introduces a parameter r_s that is the radius of a sphere occupied, on average, by one electron, assuming a uniform electron density n, divided by the effective Bohr radius, a_B^*:

$$r_s = \left(\frac{3}{4\pi n}\right)^{1/3} \cdot \frac{1}{a_B^*} \tag{8.90}$$

For GaAs with an impurity concentration $n = 10^{18}$ cm^{-13}, this gives $r_s = 0.63$. Again, because $r_s < 1$, we expect metallic behavior. This is indeed the case, and we can use our calculation of mean free path to estimate the electrical mobility and conductivity.

8.5.1.4 Calculating mobility and conductivity

Electron mobility is defined as

$$\boxed{\mu = e\tau_{el}^*/m_e^*} \tag{8.91}$$

where $1/\tau_{el}^*$ is an appropriate elastic scattering rate and m_e^* is the effective electron mass. It is usual for mobility to be quoted in CGS units of cm^2 V^{-1} s^{-1}. If we wish to calculate the mobility of electrons that are characterized on average by a Fermi wave vector k_F and a mean free path l_{k_F}, where we assume $\tau_{el}^* = \tau_{el}$, then the mobility is

$$\mu = e\tau_{el}/m_e^* = el_{k_F}/\hbar k_F \tag{8.92}$$

Conductivity is proportional to mobility and is defined as

$$\sigma = ne\mu = \frac{ne^2\tau_{el}^*}{m_e^*} \tag{8.93}$$

If we wish to calculate the conductivity of electrons which are characterized by a Fermi wave vector k_F and a mean free path l_{k_F} where we assume $\tau_{el}^* = \tau_{el}$, then the conductivity of the material is

$$\sigma = ne\mu = \frac{ne^2\tau_{el}}{m_e^*} = \frac{ne^2 l_{k_F}}{\hbar k_F} \tag{8.94}$$

For our particular example we can put in numbers for mobility of GaAs doped to $n = 10^{18}$ cm^{-3}. Using CGS units,

$$\mu_n = e\tau_{el}/m_e^* = 1.6 \times 10^{-12}/6 \times 10^{12} \times 0.07 \times 9.1 \times 10^{-28}$$
$$= 4.1 \times 10^3 \text{ cm}^2 \text{ V}^{-1} \text{ s}^{-1} \tag{8.95}$$

As shown in Fig. 8.11, the experimentally measured value of electron mobility in bulk n-type GaAs with carrier concentration $n = 10^{18}$ cm^{-3} at temperature $T = 300$ K is $\mu_n = 2 \times 10^3$–3×10^3 cm^2 V^{-1} s^{-1}. At the lower temperature of $T = 77$ K, the mobility is measured to be $\mu_n = 3 \times 10^3$–4×10^3 cm^2 V^{-1} s^{-1}. So the agreement between our very crude estimates and experiment is quite good.

As a next step to *improve* on our calculation of elastic scattering rate, $1/\tau_{el}$, we will consider how electron scattering from the static distribution of ionized impurities is

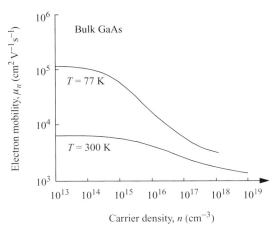

Fig. 8.11. Experimentally determined electron mobility of bulk n-type GaAs as a function of carrier density n on a logarithmic scale for the indicated temperatures, T.

modified due to the presence of mobile electron charge in the system. The coulomb potential of each ionized impurity is modified, or screened, by mobile charge carriers. Because the coulomb potential is modified, the total elastic scattering rate and its angular dependence will also change.

8.5.2   Linear screening of the coulomb potential

Previously we have assumed that a doped semiconductor has an ionized impurity distribution that is random. The static ionized charge distribution is $\rho_i(\mathbf{r})$, with net charge $Q_i = e \int d^3 r \rho_i(\mathbf{r})$. The total mobile charge attracted is exactly $-Q_i$. The mobile charge (which is considered a nearly free electron gas) is a *screening charge* and has its own distribution in space given by $\rho_s(\mathbf{r})$.

The screened potential energy from the static impurity charge and the mobile screening charge is exactly

$$V(r) = \int d^3 r' \frac{-e^2(\rho_i(\mathbf{r}') + \rho_s(\mathbf{r}'))}{4\pi\varepsilon_0 |\mathbf{r} - \mathbf{r}'|} \tag{8.96}$$

As shown schematically in Fig. 8.12, one may imagine a pile-up of electron charge density around the positive impurity ion. The mobile electron charge density *screens* the coulomb potential due to the impurity.

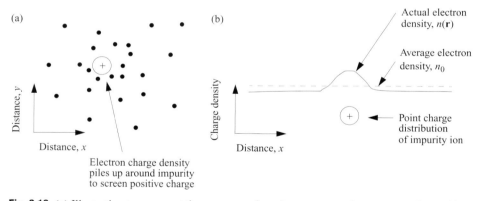

Fig. 8.12. (a) Illustration to represent the response of an electron gas to the presence of a positive charge distribution due to an impurity ion. On average, electrons spend more time in the vicinity of the impurity. (b) The electron density is greater than average near the positively charged impurity ion. The impurity ion may be modeled as a point charge.

8.5.2.1   Calculating the screened potential in real space

To calculate the screened coulomb potential, let's begin by considering an isotropic three-dimensional free-electron gas with equilibrium time-averaged electron particle density n_0. Suppose we now place a test charge (the impurity ion) into this electron gas.

The test charge will try to create a coulomb potential $\phi_{ex}(r) = -e/4\pi\varepsilon r$. However, the response of the electron gas to the presence of the test charge is to create a new electron particle density $n(r)$, which will create a new screened potential $\phi(r)$.

We make the simplifying assumption that the relationship between the energy of an electron at position $\mathbf{r}$ and its wave vector is only modified from its free-electron value by the *local potential*, so that

$$E(k) = \frac{\hbar^2 k^2}{2m} - e\phi(r) \tag{8.97}$$

The assumption of a local potential can only be true for electrons localized in space and so described (semiclassically) as wave packets. However, the electron wave packets are spread out in real space by a characteristic distance (at least $1/k_F$ for a low-temperature degenerate electron gas). To ensure that use of a local potential is an accurate approximation, we must require that $\phi(r)$ vary slowly on the scale of the wave packet size.

The new screened potential may be written in the form

$$\phi(r) = \frac{1}{r} \cdot f(r) \tag{8.98}$$

where $f(r)$ is a function we will determine using Poisson's equation $\nabla^2\phi = -\rho(r)/\varepsilon$, which relates the *local* charge density to the *local* potential. The change in equilibrium charge density is

$$\rho(r) = -e(n(r) - n_0) \tag{8.99}$$

where the averaged electron particle density is

$$n_0 = \int \frac{d^3k}{(2\pi)^3} \cdot 2 \cdot f_k = \int \frac{d^3k}{(2\pi)^3} \cdot 2 \cdot \frac{1}{e^{(E_k - \mu)/k_B T} + 1} \tag{8.100}$$

and the local particle density at position r is

$$n(r) = \int \frac{d^3k}{(2\pi)^3} \cdot 2 \cdot f_k = \int \frac{d^3k}{(2\pi)^3} \cdot 2 \cdot \frac{1}{e^{(E_k - e\phi(r) - \mu)/k_B T} + 1} \tag{8.101}$$

In the expressions for n_0 and $n(r)$, notice that f_k is the Fermi–Dirac distribution function (Eqn (7.35)) and that the factor 2 in the integral accounts for electron spin $\pm\hbar/2$. Equations (8.100) and (8.101) allow us to rewrite Eqn (8.99) as

$$\rho(r) = -e(n_0(\mu + e\phi(r)) - n_0(\mu)) \tag{8.102}$$

If the potential ϕ is small, then Eqn (8.102) may be expanded to first order to give

$$\rho(r) = -e\left(n_0(\mu) + \frac{\partial n_0}{\partial \mu} \cdot e\phi(r) - n_0(\mu)\right) = -e^2 \frac{\partial n_0}{\partial \mu} \cdot \phi(r) \tag{8.103}$$

which shows how the change in equilibrium charge density is related to the screened potential.

For an isotropic, *nondegenerate, three-dimensional electron gas at equilibrium*, the local change in number of carriers due to a local change in potential is given by a Boltzmann factor, so

$$n = n_0 e^{e\phi/k_B T} \tag{8.104}$$

and Poisson's equation becomes

$$\nabla^2 \phi(r) = \frac{-\rho(r)}{\varepsilon} = \frac{e(n - n_0)}{\varepsilon} = \frac{en_0}{\varepsilon}\left(e^{e\phi/k_B T} - 1\right) \tag{8.105}$$

This is a *nonlinear* differential equation for $\phi(r)$. To simplify the equation, we assume that the test charge is small so that one may reasonably expect that the induced potential should also be small. *If* this is true, then the exponential can be expanded to give $e^{e\phi/k_B T} \sim 1 + e\phi/k_B T + \cdots$, so that

$$\nabla^2 \phi(r) \approx \frac{e^2 n_0}{\varepsilon k_B T} \cdot \phi(r) \tag{8.106}$$

which is a *linear* differential equation for $\phi(r)$. Substituting $\phi(r) = (1/r) \cdot f(r)$ gives

$$\nabla^2 \phi(r) = \frac{e^2 n_0}{\varepsilon k_B T} \cdot \frac{1}{r} \cdot f(r) \tag{8.107}$$

In spherical coordinates, the left-hand side of Poisson's equation can be written

$$\nabla^2 \phi(r) = \nabla^2 \left(\frac{1}{r} \cdot f(r)\right) = \frac{1}{r^2}\frac{\partial}{\partial r}\left(r^2 \frac{\partial}{\partial r} \frac{1}{r} \cdot f(r)\right) \tag{8.108}$$

$$\nabla^2 \phi(r) = \frac{1}{r^2}\frac{\partial}{\partial r}\left(r^2\left(\frac{-1}{r^2} \cdot f(r) + \frac{1}{r^2}\frac{\partial f(r)}{\partial r}\right)\right) \tag{8.109}$$

$$\nabla^2 \phi(r) = \frac{1}{r^2}\frac{\partial}{\partial r}\left(r\frac{\partial f(r)}{\partial r} - f(r)\right) = \frac{1}{r^2}\left(\frac{\partial f(r)}{\partial r} + r\frac{\partial^2 f(r)}{\partial r^2} - \frac{\partial f(r)}{\partial r}\right)$$

$$= \frac{1}{r}\frac{\partial^2 f(r)}{\partial r^2} \tag{8.110}$$

so that

$$\frac{\partial^2 f(r)}{\partial r^2} = \frac{e^2 n_0}{\varepsilon k_B T} \cdot f(r) \tag{8.111}$$

Hence, the solution for $f(r)$ is simply $f(r) = e^{-q_D \cdot r}$, where

$$q_D^2 = \frac{n_0 e^2}{\varepsilon k_B T} \tag{8.112}$$

$1/q_D$ is called the Debye screening length. The Debye screening length applies to the *equilibrium, nondegenerate electron gas* and scales with carrier density as $\sqrt{1/n_0}$ and with thermal energy as $\sqrt{k_B T}$. In this case, our screened coulomb potential in real space becomes

$$\phi(r) = \frac{-e}{4\pi\varepsilon r} \cdot e^{-q_{\mathrm{D}}\cdot r} \tag{8.113}$$

where $\varepsilon = \varepsilon_0 \varepsilon_{\mathrm{r0}}$.

For a three-dimensional, *degenerate electron system* at low temperature we may obtain the Thomas–Fermi screening length by simply identifying the Fermi energy as the characteristic energy of the system $E_{\mathrm{F}} = 3k_{\mathrm{B}}T/2$. Substitution into our previous expression for q_{D} gives

$$q_{\mathrm{TF}}^2 = \frac{3n_0 e^2}{2\varepsilon E_{\mathrm{F}}} \tag{8.114}$$

Since $E_{\mathrm{F}} = \hbar^2 k_{\mathrm{F}}^2/2m$ and $n_0 = k_{\mathrm{F}}^3/3\pi^2$, this expression may be rewritten as

$$\boxed{q_{\mathrm{TF}}^2 = \frac{k_{\mathrm{F}} m e^2}{\varepsilon \pi^2 \hbar^2}} \tag{8.115}$$

The low-temperature Thomas–Fermi screening length $1/q_{\mathrm{TF}}$ scales with the characteristic Fermi wave number as $\sqrt{1/k_{\mathrm{F}}}$. In this case, our screened coulomb potential in real space becomes

$$\boxed{\phi(r) = \frac{-e}{4\pi\varepsilon r} \cdot e^{-q_{\mathrm{TF}}\cdot r}} \tag{8.116}$$

8.5.2.2 Calculating the screened potential and dielectric function in wave vector space

For the coulomb potential energy we had for a single ion at position $r = 0$

$$V_q(r) = \frac{-e^2}{4\pi\varepsilon_0 \varepsilon_{\mathrm{r0}} r} \tag{8.117}$$

This is a long-range interaction that will be screened by mobile electron charge. Suppose there is a characteristic screening length r_0. Then we approximate the screened potential energy with a function similar to Eqn (8.113) or Eqn (8.116) so that

$$V_q(r) = \frac{-e^2}{4\pi\varepsilon_0 \varepsilon_{\mathrm{r0}} r} \cdot e^{-r/r_0} \tag{8.118}$$

This is a static potential energy with no time dependence. To find $V(q)$ for this static screened potential energy, one takes the Fourier transform:

$$V(q) = \int d^3r \, V(r) \, e^{-i\mathbf{q}\cdot\mathbf{r}} \tag{8.119}$$

Leaving the integration to Exercise 8.5, the solution is

$$V(q) = \frac{-e^2}{\varepsilon_0 \varepsilon_{\mathrm{r0}} \left(q^2 + 1/r_0^2\right)} \tag{8.120}$$

We now compare this with our previous expression in terms of a dielectric function

$$V(q) = \frac{-e^2}{\varepsilon(q)q^2} = \frac{-e^2}{\varepsilon_0 \varepsilon_{r0} q^2 \left(1 + 1/q^2 r_0^2\right)} \tag{8.121}$$

where r_0 is a characteristic screening length. It is apparent that the effect of screening is to modify the dielectric function in such a way that $\varepsilon = \varepsilon(q)$. For a degenerate electron gas in the low-temperature limit one may use the Thomas–Fermi screening length such that $1/r_0^2 = q_{TF}^2$. In this case,

$$\varepsilon(q) = \varepsilon_0 \varepsilon_{r0} \left(1 + \frac{q_{TF}^2}{q^2} \right) \tag{8.122}$$

This is the Thomas–Fermi dielectric function that, *if* valid for all q, describes a statically screened, real-space potential energy of the Yukawa type

$$V_q(r) = \frac{-e^2}{4\pi \varepsilon_0 \varepsilon_{r0} r} \cdot e^{-r q_{TF}} \tag{8.123}$$

where

$$q_{TF}^2 = \frac{k_F m e^2}{\varepsilon \pi^2 \hbar^2} = \frac{k_F m e^2}{\varepsilon_0 \varepsilon_{r0} \pi^2 \hbar^2} \tag{8.124}$$

is the Thomas–Fermi wave number. The inverse of the Thomas–Fermi wave number defines the length scale for screening.

To get a feel for the value of q_{TF}, consider the semiconductor GaAs with an impurity concentration $n = 10^{18}$ cm^{-3} and a conduction-band effective electron mass of $m_e^* = 0.07\, m_0$. In this situation, $q_{TF} = 2 \times 10^6$ cm^{-1}. This may be compared with the Fermi wave vector, which has a value $k_F = (3\pi^2 n)^{1/3} = 3 \times 10^6$ cm^{-1}. The fact that $1/q_{TF} = 5$ nm and $1/k_F = 3$ nm have comparable values is not unexpected, since they are both a measure of highest spatial frequency that can be used by the electrons to screen the coulomb interaction.

In Fig. 8.13(a), the Thomas–Fermi dielectric function is plotted as a function of wave vector normalized to q_{TF}. In Fig. 8.13(b) the Thomas–Fermi statically screened real-space potential energy is shown, along with the bare coulomb potential energy as a function of distance, r.

Large values of r in the real-space potential correspond to long-wavelength excitations or equivalently small q scattering in wave vector space. At large values of r, there are many conduction electrons between the impurity and the test charge. Hence, many conduction-band electrons can respond to and effectively screen the impurity potential. Short-wavelength or high-spatial-frequency components of the potential correspond to small values of r in the real space potential. In this case, there are few electrons that can respond to screen the impurity potential. High q scattering from a real-space potential

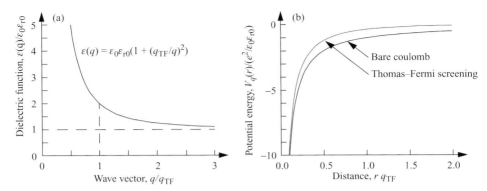

Fig. 8.13. (a) The Thomas–Fermi screened dielectric function as a function of q/q_{TF} and (b) the bare-coulomb and screened-coulomb potential for the indicated values of q.

involves the incident electron getting close to the ionized impurity. When this happens, there are fewer electrons available to screen the ion.

In general, the dielectric function will have a dynamic (or frequency-dependent) part, so $\varepsilon = \varepsilon(q, \omega)$. However, we ignore energy exchange processes (which change electron energy by $\hbar\omega$) as we consider elastic electron scattering only. There are other limitations to our model dielectric function. For example, our discussion of Debye and Thomas–Fermi screening adopted a semiclassical approximation that required the screened potential to vary slowly. This approximation is not valid in the limit of $r \rightarrow 0$ (or, equivalently, large q). Substitution of the Thomas–Fermi screened coulomb potential into Poisson's equation predicts a screened charge density proportional to $(q_{TF}^2 e^{-q_{TF} \cdot r})/r$, which diverges as $r \rightarrow 0$. This deficiency may be overcome by using a different model dielectric that does not require the screened potential to vary slowly. In one such approach, called the random phase approximation (RPA), due to Lindhard[2] one exploits the approximation that the induced charge density contributes linearly to the total potential. The Schrödinger equation is then used to calculate the electronic wave functions self-consistently in the presence of the new potential. However, for most calculations of practical interest (see Exercise 8.7), differences between the RPA and Thomas–Fermi results are relatively small, and so we will continue to use the Thomas–Fermi dielectric function to calculate elastic, ionized-impurity electron scattering rates in semiconductors.

8.5.2.3 Using the Thomas–Fermi dielectric function to calculate elastic, ionized-impurity electron scattering in GaAs

From Eqn (8.84) the elastic scattering rate is

$$\frac{1}{\tau_{el}} = \frac{2\pi m}{\hbar^3 k^3} \cdot n \cdot \left(\frac{e^2}{4\pi \varepsilon_0} \right)^2 \cdot \int_{\eta=0}^{\eta=1} \frac{d\eta}{(\varepsilon_r(q))^2 \eta^3} \tag{8.125}$$

[2] J. Lindhard, *kgl. Danske Videnskab. Selskab Mat.-Fys. Medd.* **28** no. 8(1954).

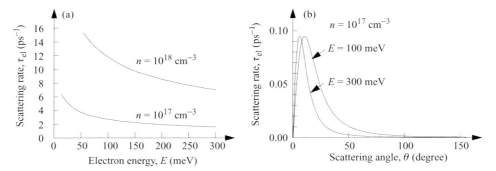

Fig. 8.14. (a) Calculated total elastic scattering rate for an electron of energy E in the conduction band of GaAs due to the presence of the indicated random ionized impurity density. The calculation uses the Thomas–Fermi dielectric function. (b) Elastic scattering rate as a function of scattered angle for an electron of initial energy $E = 100\,\text{meV}$ and $E = 300\,\text{meV}$ in the conduction band of GaAs with $n = 10^{17}\,\text{cm}^{-3}$.

where $\eta = \sin(\theta/2)$ and $q = 2k\eta$. For the Thomas–Fermi dielectric function we have a scattering rate that is given by the expression

$$\frac{1}{\tau_{\text{el}}} = \frac{2\pi m}{\hbar^3 k^3} \cdot n \cdot \left(\frac{e^2}{4\pi\varepsilon_0\varepsilon_{\text{r0}}}\right)^2 \cdot \int_{\eta=0}^{\eta=1} \frac{d\eta}{\left(1 + \frac{q_{\text{TF}}^2}{q^2}\right)^2 \eta^3} \tag{8.126}$$

Figure 8.14(a) shows the result of using Eqn (8.126) to calculate the total elastic scattering rate in GaAs for the indicated values of n-type impurity concentration. The conduction band effective electron mass is taken to be $m_{\text{e}}^* = 0.07m_0$, and the value of ε_{r0} is 13.2. The elastic scattering rate for an electron of energy $E = 200$ meV in GaAs with $n = 10^{17}\,\text{cm}^{-3}$ is about $2 \times 10^{12}\,\text{s}^{-1}$ corresponding to a scattering time of $\tau = 0.5\,\text{ps}$.

Figure 8.14(b) shows the calculated elastic scattering rate as a function of scattered angle for an electron of initial energy $E = 100$ meV and $E = 300$ meV in the conduction band of GaAs with $n = 10^{17}\,\text{cm}^{-3}$. It is clear that the coulomb potential favors small-angle scattering. This is particularly true when the electron has a large value of energy, E.

Figure 8.15 illustrates the difference in calculated elastic scattering rate as a function of scattered angle with and without Thomas–Fermi screening for the indicated electron energies in the conduction band of GaAs with $n = 10^{18}\,\text{cm}^{-3}$.

The effect of screening is to increase the dielectric constant for small scattered wave vector q, since

$$\varepsilon(q) = \varepsilon_0\varepsilon_{\text{r0}}\left(1 + \frac{q_{\text{TF}}^2}{q^2}\right) \tag{8.127}$$

This reduces the value of the integral when q is small. Small q corresponds to

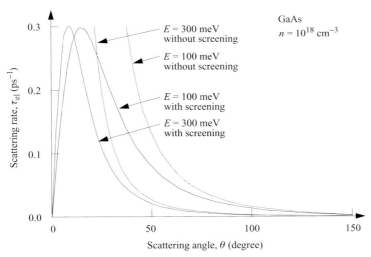

Fig. 8.15. Calculated total elastic scattering rate from random ionized impurities as a function of angle θ with and without Thomas–Fermi screening of the coulomb potential. The parameters used in the calculation are those for GaAs with n-type impurity concentration $n = 10^{18} \, \text{cm}^{-1}$.

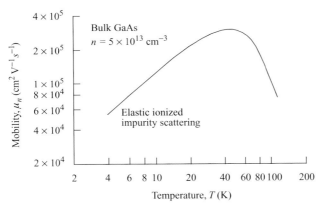

Fig. 8.16. Measured temperature dependence of mobility in bulk GaAs with an n-type impurity concentration of $n = 5 \times 10^{13} \, \text{cm}^{-3}$. The axes use a logarithmic scale. At low temperatures, elastic ionized impurity scattering dominates mobility, and mobility increases with increasing temperature dependence. At temperatures above $T = 50 \, \text{K}$, inelastic scattering from lattice vibrations dominates, causing a decrease in mobility with increasing temperature.

small-angle scattering since $q = 2k \sin(\theta/2)$. It is small-angle scattering (or long-wavelength *excitations*) that are suppressed.

Figure 8.14(a) shows that a high-energy electron scatters less than an electron of low energy. This is typical behavior for coulomb scattering, the origin of which in this case can be traced back to the $E^{-3/2}$ term in Eqn (8.85). Using this energy dependence, it is straightforward to show that, when elastic scattering from ionized impurities dominates electron dynamics, our calculations predict that mobility has a $T^{3/2}$ temperature dependence (see Exercise 8.6). Typically, elastic scattering from ionized impurities is most significant at low temperatures, and so, as shown in Fig. 8.16, mobility increases

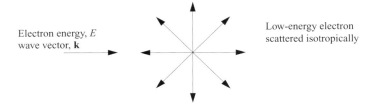

Fig. 8.17. Illustration showing a slow-velocity electron of initial energy E and wave vector $\mathbf{k}$ elastically scattered by the coulomb potential. Isotropic scattering means that the velocity of the scattered electron is independent of the initial velocity.

with increasing temperature for $T < 50$ K. However, at temperatures above $T = 50$ K, *inelastic* scattering from lattice vibrations dominates, causing a decrease in mobility with increasing temperature. In this book, we will not develop a microscopic theory of inelastic scattering.

8.5.2.4 Elastic electron scattering in the limit of small initial velocity

Recall that r_0 is a characteristic screening length that represents the range of action of the potential and that k is the magnitude of the initial wave vector. For a degenerate electron gas $1/r_0^2 = q_{TF}^2$, and for a nondegenerate electron gas $1/r_0^2 = q_D^2$. If we consider an electron with small initial velocity, then $k \to 0$, and so $kr_0 \ll 1$. The scattering amplitude is proportional to

$$V(q) = \int d^3r \; V(r) \, e^{-i\mathbf{q}\cdot\mathbf{r}} \tag{8.128}$$

In the limit $kr_0 \ll 1$, then $e^{-i\mathbf{q}\cdot\mathbf{r}} \sim 1$, since $q = 2k\sin(\theta/2)$ and $k \to 0$. Hence,

$$V(q)|_{k\to 0} = \int d^3r \; V(r) = 4\pi \int dr \; V(r) \cdot r^2 \tag{8.129}$$

and we may conclude that the scattering is isotropic, independent of the incident velocity. This is illustrated in Fig. 8.17.

8.5.2.5 Elastic electron scattering in the limit of large initial velocity

For large velocity $k \to \infty$ and so $kr_0 \gg 1$. In this limit the scattering is anisotropic into a cone $\Delta\theta \sim 1/kr_0$. This is illustrated in Fig. 8.18. Outside of this cone, the term $e^{-i\mathbf{q}\cdot\mathbf{r}}$ oscillates rapidly, and the integral of this with the slowly varying $V(r)$ is almost zero. Hence, in the limit $kr_0 \gg 1$

$$V(q)|_{k\to\infty} = 4\pi \int_{r=0}^{r=r_0} dr \; V(r) \, e^{-i\mathbf{q}\cdot\mathbf{r}} \cdot r^2 \tag{8.130}$$

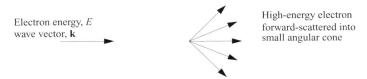

Fig. 8.18. Illustration showing a high-velocity electron of initial energy E and wave vector $\mathbf{k}$ elastically scattered by the coulomb potential. Forward-scattering means that the velocity of the scattered electron deviates little from the initial velocity of the electron. Electrons are forward-scattered into a small angular cone.

8.6 Photon emission due to electronic transitions

To understand and control the emission of photons from atoms or solids, we need to extend our knowledge to include something about the density of optical modes, light intensity, the background energy density in thermal equilibrium, Fermi's golden rule for optical transitions, the occupation factor for thermally distributed photons, and the Einstein A and B coefficients. In the next few pages, we explore these items by example. After completing this section, we will have the knowledge needed to consider the basic ingredients of a laser.

8.6.1   Density of optical modes in three dimensions

For electromagnetic plane waves characterized by wave vector $\mathbf{k}$, the density of optical states in three dimensions is

$$D_3^{\text{opt}}(k)dk = 2 \cdot 4\pi k^2 \frac{dk}{(2\pi)^3} = \frac{k^2}{\pi^2} dk \tag{8.131}$$

where the factor 2 is from the two orthogonal polarizations. This is the density of modes per unit volume in k-space. However, in a homogeneous nondispersive medium with refractive index n_r, the wave vector $k = n_r\omega/c$ and $dk = n_r d\omega/c$. Hence,

$$D_3^{\text{opt}}(\omega)d\omega = 2 \cdot 4\pi \frac{\omega^2 n_{\text{r}}^2}{c^2} \frac{dk}{d\omega} \cdot \frac{d\omega}{(2\pi)^3} = 2 \cdot 4\pi \frac{\omega^2 n_{\text{r}}^3}{c^3} \cdot \frac{d\omega}{(2\pi)^3} \tag{8.132}$$

$$D_3^{\text{opt}}(\omega)d\omega = \frac{\omega^2 n_{\text{r}}^3}{\pi^2 c^3} \cdot d\omega \tag{8.133}$$

is the mode density. We will use this density of optical modes to calculate the background photon energy density at thermal equilibrium.

Notice that the density of optical modes in a medium with refractive index $n_{\text{r}} > 1$ is larger than that of free space, where $n_{\text{r}} = 1$. The underlying reason for this is that light travels more slowly in the medium.

8.6.2   Light intensity

The Poynting vector $\mathbf{S} = \mathbf{E} \times \mathbf{H}$ (Eqn (1.94)) can be used to determine the energy flux density of a sinusoidally varying electromagnetic field. The magnitude of *average* power flux is given by

$$|\mathbf{S}_{\mathrm{av}}| = \frac{1}{2}|\mathbf{E} \times \mathbf{H}| = \frac{1}{2}|\mathbf{E}_0||\mathbf{H}_0| = \frac{1}{2}|\mathbf{E}_0| \cdot \frac{\omega \varepsilon_0}{k}|\mathbf{E}_0| = \frac{1}{2}c\varepsilon_0 \cdot |\mathbf{E}_0|^2 \qquad (8.134)$$

where the factor $1/2$ comes from taking the average. This is the average light intensity of a sinusoidally oscillating electromagnetic in field free space. It follows that the average energy density for photons in free space is

$$U(\omega) = \frac{1}{2}\varepsilon_0 \cdot |\mathbf{E}_0|^2 \qquad (8.135)$$

8.6.3   Background photon energy density at thermal equilibrium

The average value of radiation energy at frequency ω is given by the product of the density of states, the occupation factor, and the energy per photon:

$$U(\omega) = D_3^{\mathrm{opt}}(\omega) \cdot g(\omega) \cdot \hbar\omega \qquad (8.136)$$

We have already calculated $D_3^{\mathrm{opt}}(\omega)$, and the occupation factor for a system in thermal equilibrium is given by the Bose–Einstein distribution function $g(\omega)$, so

$$U(\omega) = \frac{\omega^2}{\pi^2 c^3} \cdot \frac{1}{e^{\hbar\omega/k_{\mathrm{B}}T} - 1} \cdot \hbar\omega = \frac{\hbar\omega^3}{\pi^2 c^3} \cdot \frac{1}{e^{\hbar\omega/k_{\mathrm{B}}T} - 1} \qquad (8.137)$$

$U(\omega)$ is the background radiative photon energy density per unit volume per unit frequency at thermal equilibrium. In an isotropic homogeneous medium with refractive index n_{r}, Eqn (8.137) is modified to

$$U(\omega) = \frac{\hbar\omega^3 n_{\mathrm{r}}^3}{\pi^2 c^3} \cdot \frac{1}{e^{\hbar\omega/k_{\mathrm{B}}T} - 1} \qquad (8.138)$$

The background radiative photon energy density per unit volume per unit frequency at thermal equilibrium in a dielectric medium with $n_{\mathrm{r}} > 1$ is always greater than in free space, because the density of optical modes is greater (Eqn (8.133)).

8.6.4   Fermi's golden rule for stimulated optical transitions

When deriving Fermi's golden rule we had (Eqn (8.40))

$$a_m(t) = \frac{1}{i\hbar} \int_{t'=0}^{t'=t} W_{mn} e^{i\omega_{mn}t'} dt' \qquad (8.139)$$

where $\omega_{mn} = \omega_m - \omega_n$. The simplest interaction between an atomic dipole and the oscillating electric field of a photon is by way of the dipole matrix element

$$W_{mn} = \mathbf{d}_{mn} \cdot \mathbf{E} \tag{8.140}$$

where

$$d_{mn} = e\langle m|\mathbf{r}|n\rangle = er_{mn} \tag{8.141}$$

is the dipole and

$$\mathbf{E} = \mathbf{E}_0 \cos(\omega t) = \frac{\mathbf{E}_0}{2}(e^{i\omega t} + e^{-i\omega t}) \tag{8.142}$$

is the oscillating electric field.

We wish to use Fermi's golden rule to calculate the transition rate between two states of an atomic system due to the presence of a sinusoidally oscillating electric field. For convenience, we consider an electric field in the z direction so that the dipole matrix element between state $|n\rangle$ and $|m\rangle$ changes from r_{mn} to z_{mn}. Equation (8.139) may now be written

$$a_m(t) = \frac{1}{i\hbar} \frac{e|\mathbf{E}_0|}{2} z_{mn} \int_{t'=0}^{t'=t} (e^{i\omega t'} + e^{-i\omega t'}) \cdot e^{i\omega_{mn}t'} dt' \tag{8.143}$$

$$a_m(t) = \frac{1}{\hbar} \frac{e|\mathbf{E}_0|}{2} z_{mn} \left(-\frac{e^{i(\omega+\omega_{mn})t} - 1}{\omega + \omega_{mn}} + \frac{e^{-i(\omega-\omega_{mn})t} - 1}{\omega - \omega_{mn}} \right) \tag{8.144}$$

Since $\omega + \omega_{mn} \gg \omega - \omega_{mn}$ for ω near ω_{mn}, the first term can be set to zero:

$$a_m(t) \sim \frac{1}{\hbar} \frac{e|\mathbf{E}_0|}{2} z_{mn} \left(\frac{e^{-i(\omega-\omega_{mn})t} - 1}{\omega - \omega_{mn}} \right) \tag{8.145}$$

With this approximation,

$$a_m(t) = \frac{1}{\hbar} \frac{e|\mathbf{E}_0|}{2} z_{mn} \frac{e^{-i(\omega-\omega_{mn})t/2}}{\omega - \omega_{mn}} \left(e^{-i(\omega-\omega_{mn})t/2} - e^{i(\omega-\omega_{mn})t/2} \right) \tag{8.146}$$

$$a_m(t) = \frac{i}{\hbar} e|\mathbf{E}_0|z_{mn} \frac{e^{-i(\omega-\omega_{mn})t/2}}{\omega - \omega_{mn}} \cdot \sin((\omega - \omega_{mn})t/2) \tag{8.147}$$

Hence, the probability for a transition is

$$|a_m(t)|^2 = \left(\frac{e|\mathbf{E}_0|}{\hbar} \right)^2 |z_{mn}|^2 \frac{\sin^2((\omega - \omega_{mn})t/2)}{(\omega - \omega_{mn})^2} \tag{8.148}$$

The probability of a transition to the continuum or over the complete line shape is found by integrating over all frequency ω, so that

$$|a_m(t)|^2 = \left(\frac{e|\mathbf{E}_0|}{\hbar} \right)^2 |z_{mn}|^2 \cdot \int_{-\infty}^{\infty} \frac{\sin^2((\omega - \omega_{mn})t/2)}{(\omega - \omega_{mn})^2} d\omega \tag{8.149}$$

But

$$\int_{-\infty}^{\infty} \frac{\sin^2(\alpha x)}{x^2} dx = \alpha \pi \tag{8.150}$$

So, we change variables such that $\alpha = t/2$ and $x = (\omega - \omega_{mn})$, and our expression thus becomes

$$|a_m(t)|^2 = \left(\frac{e|\mathbf{E}_0|}{\hbar}\right)^2 |z_{mn}|^2 \cdot \frac{\pi t}{2} = \frac{\pi e^2}{\varepsilon_0 \hbar^2} |z_{mn}|^2 U(\omega) \cdot t \tag{8.151}$$

where we used Eqn (8.135) to eliminate $|\mathbf{E}_0|^2$. For an isotropic energy density $U(\omega)$, averaging over the three directions of polarization and differentiating with respect to time gives the transition rate

$$\frac{d}{dt}|a_m(t)|^2 = \frac{\pi e^2}{3\varepsilon_0 \hbar^2} |z_{mn}|^2 \cdot U(\omega) = B \cdot U(\omega) \tag{8.152}$$

The factor $1/3$ comes from the average over polarization, and B is called the stimulated emission rate. The probability per unit time that an atom in state $|k\rangle$ makes a transition to any possible state $|j\rangle$ stimulated by electromagnetic radiation is

$$B = \frac{\pi e^2}{3\varepsilon_0 \hbar^2} |\langle j|\mathbf{r}|k\rangle|^2 \tag{8.153}$$

8.6.5   The Einstein A and B coefficients

We already know that electromagnetic radiation can stimulate transitions between electronic states. In addition to stimulated transitions, spontaneous transitions from a high-energy state to a lower-energy state are also possible. The existence of spontaneous emission from an excited state is *required* as a mechanism to drive the system back to thermal equilibrium. Einstein was able to show that, for a system in *thermal equilibrium*, stimulated and spontaneous transition rates are related to each other.[3]

Consider the two-energy-level atom system illustrated in Fig. 8.19 in which optical transitions take place between states $|1\rangle$ and $|2\rangle$. Under conditions of thermal equilibrium, the rate of transition from $|2\rangle$ to $|1\rangle$ must equal that from $|1\rangle$ to $|2\rangle$, so that

$$N_2(B_{21} \cdot U(\omega) + A) = N_1 B_{12} \cdot U(\omega) \tag{8.154}$$

where we have introduced a spontaneous transition rate A. Spontaneous transitions occur from $|2\rangle$ to $|1\rangle$ due to vacuum fluctuations in photon density (see Section 6.5).

---

[3] A. Einstein, *Phys. Z.* **18**, 121 (1917).

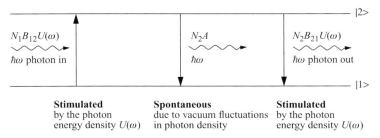

Fig. 8.19. Energy level diagram of a two-level atom system.

In thermal equilibrium, the ratio of levels is given by a Boltzmann factor, so we may write

$$\frac{B_{21}}{B_{12}} + \frac{A}{B_{12} \cdot U(\omega)} = \frac{N_1}{N_2} = e^{\hbar\omega/k_B T} \tag{8.155}$$

Substituting our expression

$$U(\omega) = \frac{\hbar\omega^3}{\pi^2 c^3} \cdot \frac{1}{e^{\hbar\omega/k_B T} - 1}$$

for black-body radiation gives

$$\frac{B_{21}}{B_{12}} + \frac{A}{B_{12}} \cdot \frac{\pi^2 c^3}{\hbar\omega^3} \cdot (e^{\hbar\omega/k_B T} - 1) = e^{\hbar\omega/k_B T} \tag{8.156}$$

which can only be true for any temperature T if $(T \to 0)$

$$\frac{A}{B_{12}} \cdot \frac{\pi^2 c^3}{\hbar\omega^3} = 1 \tag{8.157}$$

and $(T \to \infty)$

$$\frac{B_{21}}{B_{12}} = 1 \tag{8.158}$$

Hence, we obtain the *Einstein relations*

$$\boxed{\begin{aligned} A &= \frac{\hbar\omega^3}{\pi^2 c^3} \cdot B_{12} \\ B_{12} &= B_{21} \end{aligned}} \tag{8.159}$$

The stimulated and spontaneous transition rates are related to each other. Using Eqn (8.153) for the stimulated emission rate $1/\tau_{\text{stim}} = B$, we have

$$B = \frac{\pi e^2}{3\varepsilon_0 \hbar^2} |\langle j | \mathbf{r} | k \rangle|^2 \tag{8.160}$$

and it follows that the spontaneous emission rate $1/\tau_{\text{sp}} = A$ is just

$$A = \frac{e^2 \omega^3}{3\pi \varepsilon_0 \hbar c^3} |\langle j | \mathbf{r} | k \rangle|^2 \tag{8.161}$$

It can be shown (see Exercise 8.9) that since the time dependence of spontaneous light emission intensity from a number of excited atoms is $I(t) = I(t = 0)e^{-At} = I(0)e^{-t/\tau_{sp}}$, the associated spectral line has a Lorentzian line shape with FWHM $1/\tau_{sp} = A$ measured in units of rad s^{-1}.

If light emission occurs in an isotropic, homogeneous medium characterized by refractive index $n_r > 1$, then the density of optical modes contributing to $U(\omega)$ (Eqn (8.138)) increases by a factor n_r^3 and Eqn (8.35) is modified to

$$A = \frac{e^2 \omega^3 n_r^3}{3\pi \varepsilon_0 \hbar c^3} |\langle j|\mathbf{r}|k\rangle|^2 \tag{8.162}$$

8.6.5.1 Estimation of the spontaneous emission coefficient A for the hydrogen $|2p\rangle \rightarrow |1s\rangle$ transition

As a first application of quantum mechanical spontaneous emission, consider a hydrogen atom. If a hydrogen atom in free space is in an excited state with $n = 2$, then quantum mechanics predicts the atom will relax to the $n = 1$ ground state by spontaneous emission of a photon. This physical process limits the excited-state lifetime on average to a time characterized by the spontaneous emission lifetime $\tau_{sp} = 1/A$. To estimate this value, we start with our expression for the spontaneous emission coefficient given by Eqn (8.161). For the $n = 2$ to $n = 1$ transition the emission wavelength is $\lambda_{photon} = 122$ nm (Eqn (2.75)), and we calculate optical frequency using $ck = \omega = c2\pi/\lambda_{photon}$. The dipole matrix element can be estimated as $\langle j|\mathbf{r}|k\rangle \sim a_B = 0.053$ nm, where a_B is the Bohr radius of the electron in a hydrogen atom (Eqn (2.70)). Putting in the numbers gives

$$A = \frac{(2\pi)^3}{\lambda_{photon}^3} \cdot \frac{e^2}{3\pi \varepsilon_0 \hbar} \cdot a_B^2 = 1.12 \times 10^9 \text{ s}^{-1} = \frac{1}{\tau_{sp}} \tag{8.163}$$

Hence, an estimate for the spontaneous emission time is $\tau_{sp} = 0.89$ ns. A more detailed calculation gives $\langle j|\mathbf{r}|k\rangle \approx 1.12 \cdot a_B$ for the $|2p\rangle \rightarrow |1s\rangle$ transition in hydrogen, so that $\tau_{sp} = 0.71$ ns.

The electromagnetic wave produced by the transition takes energy from the excited state and converts it to electromagnetic energy. Typically, the electromagnetic field intensity decays as $e^{-t/\tau_{sp}}$, so that the *length* of a photon when $\tau_{sp} \sim 1$ ns is about 0.3 m.

As the electron makes its transition, the superposition of the $|2p\rangle$ and $|1s\rangle$ states causes the hydrogen electron probability density cloud to oscillate at difference energy $\hbar\omega = \Delta E = E_2 - E_1 = 10.2$ eV. The oscillation in expectation value of electron position creates a dipole moment, and electromagnetic radiation is emitted at wavelength $\lambda_{photon} = 2\pi c/\omega = 0.122$ μm, carrying away angular momentum of magnitude $\pm\hbar$ (quantum number ± 1).

8.6.5.2 Dipole selection rules for optical transitions

The dipole matrix element $d_{jk} = e\langle j|\mathbf{r}|k\rangle$ (Eqn (8.141)) gives rise to a set of rules for optical transitions at frequency ω between initial eigenstate $|k\rangle$ and final eigenstate $|j\rangle$. Dipole radiation requires a parity difference (even-to-odd or odd-to-even) between initial and final states to ensure oscillation in the mean position of charge. Without oscillation in the mean position of charge, there can be no dipole radiation. Clearly, the dipole matrix element $\langle even(odd)|r|odd(even)\rangle \neq 0$, whereas $\langle even(odd)|r|even(odd)\rangle = 0$ from symmetry. Hence, for quantum numbers that sequentially alternate between odd and even parity we expect $\langle j|r|k\rangle \neq 0$ for $j - k = odd$. This type of condition is often called a dipole selection rule. Other rules also apply. For example, energy conservation requires that the separation in energy between initial and final states is the energy of the photon, $\hbar\omega$.

To illustrate a practical application of these ideas, we will now use dipole selection rules to calculate the spontaneous emission lifetime for an excited state of an electron in a one-dimensional, infinite, rectangular potential well.

8.6.5.3 Spontaneous emission lifetime of an electron in a one-dimensional, rectangular potential well with infinite barrier energy

As part of the design of a laser, we wish to calculate the spontaneous emission lifetime of the first excited state for an electron confined to a one-dimensional, infinite, rectangular potential well of width $L = 12.3$ nm. Figure 8.20(a) is a sketch of a one-dimensional, rectangular potential well with infinite barrier energy showing energy eigenvalues E_1, E_2, and E_3. Figure 8.20(b) sketches the first three energy eigenfunctions

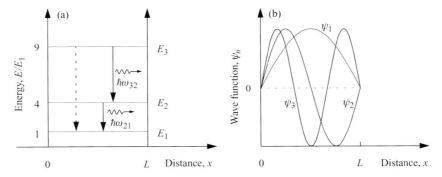

Fig. 8.20. (a) Sketch of a one-dimensional, rectangular potential well with infinite barrier energy showing the energy eigenvalues E_1, E_2, and E_3. Spontaneous emission of a photon can occur between two states when a dipole matrix element exists between the two states. This is the case for $\psi_2 \rightarrow \psi_1$ and $\psi_3 \rightarrow \psi_2$. No dipole radiation can occur between the second excited state ψ_3 and the ground state ψ_1, because the dipole matrix element is zero between states of the same parity. (b) Sketch of the energy eigenfunctions ψ_1, ψ_2, and ψ_3 for the potential shown in (a).

ψ_1, ψ_2, and ψ_3 for the potential shown in Fig. 8.20(a). In general, the eigenstates are $|j\rangle = A\sin(\pi jx/L)$, where j is a nonzero positive integer. Normalization requires $\langle j|j\rangle = 1$ giving $A = \sqrt{2/L}$. For j odd $|j\rangle$ is of even parity, and for j even $|j\rangle$ is of odd parity.

To help find a quick solution, we note that the de Broglie wavelength λ_e of an electron measured in nanometers is related to the electron energy E measured in electron volts through the relation

$$\lambda_e(\text{nm}) = \frac{1.23}{(E(\text{eV}))^{1/2}} \tag{8.164}$$

and that the wavelength of a photon λ_{photon} measured in micrometers is related to its energy E measured in electron volts through

$$\lambda_{\text{photon}}(\mu\text{m}) = \frac{1.24}{E(\text{eV})} \tag{8.165}$$

The electron wavelength of the ground state in this potential is $2L = 24.6$ nm, and the wavelength of the first excited state is $L = 12.3$ nm. This gives

$$E_1 = \left(\frac{1.23}{2L}\right)^2 = \left(\frac{1}{20}\right)^2 = 2.5 \text{ meV} \tag{8.166}$$

$$E_2 = \left(\frac{1.23}{2L}\right)^2 = \left(\frac{1}{10}\right)^2 = 10 \text{ meV} \tag{8.167}$$

Hence, the wavelength of the emitted photon is just

$$\lambda_{\text{photon}} = \frac{1.24}{E_2 - E_1}\mu\text{m} = \frac{1.24}{7.5 \times 10^{-3}}\mu\text{m} = 165.3 \ \mu\text{m} \tag{8.168}$$

We now calculate the dipole matrix element x_{jk} between two electron states $|j\rangle$ and $|k\rangle$.

$$x_{jk} = \langle j|x|k\rangle = \frac{2}{L}\int_0^L x\sin(j\pi x/L)\sin(k\pi x/L)dx \tag{8.169}$$

$$x_{jk} = \frac{1}{L}\int_0^L x(\cos((j-k)\pi x/L) - \cos((j+k)\pi x/L))dx \tag{8.170}$$

where we used the fact that $2\sin(x)\sin(y) = \cos(x-y) - \cos(x+y)$. Solving the integral by parts and introducing the integer parameter m, we have

$$\int_0^L x \cos(m\pi x/L)dx = \left[\frac{L}{m\pi}x \sin\left(\frac{m\pi x}{L}\right)\right]_0^L - \frac{L}{m\pi}\int_0^L \sin\left(\frac{m\pi x}{L}\right)dx$$

$$= \left[\left(\frac{L}{m\pi}\right)^2 \cos\left(\frac{m\pi x}{L}\right)\right]_0^L \tag{8.171}$$

For even values of m, this gives

$$\int_0^L x \cos(m\pi x/L)dx = 0 \tag{8.172}$$

and for odd values of m

$$\int_0^L x \cos(m\pi x/L)dx = -\frac{2L^2}{m^2\pi^2} \tag{8.173}$$

When $m = 0$, the integral is

$$\int_0^L x \cos(m\pi x/L)dx = \frac{L^2}{2} \tag{8.174}$$

Since j and k are integers, m is even if $|j\rangle$ and $|k\rangle$ have the same parity, m is odd if $|j\rangle$ and $|k\rangle$ are of different parity, and $m = 0$ if $j = k$.

In the situation we are considering, the state functions with odd integer values are of even parity and the state functions with even integer values are of odd parity. Hence, we may conclude the dipole matrix element

$$x_{jk} = 0 \tag{8.175}$$

for $|j\rangle$ and $|k\rangle$ of the same parity,

$$x_{jk} = -\frac{1}{L}\frac{2L^2}{\pi^2}\left(\frac{1}{(j-k)^2} - \frac{1}{(j+k)^2}\right) = \frac{-2L}{\pi^2}\left(\frac{(j+k)^2 - (j-k)^2}{(j-k)^2(j+k)^2}\right)$$

$$= \frac{-2L}{\pi^2}\frac{4jk}{(j^2-k^2)^2} \tag{8.176}$$

for $|j\rangle$ and $|k\rangle$ of different parity, and

$$x_{jk} = \frac{1}{L}\frac{2L^2}{\pi^2} = \frac{L}{2} \tag{8.177}$$

for $j = k$. As shown in Fig. 8.20(a), electron transitions involving dipole radiation of light are allowed from the first excited state to the ground state ($\psi_2 \rightarrow \psi_1$) and from the second excited state to the first excited state ($\psi_3 \rightarrow \psi_2$). However, dipole radiation from the second excited state to the ground state ($\psi_3 \rightarrow \psi_1$) does not occur, because the dipole matrix element is zero between states of the same parity.

For the transition between the first excited state $|k = 2\rangle$ and the ground state $|j = 1\rangle$, the states $|k\rangle$ and $|j\rangle$ are of different parity and so, using Eqn (8.176), we have the dipole matrix element

$$x_{12} = \frac{-16L}{9\pi^2} \tag{8.178}$$

The Einstein spontaneous emission rate given by Eqn (8.161) can be rewritten for light of wavelength λ_{photon} with units of length in nanometers and x_{12} in units of nanometers as

$$A = \frac{7.235 \times 10^{17}}{\lambda_{\text{photon}}^3} |\langle j|r|k\rangle|^2 \tag{8.179}$$

Hence, the spontaneous emission lifetime for the transition between $|2\rangle$ and $|1\rangle$ is given by

$$\tau_{\text{sp}} = \frac{1}{A} = \frac{\lambda_{\text{photon}}^3}{7.235 \times 10^{17}} \frac{1}{|x_{12}|^2} = \frac{(1.65 \times 10^5)^3}{7.235 \times 10^{17}} \cdot \frac{81\pi^4}{256 \times (12.3)^2} = 1.26 \times 10^{-3} \text{ s} \tag{8.180}$$

The frequency spectral line width is just $A = 1/\tau_{\text{sp}}$ measured in units of rad s^{-1}.

Our results would be modified if we were to consider the same transition for an electron in the conduction band of GaAs confined by a one-dimensional quantum well potential of the same width L. In this case, the conduction-band electron has a low effective electron mass of $m_e^* = 0.07m_0$, the effect of which is to increase the separation in electron energy levels by a factor of 14.3 and shorten the emission wavelength to 11.5 μm compared with our calculation for a bare electron. The Einstein spontaneous emission rate is also increased by a factor n_r^3, where n_r is the refractive index of light in the semiconductor at the emission wavelength (Eqn (8.162)). For emission wavelengths near 10 μm, the bulk refractive index for GaAs is $n_r \sim 3.3$, allowing us to estimate this contribution to an increase in spontaneous emission rate of $n_r^3 \sim 36$. The combination of reduced emission wavelength and the presence of a medium with refractive index n_r changes the spontaneous emission lifetime in Eqn (8.180) by a factor $(m_e^*/n_r m_0)^3 = (0.07/3.3)^3 = 9.5 \times 10^{-6}$, so that the spontaneous emission lifetime in the conduction band of a GaAs rectangular potential well of width $L = 12.3$ nm for wavelength $\lambda_{\text{photon}} = 11.5$ μm is $\tau_{\text{sp}} = 1.2 \times 10^{-8}$ s.

The design of a device that has efficient conversion of excited electron states to photons requires placing electrons into ψ_2 states. One way to achieve this is to use tunnel injection. Figure 8.21 shows a cross-section of a GaAs/AlGaAs conduction-band

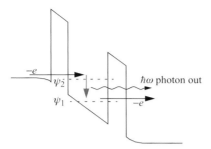

Fig. 8.21. Schematic cross-section of a GaAs/AlGaAs conduction-band profile with electron tunnel injection into ψ_2 states and optical transition to ψ_1 states.

profile with just such a tunnel injector. These ideas have been pursued to create a semiconductor laser with emission at wavelength in the $10\,\mu$m range.[4]

8.7 Example exercises

Exercise 8.1

(a) A particle of mass m is in a one-dimensional, rectangular potential well for which $V(x) = 0$ for $0 < x < L$ and $V(x) = \infty$ elsewhere. The particle is initially prepared in the ground state ψ_1 with eigenenergy E_1. Then, at time $t = 0$, the potential is very rapidly changed so that the original wave function remains the same but $V(x) = 0$ for $0 < x < 2L$ and $V(x) = \infty$ elsewhere. Find the probabilities that the particle is in the first, second, third, and fourth excited states of the system when $t \geq 0$.

(b) Consider the same situation as (a) but for the case in which at time $t = 0$ the potential is very rapidly changed so that the original wave function remains the same but $V(x) = 0$ for $0 < x < \gamma L$, where $1 < \gamma < 5$, and $V(x) = \infty$ elsewhere. Write a computer program that plots the probability of finding the particle in the ground, first, second, third and fourth excited states of the system as a function of the parameter γ.

Exercise 8.2

The coulomb potential energy in real space is

$$V(r) = \frac{-e^2}{4\pi\varepsilon_0\varepsilon_{r0}r}$$

where ε_{r0} is the low-frequency dielectric constant of the isotropic medium. By taking the Fourier transform of $V(r)$, show that the coulomb potential energy in wave-vector

---

[4] For an introduction to the quantum cascade laser, see J. Faist, F. Capasso, D. L. Sivco, C. Sirtori, A. L. Hutchinoson, and A. Y. Cho, *Science* **264**, 553 (1994).

space is

$$V(q) = \frac{-e^2}{\varepsilon_0 \varepsilon_{r0} q^2}$$

Exercise 8.3

Show that the structure factor is given by

$$s(\mathbf{q}) = \left| \sum_{j=1}^{n} e^{-i\mathbf{q}\cdot\mathbf{R}_j} \right|^2 = n$$

for a density of n sites at random positions $\mathbf{R}_j$.

Exercise 8.4

Starting from the expression for total elastic scattering rate of a particle of mass m and charge e from a density of n ionized impurities given by

$$\frac{1}{\tau_{el}} = \frac{2\pi}{\hbar} \cdot n \int \frac{d^3 q}{(2\pi)^3} \left| \frac{e^2}{\varepsilon(\mathbf{q})q^2} \right|^2 \delta(E(\mathbf{k}) - E(\mathbf{k} - \mathbf{q}))$$

show that this may be rewritten as

$$\frac{1}{\tau_{el}} = \frac{2\pi m}{\hbar^3 k^3} \cdot n \cdot \left(\frac{e^2}{4\pi \varepsilon_0} \right)^2 \cdot \int_{\eta=0}^{\eta=1} \frac{d\eta}{(\varepsilon_r(q))^2 \eta^3}$$

for an isotropic semiconductor with a density n randomly positioned ionized impurities. In this expression $\varepsilon(q) = \varepsilon_0 \varepsilon_r(q)$ and $\eta = \sin(\theta/2)$, where θ is the angle between the initial wave vector $\mathbf{k}$ and final wave vector $(\mathbf{k} - \mathbf{q})$ of the charged particle. Assume that the kinetic energy of the particle is given by $E = \hbar^2 k^2 / 2m$.

Exercise 8.5

Given that the screened coulomb potential energy in an isotropic medium is

$$V_q(r) = \frac{-e^2}{4\pi \varepsilon_0 \varepsilon_{r0} r} \cdot e^{-r/r_0}$$

Calculate $V(q)$ by taking the Fourier transform, and show that

$$V(q) = \frac{-e^2}{\varepsilon_0 \varepsilon_{r0} \left(q^2 + 1/r_0^2 \right)}$$

Exercies 8.6

Analyze the influence on current flow (for example mobility) of conduction-band electrons scattering elastically off ionized impurities in an n-doped semiconductor crystal when:

(a) The temperature of the crystal is increased. What do you expect to happen to the electron mobility as a function of temperature?

(b) The positions of n ionized impurities in a single crystal plane are correlated. Consider the case of spatial correlations arising from a net repulsion between ionized impurity sites occurring during crystal growth and the case of net attraction (resulting in spatial clustering of ionized impurities in the crystal).

Exercise 8.7

Write a computer program using f77, c++, MATLAB or similar software to calculate the total elastic scattering rate $1/\tau_{el}(E)$ of a single conduction band electron from randomly positioned ionized impurities in an n-doped semiconductor in the low-temperature limit. As a concrete example, use the parameters for GaAs (a) with an impurity concentration $n = 1 \times 10^{18}$ cm^{-3} and (b) with an impurity concentration $n = 1 \times 10^{14}$ cm^{-3}. The effective electron mass near the conduction-band minimum of GaAs is $m_e^* = 0.07 \times m_0$ (where m_0 is the bare electron mass), and the low-frequency dielectric constant is $\varepsilon_{r0} = 13.2$. Use the Thomas–Fermi dielectric function, and compare your results with the RPA dielectric function, which is

$$\varepsilon_r(q) = \varepsilon_{r0} + \frac{r_s^0}{x^3}\xi\left(x + \left(1 - \frac{x^2}{4}\right)\ln\left|\frac{x+2}{x-2}\right|\right)$$

where

$$r_s^0 = \left(\frac{3}{4\pi n}\right)^{1/3}\left(\frac{m_e^* e^2}{4\pi\varepsilon_0\hbar^2}\right)$$

$$\xi = \frac{1}{\pi^2}\left(\frac{32\pi^2}{9}\right)^{1/3}$$

$$x = \frac{q}{k_F}$$

In the equation, q is the scattered wave number and $k_F = (3\pi^2 n)^{1/3}$ is the Fermi wave number. Calculate the scattering rate for electron energies $E = 300$ meV and $E = 100$ meV above the conduction-band minimum as a function of scattering angle θ for both cases and discuss the significance and meaning of your results. What are the changes to both your results and interpretation when you weight the angular integral by a factor $(1 - \cos(\theta))$?

Exercise 8.8

A time-varying Hamiltonian $H(t')$ induces transitions from state $|k\rangle$ at time $t' = 0$ to a state $|j\rangle$ at time $t' = t$, with probability $P_{k \to j}(t)$. Use first-order time-dependent perturbation theory to show that if $P_{j \to k}(t)$ is the probability that the same Hamiltonian brings about the transition from state $|j\rangle$ to state $|k\rangle$ in the same time interval, then $P_{k \to j}(t) = P_{j \to k}(t)$.

Exercise 8.9

A single excited state of an atom decays radiatively to the ground state. Derive the time evolution of radiated power, $P(t)$ for N_0 atoms. Show that $P(t) = N_0 \hbar \omega_0 A e^{-\gamma t}$, where $\hbar \omega_0$ is the average photon energy and $1/\gamma$ is the radiative lifetime. The electric field is $\mathbf{E}(t) = e E_0 e^{-\gamma t/2} \cos(\omega_0 t)$ for $t \geq 0$ and $\mathbf{E}(t) = 0$ for $t < 0$. Derive the expression for the homogeneously broadened spectral intensity, $|\mathbf{S}(\omega)|$. In the limit $\gamma \ll \omega_0$, find an expression for $|\mathbf{S}(\omega)|$ near $\omega = \omega_0$. If there are different isotopes of the atom in the gas or Doppler shifts, how do you expect the appearance of the line shape, $|\mathbf{S}(\omega)|$, to change?

Exercise 8.10

(a) What determines the selection rules for optical transitions at frequency ω between states $|k\rangle$ and $|j\rangle$?

(b) Show that the inverse of the Einstein spontaneous emission coefficient, τ,

$$\frac{1}{A} = \frac{3\pi \varepsilon_0 \hbar c^3}{e^2 \omega^3 |\langle j|r|k\rangle|^2} = \tau$$

can be rewritten for light emission of wavelength λ from electronic transitions in a harmonic oscillator potential as

$$\frac{1}{A} = 45 \times \lambda^2(\mu m) = \tau(ns)$$

where wavelength is measured in micrometers and time τ is measured in nanoseconds.

(c) Calculate the spontaneous emission lifetime and spectral line width for an electron making a transition from the first excited state to the ground state of a harmonic oscillator potential characterized by force constant $\kappa = 3.59 \times 10^{-3} \text{ kg s}^{-2}$.

SOLUTIONS

Solution 8.1

(a) Following the solution given at the beginning of this chapter, the probability of finding the particle in an excited state when $t \geq 0$ is given by the square of the overlap integral between the ground state ψ_1 when $t < 0$ and the excited state when $t \geq 0$. After the change in potential the new state ψ is not an eigenfunction of the system. The new eigenfunctions of a rectangular potential well of width $2L$ with infinite barrier energy are $\psi_m = \sqrt{1/L} \sin(k_m x)$, where $k_m = m\pi/2L$ for $m = 1, 2, 3, \ldots$.

Since the state ψ is not an eigenfunction, it may be expressed as a sum of the new eigenfunctions, so that

$$\psi = \sum_m a_m \psi_m$$

The coefficients a_m are found by multiplying both sides by ψ_m^* and integrating over all space. This overlap integral gives the coefficients

$$a_m = \int \psi_m^* \psi \, dx$$

The effect of the overlap integral is to project out the components of the new eigenstates that contribute to the wave function ψ. The value of $|a_m|^2$ is the probability of finding the particle in the eigenstate ψ_m.

The probability that the particle is in the new first excited state when $t \geq 0$ is given by the square of the overlap integral, a_2:

$$a_2 = \langle \psi_{m=2} | \psi \rangle = \int_{x=0}^{x=L} \sqrt{\frac{1}{L}} \sin\left(\frac{2\pi x'}{2L}\right) \sqrt{\frac{2}{L}} \sin\left(\frac{\pi x'}{L}\right) dx'$$

$$a_2 = \frac{\sqrt{2}}{L} \int_{x=0}^{x=L} \left(\frac{1}{2} - \frac{1}{2}\cos\left(\frac{2\pi x'}{L}\right)\right) dx' = \frac{\sqrt{2}}{L} \left[\frac{x}{2} + \frac{1}{2} \cdot \frac{L}{2\pi} \sin\left(\frac{2\pi x'}{L}\right)\right]_0^L$$

$$a_2 = \frac{\sqrt{2}}{L}\left(\frac{L}{2} + 0\right) = \frac{1}{\sqrt{2}}$$

The probability of finding the particle in the first excited state $|\psi_m = 2\rangle$ is

$$|a_2|^2 = \frac{1}{2} = 0.50$$

We can go on to find the probability of excitation of other states. The next few are

$$|a_3|^2 = \frac{32}{25\pi^2} = 0.129\,69$$

$$|a_4|^2 = 0.0$$

$$|a_5|^2 = \frac{32}{21^2\pi^2} = 0.007\,352\,1$$

That $|a_2|^2$ has the highest probability and $|a_4|^2$ is zero is a direct consequence of symmetry imposed by the fact that in the problem the width of the potential well exactly doubled.

(b) In this part of the exercise, the new eigenfunctions of a rectangular potential well of width γL with infinite barrier energy are $\psi_m = \sqrt{2/\gamma L} \sin(k_m x)$, where $k_m = m\pi/\gamma L$ for $m = 1, 2, 3, \ldots$. We will consider the situation in which γ takes on values $1 < \gamma \leq 5$.

As in (a), the probability of finding the particle in the eigenstate ψ_m when $t \geq 0$ is given by the square of the overlap integral, a_m:

$$a_m = \langle \psi_m | \psi \rangle = \frac{2}{L}\sqrt{\frac{1}{\gamma}} \int_{x=0}^{x=L} \sin\left(\frac{m\pi x'}{\gamma L}\right) \sin\left(\frac{\pi x'}{L}\right) dx'$$

Using $2\sin(x)\sin(y) = \cos(x-y) - \cos(x+y)$ to rewrite the intergrand gives

$$a_m = \frac{1}{L}\sqrt{\frac{1}{\gamma}} \int\limits_{x=0}^{x=L} \cos\left(\left(\frac{m}{\gamma} - 1\right)\pi x'/L\right) - \cos\left(\left(\frac{m}{\gamma} + 1\right)\pi x'/L\right) dx'$$

$$a_m = \sqrt{\frac{1}{\gamma}}\left(\frac{\sin\left(\left(\frac{m}{\gamma} - 1\right)\pi\right)}{\left(\frac{m}{\gamma} - 1\right)\pi} - \frac{\sin\left(\left(\frac{m}{\gamma} + 1\right)\pi\right)}{\left(\frac{m}{\gamma} + 1\right)\pi}\right)$$

$$a_m = \sqrt{\frac{1}{\gamma}}\left(\mathrm{sinc}\left(\left(\frac{m}{\gamma} - 1\right)\pi\right) - \mathrm{sinc}\left(\left(\frac{m}{\gamma} + 1\right)\pi\right)\right)$$

Hence, the probability of finding the particle in the state $|\psi_m\rangle$ is just

$$|a_m|^2 = \frac{1}{\gamma}\left(\mathrm{sinc}\left(\left(\frac{m}{\gamma} - 1\right)\pi\right) - \mathrm{sinc}\left(\left(\frac{m}{\gamma} + 1\right)\pi\right)\right)^2$$

See the following figure.

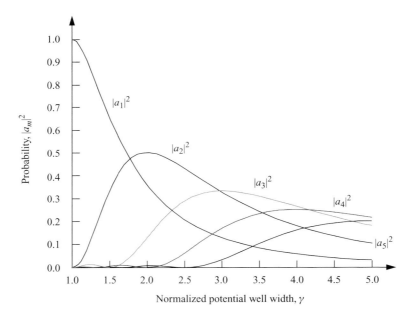

Solution 8.2

The bare coulomb potential energy in real space is

$$v(r) = \frac{-e^2}{4\pi\varepsilon_0 r}$$

where e is electron charge. To obtain the potential energy in wave-vector space, one may take the Fourier transform

$$v(q) = \int d^3 r \; v(r) \; e^{-i\mathbf{q}\cdot\mathbf{r}} = \frac{-e^2}{4\pi\varepsilon_0} \int d^3 r \frac{1}{r} \; e^{-i\mathbf{q}\cdot\mathbf{r}}$$

$$v(q) = \frac{-e^2}{4\pi\varepsilon_0} \int\limits_{r=0}^{\infty} \int\limits_{\theta=0}^{\pi} \frac{1}{r} \; e^{-iqr\cos\theta} 2\pi r^2 \sin\theta d\theta dr$$

To perform the integral over θ, it is helpful to make the change of variable $\cos\theta = x$, giving $dx = -\sin\theta d\theta$. The integral over θ becomes

$$\int\limits_{\theta=0}^{\pi} e^{-iqr\cos\theta} \sin\theta d\theta = + \int\limits_{x=-1}^{x=1} e^{-iqrx} dx = \frac{1}{iqr}(e^{iqr} - e^{-iqr}) = \frac{2\sin(qr)}{qr}$$

Substituting this result into our expression for $v(q)$ gives

$$v(q) = \frac{-e^2}{4\pi\varepsilon_0} \int\limits_{r=0}^{\infty} \frac{1}{r} 2\pi r^2 \frac{2\sin(qr)}{qr} dr = \frac{-e^2}{4\pi\varepsilon_0} \frac{4\pi}{q} \int\limits_{r=0}^{\infty} \sin(qr) dr$$

$$v(q) = \frac{-e^2}{\varepsilon_0 q} \left(\frac{-\cos(\infty)}{q} + \frac{\cos(0)}{q} \right)$$

and, hence,

$$v(q) = \frac{-e^2}{\varepsilon_0 q^2}$$

since $\cos(\infty) \equiv 0$.

Solution 8.3

The structure factor for an impurity density n of ions at positions $\mathbf{R}_j$ may be written as

$$s(\mathbf{q}) = \left| \sum_{j=1}^{n} e^{-i\mathbf{q}\cdot\mathbf{R}_j} \right|^2 = \sum_{j=1}^{n} e^{-i\mathbf{q}\cdot\mathbf{R}_j} \sum_{k=1}^{n} e^{i\mathbf{q}\cdot\mathbf{R}_k} = \sum_{j=1}^{n} 1 + \sum_{j\neq k}^{n} e^{-i\mathbf{q}\cdot(\mathbf{R}_j - \mathbf{R}_k)}$$

The second term on the right-hand side is a pair correlation that can be written in terms of sine and cosine functions.

$$s(\mathbf{q}) = n + \sum_{j\neq k}^{n} (\cos(\mathbf{q} \cdot (\mathbf{R}_j - \mathbf{R}_k)) + i \sin(\mathbf{q} \cdot (\mathbf{R}_j - \mathbf{R}_k)))$$

For a *large number* of impurities *randomly positioned* on lattice sites, the average value of the sine and cosine terms is zero, and we may write

$$s(\mathbf{q}) = n$$

It is important to remember that in small systems containing relatively few *randomly positioned* impurity sites the average over the sine and cosine terms will not become zero. In this case, there will always be some pair correlation contributing to the structure factor.

Solution 8.4

It is given that the total elastic scattering rate of an electron of mass m and charge e from a density of n ionized impurities is

$$\frac{1}{\tau_{\text{el}}} = \frac{2\pi}{\hbar} \cdot n \int \frac{d^3q}{(2\pi)^3} \left| \frac{e^2}{\varepsilon(\mathbf{q})q^2} \right|^2 \delta(E(\mathbf{k}) - E(\mathbf{k} - \mathbf{q}))$$

We wish to evaluate the total elastic scattering rate of an electron with energy E, wave vector k, and mass m, in the energy interval dE. To do this, we need to express the volume element d^3q in terms of energy E and scattering angle θ. A good way to proceed is to remind ourselves that we are working in three-dimensional k-space and then draw a diagram of the volume element:

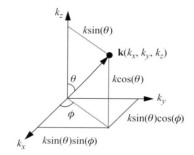

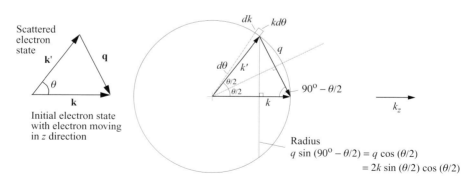

For elastic scattering in an isotropic material $q = 2k \sin(\theta/2)$. The energy of the electron is $E = \hbar^2 k^2/2m$, and momentum conservation requires $k = k' + q$. The volume element d^3q is

$$d^3q = k \sin(\theta)d\phi \cdot kd\theta \cdot dk$$

Noting that $\sin(\theta) = 2 \sin(\theta/2) \cos(\theta/2)$, we can write

$$d^3q = 2k \cdot \sin(\theta/2) \cos(\theta/2)d\phi \cdot kd\theta \cdot dk$$

Over the infinitesimal energy dE, this becomes

$$d^3q = 2k \cdot \sin(\theta/2)\cos(\theta/2)d\phi \cdot k d\theta \cdot \left(\frac{dk}{dE}\right) dE$$

Since energy $E = \hbar^2 k^2/2m$, we have $dk/dE = m/\hbar^2 k$. Substituting into our expression for d^3q and integrating over ϕ gives

$$4\pi \cdot \frac{km}{\hbar^2} \cdot \sin(\theta/2)\cos(\theta/2)d\theta dE$$

Substituting this into the equation for total elastic scattering rate gives

$$\frac{1}{\tau_{el}} = \frac{2\pi}{\hbar} \cdot n \cdot \int_{\theta=0}^{\theta=\pi} \frac{4\pi d\theta}{(2\pi)^3} \cdot \left| \frac{e^2}{\varepsilon(q)4k^2\sin^2(\theta/2)} \right|^2 \cdot \frac{km}{\hbar^2} \cdot \sin(\theta/2)\cos(\theta/2)$$

$$\frac{1}{\tau_{el}} = \frac{\pi m}{\hbar^3 k^3} \cdot n \cdot \frac{e^4}{16\pi^2} \cdot \int_{\theta=0}^{\theta=\pi} d\theta \cdot \frac{\sin(\theta/2)\cos(\theta/2)}{\varepsilon(q)^2 \sin^4(\theta/2)}$$

so that

$$\frac{1}{\tau_{el}} = \frac{2\pi m}{\hbar^3 k^3} \cdot n \cdot \left(\frac{e^2}{4\pi\varepsilon_0}\right)^2 \cdot \int_{\theta=0}^{\theta=\pi} d(\theta/2) \cdot \frac{\cos(\theta/2)}{\varepsilon_r(q)^2 \sin^3(\theta/2)}$$

where we have used the fact $\varepsilon(q) = \varepsilon_0 \varepsilon_r(q)$. We now let $\eta = \sin(\theta/2)$, so that $d\eta = \cos(\theta/2)d(\theta/2)$. This allows us to write

$$\frac{1}{\tau_{el}} = \frac{2\pi m}{\hbar^3 k^3} \cdot n \cdot \left(\frac{e^2}{4\pi\varepsilon_0}\right)^2 \cdot \int_{\eta=0}^{\eta=1} \frac{d\eta}{(\varepsilon_r(q))^2 \eta^3}$$

which is the same as Eqn (8.84).

Since $k = (2mE/\hbar^2)^{1/2}$, we see that

$$\frac{1}{\tau_{el}}(E) = \frac{2\pi m}{\hbar^3} \cdot n \cdot \left(\frac{e^2}{4\pi\varepsilon_0}\right)^2 \cdot \left(\frac{\hbar^2}{2mE}\right)^{3/2} \cdot \int_{\eta=0}^{\eta=1} \frac{d\eta}{(\varepsilon_r(2k\eta))^2 \eta^3}$$

which may be written as

$$\frac{1}{\tau_{el}}(E) = \frac{\pi}{(2m)^{1/2}} \cdot n \cdot \left(\frac{e^2}{4\pi\varepsilon_0}\right)^2 \cdot E^{-3/2} \int_{\eta=0}^{\eta=1} \frac{d\eta}{(\varepsilon_r(2k\eta))^2 \eta^3}$$

This is just Eqn (8.85).

Solution 8.5

It is given that the screened coulomb potential energy is

$$V_q(r) = \frac{-e^2}{4\pi\varepsilon_0\varepsilon_{r0}r} \cdot e^{-r/r_0}$$

Notice that the subscript q appearing in $V_q(r)$ indicates that the potential in wave-vector space is q-dependent. To calculate the potential energy in wave-vector space, we take the Fourier transform for the real-space potential energy. This gives

$$V(q) = \int d^3r \; V(r) \, e^{-i\mathbf{q}\cdot\mathbf{r}}$$

$$V(q) = \int_{\phi=0}^{\phi=2\pi} d\phi \int_{\theta=0}^{\theta=\pi} d\theta \; \sin(\theta) \int_{r=0}^{r=\infty} dr \; r^2 V(r) \, e^{-iqr\cos\theta}$$

$$V(q) = 2\pi \int_{r=0}^{r=\infty} dr \; r^2 V(r) \int_{\theta=0}^{\theta=\pi} d\theta \; \sin(\theta) \, e^{-iqr\cos\theta}$$

We now make the change of variables $x = \cos(\theta)$, so that $dx = -\sin(\theta)d\theta$, remembering that $\cos(\pi) = -1$ and $\cos(0) = 1$:

$$V(q) = 2\pi \int_{r=0}^{r=\infty} dr \; r^2 V(r) \int_{x=-1}^{x=1} dx \, e^{-iqrx} = 2\pi \int_{r=0}^{r=\infty} dr \; r^2 V(r) \cdot \left(\frac{1}{iqr} e^{iqr} - \frac{1}{iqr} e^{-iqr} \right)$$

Substituting in the function

$$V_q(r) = \frac{-e^2}{4\pi\,\varepsilon_0\varepsilon_{r0} r} \cdot e^{-r/r_0}$$

allows us to rewrite the integral

$$V(q) = \frac{-e^2}{2\varepsilon_0\varepsilon_{r0}} \cdot \frac{1}{iq} \int_{r=0}^{r=\infty} \left(e^{-\left(\frac{1}{r_0}-iq\right)r} - e^{-\left(\frac{1}{r_0}+iq\right)r} \right) dr$$

$$= \frac{-e^2}{2\varepsilon_0\varepsilon_{r0}} \cdot \frac{1}{iq} \left(\frac{1}{\frac{1}{r_0} - iq} - \frac{1}{\frac{1}{r_0} + iq} \right)$$

$$V(q) = \frac{-e^2}{2\varepsilon_0\varepsilon_{r0}} \cdot \frac{1}{iq} \left(\frac{2iq}{\frac{1}{r_0^2} + q^2} \right) = \frac{-e^2}{\varepsilon_0\varepsilon_{r0}} \left(\frac{1}{\frac{1}{r_0^2} + q^2} \right) = \frac{-e^2}{\varepsilon_0\varepsilon_{r0} q^2 \left(1 + q^2/r_0^2 \right)}$$

An alternative way to perform the integral is as follows. We start with

$$V(q) = 2\pi \int_{r=0}^{r=\infty} dr \; r^2 V(r) \cdot \left(\frac{1}{iqr} e^{iqr} - \frac{1}{iqr} e^{-iqr} \right) = 2\pi \int_{r=0}^{r=\infty} dr \; r^2 V(r) \cdot \frac{2}{qr} \sin(qr)$$

Substituting in the function

$$V_q(r) = \frac{-e^2}{4\pi\,\varepsilon_0\varepsilon_{r0} r} \cdot e^{-r/r_0}$$

gives

$$V(q) = \frac{-e^2}{\varepsilon_0 \varepsilon_{r0} q} \int\limits_{r=0}^{r=\infty} dr \, e^{-r/r_0} \cdot \sin(qr) = \frac{-e^2}{\varepsilon_0 \varepsilon_{r0} q} \cdot I$$

where we have separated out the integral

$$I = \int\limits_{r=0}^{r=\infty} dr \, e^{-r/r_0} \cdot \sin(qr)$$

We perform this integral by parts $\int U V' = U V - \int U' V$, where $U = \sin(qr)$, $U' = q \cos(qr)$, $V' = e^{-r/r_0}$, and $V = -r_0 e^{-r/r_0}$:

$$I = \left[-r_0 \sin(qr) e^{-r/r_0} \right]_{r=0}^{r=\infty} + \int\limits_{r=0}^{r=\infty} dr \, e^{-r/r_0} \cdot q r_0 \cos(qr)$$

The term in the square brackets is zero. Performing the remaining integral by parts using $U = \cos(qr)$, $U' = -q \sin(qr)$, $V' = e^{-r/r_0}$, and $V = -r_0 e^{-r/r_0}$ gives

$$I = \left[-q r_0^2 \cos(qr) e^{-r/r_0} \right]_{r=0}^{r=\infty} - \int\limits_{r=0}^{r=\infty} dr \, e^{-r/r_0} \cdot q^2 r_0^2 \cos(qr)$$

$$I = q r_0^2 - q^2 r_0^2 I$$
$$I \left(q^2 r_0^2 + 1 \right) = q r_0^2$$

$$I = \frac{q r_0^2}{(q^2 r_0^2 + 1)} = \frac{q}{(q^2 + 1/r_0^2)}$$

We now substitute in our expression for $V(q)$ to give

$$V(q) = \frac{-e^2}{\varepsilon_0 \varepsilon_{r0} q} \cdot \frac{q}{(q^2 + 1/r_0^2)} = \frac{-e^2}{\varepsilon_0 \varepsilon_{r0} (q^2 + 1/r_0^2)} = \frac{-e^2}{\varepsilon_0 \varepsilon_{r0} q^2 (1 + q^2/r_0^2)}$$

Solution 8.6

Mobility $\mu = e\tau^*/m^*$ is a measure of an appropriate electron scattering time τ^*. Conductivity is related to the mobility via the relation $\sigma = ne^2\tau^*/m^* = en\mu$.

(a) If elastic ionized impurity scattering dominates mobility in a bulk n-type semiconductor then one would expect $\tau^* \to \tau_{el}$ where τ_{el} is the elastic scattering time given by

$$\frac{1}{\tau_{el}}(E) = \frac{\pi}{(2m)^{1/2}} \cdot n \cdot \left(\frac{e^2}{4\pi \varepsilon_0} \right)^2 \cdot E^{-3/2} \cdot \int\limits_{\eta=0}^{\eta=1} \frac{d\eta}{(\varepsilon_r(2k\eta))^2 \eta^3}$$

Hence, $\tau_{el} \sim E^{3/2}$. Since, for a nondegenerate electron gas, the average energy of an electron in thermal equilibrium is proportional to the thermal energy $k_B T$, it follows that $\tau_{el} \sim (k_B T)^{3/2}$. When elastic ionized impurity scattering dominates mobility, one may surmise that the mobility will *increase* with increasing temperature as $T^{3/2}$.

(b) Coulomb scattering is weighted to small-angle (small-q) scattering. Elastic ionized impurity scattering is screened. Small-q scattering is screened most effectively. If there are spatial correlations in the positions of ionized impurities, this can influence the structure factor for a given q.

A net repulsion between impurities will tend to give rise to a long-range correlation in impurity position. Long-range order will tend to move spectral weight in the structure factor to points in the Brillouin zone of the resulting sublattice. Such order can result in a suppression in small-q scattering.

A net attraction between impurities will tend to give rise to clusters of impurity atom positions. In this case, there is not any long-range order, and scattering strength can be enhanced for high-q scattering because clusters can have large effective coulomb scattering cross-sections.

Solution 8.7

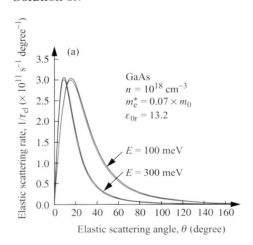
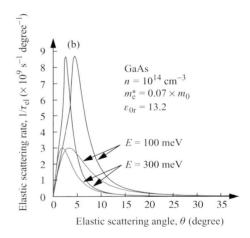

Thomas–Fermi screening predicts a lower scattering rate (lower of each curve for a given energy in the above figures) than the RPA model. This occurs because screening is overestimated in the Thomas–Fermi model. In fact, in this model, screened charge density is proportional to $(q_{TF}^2 e^{-q_{TF} \cdot r})/r$, which diverges as $r \to 0$. Notice that there is very little difference between the two models when carrier concentration is high and screening is very effective in reducing scattering in both models. The differences between Thomas–Fermi and RPA are enhanced when carrier concentration is low.

Weighting the angular integral by $(1 - \cos(\theta))$ has the effect of suppressing small-angle scattering. Such $(1 - \cos(\theta))$ weighting is used as a way to estimate the influence scattering angle has on electrical conductivity $\sigma = ne^2 \tau_{el}/m_e^*$. The intuitively obvious fact that back-scattering corresponding to $\theta = 180°$ has a much larger effect in reducing conductivity than small-angle scattering is quantified by using the $(1 - \cos(\theta))$ weighting term when calculating elastic scattering time, τ_{el}.

Solution 8.8

The probability of a transition from state $|k\rangle$ to state $|j\rangle$ at time t is

$$P_{k \to j}(t) = |c_{k \to j}(t)|^2$$

and the first-order expression for $c_{k \to j}(t)$ is

$$c_{k \to j}(t) = \frac{1}{i\hbar} \int\limits_{t'=0}^{t'=t} \langle j|H(t')|k\rangle \cdot e^{i\omega_{jk}t'} dt \qquad (1)$$

The coefficient $c_{j \to k}(t)$ for the reverse transition is given by the same expression with the indices k and j interchanged:

$$c_{k \to j}(t) = \frac{1}{i\hbar} \int\limits_{t'=0}^{t'=t} \langle k|H(t')|j\rangle \cdot e^{i\omega_{kj}t'} dt \qquad (2)$$

Since the Hamiltonian H is a Hermitian operator, it follows that $\langle k|H(t')|j\rangle = \langle j|H(t')|k\rangle^*$. Also, the change in energy due to the transition is $\hbar\omega_{kj} = E_k - E_j = -\hbar\omega_{jk}$. Therefore, the integral in Eqn (2) is the complex conjugate of the one in Eqn (1). Hence, $c_{j \to k}(t) = -(c_{k \to j}(t))^*$, giving

$$P_{j \to k}(t) = |c_{j \to k}(t)|^2 = P_{k \to j}(t)$$

The probability of transition between two states due to external stimulus is the same for transitions in either direction. This result is known as the principle of *detailed balance*.

Solution 8.9

There are N_0 excited atoms that can radiatively decay to the ground state with an average radiative lifetime $1/\gamma$. The rate equation for the population of excited atoms is $dN/dt = -\gamma N$, with solution $N(t) = N_0 e^{-\gamma t}$. Power radiated is the number of photons emitted per unit time multiplied by the energy per photon, so that $P(t) = N_0 \hbar\omega_0 \gamma e^{-\gamma t}$. We are given the electric field as a function of time $\mathbf{E}(t) = \mathbf{E}_0 e^{-\gamma t/2} \cos(\omega_0 t)$ for $t \geq 0$, $\mathbf{E}(t) = 0$ for $t < 0$. The spectral function, $|\mathbf{S}(\omega)| = \mathbf{E}^*(\omega)\mathbf{E}(\omega)/Z_0$, is the Fourier transform of the temporal electric field intensity.

We begin by calculating

$$\mathbf{E}(\omega) = \int\limits_{0}^{\infty} \mathbf{E}(t)e^{-i\omega t} dt$$

Substituting in the expression for $\mathbf{E}(t)$ and noting that $\cos(\omega_0 t) = \frac{1}{2}(e^{i\omega_0 t} + e^{-i\omega_0 t})$,

$$\mathbf{E}(\omega) = \frac{\mathbf{E}_0}{2} \int_0^\infty e^{-\gamma t/2} \cdot e^{-i\omega t}\left(e^{i\omega_0 t} + e^{-i\omega_0 t}\right)dt$$

$$\mathbf{E}(\omega) = \frac{\mathbf{E}_0}{2} \int_0^\infty \left(e^{-(\gamma/2+i\omega)t}e^{i\omega_0 t} + e^{-(\gamma/2+i\omega)t}e^{-i\omega_0 t}\right)dt$$

$$\mathbf{E}(\omega) = \frac{\mathbf{E}_0}{2} \int_0^\infty \left(e^{-(\gamma/2+i(\omega-\omega_0))t} + e^{-(\gamma/2+i(\omega+\omega_0))t}\right)dt$$

$$\mathbf{E}(\omega) = \frac{\mathbf{E}_0}{2} \left[\frac{e^{-(\gamma/2+i(\omega-\omega_0))t}}{-(\gamma/2 + i(\omega-\omega_0))} + \frac{e^{-(\gamma/2+i(\omega+\omega_0))t}}{-(\gamma/2 + i(\omega+\omega_0))}\right]_0^\infty$$

$$\mathbf{E}(\omega) = \frac{\mathbf{E}_0}{2} \left(\frac{1}{\gamma/2 + i(\omega-\omega_0)} + \frac{1}{\gamma/2 + i(\omega+\omega_0)}\right)$$

$$\mathbf{E}(\omega) = \frac{\mathbf{E}_0}{2} \left(\frac{\gamma + 2i\omega}{\omega_0^2 - \omega^2 + (\gamma/2)^2 + \frac{i\gamma\omega}{2} + \frac{i\gamma\omega_0}{2} + \frac{i\gamma\omega}{2} - \frac{i\gamma\omega_0}{2}}\right)$$

$$\mathbf{E}(\omega) = \frac{\mathbf{E}_0}{2} \left(\frac{\gamma + 2i\omega}{\omega_0^2 - \omega^2 + (\gamma/2)^2 + i\gamma\omega}\right)$$

$$\mathbf{E}(\omega) = \mathbf{E}_0 \left(\frac{\gamma/2 + i\omega}{\omega_0^2 - \omega^2 + (\gamma/2)^2 + i\gamma\omega}\right)$$

$$|\mathbf{S}(\omega)| = \frac{\mathbf{E}^*(\omega)\mathbf{E}(\omega)}{Z_0} = \frac{|\mathbf{E}_0|^2}{Z_0} \frac{(\gamma/2)^2 + \omega^2}{\left(\omega_0^2 - \omega^2 + (\gamma/2)^2\right)^2 + (\gamma\omega)^2}$$

where $Z_0 = \sqrt{\mu_0/\varepsilon_0} = 376.73\ \Omega$ is the impedance of free space.

The Lorentzian line-shape approximation may be assumed if $\gamma \to 0$. A small value of γ gives peaked $|\mathbf{S}(\omega)|$, so $\gamma^2 \to 0$ and $(\omega + \omega_0) \cong 2\omega_0$ for ω near ω_0. Hence,

$$|\mathbf{S}(\omega)| = \frac{|\mathbf{E}_0|^2}{Z_0} \frac{\omega_0^2}{(\omega_0 - \omega)^2(\omega_0 + \omega)^2 + (\gamma\omega)^2}$$

$$|\mathbf{S}(\omega)| = \frac{|\mathbf{E}_0|^2}{Z_0} \frac{\omega_0^2}{4\omega_0^2\left[(\omega_0 - \omega)^2 + (\gamma/2)^2\right]}$$

$$|\mathbf{S}(\omega)| = \frac{|\mathbf{E}_0|^2}{4Z_0} \frac{1}{(\omega_0 - \omega)^2 + (\gamma/2)^2}$$

This is in the form of a Lorentzian function. The Lorentzian function is symmetric in frequency about ω_0, but the exact solution is not. Different isotopes (distinguishable particles) or Doppler shifts give different contributions to peaks in spectrum and an asymmetric line shape.

In the following we make a comparison of an exact and a Lorentzian line shape.

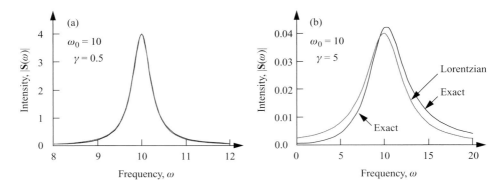

The figures above show a comparison of exact and Lorentzian line shapes for different values of γ. Note that there is very little difference between the curves when $\omega_0 = 10$, $\gamma = 0.5$, as shown in (a). γ is the full width half maximum (FWHM), and γ determines the peak value $1/\gamma^2$ at ω_0. The differences between the exact line shape and Lorentzian approximation are more apparent when $\omega_0 = 10$ and $\gamma = 5$ as shown in (b). For this large γ, heavy damping case, the exact result shows that the peak in the spectral response is shifted to a frequency above $\omega_0 = 10$, the peak value is greater than $1/\gamma^2$, the spectral line shape is asymmetric, but the FWHM is still close in value to γ.

Solution 8.10

(a) Dipole radiation requires a parity difference between initial and final states to ensure oscillation in mean position and charge. This can most easily be seen by considering the dipole matrix element $\langle even(odd)|r|odd(even)\rangle \neq 0$, whereas $\langle even(odd)|r|even(odd)\rangle = 0$ from symmetry. Hence, we expect $\langle j|r|k\rangle \neq 0$ for $j - k = odd$. Conservation of spin requires that the photon, which obeys boson statistics, must take spin quantum number ± 1 away from the system. Energy conservation requires that the separation in energy between initial and final states is the energy of the photon, $\hbar\omega$.

(b) For the harmonic oscillator, we find the matrix element using the position operator

$$\hat{x} = \left(\frac{\hbar}{2m_0\omega}\right)^{1/2}\left(\hat{b} + \hat{b}^{\dagger}\right)$$

Hence,

$$|\langle j|x|k\rangle|^2 = \frac{\hbar}{2m_0\omega}\left|\langle j|\hat{b} + \hat{b}^{\dagger}|k\rangle\right|^2$$

For spontaneous emission, we will only be considering the transition from the first excited state to the ground state. In this situation, the matrix element squared is

$$|\langle j|x|k\rangle|^2 = \frac{\hbar}{2m_0\omega}$$

Substituting into the expression for the spontaneous emission lifetime,

$$\frac{1}{A} = \tau = \frac{3\pi\varepsilon_0\hbar c^3}{e^2\omega^3|\langle j|r|k\rangle|^2} = \frac{3\pi\varepsilon_0\hbar c^3 2m_0\omega}{e^2\hbar\omega^3} = \frac{6\pi\varepsilon_0 c^3 m_0}{e^2\omega^2} = \frac{3\varepsilon_0\lambda^2 cm_0}{2\pi e^2}$$

$$\tau = \frac{3 \times 8.8 \times 10^{-12} \times \lambda^2 \times 3 \times 10^8 \times 9.1 \times 10^{-31}}{2\pi(1.6 \times 10^{-19})^2} = 45 \times 10^3 \times \lambda^2$$

which, if wavelength is measured in micrometers and time in nanoseconds, becomes

$$\tau(\text{ns}) = 45 \times \lambda^2(\mu\text{m})$$

(c) The force constant κ is related to oscillator frequency ω and particle mass m_0 through $\kappa = \omega^2 m_0$. Given that $\kappa = 3.59 \times 10^{-3}$ kg s^{-2} and m_0 is the bare electron mass, the oscillator frequency is 10 THz ($\hbar\omega = 41.36$ meV), and the emission wavelength is $\lambda = 30$ μm. This gives a spontaneous emission lifetime

$$\tau(\text{ns}) = 45 \times 900 \text{ ns} = 40 \text{ } \mu\text{s}$$

9 The semiconductor laser

9.1 Introduction

The history of the laser dates back to at least 1951 and an idea of Townes. He wanted to use ammonia molecules to amplify microwave radiation. Townes and two students completed a prototype device in late 1953 and gave it the name *maser* or *microwave amplification by stimulated emission of radiation*. In 1958 Townes and Schawlow published results of a study showing that a similar device could be made to amplify light. The device was named a *laser* which is an acronym for *light amplification by stimulated emission of radiation*. In principle, a large flux of essentially single-wavelength electromagnetic radiation could be produced by a laser. Independently, Prokhorov and Basov proposed related ideas. The first laser used a rod of ruby and was constructed in 1960 by Maiman.

In late 1962 lasing action in a current-driven GaAs *p–n* diode maintained at liquid nitrogen temperature (77 K) was reported.[1] Room-temperature operation and other improvements followed.

Soon, telephone companies recognized the potential of such components for use in communication systems. However, it took some time before useful devices and suitable glass-fiber transmission media became available. The first fiber-optic telephone installation was put in place in 1977 and consisted of a 2.4-km-long link under downtown Chicago.

Another type of laser diode suitable for use in data communication applications was inspired by the work of Iga published in 1977.[2] By the late 1990s, these vertical-cavity surface-emitting lasers (VCSELs) had appeared in volume-manufactured commercial products.

The largest number of semiconductor lasers is produced for compact disk (CD) applications and digital versatile disk (DVD) video applications. Laser diodes are also volume-manufactured for fiber-optic communication products, laser printers, and laser

---

[1] R. Hall, G. E. Fenner, J. Kingsley, T. J. Soltys, and R. O. Carlson, *Phys. Rev. Lett.* **9,** 366 (1962).
[2] H. Soda, K. Iga, C. Kitahara, and Y. Suematsu, *Jap. J. Appl. Phys.* **18,** 2329 (1977).

copiers. There is additional low-volume production of laser diodes for numerous specialty markets.

In this chapter we will be interested in various aspects of the laser, including the mechanism by which light is amplified and the role of spontaneous emission. Because laser diodes are made using semiconductors, we will investigate how this impacts laser design.

The chapter concludes with a brief discussion of why the relatively simple model we use to describe the behavior of a laser diode works so well.

9.2 Spontaneous and stimulated emission

In Chapter 8 the role emission and absorption of light have in determining transitions between two electronic states of an atom was considered. A schematic energy-level diagram for two states $|1\rangle$ and $|2\rangle$ showing stimulated and spontaneous processes is presented in Fig. 9.1(a).

In a semiconductor laser diode, optical transitions take place between conduction-band states and valence-band states. The active region where these transitions occur is typically a *direct band-gap* semiconductor, examples of which include GaAs, InP, and InGaAs. In such a semiconductor the energy minimum of the conduction band lines up with the maximum energy of the valence band in k-space. This fact is of particular importance for direct interaction of semiconductor electronic states with light of wave vector k_{opt}, because the usual dispersion relation of light $\omega = ck_{\text{opt}}/n_{\text{r}}$ is almost vertical compared with the dispersion relation for electrons in a given band $\omega = \hbar k^2/2m_{\text{r}}$, where m_{r} is an effective electron mass. Conservation of momentum during a transition from

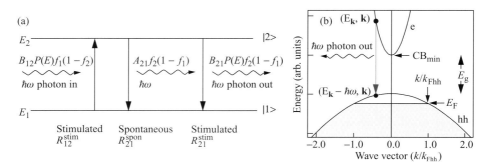

Fig. 9.1. (a) Schematic energy level diagram showing stimulated and spontaneous optical transitions between two electronic energy levels. (b) Band structure of a direct band-gap semiconductor showing valence heavy-hole band hh, conduction band e, minimum conduction-band energy CB_{min}, and band-gap energy E_{g}. The semiconductor is doped p-type, and at low temperature the Fermi energy is E_F and the Fermi wave vector is k_{Fhh}. Electrons in the conduction band can make a transition from a state characterized by wave vector $\mathbf{k}$ and energy $E_{\mathbf{k}}$ in the conduction band to wave-vector state $\mathbf{k}$ energy $E_k - \hbar\omega$ in the valence band by emitting a photon of energy $\hbar\omega$.

occupied to unoccupied electronic states via the emission or absorption of a photon requires, therefore, a vertical transition in k-space.

The simplest model of a direct band-gap semiconductor typically includes a heavy-hole valence band hh with effective hole mass m_{hh}^* and conduction band e with effective electron mass m_e^*. In GaAs, appropriate values for the effective electron masses are $m_{hh}^* = 0.5 \times m_0$ and $m_e^* = 0.07 \times m_0$. Figure 9.1(b) shows schematically spontaneous emission of a photon of energy $\hbar\omega$ accompanied by an electronic transition of a state characterized by wave vector $\mathbf{k}$ and energy $E_\mathbf{k}$ in the conduction band to wave-vector state $\mathbf{k}$ energy $E_\mathbf{k} - \hbar\omega$ in the valence band. Because the initial and final electronic states have the same value of $\mathbf{k}$, it is assumed that crystal momentum is conserved.

If we measure energy from the top of the valence band, then the energy of an electron in the conduction band with effective electron mass m_e^* is

$$E_2 = E_g + \frac{\hbar^2 k^2}{2m_e^*} \tag{9.1}$$

and the energy of an electron in the valence band with effective electron mass m_{hh}^* is

$$E_1 = -\frac{\hbar^2 k^2}{2m_{hh}^*} \tag{9.2}$$

The energy due to an electronic transition from the conduction band to the valence band is

$$\hbar\omega = E_2 - E_1 = \frac{\hbar^2 k^2}{2}\left(\frac{1}{m_e^*} + \frac{1}{m_{hh}^*}\right) + E_g \tag{9.3}$$

We can define a reduced effective electron mass m_r in such a way that

$$\frac{1}{m_r} = \frac{1}{m_e^*} + \frac{1}{m_{hh}^*} \tag{9.4}$$

In GaAs, we might use $m_e^* = 0.07 \times m_0$ and $m_{hh}^* = 0.5 \times m_0$, giving $m_r = 0.06 \times m_0$. Equation (9.4) allows Eqn (9.3) to be written

$$\hbar\omega = \frac{\hbar^2 k^2}{2m_r} + E_g \tag{9.5}$$

It follows that the three-dimensional density of electronic states coupled to vertical optical transitions of energy $\hbar\omega$ is

$$D_3(\hbar\omega) = \frac{1}{2\pi^2}\left(\frac{2m_r}{\hbar^2}\right)^{3/2}(\hbar\omega - E_g)^{1/2} \tag{9.6}$$

If Fig. 9.1(b) is representative of the physical processes involved in an optically active semiconductor, then we may wish to consider using Fermi's golden rule to calculate transition probability between electronic states. In this case, all we need to know is the matrix element coupling the initial and final states and the density of final states.

To make some rapid progress, we will proceed along this path. Later, we will discuss the source of errors in our assumptions and also explain why, in practice, our approach is so successful in describing some of the important properties of laser diodes.

First, consider a discrete two-level system inside an optical cavity that is a cube of side L, volume V, and temperature T. Possible transitions are shown in Fig. 9.1(a). For periodic boundary conditions, allowed optical modes are k states $k_{n_j} = 2\pi n_j/L$, where $j = x, y, z$ and n is an integer. The spectral density $P(E)$ of electromagnetic modes at a specific energy is found by multiplying the density of photon states $D_3^{\text{opt}}(E)$ by the Bose–Einstein occupation factor for photons $g(E)$. Since

$$D_3^{\text{opt}}(k_{\text{opt}})dk = \frac{1}{V} \cdot 2\left(\frac{L}{2\pi}\right)^3 4\pi k_{\text{opt}}^2 dk_{\text{opt}} = \left(\frac{k_{\text{opt}}}{\pi}\right)^2 dk_{\text{opt}} \tag{9.7}$$

and noting that $E = \hbar\omega = \hbar c k_{\text{opt}}$ in free space and $dE = \hbar c \cdot dk_{\text{opt}}$, we may write

$$D_3^{\text{opt}}(E) = \frac{E^2}{\pi^2 \hbar^3 c^3} \tag{9.8}$$

giving a spectral density measured in units of number of photons per unit volume per unit energy interval:

$$P(E) = D_3^{\text{opt}}(E)g(E) = \frac{E^2}{\pi^2 \hbar^3 c^3} g(E) = \frac{E^2}{\pi^2 \hbar^3 c^3} \cdot \frac{1}{e^{E/k_B T} - 1} \tag{9.9}$$

If the electromagnetic modes exist in a homogeneous dielectric medium characterized by refractive index n_r at frequency ω, then $E = \hbar\omega = \hbar c k_{\text{opt}}/n_r$. If $n_r = n_r(\omega)$, then $dk_{\text{opt}} = (1/c)(n_r + \omega \cdot (dn_r/d\omega))d\omega$. Ignoring dispersion in the refractive index ($\omega \cdot dn_r/d\omega = 0$) gives

$$P(E) = \frac{E^2 n_r^3}{\pi^2 \hbar^3 c^3} \cdot \frac{1}{e^{E/k_B T} - 1} \tag{9.10}$$

Assuming that Fermi's golden rule may be used to calculate transition rates between states $|1\rangle$ and $|2\rangle$, it makes sense to define

$$B_{21} \equiv \frac{2\pi}{\hbar} |W_{21}|^2 \tag{9.11}$$

where $|W_{21}|^2$ is the matrix element squared coupling the initial and final states. The stimulated and spontaneous rates for photons of energy $E_{21} = E_2 - E_1$ become

$$R_{12}^{\text{stim}} = B_{12} P(E_{21}) f_1 (1 - f_2) \tag{9.12}$$
$$R_{21}^{\text{stim}} = B_{21} P(E_{21}) f_2 (1 - f_1) \tag{9.13}$$
$$R_{21}^{\text{spon}} = A_{21} f_2 (1 - f_1) \tag{9.14}$$

where the Fermi–Dirac distribution function gives the probability of electron occupation f_1 at energy E_1 and the probability of an unoccupied electron state $(1 - f_2)$ at energy E_2.

When the system is in thermal equilibrium, the rates must balance, and there is only one chemical potential, so $\mu_1 = \mu_2 = \mu$. If thermal equilibrium exists between the two-level system and the electromagnetic modes of the cavity, then

$$R_{12}^{\text{stim}} = R_{21}^{\text{stim}} + R_{21}^{\text{spon}} \tag{9.15}$$

$$B_{12} P(E_{21}) f_1 (1 - f_2) = B_{21} P(E_{21}) f_2 (1 - f_1) + A_{21} f_2 (1 - f_1) \tag{9.16}$$

$$P(E_{21})(B_{12} f_1 (1 - f_2) - B_{21} f_2 (1 - f_1)) = A_{21} f_2 (1 - f_1) \tag{9.17}$$

$$P(E_{21}) = \frac{A_{21} f_2 (1 - f_1)}{B_{12} f_1 (1 - f_2) - B_{21} f_2 (1 - f_1)} = \frac{A_{21}(f_2 - f_1 f_2)}{B_{12}(f_1 - f_1 f_2) - B_{21}(f_2 - f_1 f_2)} \tag{9.18}$$

$$P(E_{21}) = \frac{A_{21}\left(\dfrac{1}{f_1} - 1\right)}{B_{12}\left(\dfrac{1}{f_2} - 1\right) - B_{21}\left(\dfrac{1}{f_1} - 1\right)} = \frac{A_{21}}{B_{12}\left(\dfrac{1/f_2 - 1}{1/f_1 - 1}\right) - B_{21}} \tag{9.19}$$

$$P(E_{21}) = \frac{A_{21}}{B_{12} e^{E_{21}/k_B T}\left(\dfrac{e^{-\mu_2/k_B T}}{e^{-\mu_1/k_B T}}\right) - B_{21}} \tag{9.20}$$

Since the system is in equilibrium, $\mu_1 = \mu_2$. Making use of Eqn (9.10) one may write

$$P(E_{21}) = \frac{A_{21}}{B_{12} e^{E_{21}/k_B T} - B_{21}} = D_3^{\text{opt}}(E_{21}) \frac{1}{e^{E_{21}/k_B T} - 1} \tag{9.21}$$

Because this relationship must hold for any temperature when the system is in equilibrium, it follows that

$$B_{12} = B_{21} \tag{9.22}$$

$$\frac{A_{21}}{B_{12}} = D_3^{\text{opt}}(E_{21}) = \frac{E_{21}^2 n_r^3}{\pi^2 \hbar^3 c^3} \tag{9.23}$$

Note that this is derived for the case in which the electron system is in thermal equilibrium with the electromagnetic modes of the cavity. This means that the complete system may be characterized by a single temperature T. Equations (9.22) and (9.23) are the Einstein relations previously discussed in Section 8.6.5.

9.2.1 Absorption and its relation to spontaneous emission

Photons of energy $E = \hbar\omega$ incident on a two-level system can cause transitions between two states $|1\rangle$ and $|2\rangle$ with energy eigenvalues $E_1 < E_2$. Absorption α may be defined as the ratio of the number of absorbed photons per second per unit volume to the number of incident photons per second per unit area. Hence,

$$\alpha = \frac{R_{\text{stim}}^{\text{net}}}{S/\hbar\omega} = \frac{R_{12}^{\text{stim}} - R_{21}^{\text{stim}}}{S/\hbar\omega} \tag{9.24}$$

where S is the magnitude of the Poynting vector and $S/\hbar\omega$ is the number of incident photons per second per unit area. It is usual for α to be measured in units of cm^{-1}. Since the absorption coefficient times the photon flux is the net stimulated rate, one may write

$$\alpha = \frac{B_{12}P(E_{21})f_1(1 - f_2) - B_{21}P(E_{21})f_2(1 - f_1)}{P(E_{21}) \cdot \dfrac{c}{n_r}} = \frac{n_r}{c} \cdot B_{12}(f_1 - f_2) \tag{9.25}$$

The ratio of spontaneous emission and absorption is

$$\frac{R_{21}^{\text{spon}}}{\alpha} = \frac{A_{21}f_2(1 - f_1)}{\dfrac{n_r}{c} \cdot B_{12}(f_1 - f_2)} = \frac{A_{21}(1 - f_1)}{\dfrac{n_r}{c} \cdot B_{12}\left(\dfrac{f_1}{f_2} - 1\right)} = \frac{A_{21}\left(\dfrac{1}{f_1} - 1\right)}{\dfrac{n_r}{c} \cdot B_{12}\left(\dfrac{1}{f_2} - \dfrac{1}{f_1}\right)} \tag{9.26}$$

$$\frac{R_{21}^{\text{spon}}}{\alpha} = \frac{A_{21}e^{(E_1 - \mu_1)/k_B T}}{\dfrac{n_r}{c} \cdot B_{12}\left(e^{(E_2 - \mu_2)/k_B T} - e^{(E_1 - \mu_1)/k_B T}\right)} = \frac{A_{21}}{\dfrac{n_r}{c} \cdot B_{12}\left(\dfrac{e^{(E_2 - \mu_2)/k_B T}}{e^{(E_1 - \mu_1)/k_B T}} - 1\right)} \tag{9.27}$$

$$\frac{R_{21}^{\text{spon}}}{\alpha} = \frac{A_{21}}{\dfrac{n_r}{c} \cdot B_{12}\left(e^{E_{21}/k_B T}e^{-(\mu_2 - \mu_1)/k_B T} - 1\right)} \tag{9.28}$$

Substituting for the ratio A_{21}/B_{12} using Eqn (9.23) gives

$$\frac{R_{21}^{\text{spon}}}{\alpha} = \frac{c}{n_r} \cdot D_3^{\text{opt}}(E_{21}) \cdot \frac{1}{e^{(E_{21} - (\mu_2 - \mu_1))/k_B T} - 1} = \frac{E_{21}^2 n_r^2}{\pi^2 c^2 \hbar^3} \cdot \frac{1}{e^{(E_{21} - \Delta\mu)/k_B T} - 1} \tag{9.29}$$

where $\Delta\mu = \mu_2 - \mu_1$ is the difference in quasi-chemical potential used to describe the distribution of electronic states at energies E_2 and E_1, respectively. The approximation made is that Eqn (9.23) (which was derived for the equilibrium condition $\Delta\mu = 0$) remains valid when $\Delta\mu \neq 0$. This is likely to be true when $\Delta\mu < k_B T$. The relationship between absorption and spontaneous emission for a system characterized by temperature T and difference in chemical potential $\Delta\mu$ given by Eqn (9.29) may be rewritten in a convenient form as

$$\alpha = \frac{\pi^2 c^2 \hbar^3}{E_{21}^2 n_r^2} \cdot R_{21}^{\text{spon}}\left(e^{(E_{21} - \Delta\mu)/k_B T} - 1\right) \tag{9.30}$$

Net optical gain exists when absorption α is negative. Since spontaneous emission is always positive, the only way the value of absorption α can change sign is if the term in parenthesis on the right-hand side of Eqn (9.30) changes sign. An easy way to see

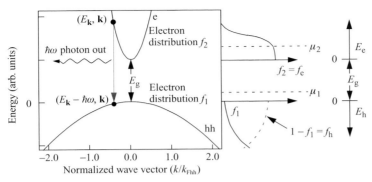

Fig. 9.2. A schematic diagram showing an electronic transition from a state of energy E_k and wave vector $\mathbf{k}$ to state of energy $E_k - \hbar\omega$ and wave vector $\mathbf{k}$, resulting in the emission of a photon of energy $\hbar\omega$. A conduction band electron has an effective electron mass m_e^* and a valence band electron has an effective electron mass m_{hh}^*. The Fermi–Dirac distribution of electron states in the conduction band is f_2, and in the valence band it is f_1. The quasi-chemical potentials are μ_2 and μ_1, respectively. Sometimes it is convenient to measure electron energy E_e from the conduction band minimum and hole energy E_h from the valence band maximum.

this is by noticing that Eqn (9.26) may be rewritten as

$$\alpha = \frac{R_{21}^{\text{spon}} \cdot \dfrac{n_r}{c} \cdot B_{12}(f_1 - f_2)}{A_{21} f_2 (1 - f_1)} \tag{9.31}$$

The denominator in Eqn (9.31) is always positive, since $0 < f_1 < 1$ and $0 < f_2 < 1$. The numerator is positive, giving positive absorption α if $f_1 > f_2$ and negative absorption (or optical gain $g_{\text{opt}} \equiv -\alpha$) if $f_1 < f_2$. The condition for optical gain is $f_2 - f_1 > 0$, or

$$\Delta\mu > E_{21} \tag{9.32}$$

This expresses the fact that the separation in quasi-chemical potentials must be greater than the photon energy for net optical gain to exist. Equation (9.32) is called the Bernard–Duraffourg condition.[3]

In a semiconductor there are not just two energy levels E_1 and E_2 to be considered, but rather a continuum of energy levels in the conduction band and valence band. This is illustrated in Fig. 9.2. Electrons in the conduction band have a Fermi–Dirac distribution f_2, and in the valence band they have a distribution f_1. As was shown in Chapter 7, *for equal carrier concentrations* in the conduction band and valence band at fixed temperature, the Fermi–Dirac distribution functions f_2 and f_1 are different since, in general, the quasi-chemical potentials are different. This occurs because the effective electron mass in each band is different, giving a different density of states, and hence

[3] M. G. A. Bernard and G. Duraffourg, *Phys. Stat. Solidi* **1**, 699 (1961).

different quasi-chemical potentials for carrier concentration n and temperature T when calculated using Eqn (7.36).

For convenience, we are going to change the way electron energy in the conduction and valence bands is measured. The energy of holes (absence of electrons) in the valence band will be measured from the valence band maxima *down*. The energy of holes is negative, and the distribution of holes is $f_h = (1 - f_1)$. The energy of electrons in the conduction band will be measured from the conduction band minimum *up*. The energy of electrons is positive, and the distribution of electrons is $f_e = f_2$. When calculating photon energy, $\hbar\omega$, for an interband transition one must remember to add the band-gap energy, E_g. Also, the condition for optical gain previously given by Eqn (9.32) becomes

$$\Delta\mu_{e-hh} > 0 \qquad (9.33)$$

In GaAs, we may use $m_e^* = 0.07 \times m_0$ and $m_{hh}^* = 0.5 \times m_0$. If temperature $T = 300\,\mathrm{K}$ and *carrier density is fixed* in value at $n = 1 \times 10^{18}\,\mathrm{cm}^{-3}$ *in each band*, then $\mu_{hh} = -55\,\mathrm{meV}$ and $\mu_e = 39\,\mathrm{meV}$ (the value of μ_e is μ_2 with energy E_g subtracted). In this situation, $\Delta\mu_{e-hh} = \mu_2 - \mu_1 - E_g = \mu_e + \mu_{hh} = -16\,\mathrm{meV}$ and, according to Eqn (9.33), GaAs is absorbing for all photon energies $\hbar\omega > E_g$. On the other hand, when $n = 2 \times 10^{18}\,\mathrm{cm}^{-3}$ in each band and $T = 300\,\mathrm{K}$, the chemical potentials are $\mu_{hh} = -36\,\mathrm{meV}$ and $\mu_e = 75\,\mathrm{meV}$. Now $\Delta\mu_{e-hh} = 39\,\mathrm{meV}$ and optical gain exists for photon energy $E_g < \hbar\omega < E_g + \Delta\mu_{e-hh}$.

The total spontaneous emission $r_{spon}(\hbar\omega)$ for photons of energy $E = \hbar\omega$ is the sum of all energy levels separated by vertical transitions of energy E. Substituting Eqn (9.23) into Eqn (9.14), using the definition of $|W_{21}|^2$ given by Eqn (9.11), and performing the sum over allowed vertical k-state transitions gives

$$r_{spon}(\hbar\omega) = \frac{2\pi}{\hbar} \frac{E_{21}^2 n_r^3}{\pi^2 \hbar^3 c^3} \sum_{k_2,k_1} |W_{21}|^2 f_2(1 - f_1)\delta(E_{21} - \hbar\omega) \qquad (9.34)$$

where the delta function ensures energy conservation. If the matrix element W_{21} is slowly varying as a function of E_{21}, it may be treated as a constant. Converting the sum to an integral and substituting $f_e = f_2$ and $f_h = (1 - f_1)$ gives

$$r_{spon}(\hbar\omega) = \frac{2\pi}{\hbar} \frac{E_{21}^2 n_r^3}{\pi^2 \hbar^3 c^3} |W_{21}|^2 \cdot \int \frac{d^3 k}{(2\pi)^3} \cdot 2 \cdot f_e f_h \cdot \delta\left(E_g + \frac{\hbar^2 k^2}{2m_r} - \hbar\omega \right) \qquad (9.35)$$

which may be written as

$$r_{spon}(\hbar\omega) = \frac{2\pi}{\hbar} \frac{\hbar^2 \omega^2 n_r^3}{\pi^2 \hbar^3 c^3} |W_{21}|^2 \cdot \frac{1}{2\pi^2} \cdot \left(\frac{2m_r}{\hbar^2} \right)^{3/2} (\hbar\omega - E_g)^{1/2} f_e f_h \qquad (9.36)$$

In this expression, we recognize $(\hbar\omega - E_g)^{1/2}$ as the energy dependence of the reduced three-dimensional density of electronic states in Eqn (9.6).

It follows from Eqn (9.30) that at equilibrium optical gain $g_{opt}(\hbar\omega)$ is related to $r_{spon}(\hbar\omega)$ through

$$g_{opt}(\hbar\omega) = -\alpha(\hbar\omega) = \hbar\left(\frac{c\pi}{n_r\omega}\right)^2 \cdot r_{spon}(\hbar\omega) \cdot \left(1 - e^{(\hbar\omega - E_g - \Delta\mu_{e-hh})/k_BT}\right) \quad (9.37)$$

This is a useful relationship that allows us to obtain the absorption function if we know the spontaneous emission. Note that if $\Delta\mu \neq 0$ then the electron–hole system is out of thermal equilibrium and our assumptions in deriving this relationship are no longer valid. For example, n_r will depend upon $\Delta\mu$.

We can also find optical gain directly by substituting Eqn (9.11) into Eqn (9.25) and performing the integral over allowed initial and final electronic density of states (Eqn (9.6)). This gives

$$g_{opt}(\hbar\omega) = \frac{2\pi n_r}{c\hbar}|W_{21}|^2 \cdot \frac{1}{2\pi^2} \cdot \left(\frac{2m_r}{\hbar^2}\right)^{3/2} (\hbar\omega - E_g)^{1/2}(f_e + f_h - 1) \quad (9.38)$$

9.3 Optical transitions using Fermi's golden rule

The matrix element W_{21} appearing in Eqn (9.36) and Eqn (9.38) remains to be evaluated. This may be done by applying Fermi's golden rule.

Consider a semiconductor illuminated with light. The interaction between an optical electric field of the form

$$\mathbf{E}_{opt} = \mathbf{E}_0 e^{i(\mathbf{k}_{opt}\cdot\mathbf{x} - \omega t)} \quad (9.39)$$

and an electron with motion in the x direction is described by the perturbation

$$W = -e|\mathbf{E}_0|x e^{i(\mathbf{k}_{opt}\cdot\mathbf{x} - \omega t)} \quad (9.40)$$

Electron states in a crystal are Bloch functions of the form given by Eqn (4.94), so the dipole matrix element coupling a conduction-band initial-state $\psi_e(x) = U_{e\mathbf{k}}(x)e^{i\mathbf{k}\cdot\mathbf{x}}$ and a heavy-hole valence-band final state $\psi_{hh}(x) = U_{hh\mathbf{k}'}(x)e^{i\mathbf{k}'\cdot\mathbf{x}}$ is

$$W_{ehh} = \langle\psi_{hh}|W|\psi_e\rangle = -e|\mathbf{E}_0| \int U_{hh\mathbf{k}'}^*(x)U_{e\mathbf{k}}(x)x e^{-i(\mathbf{k}' - \mathbf{k} - \mathbf{k}_{opt})\cdot\mathbf{x}}dx \quad (9.41)$$

The term $e^{-i(\mathbf{k}' - \mathbf{k} - \mathbf{k}_{opt})\cdot\mathbf{x}}$ in the integral rapidly oscillates, resulting in $W_{ehh} \to 0$ *except* when $\mathbf{k}' - \mathbf{k} = \mathbf{k}_{opt}$. Since electronic states have $|\mathbf{k}| \sim 3 \times 10^6 \, \text{cm}^{-1}$ (see Table 7.1) and $|\mathbf{k}_{opt}| \sim 2 \times 10^5 \, \text{cm}^{-1}$ in the semiconductor, it is reasonable to set $|\mathbf{k}_{opt}| = 0$, so that $\mathbf{k}' = \mathbf{k}$. As discussed in Section 9.2, we may assume transitions between initial and final electron states conserve crystal momentum and so have the same k-vector.

Finding the value of the matrix element in Eqn (9.41) requires detailed knowledge of the Bloch wave functions involved in the transition. The calculations can be quite

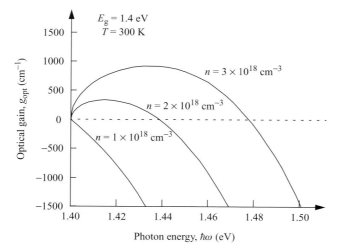

Fig. 9.3. Calculated optical gain for the indicated carrier densities, band-gap energy $E_g = 1.4$ eV, temperature $T = 300$ K, effective electron mass $m_e^* = 0.07 \times m_0$, effective hole mass $m_{hh}^* = 0.5 \times m_0$, and $g_0 = 2.64 \times 10^4$ cm^{-1} eV$^{-1/2}$.

involved,[4] and so we choose not to reproduce them here. Rather, we adopt a pragmatic approach in which the coefficients, including the matrix element squared in Eqn (9.36) and Eqn (9.38), are taken to be constants. The $\hbar^2\omega^2$ term in Eqn (9.36) is slowly varying and may be treated as a constant, since we will only be concerned with an energy range of approximately $\Delta\mu_{e-hh}$ around E_g and typically $\Delta\mu_{e-hh}/E_g \ll 1$. This allows us to write spontaneous emission (Eqn (9.36)) as

$$r_{spon}(\hbar\omega) = r_0(\hbar\omega - E_g)^{1/2} f_e f_h \tag{9.42}$$

where r_0 is a material-dependent constant. Optical gain becomes

$$g_{opt}(\hbar\omega) = g_0(\hbar\omega - E_g)^{1/2}(f_e + f_h - 1) \tag{9.43}$$

where g_0 is also a material-dependent constant. The ratio $g_0/r_0 = \pi^2 c^2/\omega^2 n_r^2$. In this simple model, the range in energy over which optical gain exists is given by the difference in chemical potential, $\Delta\mu_{e-hh}$.

Obviously, the use of constants r_0 and g_0 results in quite a crude approximation. However, it does allow us to estimate trends, such as the temperature dependence of gain. In fact, it turns out that to create a model of optical gain in a semiconductor that is even qualitatively more advanced than that described in this chapter is a very challenging task and the subject of ongoing research.

Figure 9.3 shows the result of calculating $g_{opt}(\hbar\omega)$ using Eqn (9.43) for the indicated carrier densities, band-gap energy $E_g = 1.4$ eV, temperature $T = 300$ K,

---

[4] For an introduction, see S. L. Chuang, *Physics of Optoelectronic Devices*, Wiley, New York, 1995 (ISBN 0 471 10939 8).

effective electron mass $m_{\rm e}^* = 0.07 \times m_0$, effective hole mass $m_{\rm hh}^* = 0.5 \times m_0$, and $g_0 = 2.64 \times 10^4 \ {\rm cm}^{-1} \ {\rm eV}^{-1/2}$. The parameters used are appropriate for the direct band-gap semiconductor GaAs.

According to the results shown in Fig. 9.3, there is no optical gain when carrier density $n = 1 \times 10^{18} \ {\rm cm}^{-3}$. As carrier density increases, optical gain first appears near the band-gap energy, $E_{\rm g} = 1.4 \ {\rm eV}$. When carrier density $n = 2 \times 10^{18} \ {\rm cm}^{-3}$, a peak optical gain of 330 ${\rm cm}^{-1}$ occurs for photon energy near $h\omega = 1.415 \ {\rm eV}$, and the gain bandwidth is $\Delta\mu_{\rm e-hh} = 39 \ {\rm meV}$.

9.3.1   Optical gain in the presence of electron scattering

Inelastic electron scattering has the effect of broadening electron states. Because a typical inelastic electron scattering rate $1/\tau_{\rm in}$ can be tens of ps^{-1}, corresponding to several meV broadening, this effect is significant and on the same scale as the difference in chemical potential. A Lorentzian broadening function has an energy FWHM $\gamma_{\bf k} = \hbar/\tau_{\rm in}$, so that if $\tau_{\rm in} = 25 \ {\rm fs}$ then $\gamma_{\bf k} = 26 \ {\rm meV}$. There is a subscript ${\bf k}$ in $\gamma_{\bf k}$ because, in general, scattering rate depends upon electron crystal momentum $\hbar{\bf k}$. However, in practice this fact is usually ignored and $\gamma_{\bf k}$ is treated as a constant.

To calculate optical gain in the presence of electron scattering, we first calculate the spontaneous emission using the Lorentzian broadening function to simulate the effect of electron–electron scattering. Equation (9.42) is modified to

$$r_{\rm spon}(\hbar\omega) = r_0 \int\limits_0^\infty E^{1/2} f_{\rm e} f_{\rm h} \frac{\gamma_{\bf k}/2\pi}{(E_{\rm g} + E - \hbar\omega)^2 + \left(\dfrac{\gamma_{\bf k}}{2}\right)^2} dE \qquad (9.44)$$

where the factor 2π ensures proper normalization of the Lorentzian function.

Optical gain $g_{\rm opt}$ as a function of photon energy $\hbar\omega$ is then calculated using Eqn (9.37). This ensures that optical transparency in the semiconductor occurs at a photon energy of $\Delta\mu_{\rm e-hh} + E_{\rm g}$, where $\Delta\mu_{\rm e-hh}$ is the difference in chemical potential and $E_{\rm g}$ is the band-gap energy. Optical transparency occurring at a different energy violates the concept of equilibrium in thermodynamics.

Unfortunately, even this elementary consideration is often ignored in conventional theories, which put Lorentzian broadening directly in the gain function (Eqn (9.38)).[5] As illustrated in Fig. 9.4, not only does this result in optical transparency at an energy less than $\Delta\mu_{\rm e-hh} + E_{\rm g}$, but it also predicts substantial absorption of sub-band-gap energy photons.[6] This is not observed experimentally!

---

[5] For example, see *Quantum Well Lasers*, ed. Peter Zory, Academic Press, San Diego, 1993 (ISBN 0 12 781890 1). Contributions from Corzine, Yang, Coldren, Asada, Kapon, Englemann, Shieh, and Shu all explicitly and incorrectly put Lorentzian broadening directly in the gain function.

[6] For example, see S. L. Chuang, J. O'Gorman, and A. F. J. Levi, *IEEE J. Quantum Electron.* **QE-29**, 1631 (1993) or W. W. Chow and S. W. Koch, *Semiconductor Laser Fundamentals*, Springer-Verlag, Berlin, 1999 (ISBN 3 540 64166 1).

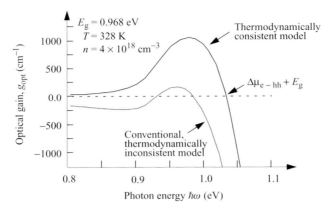

Fig. 9.4. Calculated optical modal gain, including effects of electron–electron scattering in bulk InGaAsP with room-temperature band-gap energy $E_g = 0.968$ eV and an inelastic scattering broadening factor $\gamma_k = 25$ meV.

9.4 Designing a laser diode

One may exploit the existence of optical gain in a semiconductor to make a laser diode. One might imagine constructing a p–n diode out of a direct band-gap semiconductor such as GaAs or InGaAsP. When forward-biased to pass a current, I, electrons are injected into the conduction band and holes into the valence band. The optically active region of the semiconductor is where the electrons and holes overlap in real space, so that vertical optical transitions can take place in k-space. If the density of carriers injected into the active region is great enough, then Eqn (9.32) is satisfied and optical gain exists for light at some wavelength in the semiconductor. There is, however, more to designing a useful device. Among other things, we would like to ensure that a high intensity of lasing light emission occurs at a specific wavelength.

Because a typical value of gain for an optical mode in a semiconductor laser diode is not very large ($\sim$500 cm^{-1}), and in order to precisely control emission wavelength, one typically places the active semiconductor in a high-Q optical cavity. The optical cavity has the effect of storing light at a particular wavelength, allowing it to interact with the gain medium for a longer time. In this way, relatively modest optical gain may be used to build up high light intensity in a given optical mode. Electrons contributing to injection current I are converted into lasing photons that have a single mode and wavelength. The efficiency of this conversion process is enhanced if only one high-Q optical-cavity resonance is in the same wavelength range as semiconductor optical gain. Therefore, we are interested in identifying the types of high-Q optical cavity that may be used in laser design.

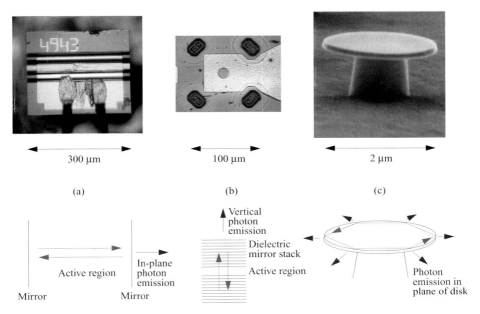

Fig. 9.5. (a) Photograph of top view of a Fabry–Perot, edge-emitting, semiconductor laser diode showing the horizontal gold metal stripe used to make electrical contact to the p-type contact of the diode. The n-type contact is made via the substrate. Two gold wire bonds attach to the large gold pad in the lower half of the picture. This particular device has a multiple-quantum-well InGaAsP active region, lasing emission at 1310 nm wavelength, and a laser threshold current of 3 mA. The sketch shows the side view of the 300-μm-long optical cavity formed by reflection at the cleaved semiconductor-air interface. (b) Photograph of top view of a VCSEL showing the gold metallization used to make electrical contact to the p-type contact of the diode. Lasing light is emitted from the small aperture in the center of the device. This VCSEL has a multiple-quantum-well GaAs active region, lasing emission at 850 nm wavelength, and a laser threshold current of 1 mA. (c) Scanning electron microscope image of a microdisk laser. The semiconductor disk is 2 μm in diameter and 0.1 μm thick. This particular device has a single-quantum-well InGaAs active region, lasing emission at 1550 nm wavelength, and an external incident optical laser threshold pump power at 980 nm wavelength of 300 μW.

9.4.1   The optical cavity

Figures 9.5(a), (b), and (c) illustrate optical cavities into which we may place the optically active semiconductor to form a Fabry–Perot laser, VCSEL,[7] and microdisk laser[8] repectively.

The Fabry–Perot optical cavity is, at least superficially, quite easy to understand, so we now consider that. Assuming that photons travel normally to the two mirror planes

[7] K. Iga, M. Oikawa, S. Misawa, J. Banno, and Y. Kokubun, *Appl. Opt.* **21**, 3456 (1982).

[8] S. L. McCall, A. F. J. Levi, R. E. Slusher, S. J. Pearton, and R. A. Logan, *Appl. Phys. Lett.* **60**, 289 (1992).

and in the z direction, we will be interested in finding expressions for the corresponding longitudinal optical resonances.

9.4.1.1 Longitudinal resonances in the z direction

The Fabry–Perot laser consists of an index-guided active gain region placed within a Fabry–Perot optical resonator.

Index guiding helps maintain the z-oriented trajectory of photons traveling perpendicular to the mirror plane. Index guiding is achieved in a buried heterostructure laser diode by surrounding the semiconductor active region with a semiconductor of lower refractive index. Usually this involves etching the semiconductor wafer to define a narrow, z-oriented active-region stripe and then planarizing the etched regions by epitaxial growth of nonactive, lower refractive index, wider band-gap semiconductor.

Optical loss for a photon inside the Fabry–Perot cavity is minimized at cavity resonances. Figure 9.6 shows a schematic diagram of a Fabry–Perot optical resonator consisting of a semiconductor active-gain medium and two mirrors with reflectivity r_1 and r_2, respectively, forming an optical cavity of length L_C. The photon round-trip time in this cavity is $t_{\text{round-trip}}$.

Suppose mirror reflectivity is such that $r_1 = r_2 = 1$. Then the Fabry–Perot cavity has an optical mode spacing given by $kL_C = \pi m$, where $m = 1, 2, 3, \ldots$ and $k = \omega n_{\text{r}}/c$. The refractive index of the dielectric is n_{r}, and c is the speed of light in vacuum. Adjacent modes are spaced in angular frequency according to

$$\Delta\omega = \frac{c(k_{m+1} - k_m)}{n_{\text{r}}} = \frac{c\pi(m+1-m)}{L_C n_{\text{r}}} = \Delta\omega = \frac{c\pi}{L_C n_{\text{r}}} = 2\pi\,\Delta f \tag{9.45}$$

This mode spacing is also called the free spectral range of the cavity.

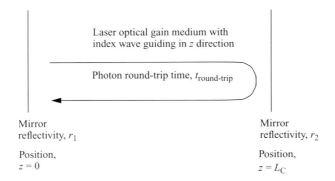

Fig. 9.6. Schematic diagram of a Fabry–Perot optical resonator consisting of a semiconductor active-gain medium and two mirrors with reflectivity r_1 and r_2, respectively, forming an optical cavity of length L_C.

Measured in units of hertz

$$\Delta f = \frac{c}{2L_C n_r} = \frac{1}{t_{\text{round-trip}}} \quad (9.46)$$

where $t_{\text{round-trip}}$ is the round-trip time for a photon in the cavity and $f = \omega/2\pi$. The mode spacing as a function of wavelength is

$$\Delta \lambda = \frac{\lambda^2}{2L_C n_r} \quad (9.47)$$

The spectral intensity as a function of frequency inside a Fabry–Perot cavity pumped by light of initial intensity I_0 in the cavity is[9]

$$I(f) = \frac{I_{\text{max}}}{1 + \left(\frac{2\mathcal{F}}{\pi}\right)^2 \sin^2\left(\frac{\pi f}{\Delta f}\right)} \quad (9.48)$$

where $\mathcal{F}$ is the finesse of the optical cavity

$$\mathcal{F} = \frac{\pi r^{1/2}}{1 - r^2} \quad (9.49)$$

and

$$I_{\text{max}} = \frac{I_0}{(1 - r)^2} \quad (9.50)$$

In these expressions, r is the round-trip attenuation factor for light amplitude in the cavity. If the only optical loss is from the two mirrors with reflectivity r_1 and r_2, respectively, then $r = r_1 r_2$. For a loss-less dielectric with refractive index n_r, the mirror reflectivity at a cleaved dielectric-to-air interface is $r_{1,2} = |(1 - n_r)/(1 + n_r)|^2$. When finesse is large ($\mathcal{F} \gg 1$), the optical line width γ_k is much smaller than Δf and $\mathcal{F} = \Delta f/\gamma_{\text{opt}}$. In this limit of $\mathcal{F} \gg 1$, the expression for the FWHM of the resonance becomes $\gamma_{\text{opt}} = \Delta f/\mathcal{F}$. The optical-$Q$ associated with the cavity is the frequency f_0 of the resonance divided by γ_{opt}, and so $Q = f_0/\gamma_{\text{opt}}$.

Consider a Fabry–Perot laser diode with cavity length $L_C = 300$ μm, an effective refractive index $n_r = 3.3$, and emission wavelength near $\lambda = 1310$ nm, corresponding to a frequency $f_0 = 229$ THz. The optical resonator has mode spacing $\Delta f = 151$ GHz or $\Delta \lambda = 0.867$ nm. The spectral intensity as a function of frequency, when $r_1 = r_2 = 0.286$, is shown in Fig. 9.7(a). Finesse is $\mathcal{F} = 0.979$. Also shown is the case in which $r_1 = 0.4$ and $r_2 = 0.8$, which has a slightly improved finesse of $\mathcal{F} = 1.98$. In this case $Q = f_0/\gamma_{\text{opt}} = 4978$ and $\gamma_{\text{opt}} = 46$ GHz.

Figure 9.7(b) shows the spectral intensity of a Fabry–Perot optical cavity of length $L_C = 3$ μm, effective refractive index $n_r = 3.3$, mirror reflectivity $r_1 = r_2 = 0.95$, and

[9] For example, see B. E. A. Saleh and M. C. Teich, *Fundamentals of Photonics*, John Wiley and Sons, New York, 1992 (ISBN 0 471 83965 5).

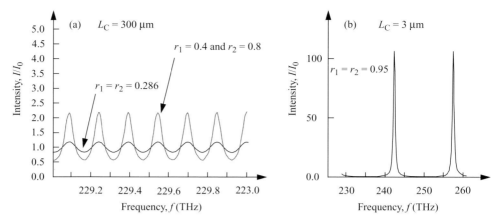

Fig. 9.7. (a) Spectral intensity as a function of frequency when $r_1 = r_1 = 0.286$ for photons inside a Fabry–Perot resonant cavity of length $L_C = 300$ μm and refractive index $n_r = 3.3$. Optical resonances are spaced by $\Delta f = 151$ GHz. The vertical axis is normalized in such a way that $I_0 = 1$ in the calculation. Also shown as the lower curve is the case in which $r_1 = 0.4$ and $r_2 = 0.8$. In this case, finesse $\mathcal{F} = 1.98$ and $\gamma_{opt} = 46$ GHz. At an optical frequency of $f = 229$ THz, this gives an optical $Q = f/\gamma_{opt} = 4978$. (b) Spectral intensity as a function of frequency when $r_1 = r_2 = 0.95$ for photons inside a Fabry–Perot resonant cavity of length $L_C = 3$ μm and refractive index $n_r = 3.3$. Optical resonances are spaced by $\Delta f = 15.1$ THz. In this case, the finesse $\mathcal{F} = \Delta f/\gamma_{opt} = 30.5$ and $\gamma_{opt} = 495$ GHz. At an optical frequency of $f = 242$ THz, this gives an optical $Q = f/\gamma_{opt} = 489$.

emission wavelength near $\lambda = 1240$ nm corresponding to a frequency $f_0 = 242$ THz. The optical resonator has resonance spacing $\Delta f = 15.1$ THz or $\Delta\lambda = 86.7$ nm. Finesse is $\mathcal{F} = \Delta f/\gamma_{opt} = 30.5$, line width is $\gamma_{opt} = 495$ GHz, and optical $Q = f_0/\gamma_{opt} = 489$.

9.4.1.2 Mode profile in an index-guided slab waveguide

We wish to calculate the optical mode profile in the x and y directions of the Fabry–Perot laser diode we have been discussing. This is an index-guided structure in which the refractive index of the active region, n_a, is greater than the refractive index, n_c, of the surrounding material. This usually small difference in refractive index acts to guide light close to the active region. Confining light to the active region is important, because only light that overlaps with the active region can experience optical gain and be amplified. We are, therefore, interested in the fraction, Γ, of a Fabry–Perot longitudinal optical resonance which overlaps with the active region.

One may proceed by using the time-independent electromagnetic wave equation, which may be derived from Maxwell's equations assuming no free charge and an electromagnetic wave traveling in the z direction:

$$\nabla^2\mathbf{E} + \varepsilon(x, y)k_0^2\mathbf{E} = 0 \tag{9.51}$$

where $k_0 = \omega/c$ is the propagation constant in free space and $\varepsilon(x, y)$ is the spatially

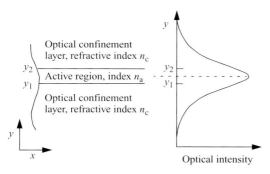

Fig. 9.8. Slab waveguide geometry showing active region of thickness $t_a = y_2 - y_1$ and optical confinement layers. The propagation of light is in the z direction, which is into the page. Optical intensity peaks in the active region, which has refractive index n_a. The refractive index of the optical confinement layer is n_c.

varying dielectric constant in the slab waveguide geometry. This has solution

$$\mathbf{E} = \mathbf{e}\phi(y)\psi(x)e^{i\beta z} \tag{9.52}$$

where $\mathbf{e}$ is the electric-field unit vector and β is the propagation constant in the dielectric.

One then solves for the optical confinement factor

$$\Gamma = \frac{\int\limits_{y_1}^{y_2} \phi^2(y)dy}{\int\limits_{-\infty}^{\infty} \phi^2(y)dy} \tag{9.53}$$

by assuming that $\varepsilon(x, y)$ varies slowly in the x direction compared with the y direction and by adopting the effective index approximation. In essence, we solve for the simple slab waveguide geometry depicted in Fig. 9.8, in which the thickness of the active region is $t_a = y_2 - y_1$.

Figure 9.9 shows the results of calculating the optical confinement factor of TE and TM modes in a slab waveguide as a function of bulk active-layer thickness. The parameters used are typical for a laser diode with emission at $\lambda_0 = 1310$ nm wavelength and an InGaAsP active region. For a given active region thickness, the optical confinement factor for TE polarization is greater than that for TM polarization. TE-polarized light propagating in the z direction has its electric field parallel to the x direction and so it is in the plane of the active-region layer.

For most Fabry–Perot laser designs that use index guiding, the ratio of active-region thickness t_a to emission wavelength λ_0 is small, and one may find the confinement factor for TE-polarized light using the approximation[10]

$$\Gamma_{TE} = 2\left(n_a^2 - n_c^2\right)\left(\frac{\pi t_a}{\lambda_0}\right)^2 \tag{9.54}$$

---

[10] W. P. Dumpke, *IEEE J. Quantum Electron.* **QE-11**, 400 (1975).

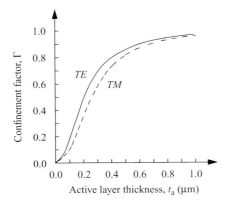

Fig. 9.9. Calculated optical confinement factor Γ for TE and TM modes in a slab waveguide as a function of bulk active-layer thickness, t_a. The lasing wave length is $\lambda = 1310$ nm the InGaAsP active layer has refractive index $n_a = n_{\text{InGaAsP}} = 3.51$, and the InP optical confinement layers have refractive index $n_c = n_{\text{InP}} = 3.22$.

In the case of TM-polarized light, the approximation for the confinement factor is

$$\Gamma_{\text{TM}} = 2\left(n_a^2 - n_c^2\right)\left(\frac{\pi n_c t_a}{n_a \lambda_0}\right)^2 \tag{9.55}$$

9.4.2   Mirror loss and photon lifetime

Previously, in Section 9.4.1.1, we used reflectivity $r_1 = r_2 = 1$ to calculate longitudinal mode frequency in a Fabry–Perot resonator. In most practical situations it will be necessary to consider the situation in which $r_1 \neq r_2 < 1$. In a typical semiconductor laser diode, the mirror facets have dielectric coatings to give power reflectivity values r_1 and r_2.

We are interested in finding the rate of loss of photons, $1/\tau_{\text{photon}}$, into regions other than the active laser region. Suppose the photon density is S. Then we wish to find S/τ_{photon}. A simple rate equation analysis shows that photon density grows exponentially in the presence of optical gain $g_{\text{opt}} = -\alpha$, so that $S = S_0 e^{-2\alpha z}$. Note the factor 2 because S is an intensity.

To convert optical gain to a rate, we introduce the rate of increase in optical intensity $G = 2g_{\text{opt}} \cdot c/n_r$, where n_r is the effective refractive index and c is the velocity of light in vacuum. For steady-state emission, the photons reflected back to the start in a single round-trip time $t_{\text{round-trip}} = 2L_c n_r/c$ must have the same density. So, if S_0 photons start out from mirror r_1 of the cavity, then $r_2 S_0 e^{(G \cdot t_{\text{round-trip}})/2}$ are reflected back from mirror r_2 to grow with another pass down the laser, and $r_1 r_2 S_0 e^{G \cdot t_{\text{round-trip}}}$ are reflected from mirror r_1 (see Fig. 9.6). Hence, in steady state,

$$S_0 = r_1 r_2 S_0 e^{G \cdot t_{\text{round-trip}}} \tag{9.56}$$

which we can rewrite by taking the logarithm of both sides:

$$G = \frac{1}{t_{\text{round-trip}}} \ln\left(\frac{1}{r_1 r_2}\right) = \frac{c}{2L_{\text{C}} n_{\text{r}}} \ln\left(\frac{1}{r_1 r_2}\right) \tag{9.57}$$

Ignoring spontaneous emission, we can write down a rate equation for the photon density

$$\frac{dS}{dt} = \left(G - \frac{1}{\tau_{\text{photon}}}\right) S = 0 \tag{9.58}$$

giving $G\tau_{\text{photon}} = 1$, so that

$$\boxed{\frac{1}{\tau_{\text{photon}}} = \frac{c}{2L_{\text{C}} n_{\text{r}}} \ln\left(\frac{1}{r_1 r_2}\right)} \tag{9.59}$$

To include the possibility of additional absorption and elastic scattering of light, it is necessary to introduce an extra photon loss term. We lump these loss-rates together as an additional internal loss-rate $1/\tau_{\text{internal}}$, so that the total photon loss-rate is

$$\frac{1}{\tau_{\text{photon}}} = \frac{1}{\tau_{\text{internal}}} + \frac{1}{\tau_{\text{mirror}}} \tag{9.60}$$

or, equivalently,

$$\boxed{\kappa = \alpha_{\text{i}} + \alpha_{\text{m}}} \tag{9.61}$$

where α_{i} is the internal photon loss-rate, α_{m} is the photon mirror loss, and κ is the total optical loss-rate.

9.4.3   The Fabry–Perot laser diode

Figure 9.10 is a sketch of a semiconductor, buried-heterostructure, Fabry–Perot laser diode. The diagram shows the bulk-active or quantum-well region exposed at one of the two cleaved-mirror faces. Carriers are injected into the region from the n-type substrate and the p-type epitaxially grown layers from below and above the p–n junction. Electrical contact to the diode is achieved by depositing a metal film and subsequent alloying into a surface layer of the semiconductor.

A laser with emission at wavelength $\lambda_0 = 1310$ nm can have a bulk active InGaAsP region. The InGaAsP composition is such that its band gap is at wavelength $\lambda_{\text{g}} = 1280$ nm. Under lasing conditions various physical effects cause the lasing wavelength to increase so that the device lases at wavelength $\lambda_0 = 1310$ nm. In a typical device, the bulk or multiple quantum-well active region is 0.12 μm thick and 0.8 μm wide. The wafer is thinned before cleaving to form the two mirror facets. Thinning the wafer to about 120-μm thickness helps to ensure that stress-induced irregularities are avoided on the cleaved mirror faces. The buried heterostructure is achieved using an etching

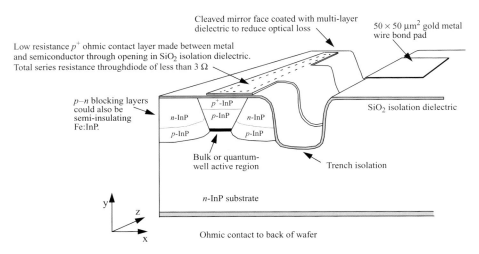

Low resistance p^+ ohmic contact layer made between metal
and semiconductor through opening in SiO$_2$ isolation dielectric.
Total series resistance throughdiode of less than 3 Ω

Cleaved mirror face coated with multi-layer
dielectric to reduce optical loss

50×50 μm^2 gold metal
wire bond pad

p–n blocking layers
could also be
semi-insulating
Fe:InP.

n-InP

p^+-InP

p-InP n-InP

p-InP p-InP

SiO$_2$ isolation dielectric

Bulk or quantum-
well active region

Trench isolation

n-InP substrate

Ohmic contact to back of wafer

Fig. 9.10. Schematic diagram of a semiconductor buried-heterostructure Fabry–Perot laser diode.
The difference in refractive index between the active region and the surrounding dielectric causes
index guiding of light propagating in the z direction.

and semiconductor regrowth process. Index guiding of the $\lambda_0 = 1310$ nm lasing mode
occurs because of the refractive index difference between the InGaAsP active layer with
$n_{\text{InGaAsP}} = 3.51$ and the InP optical confinement layers with refractive index $n_{\text{InP}} =
3.22$. For an index-guided buried heterostructure with a 0.12 μm thick and 0.8 μm
wide active region this ensures a single transverse mode and a high optical confinement
of around $\Gamma = 0.25$. The Fabry–Perot cavity length is $L_{\text{C}} = 300$ μm. A multi-layer
dielectric mirror coating is used to increase reflectivity to 0.4 on one mirror and 0.8 on
the other. This reduces optical loss and reduces laser threshold current to a value that
is typically around 3 mA.

9.4.4   Semiconductor laser diode rate equations

To understand the operation of the semiconductor buried-heterostructure Fabry–Perot
laser diode shown schematically in Fig. 9.10 and Fig. 9.11, we need to develop a model.
The simplest approach is to use rate equations.

We will assume that there can be lasing into only one optical mode of frequency ω_s.
Our calculations will not incorporate any variation in optical gain, optical loss, or carrier
density along the longitudinal (z) axis. This is equivalent to a lumped-element model.
We will adopt simple approximations for gain, spontaneous emission, and nonradiative recombination. We will also assume that the conduction-band and valence-band
electrons have the same density and that they are thermalized so that they may be
characterized by a single temperature.

When considering the rate equations one needs to be very careful to define the
parameters used. It is easy to become confused and make an error. The current, I,
injected into the diode is measured in amperes, and the volume of the active region

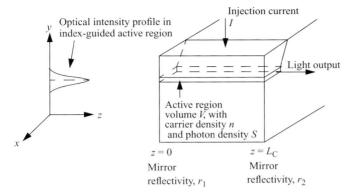

Fig. 9.11. Section through a buried-heterostructure Fabry–Perot laser diode. Index guiding ensures optical intensity is tightly confined near the active region of the device. Fabry–Perot cavity length is L_C, and mirror reflectivity is r_1 at $z = 0$ and r_2 at $z = L_C$. When current I is injected into the diode, the active region of volume V has carrier density n and photon density S.

into which electrons flow is V. The carrier density in the active region is n and is measured in either m^{-3} or cm^{-3}. The photon density in the optical mode of frequency ω_s is S and is measured in m^{-3} or cm^{-3}. The two mirrors used to form the optical cavity for photons in the device have reflectivities r_1 and r_2, respectively. Optical loss at frequency ω_s in the cavity is κ and is measured in m^{-1} or cm^{-1}. In our rate equations, we need to define the fraction of spontaneous emission, β, that feeds into the lasing mode at frequency ω_s. It turns out that this is somewhat difficult to define in an active device. In a large Fabry–Perot device with $L_C = 300$ μm, a typical value for β is in the range $10^{-4} < \beta < 10^{-5}$. In devices with smaller L_C, the value of β becomes larger. Theoretically, the maximum possible value of β is near unity.

We will use rate equations to describe rates in and out of a region of interest. The physical quantities we monitor are carrier density n, photon density S in a single optical mode of frequency ω_s, and current I as a function of time, t. Rate equations keep track of current, carrier, and photon flow in and out of the device. The challenge is to make sure that we don't miss some quantity of importance.

We begin by drawing a bucket to represent the active region of the semiconductor, see Fig. 9.12. We now imagine charge carriers supplied by a current I being poured into the bucket at a rate so that the *density* of electrons per second increases as I/eV, where e is the electron charge and V is the volume of the active region. There are losses or leaks in the bucket which represent mechanisms for removing electrons from the system. Electrons can be removed by emitting a photon or by some nonradiative process.

Because we are interested in n and S, there are two coupled rate equations that we must solve. The first equation will describe the rate of change in carrier density dn/dt in the device. Carrier density will increase as more current is injected, so we expect dn/dt to have a term proportional to current I. Carriers are removed from the active region

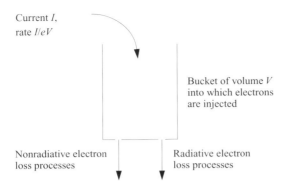

Current I,
rate I/eV

Bucket of volume V
into which electrons
are injected

Nonradiative electron
loss processes

Radiative electron
loss processes

Fig. 9.12. Bucket model for electron rates into and out of a semiconductor active region.

of the device by nonradiative and spontaneous photon-emission carrier-recombination processes. We expect to characterize this carrier loss by carrier lifetime τ_n, so that dn/dt should be proportional to $-n/\tau_n$. The negative sign reflects the fact that carriers are removed from the system. There are also carrier losses due to stimulated photon emission, in which an electron in the conduction band and a hole in the valence band are removed to create a photon in the lasing mode. This stimulated emission process influences the number of carriers via the rate $-GS$, where G is the optical gain and S is the photon density in the lasing mode.

The second equation will describe the rate of change in photon density dS/dt in the device. Photon density will increase due to the presence of optical gain, so we expect a term proportional to GS. There will also be optical losses that can be described by a total optical loss rate κ, giving a term $-\kappa S$. Finally, there is a fraction β of total spontaneous emission r_{spon} feeding into the lasing mode that makes a contribution βr_{spon}.

These considerations allow us to write down our basic coupled rate equations that describe the behavior of the laser diode:

$$\frac{dn}{dt} = \frac{I}{eV} - \frac{n}{\tau_n} - GS \tag{9.62}$$

$$\frac{dS}{dt} = (G - \kappa)S + \beta r_{\text{spon}} \tag{9.63}$$

Equations (9.62) and (9.63) are the single-mode rate equations. In Eqn (9.62), $1/\tau_n$ is the phenomenological carrier recombination rate, where we use $n/\tau_n = A_{\text{nr}}n + Bn^2 + Cn^3$, for which A_{nr} is the nonradiative recombination rate, B is the spontaneous emission rate and C is a higher-order term. The total spontaneous emission rate in the device is $r_{\text{spon}} = Bn^2$, and the function for optical gain in a device with a bulk-active region is

$$G_{\text{bulk}} = \Gamma G_{\text{slope}}(n - n_0)(1 - \varepsilon_{\text{bulk}}S) \tag{9.64}$$

where G_{slope} is the differential optical gain with respect to carrier density.

For a quantum-well active region, the optical gain function may be approximated by

$$G_{\text{QW}} = \Gamma G_{\text{const}} \cdot \ln\left(\frac{n}{n_0}\right) \cdot (1 - \varepsilon_{\text{QW}} S) \tag{9.65}$$

Note that for convenience we include the optical confinement factor Γ in the expressions for gain. The carrier density needed to achieve optical transparency at frequency ω_s is n_0. Both $\varepsilon_{\text{bulk}}$ and ε_{QW} are gain saturation terms, which become important at high optical intensities in the cavity. The gain functions we use represent fairly well the variation of peak gain with increasing carrier density.

We are now in a position to explain the steady-state carrier density and photon density characteristics of a laser diode as a function of injected current. First, we write Eqn (9.62) for the steady-state case:

$$\frac{dn}{dt} = \frac{I}{eV} - \frac{n}{\tau_n} - GS = 0 \tag{9.66}$$

This equation shows that in the steady state the rate of electron density injected into the active region is exactly balanced by the removal of electrons via the recombination rate n/τ_n and the optically stimulated recombination rate GS.

Now we note that, in the steady state, Eqn (9.63) is

$$\frac{dS}{dt} = (G - \kappa)S + \beta r_{\text{spon}} = 0 \tag{9.67}$$

which may be rewritten as

$$S = \frac{\beta r_{\text{spon}}}{(\kappa - G)} \tag{9.68}$$

It is this last equation that we can use as a starting point to explain how a laser works. The numerator βr_{spon} on the right-hand side of Eqn (9.68) shows that the optical output S of the laser amplifier is fed by a small fraction of the total spontaneous emission in the device. Because spontaneous emission is a stochastic (random) quantum mechanical process, we may view the laser as amplifying noise. Later, in Section 9.6, we will be interested in characterizing the noise and its impact on application of laser diodes for fiber-optic communications.

The denominator $(\kappa - G)$ on the right-hand side of Eqn (9.68) is the term responsible for optical amplification and lasing emission. As electrons are injected into the device, optical gain increases, and $(\kappa - G)$ approaches zero, the amplification of spontaneous emission increases. This increase in photon density in the device is so great that the stimulated carrier-recombination rate term $-GS$ in Eqn (9.66) becomes large. As $(\kappa - G)$ continues to approach zero, the net optical amplification of spontaneous emission $1/(\kappa - G)$ becomes very large, and the stimulated recombination rate $-GS$ dominates Eqn (9.66). At this point, every additional electron injected by current I

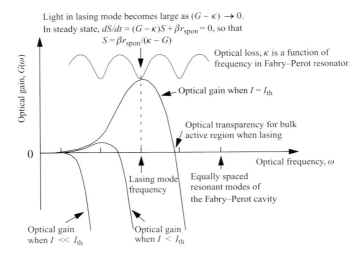

Fig. 9.13. Schematic plot of optical gain as a function of optical frequency. Lasing will occur when optical gain approaches optical loss. This will usually occur at a resonance of the Fabry–Perot resonator, the optical loss of which is also shown.

into the active region recombines very rapidly to create an additional photon in the lasing mode at frequency ω_s.

Figure 9.13 illustrates the situation we have just described. As optical gain approaches a minimum in Fabry–Perot optical-cavity loss in such a way that $(G - \kappa) \to 0$, it is clear that the lasing mode will coincide with that of the Fabry–Perot resonance of frequency ω_s nearest to peak optical gain. Because in this case only one high-Q optical-cavity resonance is in the same frequency range as semiconductor optical gain, we can expect lasing in one optical mode at frequency ω_s.

As stimulated recombination begins to dominate, light emission at frequency ω_s rapidly increases. The point at which this occurs is called the laser threshold. Associated with the laser diode threshold is a threshold current, I_{th}, and a threshold carrier density, n_{th}. For currents above the laser threshold, carrier density does not increase very much, because the stimulated recombination rate $-GS$ dominates the carrier dynamics described by Eqn (9.66). The carrier density is said to be *pinned* above threshold. Such carrier pinning results in a rapid linear increase in laser light output intensity with increasing injection current, because every extra injected electron is converted to a lasing photon.

The ideas we have discussed in the previous few paragraphs are illustrated in Fig. 9.14 and Fig. 9.15. Figure 9.14 shows total laser diode light output intensity L as a function of injected current I for a device with a threshold current I_{th}. Figure 9.15 shows carrier density n as a function of injected current I. As may be seen, carrier density is pinned to a value of approximately n_{th} when current is greater in value than I_{th}.

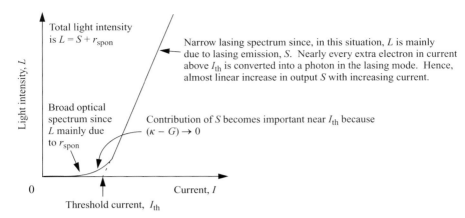

Fig. 9.14. Anticipated total light output intensity L as a function of injected current I. The device has a threshold current I_{th}.

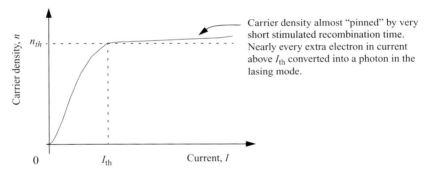

Fig. 9.15. Anticipated carrier density n as a function of injected current I. The carrier density is pinned to a value of approximately n_{th} when current is greater than threshold current I_{th}.

9.5 Numerical method of solving rate equations

In the previous section, the single-mode laser diode rate equations were used to qualitatively predict the steady-state behavior of a laser diode. While there are many practical applications that can make use of such steady-state characteristics, we are also interested in the high-speed, large-signal response of the device. Data transmission via an optical fiber medium requires a photon source in which lasing light intensity is modulated. A simple way to change lasing light emission is to change the injection current. Modulating in a one-bit digital fashion, a high level of light might correspond to binary 1 and a low level of light might correspond to binary 0. To efficiently pass data in a fiber-optic link, one would like to know how fast a laser can switch from a 1 state to a 0 state. To answer such basic questions concerning laser performance it is necessary to resort to numerical methods that are capable of predicting the large-signal dynamic response of a laser diode's carrier density and laser light emission in response to rapid changes in injection current.

Because we will be writing a computer program to solve Eqn (9.62) and Eqn (9.63), it is worth spending some time outlining how we will proceed. The basic rules of programing are to keep everything as simple as possible at first so that it is easy to fix any bugs. Always display the results graphically, and carefully document your program, so that when you return to it at a later date you can understand what you did. In this case, particular care is taken to define all of the parameters and rescale to units that are appropriate for the problem. We will scale time to nanoseconds and length to nanometers.

We will integrate the coupled rate equations using the fourth-order Runge–Kutta method to be discussed in the next section. When using this integrator, it is important to make sure that a time step size is chosen that is appropriate for the equations. In our case, a time step of around $t_{step} = 1$ ps is the correct value. Again, to maintain a simple approach, the time step is fixed. More sophisticated routines with adaptive step size are left for later consideration.

9.5.1   The Runge–Kutta method

Ordinary differential equations may always be expressed in terms of first-order differential equations. For example,

$$\frac{d^2y}{dt^2} + a(t)\frac{dy}{dt} = b(t) \tag{9.69}$$

may be written as two first-order coupled differential equations

$$\frac{dy}{dt} = f(t) \tag{9.70}$$

$$\frac{df}{dt} = b(t) - a(t)f(t) \tag{9.71}$$

We are therefore interested in the study of N coupled first-order differential equations for the functions y_j having the form

$$\frac{d}{dt}y_j(t) = f_j(t, y_1, \ldots, y_N) \tag{9.72}$$

where the right-hand side is a known function and $j = 1, 2, \ldots, N$.

After applying boundary conditions and an initial value, we use a finite step h_0 to numerically solve the equations. For example, we might use Euler's method

$$y_{n+1} = y_n + h_0 f(t_n, y_n) + 0(h_0^2) \tag{9.73}$$

to advance the solution from t_n to $t_{n+1} \equiv t_n + h_0$. It advances the solution through an interval h_0, using only derivative information contained in $f(t, y)$ at the beginning of

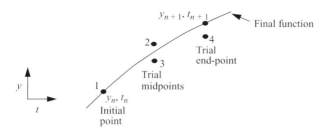

Fig. 9.16. Illustration of fourth-order Runge–Kutta method of numerically estimating integration of a function.

the interval. Unfortunately, the error in each step is $0(h_0^2)$, which is only one power of h_0 smaller than the estimate function $h_0 f(t_n, y_n)$.

One may do better than this by first taking a trial step to the midpoint of the interval and then using the value of both t and y at the midpoint to compute a more accurate step across the complete interval. If we evaluate $f(t, y)$ in such a way that first-order and some higher-order terms cancel, we can make a very accurate numerical integrator. The fourth-order Runge–Kutta method does just this. As illustrated in Fig. 9.16, each step along $f(t, y)$ is evaluated four times, once at the initial point, twice at the mid-points and once at the trial point.

The fourth-order Runge–Kutta method is summarized by the equation

$$y_{n+1} = y_n + \frac{k_1}{6} + \frac{k_2}{3} + \frac{k_3}{3} + \frac{k_4}{6} + 0(h^5) \tag{9.74}$$

where

$$k_1 = h_0 f(t_n, y_n) \tag{9.75}$$

is used to evaluate at the initial point,

$$k_2 = h_0 f\left(t_n + \frac{h_0}{2}, y_n + \frac{k_1}{2}\right) \tag{9.76}$$

is used to estimate the midpoint using $k_1/2$,

$$k_3 = h_0 f\left(t_n + \frac{h_0}{2}, y_n + \frac{k_2}{2}\right) \tag{9.77}$$

is used to estimate the midpoint using $k_2/2$, and

$$k_4 = h_0 f(t_n + h_0, y_n + k_3) \tag{9.78}$$

is used to evaluate the end point using k_3.

The single-mode rate equations Eqn (9.62) and Eqn (9.63) are two coupled first-order differential equations that can be solved using the Runge–Kutta method. However, when we do so, care has to be taken to make sure that the estimates for trial points k_1, k_2, k_3, and k_4 are updated using the most current values.

Table 9.1. *Fabry–Perot laser diode rate equation parameters*

Description	Parameter	GaAs 850 nm wavelength	InGaAsP 1310 nm wavelength	InGaAsP 1550 nm wavelength
Refractive index	n_r	3.3	4	4
Cavity length	L_C (cm)	250×10^{-4}	300×10^{-4}	500×10^{-4}
Active layer thickness	t_a (cm)	0.14×10^{-4}	0.14×10^{-4}	0.14×10^{-4}
Active layer width	w_a (cm)	0.8×10^{-4}	0.8×10^{-4}	0.8×10^{-4}
Integration time increment	t_{inc} (s)	1×10^{-12}	1×10^{-12}	1×10^{-12}
Nonradiative recombination coefficient	A_{nr} (s^{-1})	2×10^{8}	2×10^{8}	1×10^{8}
Radiative recombination coefficient	B (cm^3 s^{-1})	1×10^{-10}	1×10^{-10}	1×10^{-10}
Nonlinear recombination coefficient	C (cm^6 s^{-1})	1×10^{-29}	1×10^{-29}	5×10^{-29}
Transparency carrier density	n_0 (cm^{-3})	1×10^{18}	1×10^{18}	1×10^{18}
Optical gain-slope coefficient	G_{slope} (cm^2 s^{-1})	3.3×10^{-16}	2.5×10^{-16}	2.0×10^{-16}
Gain saturation coefficient	ε_{bulk} (cm^3)	2×10^{-18}	3×10^{-18}	5×10^{-18}
Spontaneous emission coefficient	β	1×10^{-4}	5×10^{-5}	1×10^{-5}
Optical confinement factor	Γ	0.25	0.25	0.25
Mirror 1 reflectivity	r_1 (cm^2 s^{-1})	0.3	0.32	0.32
Mirror 2 reflectivity	r_2 (cm^2 s^{-1})	0.3	0.32	0.32
Internal optical loss	α_i (cm^{-1})	20	40	50

9.5.2   Large-signal transient response

When writing a computer program to model the behavior of a laser diode with a bulk active gain region, we need to define the functions in the rate equations Eqn (9.62) and Eqn (9.63), including assigning numerical values. Table 9.1 gives some typical values for the parameters introduced in Section 9.4.4.

In Fig. 9.17, the results of calculating the large-signal response of a diode laser to a 30 mA step change in current are shown. The model uses parameters given in Table 9.1 for a laser with emission wavelength near $\lambda_0 = 1310$ nm.

There are a number of characteristic features worth mentioning. First, because the device is turned on from a zero-current state, there is a significant *turn-on delay*, t_d, associated with the fact that it takes time to inject enough carriers to bring the device to a lasing state. Second, as usual with coupled rate equations, there will be some phase delay between carrier density, n, and photon density, S. The electrons lead the photon density. This gives rise to significant photon density overshoot and *relaxation oscillations* in both the optical output, S, and the carrier density, n, as the system tries to establish steady-state conditions. This is typical of a response to a large step change in injection current, especially if current, I, passes through the threshold value, I_{th}. Relaxation oscillations limit the useful switching speed of laser diodes.

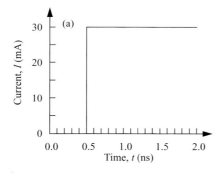

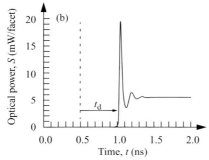

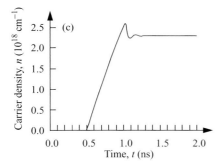

Fig. 9.17. (a) Input step current as a function of time. Current is initially zero, increasing to 30 mA at time $t = 0.5$ ns. Laser threshold is $I_{th} = 5.7$ mA. (b) Laser diode light output per mirror facet as a function of time, showing turn-on delay, t_{α}, and overshoot in optical light output. Mirror reflectivity is the same for each facet. (c) Conduction-band and valence-band carrier density as a function of time. The calculation uses parameters for an InGaAsP device as given in Table 9.1.

9.5.3 Cavity formation

The single-mode rate equations Eqn (9.62) and Eqn (9.63) can be used to learn a great deal about the high-speed performance of a laser diode. The large-signal response to a step change in current was investigated in Section 9.5.2, and it was found that turn-on delay and relaxation oscillations have to be considered when evaluating the switching speed of laser diodes. However, to some extent these effects can be mitigated by engineering and design. With this in mind, it seems appropriate to ask if there are other, more fundamental, limitations to laser diode operation and switching speed. Of course, there are and one such example is called *cavity formation*.

To explain this, we consider a photon inside a Fabry–Perot laser diode. The photon cannot know it is in a Fabry–Perot resonator or at resonance until it interacts with the mirrors. If there are no mirrors, the device is merely a light-emitting diode (LED). Hence, lasing emission into a cavity resonance requires that the photon experience at least one round-trip within the resonator. The device cannot behave as a laser until the photon cavity has formed. This takes *at least* one photon round-trip time, which, in a conventional Fabry–Perot laser diode, is about 10 ps.

You may ask what effect multiple photon round-trips have on the time evolution of lasing light emission intensity and spectra. This is not particularly easy to answer, because usually photon cavity formation is obscured by the nonlinear coupling of the optical field with the optical gain medium. Under normal conditions, it is difficult to measure the effect of multiple round-trips on the evolution of lasing light intensity and lasing spectra due to the short cavity round-trip time and charge carrier lifetime. However, by adiabatically decoupling the cavity formation (by making a large external cavity) from other processes such as charge carrier dynamics, experiments can be performed that explore this issue.[11] The results can also be predicted using time-delayed, single-mode or multi-mode rate equations. Importantly, the experiments show how a laser uses cavity formation to drive the device from an LED to a laser. Surprisingly, approximately 200 photon round-trips are needed to approach steady-state laser characteristics that are *independent* of the laser injection current (see Fig. 9.18). Obtaining pure steady-state spectral behavior requires even more photon round-trips.

9.6 Noise in laser diode light emission

The laser works by amplifying spontaneous emission. Spontaneous emission is a *fundamentally quantum mechanical* and *random process*. One therefore anticipates that the light emitted from a laser diode is inherently noisy. Because fiber-optic communication systems use intensity modulation of laser light, we will be interested in developing a model with the aim of understanding more about intensity noise.

Figure 9.19(a) shows how intensity-modulated laser light can be used to transmit bits of information. In this case, a high level of light is binary 1 and a low level of light is binary 0. Because the receiver circuitry needs time to decide between high and low light levels, data are transmitted at a well-defined rate called the bit rate, $1/\tau_{\text{bit}}$. A typical

---

[11] J. O'Gorman, A. F. J. Levi, D. Coblentz, T. Tanbun-Ek, and R. A. Logan, *Appl. Phys. Lett.* **61,** 889 (1992).

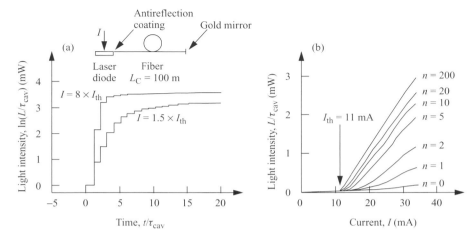

Fig. 9.18. (a) Natural logarithm of laser emission intensity as a function of time, t/τ_{cav}. Time is measured in units of the cavity round-trip time $\tau_{cav} = 0.995$ μs. Light level is normalized to the emission intensity when $0 < n < 1$, where n is the round-trip number. Inset shows the experimental arrangement. The laser diode is the active region of a 100-m-long Fabry–Perot resonator formed using a glass fiber. The diode is pumped by a step change in current, I, and the light output, L, is monitored as a function of cavity round-trip time using a high-speed photodetector. (b) Normalized light output power as a function of injection current for the indicated number of photon round-trip trips, n. When $n = 0$, the device behaves as an LED. When $n = 200$, the current–light output characteristics approach the steady-state behavior.

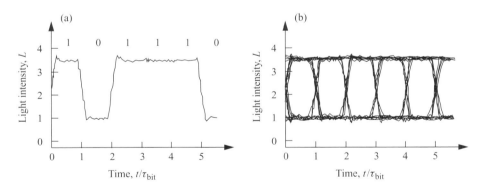

Fig. 9.19. (a) Intensity-modulated laser light can be used to transmit bits of information in which a high level of light is binary 1 and a low level of light is binary 0. Laser intensity noise is a source of bit errors. (b) A continuously sampled data stream creates an eye diagram in which the eye width and height define the area in which the receiver circuitry can minimize the probability of bit errors.

bit rate in fiber-optic communication systems is 10 Gb s^{-1}. In this case, $\tau_{bit} = 100$ ps and 10^{10} bits of information can be transmitted through the system in one second. With such large information capacity, one would like to minimize the chance of bit errors due to laser intensity noise.

If one randomly and continuously samples a data stream in time while synchronized to the bit rate, one obtains the *eye diagram* shown in Fig. 9.19(b). The opening in the center of the eye diagram is where the receiver circuitry decides between high and low light levels. The greater the amount of intensity noise in the laser signal, the smaller the region in which the receiver can make its decision and the greater the chance of errors.

Noise in the intensity of light output from a laser diode is characterized using a quantity called relative intensity noise (*RIN*). The optical intensity S contains noise, so that

$$S(t) = \langle S \rangle + \delta S(t) \tag{9.79}$$

where $\langle S \rangle$ is the time-averaged optical intensity and $\delta S(t)$ is the deviation from the average value at any given instant in time, t. The time average in the fluctuation $\delta S(t)$ is $\langle \delta S(t) \rangle = 0$. The noise $\delta S(t)$ may be characterized in the time domain by the auto-correlation function

$$g_S(\tau) = \langle \delta S(t) \delta S(t - \tau) \rangle \tag{9.80}$$

or in the frequency domain by the noise spectral density

$$\langle |\delta S(\omega)|^2 \rangle = \int_{\tau=-\infty}^{\tau=\infty} g_S(\tau) e^{-i\omega t} d\tau = \frac{1}{t'}\bigg|_{t' \to \infty} \left| \int_0^{t'} \delta S(t) e^{-i\omega t} dt \right|^2 \tag{9.81}$$

RIN is defined as the square of fluctuations in optical intensity noise at frequency ω divided by the average value of optical intensity $\langle S \rangle = S_0$ squared at frequency ω, so that

$$RIN(\omega) = \frac{\langle |\delta S(\omega)|^2 \rangle}{\langle S(\omega) \rangle^2} = \frac{\langle |\delta S(\omega)|^2 \rangle}{S_0^2(\omega)} \tag{9.82}$$

RIN is measured in units of either dB Hz^{-1} or Hz^{-1}.

RIN may be investigated theoretically using a slight modification of the single-mode rate equations Eqn (9.62) and Eqn (9.63) we have already developed. We use *Langevin* rate equations,

$$\frac{dS}{dt} = (G - \kappa)S + \beta r_{\text{spon}} + F_s(t) \tag{9.83}$$

and

$$\frac{dn}{dt} = \frac{1}{eV} - \frac{n}{\tau_n} - GS + F_e(t) \tag{9.84}$$

where S and N are the photon and carrier densities in the cavity, G is optical gain, κ is optical loss, β is the spontaneous emission factor, n/τ_n is the carrier recombination rate,

e is the charge of an electron, r_{spon} accounts for spontaneous emission into all optical modes, and V, the volume of the semiconductor active region. A source of random noise in the rate equations is included through the terms $F_s(t)$ and $F_e(t)$. These are called the Langevin noise terms. Using Eqn (9.83) and Eqn (9.84) is the simplest way to include noise into our model of a laser diode. A slightly more detailed approach also takes into account optical phase noise.[12] In the Markovian approximation, corresponding to instantaneous changes in $F_s(t)$ and $F_e(t)$, the autocorrelation and cross-correlation functions are given by

$$\langle F_e(t)F_e(t')\rangle = \left(I/eV + \frac{n}{\tau_n} + GS\right)\delta(t - t') \tag{9.85}$$

$$\langle F_s(t)F_s(t')\rangle = ((G + \kappa)S + \beta r_{spon})\delta(t - t') \tag{9.86}$$

$$\langle F_s(t)F_e(t')\rangle = -(GS - \beta r_{spon})\delta(t - t') \tag{9.87}$$

The Markovian approximation is guaranteed by use of $\delta(t - t')$. The expression $\langle F_e(t)F_e(t')\rangle$ is the square of Gaussian fluctuations around the mean value of n given by the rate equation Eqn (9.62). $\langle F_s(t)F_s(t')\rangle$ is just the square of Gaussian fluctuations around the mean value of S given by the rate equation Eqn (9.63). The cross-correlation term $\langle F_s(t)F_e(t')\rangle$ shows that the rate equations for S and n are coupled and hence correlated. The negative sign in Eqn (9.87) indicates that $F_s(t)$ and $F_e(t)$ are anticorrelated.

The mean steady-state value for S is S_0, and that for n is n_0. The magnitude of the Fourier component of S at RF angular frequency ω is $\delta S(\omega)$, and that for n is $\delta n(\omega)$. Linearizing the dynamical Eqn (9.83) and Eqn (9.84) for a constant average diode current, and neglecting gain saturation, gives

$$\delta S(\omega) = \frac{F_e(\omega)\left(\dfrac{dG}{dn}S_0 + 2\beta B n_0\right) + F_s(\omega)\left(i\omega + \dfrac{dG}{dn}S_0 + \dfrac{1}{\tau_n} + \dfrac{1}{\tau_n'}n_0\right)}{\left(i\omega\left(\dfrac{dG}{dn}S_0 + \dfrac{n}{\tau_n} + \dfrac{1}{\tau_n'}n_0\right) - \omega^2 + G\dfrac{dG}{dn}S_0 + G2\beta B n_0\right)} \tag{9.88}$$

and

$$RIN \equiv \lim(\tau \to \infty)\frac{1}{\tau}\left|\frac{\delta S(\omega)}{S_0(\omega)}\right|^2 \tag{9.89}$$

In these expressions, $F_e(\omega)$ and $F_s(\omega)$ are the Fourier components at ω of $F_e(t)$ and $F_s(t)$, respectively, and $1/\tau_n' \equiv (d/dn)(1/\tau_n)$.

---

[12] For a somewhat more complete model, see M. Ahmed, M. Yamada, and M. Saito, *IEEE J. Quantum Electron.* **QE-37**, 1600 (2001).

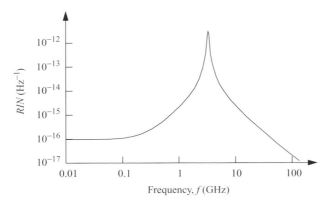

Fig. 9.20. Calculated RIN as a function of frequency, f, for a Fabry–Perot laser diode with active volume $V = 300 \times 2 \times 0.05 \ \mu m^3$. The mirror reflectivity is 0.3 per facet, and the steady-state current bias is $I_0 = 4 \times I_{th} = 7.36$ mA. The average photon number in the device is $S_0 = 9.5 \times 10^4$, and the average carrier number is $N_0 = 5.9 \times 10^7$.

The cause of the peak in RIN shown in Fig. 9.20 can be traced to the pole in Eqn (9.88). Physically it arises from fluctuations in carrier density and photon density working in phase to amplify the response to a noise fluctuation. The origin is similar to the cause of relaxation oscillation observed in the calculated average large-signal response of carriers and photons shown in Fig. 9.17.

9.7 Why our model works

It is truly remarkable that the simple model we have used throughout this chapter to describe the behavior of semiconductor lasers gives useful results. Engineers can use this, and slightly improved versions of the model, to successfully design and optimize the performance of laser diodes. The fact that this is so is largely accidental, and if we wish to understand detailed physical processes in a device our lack of knowledge becomes painfully obvious.

For example, the low-temperature energy dependence of the low-carrier-density optical absorption in GaAs is very different from the simple square-root behavior predicted by Eqn (9.43). An electron in the conduction band is attracted by way of the coulomb interaction to a hole in the valance band and can form a bound state called an exciton.[13] Absorption due to excitons can give rise to a spectrally sharp absorption peak for photons with energy *less* than the semiconductor band gap. At room temperature, this peak

---

[13] For a review of such phenomena and related issues, see the contribution by D. S. Chemla in the series *Semiconductors and Semimetals* **58**, eds. R. K. Willardson and E. R. Weber, Academic Press, New York, 1999 (ISBN 0 12 752167 5) and S. W. Koch, T. Meier, W. Hoyer, and M. Kira, *Physica*, **E14**, 45 (2002).

is broadened by thermal processes, but still makes a contribution to absorption for photon energies near the band-gap energy.

For carrier densities at levels necessary to achieve useful optical gain in a laser diode, electron scattering broadens electron energy levels in the semiconductor on an energy scale that is comparable to our measure of optical gain bandwidth, $\Delta\mu_{e-hh}$. In addition, the relatively high carrier density screens the coulomb interaction. This reduces the ability of the system to form excitons. High carrier-density can also reduce the value of the band-gap energy.

While these and other high-carrier-density effects make it quite hard to create a detailed physical model of optical gain in a semiconductor, they are also responsible for the success of our naive approach. Associated with high carrier density and room-temperature operation are energy broadening effects that quite accidently and serendip-itously conspire to turn our simple model into something that is useful. The fact remains, however, that while the model gives results that may be used to design lasers, the model itself is physically incorrect.

Relying too heavily on the model can result in misunderstanding and misinterpre-tation of device behavior. One must be careful not to draw incorrect conclusions from such a crude model.

9.8 Example exercises

Exercise 9.1
Write a computer program to calculate spontaneous emission and optical gain in a bulk direct band-gap semiconductor as a function of photon energy $\hbar\omega$ using Eqn (9.42) and Eqn (9.43). Plot your results for GaAs with carrier concentration $n = j \times 10^{18}$ cm^{-3}, where $j = 1, 2, \ldots, 10$, temperature $T = 300$ K, band-gap energy $E_g = 1.4$ eV, and constant $g_0 = 2.64 \times 10^4$ cm^{-1} eV$^{-1/2}$.

Exercise 9.2
Repeat the calculation of Exercise 9.1, but now plot difference in chemical potential, $\Delta\mu_{e-hh}$, peak optical gain, g_{peak}, and total spontaneous emission, $r_{spon-total}$ as functions of carrier density in the range $n = 10^{18}$ cm^{-3} to $n = 10^{19}$ cm^{-3}. What happens to these functions if the active region is a two-dimensional quantum well?

Exercise 9.3
Plot spontaneous emission using the Lorentzian broadening function (Eqn (9.44)) to simulate the effect of electron–electron scattering. Use the parameters

$n = 2 \times 10^{18}$ cm^{-3}, $T = 300$ K, $E_\text{g} = 1.4$ eV, $n_\text{g} = 3.3$, $\gamma_\text{k} = 15$ meV and the relationship

$$g_\text{opt}(\hbar\omega) = -\alpha_\text{opt}(\hbar\omega) = \hbar\left(\frac{c\pi}{n_\text{r}\omega}\right)^2 \cdot r_\text{spon}(\hbar\omega) \cdot \left(1 - e^{(\hbar\omega - E_\text{g} - \Delta\mu_\text{e-hh})/k_\text{B}T}\right)$$

to calculate optical gain as a function of photon energy $\hbar\omega$. Assume that $g_0 = 2.64 \times 10^4$ cm^{-1} eV$^{-1/2}$.

Exercise 9.4

A Fabry–Perot laser diode has a bulk active region 300 μm long, 0.8 μm wide, and 0.14 μm thick. The laser has emission at wavelength $\lambda_0 = 1310$ nm, internal optical loss of 40 cm^{-1}, an optical confinement factor of $\Gamma = 0.25$, a mirror reflectivity of 0.32, and a spontaneous emission factor of $\beta = 10^{-4}$. Optical transparency occurs at carrier density $n_0 = 10^{18}$ cm^{-3}, the refractive index of the semiconductor is $n_\text{r} = 4.0$, and the peak optical gain at carrier density n is $g_\text{opt} = g_\text{slope}(n - n_0)(1 - \varepsilon S)$, where $g_\text{slope} = 2.5 \times 10^{-16}$ cm^2 s^{-1}, $\varepsilon = 5 \times 10^{-18}$ cm^3, and S is the photon density.

Write a computer program that uses the Runge–Kutta method described in Section 9.5.1 to solve the rate equations for the device. Then plot: light output, L, as a function of time; carrier density, n, as a function of time; and output power as a function of carrier density for a step current of 20 mA. You may assume a nonradiative carrier recombination rate $A_\text{nr} = 2 \times 10^8$ s^{-1}, a radiative carrier recombination rate coefficient $B = 1 \times 10^{-10}$ cm^3 s^{-1}, and a nonlinear carrier recombination rate coefficient $C = 1 \times 10^{-29}$ cm^6 s^{-1}.

Exercise 9.5

Modify the computer program used in Exercise 9.4 to find the steady-state laser light output, L, and carrier density, n, as functions of diode injection current, I. Use the device parameters given in Exercise 9.4. Plot L on both a linear and a logarithmic scale as a function of I, and determine the laser diode threshold current, I_th. Plot carrier density, n, as a function of current, I, and determine the carrier density at laser threshold n_th.

SOLUTIONS

Solution 9.1

The following figures show the results of calculating spontaneous emission and optical gain in a bulk direct band-gap semiconductor as a function of photon energy $\hbar\omega$ using Eqn (9.42) and Eqn (9.43). In each figure, carrier concentration $n = j \times 10^{18}$ cm^{-3}, where $j = 1, 2, \ldots, 10$, temperature $T = 300$ K, and band-gap energy $E_\text{g} = 1.4$ eV. Optical gain $g_\text{opt}(\hbar\omega)$ is calculated using the constant $g_0 = 2.64 \times 10^4$ cm^{-1} eV$^{-1/2}$. Because $g_0/r_0 = \pi^2 c^2/\omega^2 n_\text{r}^2$, it is not necessary to specify r_0.

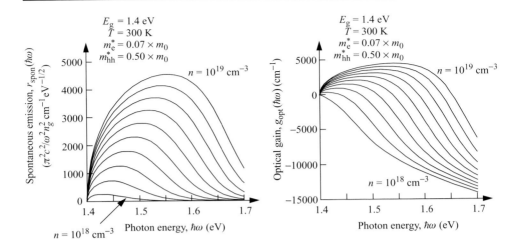

The computer program used to calculate the above figures calls the algorithm (in this case a function called mu which in turn uses the function fermi) developed in Exercise 7.3 to determine the chemical potential for electrons and holes. The following lists an example program to generate the above figures written in the MATLAB language.

Listing of MATLAB program for Exercise 9.1

```
%Chapt9Exercise1.m
%plot gain and spontaneous emission as function of photon energy
%for carrier 10 different carrier densities
%uses function mu.m and fermi.m
%carrier density n(m-3), temperature kelvin(K)
%
clear
clf;
    hbar=1.05457159e-34;          %Planck's constant (J s)
    kB=8.61734e-5;                %Boltzmann constant (eV K-1)

    m0=9.109382e-31;              %bare electron mass (kg)
    me=0.07*m0;                   %effective electron mass (kg)
    mhh=0.5*m0;                   %effective heavy hole mass (kg)
    mr=1/(1/me+1/mhh);            %reduced electron mass
    rerr=1e-3;                    %relative error

    Eg=1.4                        %GaAs band gap energy (eV)
    kelvin=300.0;                 %temperature (K)
    kBT=kB*kelvin;                %thermal energy (eV)
    beta=1/kBT;                   %inverse thermal energy (eV-1)

    for k=1:1:10

    n=k*1.e18;                    %carrier density (cm-3)
```

```
    ncarrier=n*1e6;                         %convert carrier density to (m-3)
    muhh=mu(mhh,ncarrier,kelvin,rerr)       %call mu chemical potential function for holes (eV)
    mue=mu(me,ncarrier,kelvin,rerr)         %call mu chemical potential function for electrons (eV)
    deltamu=mue+muhh                        %difference in chemical potential (eV)

    const=2.64e4;                           %GaAs constant gives gain 330 cm-1 at n = 2e18 cm-3

    deltae=0.001;

    for j=1:300
        Energy(j)=j*deltae;                 %photon energy - Eg
        Ehh=(Energy(j))/(1+mhh/me);         %energy in hole band
        Ee=(Energy(j))/(1+me/mhh);          %energy in conduction band
        fhh=fermi(beta,Ehh,muhh);           %call Fermi function for holes
        fe=fermi(beta,Ee,mue);              %call Fermi function for electrons
        gain(j)=const*(Energy(j)^0.5)*(fe+fhh-1);
        rspon(j)=(const)*(Energy(j)^0.5)*(fe*fhh);
    end

    figure(1)
    hold on;
    plot(Energy+Eg, gain);
    xlabel( 'Photon energy, hw (eV)' );
    ylabel( 'Optical gain, g (cm-1)' );
    title([ 'n(min)=',num2str(1),'x 10^{18} cm^{-3}, n(max)=',num2str(k),'x 10^{18} cm^{-3},
    me=',num2str(me / m0),', mhh=',num2str(mhh / m0),', T=',num2str(kelvin),'K, Eg=',num2str(Eg),
    'eV' ]);
        grid on;
        hold off;

    figure(2)
    hold on;
    plot(Energy+Eg,rspon,'r');
    xlabel('Photon energy, hw (eV)');
    ylabel('Spontaneous emission, rsp (arb.)');
    title([ 'n(min)=',num2str(1),'x 10^{18} cm^{-3}, n(max)=',num2str(k),'x 10^{18} cm^{-3},
    me=',num2str(me/m0),', mhh=',num2str(mhh / m0),', T=',num2str(kelvin),'K, Eg=',num2str(Eg),
    'eV' ]);
        grid on;
        end
        hold off;
```

Solution 9.2

Here, we basically repeat the calculation of Exercise 9.1, but now we plot difference in chemical potential, $\Delta \mu_{e-hh}$, peak optical gain, g_{peak}, and total spontaneous emission, $r_{spon-total}$, as functions of carrier density in the range $n = 10^{18}$ cm^{-3} to $n = 10^{19}$ cm^{-3}. The following figures show the results of performing the calculations. Notice how peak optical gain increases essentially linearly with increasing carrier concentration, n. This justifies the use of the linear approximation for optical gain given by Eqn (9.64).

The total spontaneous emission increases faster than linearly with increasing carrier concentration, supporting the use of the approximation $r_{\text{spon-total}} = Bn^2$.

If the active region is a two-dimensional quantum well in place of the bulk, three-dimensional gain medium we have been considering, the energy dependence of the electronic density of states is altered from a $E^{1/2}$ behavior to a constant. This will change how the difference in chemical potential, $\Delta\mu_{\text{e-hh}}$, peak optical gain, g_{peak}, and total spontaneous emission, $r_{\text{spon-total}}$, depend upon carrier density. Peak gain as a function of n can be approximated by the logarithmic function given by Eqn (9.65).

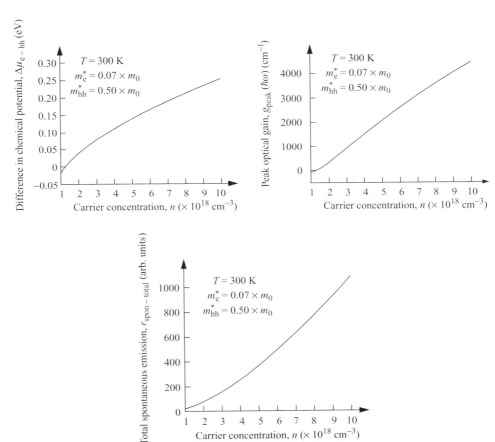

Solution 9.3

In this exercise we use a Lorentzian broadening function to simulate the effect of electron–electron scattering, and we plot spontaneous emission (Eqn (9.44)) for the parameters $n = 2 \times 10^{18}$ cm^{-3}, $T = 300$ K, $E_{\text{g}} = 1.4$ eV, $n_{\text{g}} = 3.3$, $\gamma_{\mathbf{k}} = 15$ meV. We then use the relationship

$$g_{\text{opt}}(\hbar\omega) = -\alpha_{\text{opt}}(\hbar\omega) = \hbar\left(\frac{c\pi}{n_{\text{r}}\omega}\right)^2 \cdot r_{\text{spon}}(\hbar\omega) \cdot \left(1 - e^{(\hbar\omega - E_{\text{g}} - \Delta\mu_{\text{e-hh}})/k_{\text{B}}T}\right)$$

to calculate optical gain as a function of photon energy $\hbar\omega$, assuming that $g_0 = 2.64 \times 10^4$ cm^{-1} eV$^{-1/2}$.

The computer program used to do this exploits what was developed in Exercise 9.1. The additional complication is the inclusion of the Lorentzian broadening function. The results shown in the following figures are indicative of what happens to the spontaneous emission and gain spectra when $\gamma_\mathbf{k}$ is included. The peak values of both r_{spon} and g_{opt} decrease and there is a low-energy tail that extends into the band-gap region.

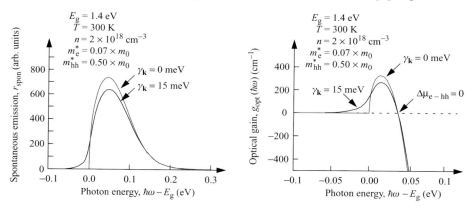

Solution 9.4

In this exercise we are asked to write a computer program that simulates the large-signal response of a Fabry–Perot laser diode using the parameters provided.

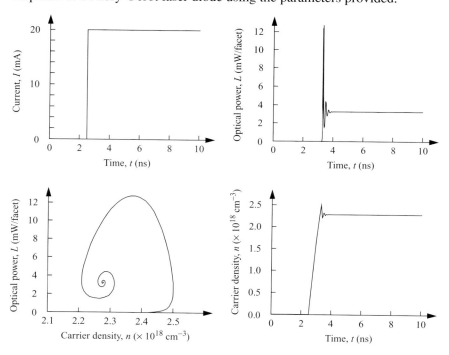

The above figures plot: light output, L, from one mirror facet as a function of time; carrier density, n, as a function of time, t; and output power as a function of carrier density for a step-current of 20 mA. Notice the relaxation oscillations in both the L–t and n–t plots. These oscillations arise from the response of the system to a large-signal step change in current. There is a phase lag between the photons and the carriers during the transient that can be seen in the figure that plots L–n.

The computer program is written in two parts so that the Runge–Kutta integrator can be called from the main program.

Listing of MATLAB program for Exercise 9.4

```
%Chapt9Exercise4.m
%solves single-mode laser diode rate equations
%calls runge4.m which is a 4th order Runge-Kutta integrator for fixed time increment
%plots carrier density and photon density as function of time.
%gain=g=gamma*gslope*(n-n0)*(1-epsi*s)
%carrier density, n, current, I, and photon density, s, are related via
%fcn=n'=(I/ev)-(n/tau_n)-(g*s)
%fcs=s'=(g-K)*s+beta*Rsp

clear;
clf;

%declare constants
hconstjs=6.626069e-34;          %Planck's constant (J s)
echargec=1.60217646e-19;        %electron charge (C)
vlightcm=2.99792458e10;         %velocity of light (cm s-1)

nsteps=10000;                   %number of time-steps
%****************** start the main program *******************************/
%assign values to parameters

ngroup=4;                       %refractive index
clength=3.00E-02;               %cavity length (cm)
thick=1.40E-05;                 %thickness of active region (cm)
width=8.00E-05;                 %width of active region (cm)

tincrement=1.00E-12;            %time increment (s)
initialn=0.00E+18;              %initial value of carrier density (cm-3)

Anr=2.00E+08;                   %non-radiative recombination rate (s-1)
Bcons=1.00E-10;                 %radiative recombination coefficent (cm3 s-1)
Ccons=1.00E-29;                 %non-linear recombination coefficient (cm6 s-1)

n0density=1.00E+18;             %transparency carrier density (cm-3)
gslope=2.50E-16;                %optical gain-slope coefficient (cm2 s-1)
epsi=5.00E-18;                  %gain compression (cm3)
beta=1.00E-04;                  %spontaneous emission coefficient
gamma_cons=0.25;                %optical confinement factor
wavelength=1.31;                %optical emission wavelength (um)
```

```
mirrone=0.32;          %optical reflectivity from first mirror
mirrtwo=0.32;          %optical reflectivity from second mirror
alfa_i=40;             %internal optical loss (cm-1)

pstart1=2.5;           %start of first current step (ns)
pstart2=3.5;           %start of second current step (ns)

rioff1=0;              %off value of current (mA)
rion1=20;              %first step in current (mA)
rion2=20;              %second step in current (mA)

%initialize sdensity, ndensity and calculate total optical loss (cm-1)
sdensity=0.0;
ndensity=initialn * 1.0e-21;

alfa_m=log ( 1.0/(mirrone * mirrtwo) ) / ( 2.0 * clength );
alfasum=alfa_m + alfa_i;

%rescale integrator: time[ns], current[mA], length[nm] and 1nm3=1e-21cm3
echarge=echargec*1.0e9*1.0e3;      %electron charge
vlight=vlightcm/1.0e9/1.0e-7;      %velocity of light (m/s) to (nm s-1)

tincrement=tincrement*1.0e9;
gslope=gslope*1.0e14;
n0density=n0density*1.0e-21;

%length scale in nm.
 clength=clength / 1.0e-7;
 width=width / 1.0e-7;
 thick=thick / 1.0e-7;

%recombination constants
 Anr=Anr / 1.0e9;
 Bcons=Bcons / (1.0e9*1.0e-21);
 Ccons=Ccons / (1.0e9*1.0e-42);

epsi=epsi / 1.0e-21;
gslope=gslope*vlight / ngroup;
alfasum=alfasum*1.0e-7;
kappa=alfasum*vlight / ngroup;
volume=clength*width*thick;

%scale light output to mW from mirror one
tmp=hconstjs*(vlightcm*1.0e-2)/(wavelength*1.0e-6);    %use m (J)
tmp=tmp*alfa_m*vlightcm;                               %use cm (s-1)
tmp=tmp*1000.0*volume/ngroup;                          %use mW and nm (nm3)
lightout=tmp*(1.0-mirrone)/(2.0-mirrone-mirrtwo);      %output mirror one

data(1)=gamma_cons;
data(2)=gslope;
data(3)=n0density;
data(4)=epsi;
data(5)=kappa;
```

```
data(6)=Bcons;
data(7)=Anr;
data(8)=Ccons;
data(9)=volume;
data(10)=beta;

%***************** the main integration loop *****************************

for i = 1 : 1 : nsteps
   time(i) = ( double(i-1)+1.0* tincrement;
   current(i) = rioff1;

   if time(i)>pstart1
      current(i)=rion1;
   end

   if time(i)>pstart2
      current(i)=rion2;
   end

   [sdensity, ndensity]=runge4( current(i), sdensity, ndensity, tincrement, data);
   carrier(i)=ndensity*1.0e21/11.0e18;
   photon(i)=(lightout*sdensity);
end;

photon_number=photon( nsteps - 1 )*volume / lightout      %number of photon at last data point
carrier_number=carrier( nsteps - 1 )*volume               %number of carriers at last data point

%start plotting the time-evolution of corrent, light output and carrier density.
%injected current change in time.
figure(1);
hold on;
axis([0 time(nsteps) 0 max(current)*1.05]);
temp=['Injected current as function of time'];
title(temp, 'fontsize', 12);
xlabel(['Time, t (ns)'],'fontsize', 12);
ylabel(['Current, I (mA)'],'fontsize', 12);
grid on;
plot(time,current,'r-');
hold off;

%Evolution of light output from mirror one.
figure(2);
hold on;
axis([0 time(nsteps) 0 max(photon)*1.05]);
temp=['Optical power from mirror-one as function of time'];
title(temp, 'fontsize', 12);
xlabel(['Time, t (ns)'],'fontsize', 12);
ylabel(['Power, L (mW/facet)'],'fontsize', 12);
grid on;
plot(time,photon,'r-');
hold off;
```

```
%Evolution of carreir density.
figure(3);
hold on;
axis([0 time(nsteps) min(carrier) max(carrier)*1.05]);
temp=['Carrier density as function of time'];
title(temp, 'fontsize', 12);
grid on;
xlabel(['Time, t (ns)'],'fontsize', 12);
ylabel(['Carrier density, n (10^1^8cm^-^3)'],'fontsize', 12);
plot(time,carrier,'r-');
hold off;

%Light output as function of carreir density.
figure(4);
hold on;
axis([min(carrier) max(carrier)*1.05 max(photon)*1.05]);
temp=['Light output as function of carrier density'];
title(temp, 'fontsize', 12);
grid on;
xlabel(['Carrier density, n (10^1^8cm^-^3)'],'fontsize', 12);
ylabel(['Power, L (mW/facet)'],'fontsize', 12);
plot(carrier,photon,'r-');
hold off;

%main ends here.
```

Listing of runge4 function for MATLAB program used in Exercise 9.4

```
%*******************Runge Kutta 4-step algorithm.*************************
function [ret_s, ret_n]=runge4(current, sdensity, ndensity, tincrement, data)
%runge kutta 4 step algorithm
echargec =1.60217646e-19;              %electron charge (Coulomb)
echarge =echargec*1.0e9*1.0e3;         %electron charge
vlightcm =2.9979458e10;                 velocity of light [cm s-1]
vlight=vlightcm/1.0e9/1.0e-7;           velocity of light [nm s-1]

%start reading parameters from the main program
gamma_cons=data(1);
gslope=data(2);
n0density=data(3);
epsi=data(4);
kappa=data(5);
Bcons=data(6);
anr=data(7);
Ccons=data(8);
volume=data(9);
beta=data(10);
%finish reading parameters from the main program.
```

%Runge-Kutta 4 step integrator.
temps=sdensity;
tempn=ndensity;

%photon and carrier number in the 1st step of Runge-Kutta method.
gn=gamma_cons*gslope*(ndensity - n0density)*(1.0 - epsi*sdensity);
netgain=gn-kappa;

Rsp=Bcons*ndensity*ndensity;
temp=Anr*ndensity + Rsp + Ccons*ndensity*ndensity*ndensity;
ret_val=current / echarge / volume - temp - gn*sdensity;
nk1=tincrement*ret_val;

ret_val=netgain*sdensity+ beta*Rsp;
sk1=tincrement*ret_val;

ndensity=nk1*0.5+tempn;
sdensity=sk1*0.5+temps;

%photon and carrier number in the 2nd step of Runge-Kutta method.
gn=gamma_cons*gslope*(ndensity - n0density)*(1.0 - epsi*sdensity);
netgain=gn-kappa;

Rsp=Bcons*ndensity*ndensity;
temp=Anr*ndensity + Rsp + Ccons*ndensity*ndensity*ndensity;
ret_val=current / echarge / volume - temp - gn*sdensity;
nk2=tincrement*ret_val;

ret_val=netgain*sdensity+ beta*Rsp;
sk2=tincrement*ret_val;

ndensity=nk2*0.5+tempn;
sdensity=sk2*0.5+temps;

%photon and carrier number in the 3rd step of Runge-Kutta method.
gn=gamma_cons*gslope*(ndensity - n0density)*(1.0 - epsi*sdensity);
netgain=gn-kappa;

Rsp=Bcons*ndensity*ndensity;
temp=Anr*ndensity + Rsp + Ccons*ndensity*ndensity*ndensity;
ret_val=current / echarge / volume - temp - gn*sdensity;
nk3=tincrement*ret_val;

ret_val=netgain*sdensity + beta*Rsp;
sk3=tincrement*ret_val;

ndensity=nk3+tempn;
sdensity=sk3+temps;

%photon and carrier number in the 4th step of Runge-Kutta method.
%double gain(double freq, double sdensity, double ndensity, double Ef, double En, double beta)
gn=gamma_cons*gslope*(ndensity - n0density)*(1.0 - epsi*sdensity);
netgain=gn-kappa;

Rsp=Bcons*ndensity*ndensity;
temp=Anr*ndensity + Rsp + Ccons*ndensity*ndensity*ndensity;
ret_val = current / echarge / volume - temp - gn*sdensity;
nk4=tincrement*ret_val;

ret_val=netgain*sdensity + beta*Rsp;
sk4=tincrement*ret_val;

ret_s=temps+(sk1+2.0*sk2+2.0sk3+sk4)/6.0;
ret_n=tempn+(nk1+2.0*nk2+2.0nk3+nk4)/6.0;

sdensity=ret_s;
ndensity=ret_n;

%Runge-Kutta function ends here.

Solution 9.5

A simple way to find the steady-state $L-I$ and $n-I$ characteristsics of a laser diode is to modify the computer program used in Exercise 9.4. One calculates the time domain response to a fixed current I and waits until steady-state conditions are reached. After recording the value of L and n for that value of I, one increases the current and repeats the calculaton using the previous value of n as an initial condition.

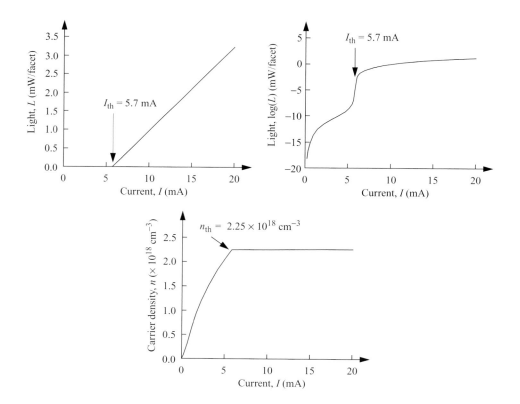

Of course, one has to know that the steady state has been achieved before recording the values of L and n. The most straightforward approach is to simulate the device for such a long time interval that you are sure that there are no transient effects left in the system. For example, the time interval can be fixed at, say, 2 ns. This time is longer than any other time scale in the system, and so should provide good steady-state results for each value of current I. There are more efficient ways to determine if the steady state has been reached, but they are more complicated and involve sampling a number of values for photon intensity and determining if they are the same to within a predetermined margin.

In the figures above light output L from one mirror facet is plotted on both linear and logarithmic scales as a function of I. The laser diode threshold current, I_{th}, is determined by linearly extrapolating the high-slope portion of the curve to $L = 0$. This is best done by using the linear L–I plot. As may be seen, threshold current is $I_{th} = 5.7$ mA. The $\log(L)$–I curve is particularly useful for showing the behavior of below threshold light level. Clearly, for this device, laser threshold is associated with a very rapid change in optical output power near $I = I_{th}$.

The n–I plot shows that carrier density at threshold is $n_{th} = 2.25 \times 10^{18}$ cm^{-3} and that it does not increase significantly with increasing current, I. One says carrier density is *pinned* above threshold.

10 Time-independent perturbation

Time-independent perturbation

10.1 Introduction

Often there are situations in which the solutions to the time-independent Schrödinger equation are known for a particular potential but not for a similar but different potential. Time-independent perturbation theory provides a means of finding approximate solutions using an expansion in the known eigenfunctions.

As an example, consider the one-dimensional, rectangular potential well with infinite barrier energy shown in Fig. 10.1. The width of the well is L, and the potential is $V(x) = 0$ for $0 < x < L$ and $V(x) = \infty$ for $0 > x > L$.

The time-independent Schrödinger equation for a particle of mass m in the potential $V(x)$ is

$$H\psi_n(x) = \frac{-\hbar^2}{2m} \frac{\partial^2}{\partial x^2} \psi_n(x) + V(x)\psi_n(x) = E_n\psi_n(x) \tag{10.1}$$

The solutions to this equation have eigenfunctions

$$\psi_n(x) = \sqrt{\frac{2}{L}} \sin(k_n x) \tag{10.2}$$

and eigenenergy

$$E_n = \frac{\hbar^2 k_n^2}{2m} \tag{10.3}$$

where

$$k_n = \frac{n\pi}{L} \tag{10.4}$$

and the value of n takes an integer value so that $n = 1, 2, 3, \ldots$.

We now suppose that the potential $V(x)$ is deformed by the presence of an additional term $W(x)$. In Fig. 10.1, $W(x)$ is shown as a small curved bump. The challenge is to find the new eigenstates and eigenvalues of the system. One approach is to use time-independent perturbation theory.

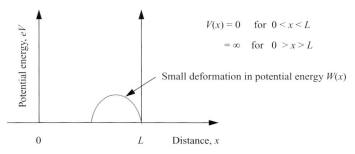

Fig. 10.1. Sketch of a one-dimensional, rectangular potential well with infinite barrier energy. The width of the well is L, and the potential is $V(x) = 0$ for $0 < x < L$ and $V(x) = \infty$ for $0 > x > L$. The known eigenfunction for this potential can be used to obtain approximate solutions in the presence of a "small" deformation in the potential.

10.2 Time-independent nondegenerate perturbation

To develop time-independent nondegenerate perturbation theory, we assume a Hamiltonian of the form

$$\boxed{H = H^{(0)} + W} \tag{10.5}$$

where $H^{(0)}$ is the unperturbed Hamiltonian and W is the perturbation. Formally, the solution for the total Hamiltonain H has eigenfunctions ψ_n and eigenenergies E_n for which

$$H\psi_n = E_n\psi_n \tag{10.6}$$

The solution for the Hamiltonian $H^{(0)}$ has eigenfunctions $\psi_m^{(0)}$ and eigenenergies $E_m^{(0)}$ for which

$$H^{(0)}\psi_m^{(0)} = E_m^{(0)}\psi_m^{(0)} \tag{10.7}$$

For the situations we will consider, $H^{(0)}$, $\psi_m^{(0)}$, and $E_m^{(0)}$ are known. Because the perturbation W is assumed to be small, it should be possible to expand ψ_n and E_n as a power series in W.

The perturbed eigenfunctions and eigenvalues are written

$$\begin{aligned}
\psi &= \psi^{(0)} + \lambda\psi^{(1)} + \lambda^2\psi^{(2)} + \cdots \\
E &= E^{(0)} + \lambda E^{(1)} + \lambda^2 E^{(2)} + \cdots
\end{aligned} \tag{10.8}$$

where $\lambda = 1$ and is a dummy variable that we employ for keeping track of the order of the terms in the power series that we will use. Hence,

$$\left(H^{(0)} + \lambda W\right)\left(\psi^{(0)} + \lambda\psi^{(1)} + \cdots\right) = \left(E^{(0)} + \lambda E^{(1)} + \cdots\right)\left(\psi^{(0)} + \lambda\psi^{(1)} + \cdots\right) \tag{10.9}$$

Equating equal powers of λ gives

$$\left(H^{(0)} - E^{(0)}\right)\psi^{(0)} = 0 \tag{10.10}$$

$$\left(H^{(0)} - E^{(0)}\right)\psi^{(1)} = \left(E^{(1)} - W\right)\psi^{(0)} \tag{10.11}$$

$$\left(H^{(0)} - E^{(0)}\right)\psi^{(2)} = \left(E^{(1)} - W\right)\psi^{(1)} + E^{(2)}\psi^{(0)} \tag{10.12}$$

$$\left(H^{(0)} - E^{(0)}\right)\psi^{(3)} = \left(E^{(1)} - W\right)\psi^{(2)} + E^{(2)}\psi^{(1)} + E^{(3)}\psi^{(0)} \tag{10.13}$$

Equation (10.10) is the zero-order solution. Equation (10.11) has the first-order correction. Equation (10.12) is to second order and so on.

By now it is clear that this theory has an important flaw. There is no self-consistent method to decide when to terminate the perturbation series. Often, this is not a limitation, because the high-spatial-frequency components of a potential usually decrease with increasing frequency. However, this is not always the case, and so one must proceed with caution.

10.2.1 The first-order correction

Because $\psi^{(0)}$ forms a complete set, we can expand each correction term in $\psi^{(0)}$. For $\psi^{(1)}$, this gives

$$\psi^{(1)} = \sum_n a_n^{(1)} \psi_n^{(0)} \tag{10.14}$$

where it is worth noting that the unperturbed solution gives

$$H^{(0)}\psi_m^{(0)} = E_m^{(0)}\psi_m^{(0)} \tag{10.15}$$

The first-order corrected solution for the perturbed m-th eigenvalue and eigenfunction is (Eqn (10.11))

$$\left(H^{(0)} - E_m^{(0)}\right)\psi^{(1)} = \left(E^{(1)} - W\right)\psi_m^{(0)} \tag{10.16}$$

Substituting for $\psi^{(1)}$ gives

$$\left(H^{(0)} - E_m^{(0)}\right)\sum_n a_n^{(1)}\psi_n^{(0)} = \left(E^{(1)} - W\right)\psi_m^{(0)} \tag{10.17}$$

We now multiply both sides by $\psi_k^{(0)*}$, integrate, and make use of the orthonormal property of eigenfunctions so that $\langle i|j \rangle = \delta_{ij}$

$$\int \psi_k^{(0)*} H^{(0)} \sum_n a_n^{(1)}\psi_n^{(0)} - E_m^{(0)} \int \psi_k^{(0)*} \sum_n a_n^{(1)}\psi_n^{(0)} = E^{(1)}\delta_{km} - W_{km} \tag{10.18}$$

$$\boxed{a_k^{(1)}\left(E_k^{(0)} - E_m^{(0)}\right) = E^{(1)}\delta_{km} - W_{km}} \tag{10.19}$$

where W_{km} is the matrix element $\int \psi_k^{(0)*} W \psi_m^{(0)} = \langle k|W|m \rangle$. Thus, for $k \neq m$

$$a_k^{(1)} = \frac{W_{km}}{E_m - E_k} \qquad k \neq m \tag{10.20}$$

and for $k = m$

$$E^{(1)} = W_{mm} \qquad k = m \tag{10.21}$$

$E^{(1)}$ is the first-order correction to eigenvalue E_m under the perturbation W.

We must now evaluate $a_m^{(1)}$. This is done by requiring that $\psi = \psi^{(0)} + \psi^{(1)}$, the first-order corrected wave function, is normalized to unity:

$$\int \psi^* \psi = \int \left(\psi_m^{(0)} + \lambda \psi^{(1)} \right)^* \left(\psi_m^{(0)} + \lambda \psi^{(1)} \right) = 1 \tag{10.22}$$

$$\int \psi^* \psi = \int \left(\psi_m^{(0)} + \lambda \sum_i a_i^{(1)} \psi_i^{(0)} \right)^* \left(\psi_m^{(0)} + \lambda \sum_j a_j^{(1)} \psi_j^{(0)} \right) \tag{10.23}$$

$$\int \psi^* \psi = 1 + \lambda a_m^{(1)} + \lambda a_m^{(1)*} + \lambda^2 \sum_i a_i^{(1)} a_i^{(1)*} \tag{10.24}$$

Neglecting the second-order term gives $a_m^{(1)} = 0$ as a solution. Hence, we may conclude that the eigenfunction and eigenvalue to first order are

$$\begin{aligned} \psi &= \psi_m^{(0)} + \sum_{k \neq m} \frac{W_{km}}{E_m - E_k} \psi_k^{(0)} \\ E &= E_m^{(0)} + W_{mm} \end{aligned} \tag{10.25}$$

10.2.2 The second-order correction

The second-order correction to the eigenfunction $\psi^{(2)}$ may be expanded as

$$\psi^{(2)} = \sum_n a_n^{(2)} \psi_n^{(0)} \tag{10.26}$$

Again we note that the unperturbed solution gave

$$H^{(0)} \psi_m^{(0)} = E_m^{(0)} \psi_m^{(0)} \tag{10.27}$$

The second-order solution for the perturbed m-th eigenvalue and eigenfunction is (Eqn (10.12))

$$\left(H^{(0)} - E_m^{(0)} \right) \psi^{(2)} = \left(E^{(1)} - W \right) \psi^{(1)} + E^{(2)} \psi_m^{(0)} \tag{10.28}$$

Substituting our expansions for $\psi^{(1)}$ and $\psi^{(2)}$ in terms of $\psi^{(0)}$ gives

$$H^{(0)} \sum_n a_n^{(2)} \psi_n^{(0)} - E_m^{(0)} \sum_n a_n^{(2)} \psi_n^{(0)} = E^{(1)} \sum_n a_n^{(1)} \psi_n^{(0)} - W \sum_n a_n^{(1)} \psi_n^{(0)} + E^{(2)} \psi_m^{(0)}$$

(10.29)

Multiplying both sides by $\psi_k^{(0)*}$ and integrating gives

$$\int \psi_k^{(0)*} H^{(0)} \sum_n a_n^{(2)} \psi_n^{(0)} - E_m^{(0)} \int \psi_k^{(0)*} \sum_n a_n^{(2)} \psi_n^{(0)}$$

$$= E^{(1)} \int \psi_k^{(0)*} \sum_n a_n^{(1)} \psi_n^{(0)} - \int \psi_k^{(0)*} W \sum_n a_n^{(1)} \psi_n^{(0)} + E^{(2)} \int \psi_k^{(0)*} \psi_m^{(0)} \quad (10.30)$$

Hence,

$$\boxed{a_k^{(2)}\left(E_k^{(0)} - E_m^{(0)}\right) = a_k^{(1)} E^{(1)} - \sum_n a_n^{(1)} W_{kn} + E^{(2)} \delta_{mk}}$$

(10.31)

10.2.2.1 Second-order correction to eigenvalues ($k = m$)

For $k = m$, we have

$$E^{(2)} = \sum_n a_n^{(1)} W_{mn} - a_m^{(1)} E^{(1)}$$

$$= \sum_{n \neq m} a_n^{(1)} W_{mn} + a_m^{(1)} W_{mm} - a_m^{(1)} E^{(1)} \quad (10.32)$$

But from our first-order perturbation results we had $E^{(1)} = W_{mm}$, so the second and third terms on the right-hand side cancel and

$$E^{(2)} = \sum_{n \neq m} a_n^{(1)} W_{mn} \quad (10.33)$$

Substituting for $a_n^{(1)}$ from our first-order perturbation results gives

$$\boxed{E^{(2)} = \sum_{n \neq m} \frac{|W_{mn}|^2}{E_m - E_n}}$$

(10.34)

10.2.2.2 Second-order coefficients $a_k^{(2)}$

There are two situations we should consider – the values of $a_k^{(2)}$ when $k \neq m$ and when $k = m$. For the case in which $k \neq m$, we use our previous results

$$a_n^{(1)} = \frac{W_{nm}}{E_m - E_n} \quad (10.35)$$

and

$$E^{(1)} = W_{mm} \tag{10.36}$$

Substituting into

$$a_k^{(2)}\left(E_k^{(0)} - E_m^{(0)}\right) = a_k^{(1)}E^{(1)} - \sum_n a_n^{(1)}W_{kn} + E^{(2)}\delta_{mk} \tag{10.37}$$

gives

$$a_k^{(2)}\left(E_m^{(0)} - E_k^{(0)}\right) = \sum_n a_n^{(1)}W_{kn} - a_k^{(1)}E^{(1)} \tag{10.38}$$

$$a_k^{(2)}\left(E_m^{(0)} - E_k^{(0)}\right) = \sum_n \frac{W_{nm}W_{kn}}{\left(E_m^{(0)} - E_n^{(0)}\right)} - \frac{E^{(1)}W_{km}}{\left(E_m^{(0)} - E_k^{(0)}\right)} \tag{10.39}$$

$$\boxed{a_k^{(2)} = \sum_n \frac{W_{nm}W_{kn}}{(E_m - E_n)(E_m - E_k)} - \frac{W_{mm}W_{km}}{(E_m - E_k)^2}} \qquad k \neq m \tag{10.40}$$

For the case in which $k = m$, we find $a_m^{(2)}$ by using normalization of the corrected wave function

$$\int \psi^*\psi = \int \left(\psi_m^{(0)} + \lambda\psi^{(1)} + \lambda^2\psi^{(2)}\right)^*\left(\psi_m^{(0)} + \lambda\psi^{(1)} + \lambda^2\psi^{(2)}\right) = 1 \tag{10.41}$$

$$\int \psi^*\psi = \int \left(\psi_m^{(0)} + \lambda\sum_i a_i^{(1)}\psi_i^{(0)} + \lambda^2\sum_i a_i^{(2)}\psi_i^{(0)}\right)^*$$
$$\times \left(\psi_m^{(0)} + \lambda\sum_j a_j^{(1)}\psi_j^{(0)} + \lambda^2\sum_j a_j^{(2)}\psi_j^{(0)}\right) \tag{10.42}$$

$$\int \psi^*\psi = 1 + \lambda a_m^{(1)} + \lambda a_m^{(1)*} + \lambda^2\sum_n a_n^{(1)*}a_n^{(1)} + \lambda^2 a_m^{(2)*} + \lambda^2 a_m^{(2)} \tag{10.43}$$

We now use the fact that $a_m^{(1)} = 0$, so that

$$2\lambda^2 a_m^{(2)} = -\lambda^2\sum_n a_n^{(1)*}a_n^{(1)} \tag{10.44}$$

$$a_m^{(2)} = -\frac{1}{2}\sum_n \left|a_n^{(1)}\right|^2 \tag{10.45}$$

and since $a_m^{(1)} = 0$, we have for $k = m$

$$\boxed{a_m^{(2)} = -\frac{1}{2}\sum_{n\neq m} \frac{|W_{mn}|^2}{(E_m - E_n)^2}} \qquad k = m \tag{10.46}$$

One may now write down the eigenfunction and eigenvalue of the perturbed system to second order. We have

$$E = E^{(0)} + E^{(1)} + E^{(2)} \tag{10.47}$$

$$E = E_m^{(0)} + W_{mm} + \sum_{n \neq m} \frac{|W_{mn}|^2}{E_m - E_n} \tag{10.48}$$

and

$$\psi = \psi^{(0)} + \psi^{(1)} + \psi^{(2)} \tag{10.49}$$

$$\psi = \psi_m^{(0)} + \sum_k a_k^{(1)} \psi_k^{(0)} + \sum_k a_k^{(2)} \psi_k^{(0)} \tag{10.50}$$

$$\psi = \psi_m^{(0)} + \sum_{k \neq m} \frac{W_{mk}}{E_m - E_k} \psi_k^{(0)} + \sum_{k \neq m} \left(\left(\sum_{n \neq m} \frac{W_{kn} W_{mn}}{(E_m - E_n)(E_m - E_k)} \right. \right.$$
$$\left. \left. - \frac{W_{mm} W_{km}}{(E_m - E_k)^2} \right) \times \psi_k^{(0)} - \frac{1}{2} \frac{|W_{mn}|^2}{(E_m - E_n)^2} \psi_m^{(0)} \right) \tag{10.51}$$

This completes the formal aspects of time-independent nondegenerate perturbation theory up to second order in both the eigenvalues and eigenfunctions. It is now time to get some practice using this formalism. In the following section we will consider the effect a perturbing potential has on the one-dimensional harmonic oscillator.

10.2.3   Harmonic oscillator subject to perturbing potential in x

We start by considering a particle of mass m and charge e moving in a one-dimensional harmonic potential that is subject to a constant small electric field, $\mathbf{E}$, in the x direction. This is a good choice, because we have already found the exact solution in Chapter 6. Hence, a meaningful comparison between the different approaches can be made.

To find the new energy eigenvalues and eigenfunctions for the perturbed system, one starts by writing down the Hamiltonian

$$H = \frac{p^2}{2m} + \frac{\kappa}{2} x^2 + W \tag{10.52}$$

where spring constant $\kappa = m\omega^2$ and perturbation is $W = -e|\mathbf{E}|\hat{x}$. This perturbation involves the position operator $\hat{x}$, and so we will be interested in finding matrix elements $W_{nm} = -e|\mathbf{E}|x_{nm}$.

From our previous work on the harmonic oscillator, we know that the position operator may be written as a linear combination of creation and annihilation operators in such a way that $x = (\hbar/2m\omega)^{1/2}(\hat{b}^\dagger + \hat{b})$. It follows that the matrix elements $\langle n|\hat{b}^\dagger|m \rangle = (m+1)^{1/2}\delta_{m=n-1}$ and $\langle n|\hat{b}|m \rangle = m^{1/2}\delta_{m=n+1}$, so that matrix element $W_{nm} = -e|\mathbf{E}|x_{nm}$ has only two nonzero values:

$$W_{n,n+1} = -e|\mathbf{E}| \left(\frac{\hbar}{2m\omega} \right)^{1/2} (n+1)^{1/2} \tag{10.53}$$

and

$$W_{n,n-1} = -e|\mathbf{E}|\left(\frac{\hbar}{2m\omega}\right)^{1/2} n^{1/2} \tag{10.54}$$

The unperturbed energy levels of the harmonic oscillator are

$$E_n^{(0)} = \hbar\omega\left(n + \frac{1}{2}\right) \tag{10.55}$$

The first-order correction is given by

$$E^{(1)} = W_{mm} = -e|\mathbf{E}|x_{mm} = 0 \tag{10.56}$$

This is zero, since the matrix element $\langle m|x|m\rangle = x_{mm} = 0$. However, the second-order correction is

$$E^{(2)} = \sum_{n \neq m} \frac{|W_{mn}|^2}{E_m - E_n} = \frac{e^2|\mathbf{E}|^2\hbar}{2m\omega}\left(\frac{x_{n,n+1}^2}{-\hbar\omega} + \frac{x_{n,n-1}^2}{\hbar\omega}\right) \tag{10.57}$$

$$E^{(2)} = \frac{e^2|\mathbf{E}|^2}{2m\omega^2}(-(n+1) + n) = \frac{-e^2|\mathbf{E}|^2}{2m\omega^2} = \frac{-e^2|\mathbf{E}|^2}{2\kappa} \tag{10.58}$$

The new energy levels of the oscillator are to second order:

$$E = \hbar\omega\left(n + \frac{1}{2}\right) - \frac{e^2|\mathbf{E}|^2}{2\kappa} \tag{10.59}$$

which is the same as the exact result. Physically the particle oscillates at the same frequency, ω, as the unperturbed case, but it is displaced a distance $e|\mathbf{E}|/m\omega^2$, and the new energy levels are shifted by $-e^2|\mathbf{E}|^2/2m\omega^2$ (see Fig. 10.2).

The fact that perturbation theory gives the same result we previously obtained by an exact calculation is a validation of our approach.

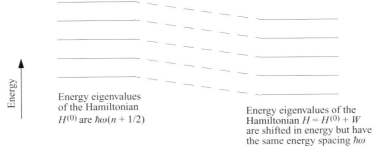

Fig. 10.2. Illustration of energy eigenvalues of the one-dimensional harmonic oscillator with Hamiltonian $H^{(0)}$ (left) and subject to perturbation $W = -e|\mathbf{E}|x$ in the potential (right). In the presence of the perturbation, the energy levels are shifted but the energy-level spacing remains the same.

10.2.4   Harmonic oscillator subject to perturbing potential in x^2

Following our success in calculating the energy levels for a one-dimensional harmonic oscillator subject to a perturbation linear in x, we now consider the case of a perturbation in x^2.

As usual, we start by writing down the Hamiltonian for the complete system:

$$H = \frac{p^2}{2m} + \frac{\kappa}{2}x^2 + W \tag{10.60}$$

where spring constant $\kappa = m\omega^2$ and the perturbation is $W = \kappa\xi x^2/2$.

In this case, the effect of the perturbation is to change the spring constant of the harmonic oscillator. If we set $\omega'^2 = \omega^2(1 + \xi)$, then

$$H = \frac{p^2}{2m} + \frac{\kappa}{2}x^2 + \frac{\kappa}{2}\xi x^2 = \frac{p^2}{2m} + \frac{m\omega^2}{2}(1 + \xi)x^2 = \frac{p^2}{2m} + \frac{m\omega'^2}{2}x^2 \tag{10.61}$$

The right-hand side of Eqn (10.61) is another harmonic oscillator. One may, therefore, solve this exactly. The eigenvalues are

$$E_n = \left(n + \frac{1}{2}\right)\hbar\omega' = \left(n + \frac{1}{2}\right)\hbar\omega(1 + \xi)^{1/2}$$

$$= \left(n + \frac{1}{2}\right)\hbar\omega\left(1 + \frac{\xi}{2} - \frac{\xi^2}{8} + \cdots\right) \tag{10.62}$$

where the expansion $(1 + x)^n = 1 + nx + n(n - 1)x^2/2! + \cdots$ has been used. The same result may be found using perturbation theory.

The perturbation W may be written

$$W = \frac{\kappa}{2}\xi x^2 \tag{10.63}$$

$$W = \frac{\kappa}{2}\xi\frac{\hbar}{2m\omega}\left(\hat{b}^\dagger + \hat{b}\right)^2 = \frac{\xi}{4}\hbar\omega\left(\hat{b}^\dagger + \hat{b}\right)^2 \tag{10.64}$$

$$W = \frac{\xi}{4}\hbar\omega\left(\hat{b}^{\dagger 2} + \hat{b}^2 + \hat{b}\hat{b}^\dagger + \hat{b}^\dagger\hat{b}\right) \tag{10.65}$$

and, since $\hat{b}\hat{b}^\dagger = \hat{b}^\dagger\hat{b} + 1$

$$W = \frac{\xi}{4}\hbar\omega\left(\hat{b}^{\dagger 2} + \hat{b}^2 + 2\hat{b}^\dagger\hat{b} + 1\right) \tag{10.66}$$

As usual, the nonzero matrix elements are easy to find. They are

$$\langle\phi_n|W|\phi_n\rangle = \frac{1}{2}\xi\left(n + \frac{1}{2}\right)\hbar\omega \tag{10.67}$$

$$\langle\phi_{n+2}|W|\phi_n\rangle = \frac{1}{4}\xi((n + 1)(n + 2))^{1/2}\hbar\omega \tag{10.68}$$

$$\langle\phi_{n-2}|W|\phi_n\rangle = \frac{1}{4}\xi(n(n - 1))^{1/2}\hbar\omega \tag{10.69}$$

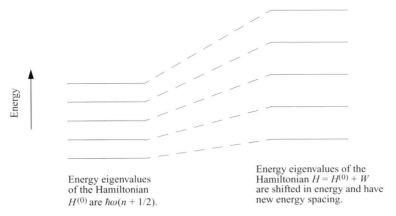

Fig. 10.3. Illustration of energy eigenvalues of the one-dimensional harmonic oscillator with Hamiltonian $H^{(0)}$ (left) and subject to perturbation $W = \kappa \xi x^2 / 2$ in the potential (right). In the presence of the perturbation, both the energy levels and the energy-level spacing are changed. In the figure it is assumed that $\xi < 1$.

Now we use these results to evaluate the energy terms up to second order:

$$E = E^{(0)} + E^{(1)} + E^{(2)} \tag{10.70}$$

$$E_n = E_n^{(0)} + W_{nn} + \sum_{m \neq n} \frac{|W_{nm}|^2}{E_n - E_m} \tag{10.71}$$

$$E_n = E_n^{(0)} + \frac{\xi}{2}\left(n + \frac{1}{2}\right)\hbar\omega - \frac{\xi^2}{16}(n+1)(n+2)\frac{\hbar\omega}{2} + \frac{\xi^2}{16}n(n-1)\frac{\hbar\omega}{2} + \cdots \tag{10.72}$$

$$E_n = E_n^{(0)} + \left(n + \frac{1}{2}\right)\hbar\omega\frac{\xi}{2} - \left(n + \frac{1}{2}\right)\hbar\omega\frac{\xi^2}{8} + \cdots \tag{10.73}$$

$$E_n = \left(n + \frac{1}{2}\right)\hbar\omega\left(1 + \frac{\xi}{2} - \frac{\xi^2}{8} + \cdots\right) \tag{10.74}$$

which is the result we had before. The energy-level diagram in Fig. 10.3 illustrates the result.

Again, perturbation theory is in good agreement with an alternative exact approach, thereby justifying our method.

10.2.5 Harmonic oscillator subject to perturbing potential in x^3

We consider a one-dimensional harmonic oscillator subject to a perturbation in x^3. As before, the way to proceed is by writing down the Hamiltonian for the complete system:

$$H = \frac{p^2}{2m} + \frac{\kappa}{2}x^2 + W \tag{10.75}$$

Here, the spring constant $\kappa = m\omega^2$, and the perturbation is

$$W = \xi x^3 \hbar\omega(m\omega/\hbar)^{3/2}$$

Because we have factored out $(\hbar/m\omega)^{1/2}$, the position operator may be written as $x = (1/2)^{1/2}(\hat{b}^\dagger + \hat{b})$, and since we are interested in x^3, we need to find $x^3 \propto (\hat{b}^\dagger + \hat{b})^3$:

$$\left(\hat{b}^\dagger + \hat{b}\right)^3 = \left(\hat{b}^{\dagger 2} + \hat{b}\hat{b}^\dagger + \hat{b}^\dagger\hat{b} + \hat{b}^2\right)\left(\hat{b}^\dagger + \hat{b}\right) \tag{10.76}$$

$$\left(\hat{b}^\dagger + \hat{b}\right)^3 = \left(\hat{b}^{\dagger 3} + \hat{b}\hat{b}^\dagger\hat{b}^\dagger + \hat{b}^\dagger\hat{b}\hat{b}^\dagger + \hat{b}\hat{b}\hat{b}^\dagger + \hat{b}^\dagger\hat{b}^\dagger\hat{b} + \hat{b}\hat{b}^\dagger\hat{b} + \hat{b}^\dagger\hat{b}\hat{b} + \hat{b}^3\right) \tag{10.77}$$

Making use of the operator $\hat{n} = \hat{b}^\dagger\hat{b}$ and $\hat{n} + 1 = \hat{b}\hat{b}^\dagger$, and substituting into the equation

$$\left(\hat{b}^\dagger + \hat{b}\right)^3 = \left(\hat{b}^{\dagger 3} + (n+1)\hat{b}^\dagger + n\hat{b}^\dagger + \hat{b}(n+1) + \hat{b}^\dagger n + (n+1)\hat{b} + n\hat{b} + \hat{b}^3\right) \tag{10.78}$$

Now, exploiting the commutation relations $[n, \hat{b}] = n\hat{b} - \hat{b}n = -\hat{b}$ and $[n, \hat{b}^\dagger] = n\hat{b}^\dagger - \hat{b}^\dagger n = \hat{b}^\dagger$ one finds

$$\left(\hat{b}^\dagger + \hat{b}\right)^3 = \left(\hat{b}^{\dagger 3} + n\hat{b}^\dagger + \hat{b}^\dagger + n\hat{b}^\dagger + (n+1)\hat{b} + \hat{b} + n\hat{b}^\dagger - \hat{b}^\dagger \right.$$
$$\left. + (n+1)\hat{b} + (n+1)\hat{b} - \hat{b} + \hat{b}^3\right)$$

$$\left(\hat{b}^\dagger + \hat{b}\right)^3 = \left(\hat{b}^{\dagger 3} + 3n\hat{b}^\dagger + 3(n+1)\hat{b} + \hat{b}^3\right) \tag{10.79}$$

Therefore, the perturbation is

$$W = \frac{\xi\hbar\omega}{2^{3/2}}\left(\hat{b}^{\dagger 3} + 3\hat{n}\hat{b}^\dagger + 3(\hat{n}+1)\hat{b} + \hat{b}^3\right) \tag{10.80}$$

Hence, the only nonzero matrix elements of the perturbation W have the effect of mixing $|\phi_n\rangle$ with states $|\phi_{n+1}\rangle$, $|\phi_{n-1}\rangle$, $|\phi_{n+3}\rangle$ and $|\phi_{n-3}\rangle$. The matrix elements are

$$\langle\phi_{n+3}|W|\phi_n\rangle = \xi\left(\frac{(n+3)(n+2)(n+1)}{8}\right)^{1/2}\hbar\omega \tag{10.81}$$

$$\langle\phi_{n-3}|W|\phi_n\rangle = \xi\left(\frac{n(n-1)(n-2)}{8}\right)^{1/2}\hbar\omega \tag{10.82}$$

$$\langle\phi_{n+1}|W|\phi_n\rangle = 3\xi\left(\frac{n+1}{2}\right)^{3/2}\hbar\omega \tag{10.83}$$

$$\langle\phi_{n-1}|W|\phi_n\rangle = 3\xi\left(\frac{n}{2}\right)^{3/2}\hbar\omega \tag{10.84}$$

Using these results to evaluate the energy terms up to second order gives

$$E_n = E_n^{(0)} + W_{nn} + \sum_{m \neq n}\frac{|W_{nm}|^2}{E_n - E_m} \tag{10.85}$$

$$E_n = \left(n + \frac{1}{2}\right)\hbar\omega + 0 + \sum_{m \neq n}\frac{|W_{nm}|^2}{E_n - E_m} \tag{10.86}$$

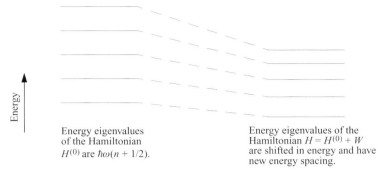

Fig. 10.4. Illustration of energy eigenvalues of the one-dimensional harmonic oscillator with Hamiltonian $H^{(0)}$ (left) and subject to perturbation $W = \xi x^3 \hbar\omega(m\omega/\hbar)^{3/2}$ in the potential (right). In the presence of the perturbation, both the energy levels and the energy-level spacing are changed. The effect of W is to lower the unperturbed energy levels whatever the sign of ξ. In addition, the difference between adjacent energy eigenvalues decreases with increasing energy.

$$E_n = \left(n + \frac{1}{2}\right)\hbar\omega - \frac{15}{4}\xi^2\left(n + \frac{1}{2}\right)^2\hbar\omega - \frac{7}{16}\xi^2\hbar\omega + \cdots \tag{10.87}$$

As shown in Fig. 10.4, the effect of W is to *lower* the unperturbed energy levels. The energy levels are lowered *whatever the sign of* ξ. In addition, the difference between adjacent energy eigenvalues decreases with increasing energy (Exercise 10.3).

10.3 Time-independent degenerate perturbation

Often, the potential in which a particle moves contains symmetry that results in eigenvalues that are degenerate. If the symmetry producing this degeneracy is destroyed by the perturbation W, the degenerate state separates, or splits, into distinct energy levels. This is illustrated schematically in Fig. 10.5.

We assume the Hamiltonian that describes the system is of the form

$$H = H^{(0)} + W \tag{10.88}$$

where $H^{(0)}$ is the unperturbed Hamiltonian with degenerate energy levels and W is the perturbation. Formally, the solution for the Schrödinger equation with total Hamiltonain H has eigenfunctions ψ_n and eigenenergies E_n, so that

$$H\psi_n = E_n\psi_n \tag{10.89}$$

The solution for the Schrödinger equation using the unperturbed Hamiltonian $H^{(0)}$ has eigenfunctions $\psi_m^{(0)}$ and degenerate eigenenergies $E_m^{(0)}$ so that

$$H^{(0)}\psi_m^{(0)} = E_m^{(0)}\psi_m^{(0)} \tag{10.90}$$

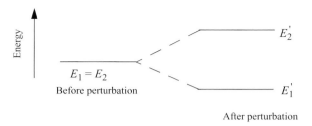

Fig. 10.5. Energy-level diagram illustrating degenerate energy eigenvalues split into separate energy levels by the presence of a perturbing potential.

10.3.1 A two-fold degeneracy split by time-independent perturbation

If we expand the *first-order* corrected wave functions in terms of the unperturbed wave functions, we obtain

$$\psi_m^{(1)} = \sum_n a_{mn}^{(1)} \psi_n^{(0)} \tag{10.91}$$

and

$$a_{mn}^{(1)} = \frac{W_{mn}}{E_m^{(0)} - E_n^{(0)}} \tag{10.92}$$

for the values of m and n, for which we have degeneracy $E_m^{(0)} - E_n^{(0)} = 0$ and for which a_{mn} is infinite. We get around this problem by diagonalizing (finding solutions) the Hamiltonian H. We adopt the matrix formulation and use only N terms to keep the problem tractable.

10.3.2 Matrix method

In the matrix method, the Hamiltonian is treated as a finite matrix equation. Diagonalization of this matrix gives the solution to how a group of initially unperturbed states interact via the perturbation *with one another*.

The matrix form of the Schrödinger equation $H|a\rangle = E|a\rangle$ describing the interacting system may be written as

$$\begin{bmatrix} H_{11} & H_{12} & H_{13} & \cdots & \\ H_{21} & H_{22} & H_{23} & \cdots & \\ H_{31} & H_{32} & H_{33} & \cdots & \\ \vdots & \vdots & \vdots & \ddots & \\ & & & & H_{NN} \end{bmatrix} \begin{bmatrix} a_1 \\ a_2 \\ a_3 \\ \vdots \\ a_N \end{bmatrix} = E \begin{bmatrix} a_1 \\ a_2 \\ a_3 \\ \vdots \\ a_N \end{bmatrix} \tag{10.93}$$

In Eqn (10.93), the matrix elements are $H_{mn} = \langle m|H|n\rangle$.

The approximation in the matrix method arises because *we only use N terms*. This method is good for the problem of degenerate energy levels split by a perturbation. Hence, it is sometimes called degenerate perturbation theory.

This approximation works because, first, the larger the energy separation between states the weaker the effect of the perturbation and, second, the smaller the matrix element $W_{mn} = \langle m|W|n \rangle$ the weaker the effect of the perturbation.

The matrix equation (Eqn (10.93)) may be rewritten

$$\sum_{n=1}^{N} [H_{mn} - E\delta_{mn}]a_n = 0, \quad m = 1, 2, \ldots, N \tag{10.94}$$

which has a nontrivial solution if the characteristic determinant vanishes, giving the *secular equation*

$$\begin{vmatrix} H_{11} - E & H_{12} & \cdots \\ H_{21} & H_{22} - E & \cdots \\ \vdots & \vdots & \ddots \\ & & & H_{NN} - E \end{vmatrix} = 0 \tag{10.95}$$

10.3.2.1 Matrix method for two states

As a simple example of the matrix method, consider two states. The secular equation is

$$\begin{vmatrix} H_{11} - E & H_{12} \\ H_{21} & H_{22} - E \end{vmatrix} = 0 \tag{10.96}$$

$$(H_{11} - E)(H_{22} - E) - H_{12}H_{21} = 0 \tag{10.97}$$

$$E^2 - EH_{22} - EH_{11} + H_{11}H_{22} - H_{12}H_{21} = 0 \tag{10.98}$$

$$E^2 - (H_{22} + H_{11})E + H_{11}H_{22} - H_{12}H_{21} = 0 \tag{10.99}$$

This equation is of the form $ax^2 + bx + c = 0$, and so it has a solution $x = (-b \pm \sqrt{b^2 - 4ac})/2a$. Hence, the two new energy eigenvalues are

$$E = \frac{(H_{11} + H_{22}) \pm \sqrt{(H_{11} + H_{22})^2 - 4(H_{11}H_{22} - H_{12}H_{21})}}{2} \tag{10.100}$$

$$E = \frac{(H_{11} + H_{22})}{2} \pm \frac{\left(H_{11}^2 + H_{22}^2 + 2H_{11}H_{22} - 4H_{11}H_{22} + 4H_{12}H_{21}\right)^{1/2}}{2} \tag{10.101}$$

$$\boxed{E_{\pm} = \frac{(H_{11} + H_{22})}{2} \pm \left(\frac{1}{4}(H_{11} - H_{22})^2 + H_{12}H_{21}\right)^{1/2}} \tag{10.102}$$

or

$$E_{\pm} = \frac{(H_{11} + H_{22})}{2} \pm \left(\frac{1}{4}(H_{11} - H_{22})^2 + |H_{12}|^2 \right)^{1/2} \tag{10.103}$$

We now determine the coefficients a_1 and a_2. Equation (10.95) may be written as

$$\begin{bmatrix} H_{11} - E & H_{12} \\ H_{21} & H_{22} - E \end{bmatrix} \begin{bmatrix} a_1 \\ a_2 \end{bmatrix} = 0 \tag{10.104}$$

Multiplying out the matrix gives two equations:

$$(H_{11} - E)a_1 + H_{12}a_2 = 0 \tag{10.105}$$

and

$$H_{21}a_1 + (H_{22} - E)a_2 = 0 \tag{10.106}$$

In addition to these two equations, there is the constraint that ψ is normalized. Hence, we require

$$|a_1|^2 + |a_2|^2 = 1 \tag{10.107}$$

Solving Eqn (10.105) is achieved by writing

$$(H_{11} - E)a_1 = -H_{12}a_2 \tag{10.108}$$

and then squaring both sides to give

$$(H_{11} - E)^2|a_1|^2 = H_{12}^2|a_2|^2 \tag{10.109}$$

Using the fact that $|a_2|^2 = 1 - |a_1|^2$ (Eqn (10.107)), one may write

$$(H_{11} - E)^2|a_1|^2 = H_{12}^2\left(1 - |a_1|^2\right) \tag{10.110}$$

$$\left((H_{11} - E)^2 + H_{12}^2\right)|a_1|^2 = H_{12}^2 \tag{10.111}$$

Hence,

$$|a_1|^2 = \frac{|H_{12}|^2}{(H_{11} - E)^2 + |H_{12}|^2} \tag{10.112}$$

If we let

$$D^2 = |H_{12}|^2 + (H_{11} - E)^2 \tag{10.113}$$

then

$$a_1 = \frac{H_{12}}{D} \tag{10.114}$$

and substituting this into Eqn (10.105) gives

$$a_2 = \frac{-(H_{11} - E)a_1}{H_{12}} = \frac{E - H_{11}}{D} \tag{10.115}$$

10.3.3   The two-dimensional harmonic oscillator subject to perturbation in xy

The unperturbed Hamiltonian for a particle of mass m in a two-dimensional harmonic potential is

$$H^{(0)} = \frac{p_x^2 + p_y^2}{2m} + \frac{\kappa}{2}(x^2 + y^2) \tag{10.116}$$

where κ is the spring constant. Figure 10.6 illustrates the potential $V(x, y)$.

The unperturbed Hamiltonian given by Eqn (10.116) can be rewritten in terms of raising and lowering operators:

$$H^{(0)} = \hbar\omega\left(\hat{b}_x^\dagger\hat{b}_x + \frac{1}{2} + \hat{b}_y^\dagger\hat{b}_y + \frac{1}{2}\right) = \hbar\omega(\hat{b}_x^\dagger\hat{b}_x + \hat{b}_y^\dagger\hat{b}_y + 1) \tag{10.117}$$

where

$$\hat{x} = \left(\frac{\hbar}{2m\omega}\right)^{1/2}(\hat{b}_x + \hat{b}_x^\dagger) \text{ and } \hat{y} = \left(\frac{\hbar}{2m\omega}\right)^{1/2}(\hat{b}_y + \hat{b}_y^\dagger)$$

The eigenstates of $H^{(0)}$ are of the form

$$\phi_{nm} = \phi_n(x)\phi_m(y) = |nm\rangle \tag{10.118}$$

and the eigenenergy is

$$E_{nm} = \hbar\omega(n + m + 1) \tag{10.119}$$

which is $(n + m + 1)$-fold degenerate. For example, states with energy $2\hbar\omega$ are two-fold degenerate, since $E_{10} = E_{01} = 2\hbar\omega$. The corresponding eigenstates are $|10\rangle$ and $|01\rangle$.

Let's consider the effect of the perturbing potential $W = \kappa'xy$ on this degeneracy and find the two new wave functions and eigenvalues that diagonalize W. The new wave functions are linear combinations of the unperturbed wave function, so that

$$\psi_1 = a_1\phi_{10} + a_2\phi_{01} \tag{10.120}$$

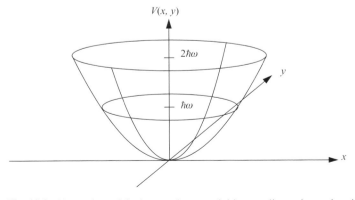

Fig. 10.6. Illustration of the harmonic potential in two dimensions, showing quantization energy $\hbar\omega$ and $2\hbar\omega$.

and

$$\psi_2 = a_1 \phi_{10} - a_2 \phi_{01} \tag{10.121}$$

The submatrix W in the basis $\{\phi_{10}, \phi_{01}\}$ is

$$W = \kappa' \begin{bmatrix} \langle 10|xy|10 \rangle & \langle 10|xy|01 \rangle \\ \langle 01|xy|10 \rangle & \langle 01|xy|01 \rangle \end{bmatrix} = \kappa' \begin{bmatrix} W_{11} & W_{12} \\ W_{21} & W_{22} \end{bmatrix} \tag{10.122}$$

The next task is to evaluate the matrix elements. For example

$$W_{12} = \langle 10|xy|01 \rangle = \frac{\hbar}{2m\omega} \langle 10|(\hat{b}_x + \hat{b}_x^\dagger)(\hat{b}_y + \hat{b}_y^\dagger)|01 \rangle \tag{10.123}$$

$$W_{12} = \frac{\hbar}{2m\omega} \langle 10|\hat{b}_x\hat{b}_y + \hat{b}_x^\dagger\hat{b}_y + \hat{b}_x\hat{b}_y^\dagger + \hat{b}_x^\dagger\hat{b}_y^\dagger|01 \rangle = \frac{\hbar}{2m\omega} \langle 10|\hat{b}_x^\dagger\hat{b}_y|01 \rangle \tag{10.124}$$

$$W_{12} = \frac{\hbar}{2m\omega} \tag{10.125}$$

and

$$\langle 10|xy|10 \rangle = \frac{\hbar}{2m\omega} \langle 10|\hat{b}_x\hat{b}_y + \hat{b}_x^\dagger\hat{b}_y + \hat{b}_x\hat{b}_y^\dagger + \hat{b}_x^\dagger\hat{b}_y^\dagger|10 \rangle = 0 \tag{10.126}$$

Hence,

$$W = \frac{\hbar\kappa'}{2m\omega} \begin{bmatrix} 0 & 1 \\ 1 & 0 \end{bmatrix} \tag{10.127}$$

and the secular equation

$$\begin{vmatrix} H_{11} - E & H_{12} \\ H_{21} & H_{22} - E \end{vmatrix} = 0 \tag{10.128}$$

may be written as

$$\begin{vmatrix} -E & \dfrac{\hbar\kappa'}{2m\omega} \\ \dfrac{\hbar\kappa'}{2m\omega} & -E \end{vmatrix} = 0 \tag{10.129}$$

which has the solutions

$$E = \pm\frac{\hbar\kappa'}{2m\omega} = \pm\frac{\Delta E}{2} \tag{10.130}$$

We therefore find that the perturbation separates the first excited state by an amount

$$\frac{2\hbar\kappa'}{2m\omega} = \frac{\hbar\kappa'}{m\omega} = \Delta E \tag{10.131}$$

The lifting of the first excited-state degeneracy by the perturbation $W = \kappa'xy$ is illustrated in Fig. 10.7.

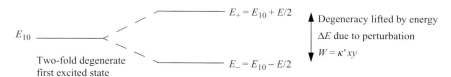

Fig. 10.7. The two-fold degeneracy of the first excited state of the two-dimensional harmonic oscillator is split into two energy eigenvalues by the perturbation $W = \kappa' xy$.

The new wave functions are obtained by substituting these values into the matrix equation

$$\sum_{n=1}^{N}[H_{nm} - E\delta_{nm}]a_n = 0 \tag{10.132}$$

For our case,

$$\begin{bmatrix} -E & \dfrac{\Delta E}{2} \\ \dfrac{\Delta E}{2} & -E \end{bmatrix} \begin{bmatrix} a_1 \\ a_2 \end{bmatrix} = 0 \tag{10.133}$$

If $E = \Delta E/2$, then there is a solution $a_1 = a_2$. On the other hand, if $E = -\Delta E/2$, then there is a solution $a_1 = -a_2$. From this, we obtain symmetric and antisymmetric eigenfunction states and eigenvalues. The symmetric state

$$\psi_1 = \frac{1}{\sqrt{2}}(\phi_{10} + \phi_{01}) \tag{10.134}$$

has eigenenergy

$$E_+ = \frac{\Delta E}{2} \tag{10.135}$$

and the antisymmetric state

$$\psi_2 = \frac{1}{\sqrt{2}}(\phi_{10} - \phi_{01}) \tag{10.136}$$

has the lower eigenenergy

$$E_- = -\frac{\Delta E}{2} \tag{10.137}$$

10.4 Example exercises

Exercise 10.1

A particle of mass m moves in a one-dimensional, infinitely deep potential well having a parabolic bottom,

$$V(x) = \infty \text{ for } |x| \geq L$$

and

$$V(x) = \xi x^2/L^2 \text{ for } -L < x < L$$

where ξ is small compared with the ground-state energy. Treat the term ξ as a perturbation on the square potential well (denoting the unperturbed states as $\phi_0, \phi_1, \phi_2, \ldots$ in order of increasing energy), and calculate, to first order in ξ only, the energy and the amplitudes A_0, A_1, A_2, A_3 of the first four perturbed states.

Exercise 10.2

Calculate the energy levels of an anharmonic oscillator with potential of the form

$$V(x) = \frac{\kappa}{2}x^2 + \xi x^3 \hbar \omega$$

where κ is the spring constant for a harmonic potential. Show that the difference between two adjacent perturbed levels is $E_n - E_{n-1} = \hbar\omega(1 - 15\xi^2(\hbar/m\omega)^3 n/2)$. A heterodi-atomic molecule can absorb or emit electromagnetic waves the frequency of which coincides with the vibrational frequency of anharmonic oscillations of the molecule about its equilibrium position. For a molecule initially in the ground state, what do you expect to observe in the absorption spectrum of the molecule?

Exercise 10.3

The potential function of a one-dimensional oscillator of mass m and angular frequency ω is $V(x) = \kappa x^2/2 + \xi x^4$, where κ is the spring constant for a harmonic potential and the second term is small compared with the first.

(a) Show that, to first order, the effect of the anharmonic term is to change the energy of the ground state by $3\xi(\hbar/2m\omega)^2$.

(b) What would be the first-order effect of an additional x^3 term in the potential?

Exercise 10.4

A particle of mass m and charge e oscillates in a one-dimensional harmonic potential with angular frequency ω.

(a) Show, using perturbation theory, that the effect of an applied uniform electric field $\mathbf{E}$ is to lower all the energy levels by $e^2|\mathbf{E}|^2/2m\omega^2$.

(b) Compare this with the classical result.

(c) Use perturbation theory to calculate the new ground-state wave function.

Exercise 10.5

The potential seen by an electron with effective mass m_e^* in a GaAs quantum well is approximated by a one-dimensional rectangular potential well of width $2L$ in such a way that

$$V(x) = 0 \qquad \text{for } 0 < x < 2L$$

and

$V(x) = \infty$ elsewhere

(a) Find the eigenvalues E_n, eigenfunctions ψ_n, and parity of ψ_n.

(b) The system is subject to a perturbation in the potential energy so that $V(x) = e|\mathbf{E}|x$, where $\mathbf{E}$ is a constant electric field in the x direction. Find the value of the new energy eigenvalues to first order (the linear Stark effect) for a quantum well of a width of 10 nm subject to an electric field of 10^5 V cm^{-1}. Compare the change in energy value with thermal energy at room temperature.

(c) Find the expression for the second-order correction to the energy eigenvalues for the perturbation in (b).

Exercise 10.6

In this chapter we solved for the first excited state of a two-dimensional harmonic oscillator subject to perturbation $W = \kappa'xy$. How do the three-fold degenerate energy $E = 3\hbar\omega$ and the four-fold degenerate energy $E = 4\hbar\omega$ separate due to the same perturbation?

Exercise 10.7

(a) An electron moves in a one-dimensional box of length X. Apply the periodic boundary condition $\phi(x) = \phi(x + X)$ to find the electron eigenfunction and eigenvalues.

(b) Now apply a weak periodic potential $V(x) = V(x + L)$ to the system, where $X = NL$ and N is a large positive integer. Using nondegenerate perturbation theory, find the first-order correction to the wave functions and the second-order correction to the eigenenergies.

(c) When wave vector k is close to $n\pi/L$, where n is an integer, the result in (b) is no longer valid. Use two-state degenerate perturbation theory to find the corrected energy values for $k = n\pi((1 + \Delta)/L)$ and $k' = n\pi((1 - \Delta)/L)$, where Δ is small compared with π/L.

(d) Use the results of (b) and (c) to draw the electron dispersion relation, $E(k)$.

(e) If we choose the lowest-frequency Fourier component of the perturbative periodic potential in part (b), then $V(x) = V_1 \cos(\pi x/L)$. Repeat (b), (c), and (d) using this potential.

Hint: $V(x) = V_0 + \sum_{n \neq 0} V_n e^{i2\pi nx/L}$, and choose $V_0 = 0$.

Exercise 10.8

(a) What is the effect of applying a uniform electric field on the energy spectrum of an atom?

(b) If spin effects are neglected, the four states of the hydrogen atom with quantum number $n = 2$ have the same energy, E^0. Show that when an electric field $\mathbf{E}$

is applied to hydrogen atoms in these states, the resulting first-order energies are $E^0 \pm 3a_B e|\mathbf{E}|$, E^0, E^0.

Treat the z-directed electric field as a perturbation on the separable, orthonormal, unperturbed electron wave functions $\psi_{nlm}(r, \theta, \phi) = R_n(r)\Theta_l(\theta)\Phi_m(\phi)$, where r, θ, and ϕ are the standard spherical coordinates. You may use the unperturbed wave functions

$$\psi_{200} = \frac{2}{(2a_B)^{3/2}} \cdot \left(1 - \frac{r}{2a_B}\right) \cdot e^{-r/2a_B} \cdot \left(\frac{1}{4\pi}\right)^{1/2}$$

$$\psi_{210} = \frac{1}{\sqrt{3}(2a_B)^{3/2}} \cdot \frac{r}{a_B} \cdot e^{-r/2a_B} \cdot \frac{1}{2}\left(\frac{3}{\pi}\right)^{1/2} \cos(\theta)$$

SOLUTIONS

Solution 10.1

The eigenfunctions for a rectangular potential well of width $2L$ centered at $x = 0$ and infinite barrier energy may be expressed in terms of sine functions. Hence,

$$\phi_n^{(0)} = \frac{1}{\sqrt{L}} \sin\left(\frac{(n+1)\pi(x+L)}{2L}\right) = \frac{1}{\sqrt{L}} \sin(k_n x)$$

where the index $n = 0, 1, 2, \ldots$ labels the eigenstate, and

$$k_n = \frac{(n+1)\pi}{2L}$$

So, the first few eigenfunctions are

$$\phi_0^{(0)} = \frac{1}{\sqrt{L}} \sin\left(\frac{\pi(x+L)}{2L}\right) = \frac{1}{\sqrt{L}} \cos\left(\frac{\pi x}{2L}\right)$$

$$\phi_1^{(0)} = \frac{1}{\sqrt{L}} \sin\left(\frac{\pi(x+L)}{L}\right) = \frac{-1}{\sqrt{L}} \sin\left(\frac{\pi x}{L}\right)$$

$$\phi_2^{(0)} = \frac{1}{\sqrt{L}} \sin\left(\frac{3\pi(x+L)}{2L}\right) = \frac{-1}{\sqrt{L}} \cos\left(\frac{3\pi x}{2L}\right)$$

$$\phi_3^{(0)} = \frac{1}{\sqrt{L}} \sin\left(\frac{2\pi(x+L)}{L}\right) = \frac{1}{\sqrt{L}} \sin\left(\frac{2\pi x}{L}\right)$$

In general, the eigenvalues are

$$E_n^{(0)} = \frac{\hbar^2 k_n^2}{2m} = \frac{\hbar^2 \pi^2 (n+1)^2}{8mL^2}$$

So, the first few eigenvalues are

$$E_0^{(0)} = \frac{\hbar^2 \pi^2}{8mL^2}$$

$$E_1^{(0)} = \frac{\hbar^2 \pi^2}{2mL^2}$$

$$E_2^{(0)} = \frac{9\hbar^2\pi^2}{8mL^2}$$

$$E_3^{(0)} = \frac{2\hbar^2\pi^2}{mL^2}$$

In the presence of the perturbation, $W(x) = \xi x^2/L^2$, the energy eigenvalues and eigenfunctions are to first order given by

$$E = E_m^{(0)} + W_{mm}$$

$$\psi = \psi_m^{(0)} + \sum_{k \neq m} \frac{W_{km}}{E_m - E_k} \psi_k^{(0)}$$

where the first-order correction to energy eigenvalues are the diagonal matrix elements

$$E_m^{(1)} = W_{mm} = \left\langle \phi_m \left| \xi \frac{x^2}{L^2} \right| \phi_m \right\rangle$$

$$W_{mm} = \frac{\xi}{L^3} \int_{-L}^{L} \sin\left(\frac{(m+1)\pi(x+L)}{2L}\right) x^2 \sin\left(\frac{(m+1)\pi(x+L)}{2L}\right) dx$$

Using $2\sin(x)\sin(y) = \cos(x - y) - \cos(x + y)$, this integral may be written

$$W_{mm} = \frac{\xi}{2L^3} \int_{-L}^{L} x^2 \left(1 - \cos\left(\frac{(m+1)\pi(x+L)}{L}\right)\right) dx$$

$$W_{mm} = \frac{\xi}{2L^3} \left(\frac{2L^3}{3} - \int_{-L}^{L} x^2 \cos\left(\frac{(m+1)\pi(x+L)}{L}\right) dx\right)$$

Solving the integral by parts $\int U V' dx = UV - \int U'V dx$, we set

$$U = x^2$$

$$U' = 2x$$

$$V' = \cos\left(\frac{(m+1)\pi(x+L)}{L}\right)$$

$$V = \frac{L}{(m+1)\pi} \sin\left(\frac{(m+1)\pi(x+L)}{L}\right)$$

$$\int U V' dx = \frac{Lx^2}{(m+1)\pi} \sin\left(\frac{(m+1)\pi(x+L)}{L}\right)\Bigg|_{-L}^{L}$$

$$- \frac{2L}{(m+1)\pi} \int_{-L}^{L} x \sin\left(\frac{(m+1)\pi(x+L)}{L}\right) dx$$

The first term on the right-hand side is zero for all integer values of m. Solving the integral by parts again, we set

$$U = \frac{2Lx}{(m+1)\pi}$$

$$U' = \frac{2L}{(m+1)\pi}$$

$$V' = \sin\left(\frac{(m+1)\pi(x+L)}{L}\right)$$

$$V = \frac{-L}{(m+1)\pi}\cos\left(\frac{(m+1)\pi(x+L)}{L}\right)$$

$$-\int UV'dx = \left.\frac{2L^2x}{(m+1)^2\pi^2}\cos\left(\frac{(m+1)\pi(x+L)}{L}\right)\right|_{-L}^{L}$$

$$+\frac{2L^2}{(m+1)^2\pi^2}\int_{-L}^{L}\cos\left(\frac{(m+1)\pi(x+L)}{L}\right)dx$$

$$-\int UV'dx = \frac{4L^3}{(m+1)^2\pi^2} + 0$$

so that

$$W_{mm} = \frac{\xi}{2L^3}\left(\frac{2L^3}{3} - \frac{4L^3}{(m+1)^2\pi^2}\right)$$

and the first-order corrected eigenvalues are

$$E_m = E_m^{(0)} + W_{mm} = \frac{\hbar^2\pi^2(n+1)^2}{8mL^2} + \xi\left(\frac{1}{3} - \frac{2}{(m+1)^2\pi^2}\right)$$

Notice that in the limit $m \to \infty$ the first-order correction to energy eigenvalues is $W_{mm} \to \xi/3$. This limit is easy to understand, since for states with large m the probability of finding the particle somewhere in the range $-L < x < L$ is uniform. In this case, the energy shift is given by the average value of the perturbation in the potential:

$$\langle V(x)\rangle = \frac{1}{L^2}\int_{-L}^{L}\xi x^2 dx \bigg/ \int_{-L}^{L}dx = \frac{1}{3}\xi$$

One may now interpret the term $-2\xi/(m+1)^2\pi^2$, which is significant for states with small values of m. This decrease in energy shift compared with the average value of the potential is due to the fact that the probability of finding the particle somewhere in the range $-L < x < L$ is nonuniform. For low values of m, particle probability tends to be greater near to $x = 0$ compared with $|x| = L$. Because the perturbing potential is zero at $x = 0$, the first-order energy shift for states with small values of m is always smaller than for states with large values of m.

The first few energy levels to first order are

$$E_0 = \frac{\hbar^2\pi^2}{8mL^2} + \xi\left(\frac{1}{3} - \frac{2}{\pi^2}\right)$$

$$E_1 = \frac{\hbar^2\pi^2}{2mL^2} + \xi\left(\frac{1}{3} - \frac{1}{2\pi^2}\right)$$

$$E_2 = \frac{9\hbar^2\pi^2}{8mL^2} + \xi\left(\frac{1}{3} - \frac{2}{9\pi^2}\right)$$

$$E_3 = \frac{2\hbar^2\pi^2}{mL^2} + \xi\left(\frac{1}{3} - \frac{2}{16\pi^2}\right)$$

To find the eigenfunctions in the presence of the perturbation, we need to evaluate the matrix elements

$$W_{km} = \left\langle \phi_k \left| \xi\frac{x^2}{L^2} \right| \phi_m \right\rangle = \frac{\xi}{L^3}\int\limits_{-L}^{L} \sin\left(\frac{(k+1)\pi(x+L)}{2L}\right)x^2\sin\left(\frac{(m+1)\pi(x+L)}{2L}\right)dx$$

$$W_{km} = \frac{\xi}{2L^3}\int\limits_{-L}^{L} x^2\left(\cos\left(\frac{(k-m)\pi(x+L)}{2L}\right) - \cos\left(\frac{(k+m)\pi(x+L)}{2L}\right)\right)dx$$

$$W_{km} = \frac{8\xi}{\hbar^2}\left(\frac{1}{(m-n)^2} - \frac{1}{(m+n)^2}\right)$$

$$\psi_1 = \phi_1 + \sum_{m\neq 1}\frac{W_{m1}}{E_1 - E_m}\phi_m = \phi_1 + \sum_{m(odd)\neq 1}\frac{\dfrac{8\xi}{\pi^2}\left(\dfrac{1}{(m-n)^2} - \dfrac{1}{(m+n)^2}\right)}{\dfrac{\hbar^2\pi^2}{8mL^2}(1-m)^2}\phi_m$$

$$\psi_2 = \phi_2 + \sum_{m\neq 2}\frac{W_{m2}}{E_2 - E_m}\phi_m = \phi_2 + \sum_{m(even)\neq 2}\frac{\dfrac{8\xi}{\pi^2}\left(\dfrac{1}{(m-n)^2} - \dfrac{1}{(m+n)^2}\right)}{\dfrac{\hbar^2\pi^2}{8mL^2}(4-m)^2}\phi_m$$

$$\psi_3 = \phi_3 + \sum_{m\neq 3}\frac{W_{m3}}{E_3 - E_m}\phi_m = \phi_3 + \sum_{m(odd)\neq 3}\frac{\dfrac{8\xi}{\pi^2}\left(\dfrac{1}{(m-n)^2} - \dfrac{1}{(m+n)^2}\right)}{\dfrac{\hbar^2\pi^2}{8mL^2}(9-m)^2}\phi_m$$

Solution 10.2

The Hamiltonian for the one-dimensional harmonic oscillator subject to perturbation W is

$$H = \frac{p^2}{2m} + \frac{\kappa}{2}x^2 + W$$

where $\kappa = m\omega^2$. From Section 10.2.5 we have for a perturbation $W = \xi x^3 \hbar\omega(m\omega/\hbar)^{3/2}$

$$E_n = \left(n + \frac{1}{2}\right)\hbar\omega - \frac{15}{4}\xi^2\left(n + \frac{1}{2}\right)^2\hbar\omega - \frac{7}{16}\xi^2\hbar\omega$$

which, expanding the square, may be written

$$E_n = \left(n + \frac{1}{2}\right)\hbar\omega - \frac{15}{4}\xi^2\left(n^2 + n + \frac{1}{4}\right)\hbar\omega - \frac{7}{16}\xi^2\hbar\omega$$

Hence,

$$E_{n-1} = \left(n - \frac{1}{2}\right)\hbar\omega - \frac{15}{4}\xi^2\left(n^2 - n + \frac{1}{4}\right)\hbar\omega - \frac{7}{16}\xi^2\hbar\omega$$

so that

$$E_n - E_{n-1} = \hbar\omega - \frac{15}{2}\xi^2 n$$

Because, in this exercise, W did not explicitly contain the factor $(m\omega/\hbar)^{3/2}$, we need to put this back in. Since ξ appears as a squared term, the inverse of the factor $(m\omega/\hbar)^{3/2}$ is also squared to give

$$E_n - E_{n-1} = \hbar\omega\left(1 - \frac{15}{2}\xi^2\left(\frac{\hbar}{m\omega}\right)^3 n\right)$$

The absorption spectrum of an anharmonic diatomic molecule initially in the ground state will consist of a series of absorption lines with energy separation between adjacent lines that decreases with increasing energy. The absorption lines will be at energy

$$E_n - E_0 = n\hbar\omega - \frac{15}{4}\xi^2\left(n^2 + n + \frac{1}{4}\right)\hbar\omega - \frac{1}{16}\xi^2\hbar\omega$$

and the wavelength is given by $\lambda = \hbar c/(E_n - E_0)$.

Solution 10.3

(a) Using the position operator $x = (\hbar/2m\omega)^{1/2}(\hat{b}^\dagger + \hat{b})$, one may express the perturbation as

$$W = \xi x^4 = \xi\left(\frac{\hbar}{2m\omega}\right)^2(\hat{b}^\dagger + \hat{b})^4$$

$$W = \xi\left(\frac{\hbar}{2m\omega}\right)^2(\hat{b}^\dagger + \hat{b})(\hat{b}^{\dagger 3} + \hat{b}\hat{b}^\dagger\hat{b}^\dagger + \hat{b}^\dagger\hat{b}\hat{b}^\dagger + \hat{b}\hat{b}\hat{b}^\dagger + \hat{b}^\dagger\hat{b}^\dagger\hat{b} + \hat{b}\hat{b}^\dagger\hat{b} + \hat{b}^\dagger\hat{b}\hat{b} + \hat{b}^3)$$

$$W = \xi\left(\frac{\hbar}{2m\omega}\right)^2(\hat{b}^{\dagger 4} + \hat{b}^\dagger\hat{b}\hat{b}^\dagger\hat{b}^\dagger + \hat{b}^\dagger\hat{b}^\dagger\hat{b}\hat{b}^\dagger + \hat{b}^\dagger\hat{b}\hat{b}\hat{b}^\dagger + \hat{b}^\dagger\hat{b}^\dagger\hat{b}^\dagger\hat{b} + \hat{b}^\dagger\hat{b}\hat{b}^\dagger\hat{b} + \hat{b}^\dagger\hat{b}^\dagger\hat{b}\hat{b}$$

$$+ \hat{b}^\dagger\hat{b}^3 + \hat{b}\hat{b}^{\dagger 3} + \hat{b}\hat{b}\hat{b}^\dagger\hat{b}^\dagger + \hat{b}\hat{b}^\dagger\hat{b}\hat{b}^\dagger + \hat{b}\hat{b}\hat{b}\hat{b}^\dagger + \hat{b}\hat{b}^\dagger\hat{b}^\dagger\hat{b} + \hat{b}\hat{b}\hat{b}^\dagger\hat{b} + \hat{b}\hat{b}^\dagger\hat{b}\hat{b} + \hat{b}^4)$$

Energy eigenvalues in first-order perturbation theory couple the same state, so only symmetric terms with two $\hat{b}^\dagger$ and two $\hat{b}$ will contribute.

$$W_{nn} = \xi \left(\frac{\hbar}{2m\omega}\right)^2 \left((n+1)(n+2) + (n+1)^2 + n^2 + n(n-1) + 2n(n+1)\right)$$

$$W_{nn} = \xi \left(\frac{\hbar}{2m\omega}\right)^2 (6n^2 + 6n + 3)$$

so, to first order,

$$E_n = \hbar\omega\left(n + \frac{1}{2}\right) + \xi \left(\frac{\hbar}{2m\omega}\right)^2 (6n^2 + 6n + 3)$$

and the new ground state is

$$E_n = \hbar\omega\left(n + \frac{1}{2}\right) + 3\xi \left(\frac{\hbar}{2m\omega}\right)^2$$

(b) There is no first-order correction for a perturbation in x^3, because it cannot couple to the same state (there are always an odd number of operators $\hat{b}^\dagger$ or $\hat{b}$).

Solution 10.4

(a) The solution follows that already given in this chapter.

(b) The new energy levels of the oscillator are to second order

$$E = \hbar\omega\left(n + \frac{1}{2}\right) - \frac{e^2|\mathbf{E}|^2}{2\kappa}$$

which is the same as the exact result. Physically, the particle oscillates at the same frequency, ω, as the unperturbed case, but it is displaced a distance of $e|\mathbf{E}|/m\omega^2$, and the new energy levels are shifted by $-e^2|\mathbf{E}|^2/2m\omega^2$.

(c) The ground-state wave function of the unperturbed harmonic oscillator is

$$\psi_0(x) = \left(\frac{m\omega}{\pi\hbar}\right)^{1/4} e^{-x^2 m\omega/2\hbar}$$

After the perturbation, it is

$$\psi_0(x) = \left(\frac{m\omega}{\pi\hbar}\right)^{1/4} e^{-\frac{m\omega}{2\hbar}\left(x - \frac{e|\mathbf{E}|}{m\omega^2}\right)^2}$$

The result using second-order perturbation theory

$$\psi = \psi_m^{(0)} + \sum_{k \neq m} \frac{W_{mk}}{E_m - E_k} \psi_k^{(0)}$$

$$+ \sum_{k \neq m}\left(\left(\sum_{n \neq m} \frac{W_{kn}W_{mn}}{(E_m - E_n)(E_m - E_k)} - \frac{W_{mm}W_{km}}{(E_m - E_k)^2}\right)\psi_k^{(0)} - \frac{1}{2}\frac{|W_{mn}|^2}{(E_m - E_n)^2}\psi_m^{(0)}\right)$$

only approximates the exact solution.

Solution 10.5

(a) It is given that the potential seen by an electron with effective mass m_e^* in a GaAs quantum well is approximated by a one-dimensional, rectangular potential well of width $2L$ in such a way that

$$V(x) = 0$$

for $0 < x < 2L$ and

$$V(x) = \infty$$

elsewhere. We find the eigenfunctions and eigenvalues by solving the time-independent Schrödinger equation

$$H^{(0)} \psi_n^{(0)} = E_n^{(0)} \psi_n^{(0)}$$

where the Hamiltonian for the electron in the potential is

$$H^{(0)} = \frac{p^2}{2m_e^*} + V(x)$$

The solution for the eigenfunctions is

$$\psi_n^{(0)} = \frac{1}{\sqrt{L}} \sin\left(\frac{n\pi x}{2L}\right) \quad n = 1, 2, \ldots$$

and the parity of the eigenfunctions is even for odd-integer and odd for even-integer values of n.

The solution for the eigenvalues is

$$E_n^{(0)} = \frac{\hbar^2 k_n^2}{2m_e^*} = \frac{\hbar^2 n^2 \pi^2}{8m_e^* L^2}$$

where

$$k_n = \frac{n2\pi}{2.2L} = \frac{n\pi}{2L}$$

(b) First-order correction to energy eigenvalues is

$$E_n^{(1)} = \langle n|V|n\rangle = \langle n|e|\mathbf{E}|x|n\rangle = V_{nn} = \frac{e|\mathbf{E}|}{L} \int\limits_{x=0}^{x=2L} x \sin^2\left(\frac{n\pi x}{2L}\right) dx$$

Using the relation $2\sin(x)\sin(y) = \cos(x - y) - \cos(x + y)$ with $x = y = n\pi x/2L$ allows us to rewrite the integrand, giving

$$V_{nn} = \frac{e|\mathbf{E}|}{2L} \int\limits_{x=0}^{x=2L} x\left(1 - \cos\left(\frac{n\pi x}{L}\right)\right) dx$$

$$V_{nn} = \frac{e|\mathbf{E}|}{2L}\left[\frac{x^2}{2}\right]_{x=0}^{x=2L} = e|\mathbf{E}|L$$

which is the *linear Stark effect*. The energy-level shift for an electric field $\mathbf{E} = 10^5$ V cm^{-1} in the x direction across a well of width $2L = 10$ nm is

$$\Delta E = 10^7 \times 5 \times 10^{-9} = 50 \text{ meV}$$

At temperature $T = 300$ K and for an energy splitting $\Delta E > k_B T = 25$ meV one may assume that the Stark effect produces a large-enough change in energy eigenvalue to be of potential use in a room-temperature device.

(c) The new energy levels to second order are found using

$$E_n = E_n^{(0)} + E_n^{(1)} + E_n^{(2)}$$

where

$$E_k^{(1)} = V_{kk}$$

and

$$E_k^{(2)} = \sum_{j \neq k} \frac{V_{kj} V_{jk}}{E_k^{(0)} - E_j^{(0)}}$$

The second-order matrix elements are the off-diagonal terms

$$V_{kj} = \langle k|V|j \rangle = \langle k|e|\mathbf{E}|x|j \rangle = \frac{e|\mathbf{E}|}{L} \int\limits_{x=0}^{x=2L} x \sin\left(\frac{k\pi x}{2L}\right) \sin\left(\frac{j\pi x}{2L}\right) dx$$

Using the relation $2\sin(x)\sin(y) = \cos(x - y) - \cos(x + y)$ with $x = k\pi x/2L$ and $y = j\pi x/2L$ allows us to rewrite the integrand, giving

$$V_{kj} = \frac{e|\mathbf{E}|}{2L} \int\limits_{x=0}^{x=2L} x \left(\cos\left(\frac{(k - j)\pi x}{2L}\right) - \cos\left(\frac{(k + j)\pi x}{2L}\right) \right) dx$$

For $(k \pm j)$ odd we integrate by parts, using $UV'dx = UV - \int U'V dx$ with $U = x$ and $V' = \cos((k \pm j)\pi x/2L)$. This gives

$$V_{kj} = \frac{e|\mathbf{E}|}{2L} \left(\frac{4L^2(\cos((k - j)\pi) - 1)}{\pi^2(k - j)^2} - \frac{4L^2(\cos((k + j)\pi) - 1)}{\pi^2(k + j)^2} \right)$$

$$V_{kj} = \frac{-4e|\mathbf{E}|L}{\pi^2} \left(\frac{1}{(k + j)^2} - \frac{1}{(k - j)^2} \right) = \frac{-16e|\mathbf{E}|L}{\pi^2} \frac{kj}{(k^2 - j^2)^2}$$

For $(k \pm j)$ even, symmetry requires that

$$V_{kj} = 0$$

So, the perturbing potential only mixes states of different parity.

Solution 10.6

A two-dimensional harmonic oscillator with motion in the x–y plane is subject to perturbation $W = \kappa'xy$. We are asked to find how the three-fold degenerate energy $E = 3\hbar\omega$ and the four-fold degenerate energy $E = 4\hbar\omega$ separate due to this perturbation.

Separation of variables x and y allows us to write the unperturbed Hamiltonian as

$$H^{(0)} = \frac{\hbar\omega}{2m}\left(p_x^2 + p_y^2\right) + \frac{\kappa}{2}\left(x^2 + y^2\right) = \hbar\omega\left(\hat{b}_x^\dagger\hat{b}_x + \hat{b}_y^\dagger\hat{b}_y + 1\right)$$

where $\hat{x} = \sqrt{\hbar/2m\omega}\cdot(\hat{b}_x + \hat{b}_x^\dagger)$ and $\hat{y} = \sqrt{\hbar/2m\omega}\cdot(\hat{b}_y + \hat{b}_y^\dagger)$. The eigenstates are of the form $\psi_{nm}^{(0)} = \phi_n^{(0)}(x)\phi_m^{(0)}(y) = |nm\rangle$, and the energy eigenvalues are

$$E_{nm} = \hbar\omega\,(n + m + 1)$$

where n and m are positive integers. States with eigenenergy $3\hbar\omega$ are E_{02}, E_{11}, and E_{20}, and so they are three-fold degenerate. To find the effect of the perturbation $W = \kappa'xy$, we start by writing the total Hamiltonian:

$$H = H^{(0)} + W = \hbar\omega\left(\hat{b}_x^\dagger\hat{b}_x + \hat{b}_y^\dagger\hat{b}_y + 1\right) + \frac{\hbar\kappa'}{2m\omega}\left(\hat{b}_x + \hat{b}_x^\dagger\right)\left(\hat{b}_y + \hat{b}_y^\dagger\right)$$

This has eigenfunction solutions that are linear combinations of the unperturbed eigenstates so that

$$\Psi_1 = a_1\psi_1 + a_2\psi_2 + a_3\psi_3$$

The coefficients a_n may be found by writing the Schrödinger equation in matrix form

$$\begin{bmatrix} \langle 02|H|02\rangle & \langle 02|H|11\rangle & \langle 02|H|20\rangle \\ \langle 11|H|02\rangle & \langle 11|H|11\rangle & \langle 11|H|20\rangle \\ \langle 20|H|02\rangle & \langle 20|H|11\rangle & \langle 20|H|20\rangle \end{bmatrix}\begin{bmatrix} a_1 \\ a_2 \\ a_3 \end{bmatrix} = E\begin{bmatrix} a_1 \\ a_2 \\ a_3 \end{bmatrix}$$

The diagonal terms have a value that is the unperturbed eigenvalue $3\hbar\omega$. To show this, consider

$$\langle 02|H|02\rangle = \hbar\omega\left\langle 02\left|\hat{b}_x^\dagger\hat{b}_x + \hat{b}_y^\dagger\hat{b}_y + 1\right|02\right\rangle + \frac{\hbar\kappa'}{2m\omega}\left\langle 02\left|\hat{b}_x\hat{b}_y + \hat{b}_x^\dagger\hat{b}_y + \hat{b}_x\hat{b}_y^\dagger + \hat{b}_x^\dagger\hat{b}_y^\dagger\right|02\right\rangle$$

The first term on the right-hand side has value $3\hbar\omega$, and the second term on the right-hand side is zero. Because the perturbation W is linear in x and y, only the off-diagonal terms adjacent to the diagonal are finite. For example,

$$\langle 11|H|02\rangle = \hbar\omega\left\langle 11\left|\hat{b}_x^\dagger\hat{b}_x + \hat{b}_y^\dagger\hat{b}_y + 1\right|02\right\rangle + \frac{\hbar\kappa'}{2m\omega}\left\langle 11\left|\left(\hat{b}_x + \hat{b}_x^\dagger\right)\left(\hat{b}_y + \hat{b}_y^\dagger\right)\right|02\right\rangle$$

The first term on the right-hand side is zero, leaving

$$\langle 11|H|02\rangle = \frac{\hbar\kappa'}{2m\omega}\left\langle 11\left|\hat{b}_x\hat{b}_y + \hat{b}_x^\dagger\hat{b}_y + \hat{b}_x\hat{b}_y^\dagger + \hat{b}_x^\dagger\hat{b}_y^\dagger\right|02\right\rangle = \frac{\hbar\kappa'}{2m\omega}\left\langle 11\left|\hat{b}_x^\dagger\hat{b}_y\right|02\right\rangle$$

Recalling that $|\hat{b}^\dagger n\rangle = (n + 1)^{1/2}|n + 1\rangle$ (Eqn (6.73)) and $|\hat{b}n\rangle = n^{1/2}|n - 1\rangle$

(Eqn (6.74)) allows us to conclude that

$$\langle 11|H|02\rangle = \frac{\hbar\kappa'}{2m\omega}\langle 11|\hat{b}_x^\dagger\hat{b}_y|02\rangle = \frac{\sqrt{2}\hbar\kappa'}{2m\omega}\langle 11|\hat{b}_x^\dagger|01\rangle = \frac{\sqrt{2}\hbar\kappa'}{2m\omega}\langle 11|11\rangle = \frac{\sqrt{2}\hbar\kappa'}{2m\omega}$$

It follows that

$$
\begin{bmatrix}
3\hbar\omega & \dfrac{\sqrt{2}\hbar\kappa'}{2m\omega} & 0 \\
\dfrac{\sqrt{2}\hbar\kappa'}{2m\omega} & 3\hbar\omega & \dfrac{\sqrt{2}\hbar\kappa'}{2m\omega} \\
0 & \dfrac{\sqrt{2}\hbar\kappa'}{2m\omega} & 3\hbar\omega
\end{bmatrix}
\begin{bmatrix} a_1 \\ a_2 \\ a_3 \end{bmatrix} = E \begin{bmatrix} a_1 \\ a_2 \\ a_3 \end{bmatrix}
$$

All we need to do now is find the eigenvalues of the matrix. The solutions are

$$E_1 = 3\hbar\omega$$
$$E_2 = \frac{3m\hbar\omega^2 - \hbar\kappa'}{m\omega}$$
$$E_3 = \frac{3m\hbar\omega^2 + \hbar\kappa'}{m\omega}$$

The corresponding eigenfunctions are given by the coefficients

$$
\begin{bmatrix} a_1 \\ a_2 \\ a_3 \end{bmatrix} = \frac{1}{\sqrt{2}} \begin{bmatrix} -1 \\ 0 \\ 1 \end{bmatrix}
$$

$$
\begin{bmatrix} a_1 \\ a_2 \\ a_3 \end{bmatrix} = \frac{1}{2} \begin{bmatrix} 1 \\ -\sqrt{2} \\ 1 \end{bmatrix}
$$

$$
\begin{bmatrix} a_1 \\ a_2 \\ a_3 \end{bmatrix} = \frac{1}{2} \begin{bmatrix} 1 \\ \sqrt{2} \\ 1 \end{bmatrix}
$$

We may follow a similar procedure to find how the four-fold degenerate levels of a two-dimensional harmonic oscillator with motion in the x–y plane subject to perturbation $W = \kappa'xy$ change. The perturbed Schrödinger equation matrix is

$$
\begin{bmatrix}
4\hbar\omega & \dfrac{\sqrt{3}\hbar\kappa'}{2m\omega} & 0 & 0 \\
\dfrac{\sqrt{3}\hbar\kappa'}{2m\omega} & 4\hbar\omega & \dfrac{\hbar\kappa'}{m\omega} & 0 \\
0 & \dfrac{\hbar\kappa'}{m\omega} & 4\hbar\omega & \dfrac{\sqrt{3}\hbar\kappa'}{2m\omega} \\
0 & 0 & \dfrac{\sqrt{3}\hbar\kappa'}{2m\omega} & 4\hbar\omega
\end{bmatrix}
\begin{bmatrix} a_1 \\ a_2 \\ a_3 \\ a_4 \end{bmatrix} = E \begin{bmatrix} a_1 \\ a_2 \\ a_3 \\ a_4 \end{bmatrix}
$$

which has eigenvalues

$$E_1 = \frac{8m\hbar\omega^2 - 3\hbar\kappa'}{2m\omega}$$

$$E_2 = \frac{8m\hbar\omega^2 - \hbar\kappa'}{2m\omega}$$

$$E_3 = \frac{8m\hbar\omega^2 + \hbar\kappa'}{2m\omega}$$

$$E_4 = \frac{8m\hbar\omega^2 + 3\hbar\kappa'}{2m\omega}$$

The corresponding eigenfunctions are given by the coefficients

$$\begin{bmatrix} a_1 \\ a_2 \\ a_3 \\ a_4 \end{bmatrix} = \frac{1}{2\sqrt{2}} \begin{bmatrix} -1 \\ \sqrt{3} \\ -\sqrt{3} \\ 1 \end{bmatrix}$$

$$\begin{bmatrix} a_1 \\ a_2 \\ a_3 \\ a_4 \end{bmatrix} = \frac{\sqrt{3}}{2\sqrt{2}} \begin{bmatrix} 1 \\ -1/\sqrt{3} \\ -1/\sqrt{3} \\ 1 \end{bmatrix}$$

$$\begin{bmatrix} a_1 \\ a_2 \\ a_3 \\ a_4 \end{bmatrix} = \frac{\sqrt{3}}{2\sqrt{2}} \begin{bmatrix} -1 \\ -1/\sqrt{3} \\ 1/\sqrt{3} \\ 1 \end{bmatrix}$$

$$\begin{bmatrix} a_1 \\ a_2 \\ a_3 \\ a_4 \end{bmatrix} = \frac{1}{2\sqrt{2}} \begin{bmatrix} 1 \\ \sqrt{3} \\ \sqrt{3} \\ 1 \end{bmatrix}$$

Solution 10.7

(a) Here, we are interested in an electron of mass m that in the unperturbed state is free to move in a one-dimensional box of length X with periodic boundary conditions. In this situation, the potential is zero and the solution to the time-independent Schrödinger equation gives eigenfunctions that are plane waves of the form $\psi_k(x) = Ae^{ikx}$. Applying the periodic boundary conditions to the eigenstates, $\psi(x) = \psi(x + X)$, gives $e^{i\mathbf{k} \cdot X} = 1$ and $k_n = 2n\pi/X$ for integer n in such a way that $n = 1, 2, 3, \ldots$. Hence, the energy eigenvalues are $E = \hbar^2 k_n^2/2m$.

(b) To find solutions to the time-independent Schrödinger equation in the presence of a periodic potential $V(x) = V(x + L)$, where $X = NL$ and N is a large integer, we will treat the potential as a perturbation. The potential can be decomposed into its

Fourier components, so that

$$V(x) = V_0 + \sum_{n\neq 0} V_n e^{i2\pi nx/L} = V_0 + \sum_{n\neq 0} V_n e^{i2\pi nNx/X}$$

Setting $V_0 = 0$ and recalling that $(1/2\pi) \int e^{-i(k-k')x} dx = \delta(k - k')$, the first-order correction to the eigenenergies is

$$E_n^{(1)} = W_{nn} = \frac{2n^2\pi^2}{mX^2} + \int_{-\infty}^{\infty} A^2 \sum_{k\neq 0} V_k e^{i2\pi kNx/X} dx = 0$$

The new eigenstates are to first order

$$\phi_n = \phi_n^{(0)} + \sum_{m\neq n} \frac{W_{mn}}{E_m^{(0)} - E_n^{(0)}} \phi_m^{(0)}$$

$$\phi_n = A e^{i2\pi nx/X} + \sum_{m\neq n} \frac{mX^2 A^2 \int_{-\infty}^{\infty} \sum_{k\neq 0} V_k e^{i2\pi kNx/X} e^{i2\pi(n-m)x/X} dx}{2\pi^2(m^2 - n^2)} A e^{i2m\pi x/X}$$

where nonzero contributions occur for $n - m = -kN$, allowing us to write

$$\phi_n = A e^{i2\pi nx/X} + \sum_{k\neq 0} \frac{mN^2 A^3 V_k}{\pi kN(2n - kN)} e^{i2\pi(n+kN)x/X}$$

The second-order correction to the eigenenergy is

$$E_n^{(2)} = \sum_{m\neq n} \frac{|W_{mn}|^2}{E_m^{(0)} - E_n^{(0)}} \phi_m^{(0)} = \sum_{m\neq n} \frac{mX^2 A^4 \left| \int_{-\infty}^{\infty} \sum_{k\neq 0} V_k e^{i2\pi kNx/X} e^{i2\pi(n-m)x/X} dx \right|^2}{2\pi^2(m^2 - n^2)}$$

which is nonzero if $n - m = -kN$, so

$$E_n^{(2)} = \sum_{k\neq 0} \frac{mN^2 A^4 V_k^2}{\pi kN(2n - kN)}$$

(c) When $k \to n\pi/L$, the denominator in the previous equation approaches zero. Nondegenerate perturbation theory can no longer be used. To gain some insight into the situation, we use two-state degenerate perturbation theory to find the corrected energy values for $k = n\pi((1 + \Delta)/L)$ and $k' = n\pi((1 - \Delta)/L)$, where Δ is small compared with π/L. In this case, the perturbed Hamiltonian matrix is

$$\begin{bmatrix} \dfrac{\hbar^2 k^2}{2m} & A^2 V_n \\ A^2 V_n & \dfrac{\hbar^2 k^2}{2m} \end{bmatrix} \begin{bmatrix} a_1 \\ a_2 \end{bmatrix} = E \begin{bmatrix} a_1 \\ a_2 \end{bmatrix}$$

The new eigenenergies (the eigenvalues of the matrix) are

$$E_{1,2} = \frac{\hbar^2(k^2 + k'^2) \pm \sqrt{16m^2 V_n^2 A^4 + \hbar^4(k^2 + k'^2)^2}}{4m}$$

or

$$E_{1,2} = \frac{\hbar^2 n^2 \pi^2(1 + \Delta^2) \pm 2L^2 \sqrt{\dfrac{\hbar^4 n^4 \pi^4 \Delta^2}{L^4} + A^4 m^2 V_n^2}}{2L^2 m}$$

(d) The following figure sketches the dispersion relation we might anticipate for electrons perturbed by a periodic potential. The parabolic dispersion of a free electron is modified to include band gaps of value E_g at wave vectors $\pm n\pi/L$.

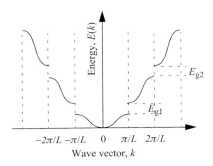

(e) If we choose the lowest-frequency Fourier component of the perturbative periodic potential in part (b), then $V(x) = V_1 \cos(\pi x/L) = -2U_G \cos(Gx)$, where $G = 2\pi/L$ is the reciprocal lattice vector. The following figure sketches the potential.

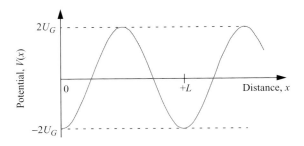

Each oscillation of $V(x)$ is one "cell" of the crystal. If the electron is displaced by nL, where n is an integer, and L is the oscillation period, then the electron must find itself in an identical environment. A perfectly periodic crystal looks the same to a particle displaced by nL.

The unperturbed wave function is $\psi_k(x) = A_0 e^{ikx}$. The wave functions are plane waves in the absence of the periodic potential perturbation. We impose a periodic boundary condition $\psi_k(x + X) = \psi_k(x)$, where $X = NL$ is the length of the crystal.

Hence, we can now normalize

$$\psi_k(x) = |k\rangle = \frac{1}{\sqrt{X}} e^{ikx}$$

where $k = (n/N)G$ and $n = 0, \pm1, \pm2, \ldots$.

$$H_{11} = H_{22} = E_0 = \frac{\hbar^2 k^2}{2m}$$

is the unperturbed energy.

To evaluate H_{12} and H_{21} we find the matrix element

$$H_{12} = \langle k_1 | V | k_2 \rangle$$

$$H_{12} = \int_0^X \psi_{k_1}^*(x) V \psi_{k_2}(x) dx = \int_0^X \frac{1}{\sqrt{X}} e^{-ik_1 x} \cdot -2U_G \cos(Gx) \cdot \frac{1}{\sqrt{X}} e^{ik_2 x} dx$$

$$H_{12} = -\frac{U_G}{X} \int_0^X e^{-ik_1 x} \left(e^{Gx} + e^{-Gx} \right) e^{ik_2 x} dx = -\frac{U_G}{X} \int_0^X \left(e^{i(k_2 - k_1 + G)x} + e^{i(k_2 - k_1 - G)x} \right) dx$$

We note that the definition of the delta function is $\delta(x - x') = (1/2\pi) \int_{-\infty}^{\infty} e^{iy(x-x')} dy$.

Our periodic boundary condition requires that $H_{12} = 0$, unless, e.g., $k_2 - k_1 = \pm G = \pm 2\pi/L$, in which case the integral is a Dirac delta function. Hence, the matrix element connecting two arbitrary plane-wave states $|k_1\rangle$ and $|k_2\rangle$ is zero unless

$$H_{12} = H_{21} = -U_G \cdot \delta(k_2 - k_1 = \pm G)$$

For the case in which k is near $\pm G/2$, it is useful if we rewrite

$$k_- = k_1 = -\frac{G}{2} + k'$$

$$k_+ = k_2 = \frac{G}{2} + k'$$

so that

$$\psi(x) = e^{ik'x} \cdot \left(a_+ e^{i\frac{G}{2}x} + a_- e^{-i\frac{G}{2}x} \right)$$

is the perturbed wave function and $|k_+\rangle$ and $|k_-\rangle$ are the unperturbed wave functions.

If we consider the limit $k = \pm G/2$, then we set $k' = 0$ and $k_1 = -k_2 = -G/2$. There is a degeneracy at the unperturbed energy

$$H_{11} = H_{22} = E_0 = \frac{\hbar^2 k^2}{2m} = \frac{\hbar^2 G^2}{8m}$$

$$H_{21} = H_{12} = \left\langle -\frac{G}{2} \middle| V \middle| \frac{G}{2} \right\rangle = -U_G$$

The secular equation for this problem is

$$\begin{vmatrix} H_{11} - E & H_{12} \\ H_{21} & H_{22} - E \end{vmatrix} = 0$$

which has solution

$$E_{\pm} = \frac{(H_{11} + H_{22})}{2} \pm \left(\frac{1}{4}(H_{11} - H_{22})^2 + H_{12}H_{21} \right)^{1/2}$$

Since $H_{11} = H_{22}$ and $H_{12}H_{21} = U_G^2$

$$E_{\pm} = E_0 \pm U_G$$

The following figure illustrates the effect the lowest-frequency Fourier component of the periodic potential has on electron dispersion.

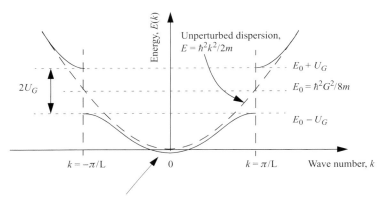

Energy shift due to first-order correction is zero, but there is a shift in energy in second order.

The following few sketches illustrate the effect the lowest-frequency Fourier component of the periodic potential has on electron dispersion at $G/2$.

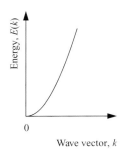

Free electron $E = \hbar k/2m$ parabolic dispersion relation. Apply periodic perturbation $V(x) = -2G\cos(Gx)$, where $G = 2/L$ with L the "lattice" constant and G the reciprocal lattice vector.

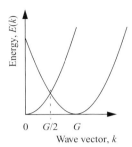

Bloch's theorem for periodic potential.
$\psi_k(x + L) = \psi_k(x) \exp(ikL)$
$G = 2/L$
Hence, $E_k = E_{k+G}$. In general $E_k = E_{k+nG}$
where $n = 0, 1, 2, 3 \ldots$

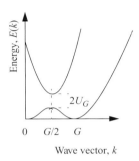

Degeneracy of two parabolas at $k = G/2$
is split by energy $2U_G$.

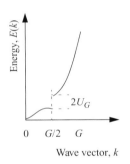

Net result of perturbation on free-electron
parabolic dispersion relation.

Solution 10.8

(a) We are asked to predict the effect that applying a uniform electric field has on the energy spectrum of an atom. It is reasonable to expect that the atom is in its lowest-energy ground state. Using perturbation theory, the perturbation due to an electric field applied in the z direction is $W = e|\mathbf{E}|z$, and the matrix element between the ground state and the first excited state is $W_{01} = e|\mathbf{E}|\langle 0|z|1\rangle = e|\mathbf{E}|z_{01}$. The matrix element is the expectation value of the position operator, and so it is at most of the order of the size of an atom $\sim 10^{-8}$ cm. Maximum electric fields in a laboratory are $\sim 10^6$ V cm^{-1} on a macroscopic scale. Transitions can only take place between energy levels that differ at most by the potential difference induced by the field – i.e., $\sim 10^{-2}$ eV. This will not be enough to induce transitions between principal quantum numbers n, but may break the degeneracy between states with different l belonging to the same n. The simple theory of the atom in Section 2.2.3.2 indicated that the degeneracy of an electronic state ψ_{nlm} in the hydrogen atom is determined by the principal quantum number n to be n^2 since

$$\sum_{l=0}^{l=n-1} (2l + 1) = n^2$$

In our assessment of the influence an electric field has on the energy levels of an atom, it was assumed that the electric field at the atom is the same as the macroscopic field. However, on a microscopic scale, electric fields can be dramatically enhanced locally depending on the exact geometry.

(b) If we ignore the effects of electron spin, the time-independent Schrödinger equation for an electron in the hydrogen atom is

$$H^{(0)}\psi = E^{(0)}\psi$$

which has the solutions

$$\psi_{nlm}(r, \theta, \phi) = R_n(r)\Theta_l(\theta)\Phi_m(\phi) = |nlm\rangle$$

The principal quantum number n specifies the energy of a state, the orbital quantum number is $l = 0, 1, 2, \ldots, (n - 1)$, and the azimuthal quantum number is $m = \pm l, \ldots, \pm 2, \pm 1, 0$.

The electron states specified by nlm are n^2 degenerate, and each state has definite parity. For $n = 2$, there are four states with the same energy. They are $|200\rangle$, $|21 - 1\rangle$, $|210\rangle$, and $|211\rangle$.

We now apply an electric field $\mathbf{E}$ and seek solutions to the time-independent Schrödinger equation

$$\left(H^{(0)} + W\right)\psi = E\psi$$

where, taking the electric field to be in the z direction, we have

$$W = e|\mathbf{E}|\hat{\mathbf{z}}$$

Because the degeneracy of the $n = 2$ state is 4, we wish to find solutions to the 4×4 matrix

$$\sum_{k=1}^{N=4}[H_{jk} - E\delta_{jk}]a_k = 0$$

The solutions are given by the secular equation

$$\begin{vmatrix} \langle 200|W|200\rangle - E & \langle 200|W|21-1\rangle & \langle 200|W|210\rangle & \langle 200|W|211\rangle \\ \langle 21-1|W|200\rangle & \langle 21-1|W|21-1\rangle - E & \langle 21-1|W|210\rangle & \langle 21-1|W|211\rangle \\ \langle 210|W|200\rangle & \langle 210|W|21-1\rangle & \langle 210|W|210\rangle - E & \langle 210|W|211\rangle \\ \langle 211|W|200\rangle & \langle 211|W|21-1\rangle & \langle 211|W|210\rangle & \langle 211|W|211\rangle - E \end{vmatrix} = 0$$

The diagonal matrix elements are zero, because the odd parity of z forces the integrand to odd parity:

$$e|\mathbf{E}|\langle nlm|\hat{\mathbf{z}}|nlm\rangle = e|\mathbf{E}| \int \psi_{nlm}^* \hat{\mathbf{z}} \psi_{nlm} d^3r = 0$$

The perturbation is in the z direction which in spherical coordinates only involves r and θ via the relation $z = r\cos(\theta)$. Hence, because the eigenfunctions are separable into orthonormal functions of r, θ, and ϕ in such a way that

$$\psi_{nlm}(r,\theta,\phi) = R_n(r)\Theta_l(\theta)\Phi_m(\phi)$$

it follows that the matrix elements between states with different m for a z-directed perturbation are zero. This is because states described by $\Phi_m(\phi)$ are orthogonal, and the perturbation in z has no ϕ dependence. Hence, the only possible nonzero off-diagonal matrix elements are those that involve states with the *same* value of m:

$$\langle 200|W|210\rangle = \langle 210|W|200\rangle = e|\mathbf{E}|\psi_{210}^* \hat{\mathbf{z}} \psi_{200} d^3r$$

The wave functions are

$$\psi_{200} = \frac{2}{(2a_B)^{3/2}} \cdot \left(1 - \frac{r}{2a_B}\right) \cdot e^{-r/2a_B} \cdot \left(\frac{1}{4\pi}\right)^{1/2}$$

which has even parity, and

$$\psi_{210} = \frac{1}{\sqrt{3}(2a_B)^{3/2}} \cdot \frac{r}{a_B} \cdot e^{-r/2a_B} \cdot \frac{1}{2}\left(\frac{3}{\pi}\right)^{1/2} \cos(\theta)$$

which has angular dependence and odd parity. Using $z = r\cos(\theta)$, we now calculate the matrix element

$$e|\mathbf{E}| \int \psi_{210}^* \hat{\mathbf{z}} \psi_{200} d^3r = \frac{ea_B|\mathbf{E}|}{32\pi} \int_{-\infty}^{\infty} r^4\left(2 - \frac{r}{a_B}\right) \cdot e^{-r/2a_B} dr \int_0^\pi \cos^2(\theta)\sin(\theta)d\theta \int_0^{2\pi} d\phi$$

$$e|\mathbf{E}| \int \psi_{210}^* \hat{\mathbf{z}} \psi_{200} d^3r = -3e|\mathbf{E}|a_B$$

To solve the integral, we used

$$\int_0^\infty r^n e^{-r/a_B} dr = n! a_B^{n+1}$$

The secular equation may now be written

$$\begin{vmatrix} -E & 0 & -3e|\mathbf{E}|a_B & 0 \\ 0 & -E & 0 & 0 \\ -3e|\mathbf{E}|a_B & 0 & -E & 0 \\ 0 & 0 & 0 & -E \end{vmatrix} = 0$$

This has four solutions $\pm 3a_B e|\mathbf{E}|$, 0, 0, so that the new first-order corrected energy levels are $E = E^{(0)} \pm 3a_B e|\mathbf{E}|$, $E^{(0)}$, $E^{(0)}$. This partial lifting of degeneracy in hydrogen due to application of an electric field is called the Stark effect.

The effect of applying an electric field is to break the symmetry of the potential and partially lift the degeneracy of the state. For the states for which degeneracy is lifted, it is as if the atom has an electric dipole moment of magnitude $3e|\mathbf{E}|a_B$.

To find the wave functions for the perturbed states, we need only consider the 2×2 matrix that relates $|200\rangle$ and $|210\rangle$:

$$\begin{bmatrix} 0 & -3e|\mathbf{E}|a_B \\ -3e|\mathbf{E}|a_B & 0 \end{bmatrix}$$

This has eigenfunctions

$$\psi_+ = \frac{1}{\sqrt{2}}(\psi_{210} - \psi_{200}) \text{ with eigenvalue } E_+ = E^{(0)} + 3e|\mathbf{E}|a_B$$

and

$$\psi_- = \frac{1}{\sqrt{2}}(\psi_{210} - \psi_{200}) \text{ with eigenvalue } E_- = E^{(0)} - 3e|\mathbf{E}|a_B$$

Because the eiegnfunctions $|200\rangle$ and $|210\rangle$ are of different parity, the eigenfunctions ψ_+ and ψ_- are of mixed parity.

Appendix A

Physical values

SI-MKS[1,2]

Speed of light in free space	$c = 2.99792458 \times 10^8$ m s^{-1}
Planck's constant	$\hbar = 6.58211889(26) \times 10^{-16}$ eV s
	$\hbar = 1.054571596(82) \times 10^{-34}$ J s
Electron charge	$e = 1.602176462(63) \times 10^{-19}$ C
Electron mass	$m_0 = 9.10938188(72) \times 10^{-31}$ kg
Neutron mass	$m_{\mathrm{n}} = 1.67492716(13) \times 10^{-27}$ kg
Proton mass	$m_{\mathrm{p}} = 1.67262158(13) \times 10^{-27}$ kg
Boltzmann constant	$k_{\mathrm{B}} = 1.3806503(24) \times 10^{-23}$ J K^{-1}
	$k_{\mathrm{B}} = 8.617342(15) \times 10^{-5}$ eV K^{-1}
Permittivity of free space	$\varepsilon_0 = 8.8541878 \times 10^{-12}$ F m^{-1}
Permeability of free space	$\mu_0 = 4\pi \times 10^{-7}$ H m^{-1}
Speed of light in free space	$c = 1/\sqrt{\varepsilon_0 \mu_0}$
Avagadro's number	$N_{\mathrm{A}} = 6.02214199(79) \times 10^{23}$ mol^{-1}
Bohr radius	$a_{\mathrm{B}} = 0.52917721(19) \times 10^{-10}$ m

$$a_{\mathrm{B}} = \frac{4\pi \varepsilon_0 \hbar^2}{m_0 e^2}$$

Inverse fine-structure constant $\alpha^{-1} = 137.0359976(50)$

$$\alpha^{-1} = \frac{4\pi \varepsilon_0 \hbar c}{e^2}$$

A.1 Constants in quantum mechanics

In this book, we have assumed that Planck's constant, the speed of light in free space, the value of the electron charge, etc., may be treated as constants. Notice, however, that this Appendix is titled "Physical values" and not "Physical constants". The reason for this is that we have no way to prove theoretically or experimentally that, for example, Planck's constant does in fact have the same value in all parts of space or that it has been the same over all time. On the contrary, there seems to be some tentative evidence

---

[1] See http://physics.nist.gov/constants.
[2] The number in parentheses is one standard deviation uncertainty in the last two digits.

that the values of these constants may have changed with time.[3] However, because of the way SI defines some units of measure and the use of quantum mechanics to find relationships between quantities, some physical values are absolute (there is no uncertainty in their value). Examples include: the speed of light, c; the permittivity of free space, ε_0; and the permeability of free space, μ_0.

A.2 The MKS and SI units of measurement

To ensure uniform standards in commercial transactions involving weights and measures and to facilitate information exchange within the science and technology community, it is useful to agree on a system of units of measurement. The foundation of our measurement system may be traced to the creation of the decimal metric method during the time of the French Revolution (1787–1799). An important step in the development of our standardized system of measurement occurred in 1799 when physical objects representing the meter and the kilogram where placed in the Archives de la République in Paris, France. In 1874, the British Association for the Advancement of Science (BAAS) introduced a system of measurement called CGS in which the centimeter is used for measurement of distance, the gram for mass, and the second for time. This was rapidly adopted by the main-stream experimental physical science community. Then, in 1901, Giorgi suggested that if the metric system of meters for distance, kilograms for mass, and seconds for time were used instead of centimeter, gram, and second, a consistent system of electromagnetic units could be developed. This meter, kilogram, second approach is called the MKS system, and it was adopted by the International Electro-technical Commission in 1935.

Of course, there are other units that should also be defined, such as force, energy, and power. Recognizing this need, in 1948 the General Conference on Weights and Measures introduced a number of units, including the newton for force, the joule for energy, and the watt for power. The newton, joule, and watt are named after scientists. The first letter of each name serves as the abbreviation for the unit and is written in upper case.

Internationally agreed upon units of measure continue to evolve over time. In 1960, the General Conference on Weights and Measures extended the MKS scheme using seven basic units of measure from which all other units may be derived. Called Système Internationale d'Unités, it is commonly known as SI.

---

[3] J. K. Webb, M. T. Murphy, V. V. Flambaum, V. A. Dzuba, J. D. Barrow, C. W. Churchill, J. X. Prochaska, and A. M. Wolfe, *Phys. Rev. Lett.* **87**, 091301/1-4(2001).

The seven basic units of measure are:

m   The meter for length
kg   The kilogram for mass
s   The second for time
A   The ampere for current
cd   The candela for light intensity
mol   The mol for the amount of a substance
K   The kelvin for thermodynamic temperature

The definition of these and derived units often seems a little arbitrary and, in some cases, has changed with time.

The *meter* (symbol m) is the basic unit of length in SI. One meter is equal to approximately 39.37 inches. The meter was first defined by the French Academy of Sciences in 1791 as $1/10^7$ of the quadrant of the Earth's circumference running from the North Pole through Paris to the equator. In 1889, the International Bureau of Weights and Measures defined the meter as the distance between two lines on a particular standard bar of 90 percent platinum and 10 percent iridium. In 1960, the definition of the meter changed again and it was defined as being equal to 1650763.73 wavelengths of the orange-red line in the spectrum of the krypton-86 atom in a vacuum. The definition of the meter changed yet again in 1983, when the General Conference on Weights and Measures defined the meter as the distance traveled by light in a vacuum in 1/299792458 of one second. This definition of the meter in terms of the speed of light has the consequence that the speed of light is an *exact* quantity in SI.

The *kilogram* (symbol kg) is the basic unit of mass in SI. It is equal to the mass of a particular platinum–iridium cylinder kept at the International Bureau of Weights and Measures laboratory (Bureau International des Poids at Mesures or BIPM) at Sèvres, near Paris. Today, the BIPM kilogram is the only SI unit that is based on a physical object (sometimes called an artifact). The cylinder was supposed to have a mass equal to a cubic decimeter of water at its maximum density. The cylinder was later discovered to be 28 parts per million too large. Unfortunately, since the same mass of water at maximum density defines the liter, one liter is 1000.028 cm³. Compounding this lack of elegance, in 1964 the General Conference on Weights and Measures redefined the liter to be a cubic decimeter while, at the same time recommending that the unit not be used in work requiring great precision. To make connection to the imperial unit of weight, the pound is now defined as being exactly 0.45359237 kg.

The *second* (symbol s) is the basic unit of time in SI. The second was defined as 1/86400 of the mean solar day, which is the average period of rotation of the Earth on its axis relative to the Sun. In 1956, the International Committee on Weights and Measures defined the ephemeris second as 1/31556925.9747 of the length of the tropical year for 1900. This definition was ratified by the General Conference on Weights and Measures in 1960. In 1964, the International Committee on Weights and

Measures suggested a second be defined as 9192631770 periods of radiation from the transition between the two hyperfine levels of the ground state of the cesium-133 atom when unperturbed by external fields. In 1967, this became the official definition of the second in SI.

The *ampere* (symbol A) is the basic unit of electric current in SI. The ampere corresponds to the flow of one coulomb of electric charge per second. A flow of one ampere is produced in a resistance of one ohm by a potential difference of one volt. Since 1948, the ampere has been defined as the constant current that, if maintained in two straight parallel conductors of infinite length of negligible circular cross-section and placed one meter apart in a vacuum, would produce between these conductors a force equal to 2×10^{-7} newton per meter of length.

The *candela* (symbol cd) is the unit of luminous intensity in SI. The candela replaces the international candle. One candela is 0.982 international candles. One candela is defined as the luminous intensity in a given direction of a source that emits monochromatic radiation of frequency 540×10^{12} Hz and has a radiant intensity in that same direction of $1/683$ watt per steradian.

The *mol* (symbol mol) is defined as the amount of substance containing the same number of chemical units (atoms, molecules, ions, electrons, or other specified entities or groups of entities) as exactly 12 grams of carbon-12. The number of units in a mol, also known as Avogadro's number, is 6.022141×10^{23}.

The *kelvin* (symbol K) is the unit for thermodynamic temperature in SI. Absolute zero temperature (0 K) is the temperature at which a thermodynamic system has the lowest energy. It corresponds to -273.16 on the Celsius scale and to -459.67 on the Fahrenheit scale. The kelvin is defined as $1/273.16$ of the triple point of pure water. The triple point of pure water is the temperature at which the liquid, solid, and gaseous forms can be maintained simultaneously.

Another absolute temperature scale used by engineers in the United States of America is the Rankine scale. Although the zero point of the Rankine scale is also absolute zero, each rankine is $5/9$ of the kelvin.

The *newton* (symbol N) is the unit of force in SI. It is defined as that force necessary to provide a mass of one kilogram with an acceleration of one meter per second per second. One newton is equal to a force of 100000 dynes in the CGS system, or a force of about 0.2248 pound in the foot-pound-second.

The *joule* (symbol J) is the unit of energy in SI. It is equal to the work done by a force of one newton acting through one meter. It is also equal to one watt-second, which is the energy dissipated in one second by a current of one ampere through a resistance of one ohm. One joule is equal to 10^7 ergs in CGS.

The *watt* (symbol W) is the unit of power in SI. One watt is equal to one joule of work per second, $1/746$ horsepower, or the power dissipated in an electrical conductor carrying one amp over a potential drop of one volt.

A.3 Other units of measurement

While SI has been in use for many years, there are still a number of other schemes that you may find in the scientific literature. One of the more common is CGS (centimeter, gram, second). For reference, the list of physical values is given below.

CGS

Speed of light	$c = 2.99792458 \times 10^{10}$ cm s^{-1}
Planck's constant	$\hbar = 6.58211889(26) \times 10^{-16}$ eV s
	$\hbar = 1.054571596(82) \times 10^{-27}$ erg s
Electron charge	$e = 1.602176462(63) \times 10^{12}$ erg eV^{-1}
Electron mass	$m_0 = 9.10938188(72) \times 10^{-28}$ g
Neutron mass	$m_\mathrm{n} = 1.67492716(13) \times 10^{-24}$ g
Proton mass	$m_\mathrm{p} = 1.67262158(13) \times 10^{-24}$ g
Boltzmann constant	$k_\mathrm{B} = 1.3806503(24) \times 10^{-16}$ erg K^{-1}
	$k_\mathrm{B} = 8.617342 \times 10^{-5}$ eV K^{-1}
Bohr radius	$a_\mathrm{B} = 0.5291772083(19) \times 10^{-8}$ cm

$$a_\mathrm{B} = \frac{\hbar c}{m_0 e^2}$$

Inverse fine-structure constant $\alpha^{-1} = 137.0359976(50)$

$$\alpha^{-1} = \frac{\hbar c}{e^2}$$

In addition to internationally agreed upon units of measurement, theorists sometimes adopt a short-hand notation in which Planck's constant is unity. You may also find occasions when the speed of light is set to unity.

A.4 Use of quantum mechanical effects to define units of measure

Over the years there has been a trend to try to use the properties of atoms and quantum mechanical phenomena to define units of measure. In the 1960s the second was defined using radiation from the transition between the two hyperfine levels of the ground state of the cesium-133 atom. Such radiation is, of course, quantum mechanical in origin. More recently, the macroscopic quantum phenomena of both Josephson tunneling and the quantum Hall effect have been used to precisely define voltage and resistance values, respectively. In 1988, the International Committee on Weights and Measures adopted the Josephson constant, K_J, and the von Klitzing constant, R_K, for electrical measurements. The Josephson constant ($K_\mathrm{J} = e/\pi\hbar = 483597.898(19)$ GHz V^{-1}) relates frequency and voltage by way of the ac Josephson effect. The von Klitzing constant ($R_\mathrm{K} = 2\pi\hbar/e^2 = 25,812.807572(95)\,\Omega$) is the unit of resistance in the integer quantum Hall effect. This leads directly to an expression for Planck's constant, $\hbar = 2/\pi K_J^2 R_K$.

In the future, it may be possible to use the Einstein expressions for energy $E = mc^2$ and $E = \hbar\omega$ to eliminate the need for a physical BIPM artifact in France to define the kilogram mass in SI. If this approach were used, then Planck's constant would become an exact quantity. The definition of *one kilogram* would define Planck's constant as $\hbar = c^2/\omega$ in SI.

Appendix B

Coordinates and trigonometry

B.1 Coordinates

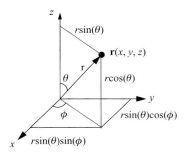

The position of a point particle in three-dimensional Euclidian space is given by the vector $\mathbf{r}(x, y, z)$ in Cartesian coordinates, where x, y, and z are measured in the $\hat{\mathbf{x}}$, $\hat{\mathbf{y}}$, and $\hat{\mathbf{z}}$ unit-vector directions, respectively. In spherical coordinates, the position vector is $\mathbf{r}(r, \theta, \phi)$, where r, θ, and ϕ are the radial and two-angular coordinates, respectively. The relationship between the two coordinate systems is given by

$$x = r \sin(\theta) \cos(\phi)$$
$$y = r \sin(\theta) \sin(\phi)$$
$$z = r \cos(\theta)$$

The value of r is related to Cartesian coordinates through

$$r^2 = x^2 + y^2 + z^2$$

B.2 Useful trigonometric relations

Functions of real variable x

$$\sin(x) = \frac{1}{2i} \left(e^{ix} - e^{-ix} \right)$$

$$\cos(x) = \frac{1}{2} \left(e^{ix} + e^{-ix} \right)$$

Hyperbolic functions of complex variable z

$$\sinh(z) = \frac{1}{2} \left(e^{z} - e^{-z} \right)$$

$$\cosh(z) = \frac{1}{2} \left(e^{z} + e^{-z} \right)$$

$$e^{ix} = \cos(x) + i\sin(x)$$
$$e^{-ix} = \cos(x) - i\sin(x)$$
$$\cos^2(x) + \sin^2(x) = 1 \qquad \cosh^2(z) - \sinh^2(z) = 1$$

$$2\sin(x)\cos(y) = \sin(x+y) + \sin(x-y)$$
$$2\cos(x)\cos(y) = \cos(x+y) + \cos(x-y)$$
$$2\sin(x)\sin(y) = \cos(x-y) - \cos(x+y)$$

$$\sin(x \pm y) = \sin(x)\cos(y) \pm \cos(x)\sin(y)$$
$$\cos(x \pm y) = \cos(x)\cos(y) \mp \sin(x)\sin(y)$$

$$\tan(x \pm y) = \frac{\tan(x) \pm \tan(y)}{1 \mp \tan(x)\tan(y)}$$

$$1 + \tan^2(x) = \sec^2(x) = \frac{1}{\cos^2(x)}$$

$$1 + \cot^2(x) = \csc^2(x) = \frac{1}{\sin^2(x)}$$

$$\operatorname{sech}(z) = \frac{1}{\cosh(z)} \qquad\qquad \operatorname{cosech}(z) = \frac{1}{\sinh(z)}$$

$$\sin^{-1}(z) = -i\ln\left(iz + \left(1 - z^2\right)^{1/2}\right) \qquad \sinh^{-1}(z) = \ln\left(z + \left(z^2 + 1\right)^{1/2}\right)$$

$$\cos^{-1}(z) = -i\ln\left(z + \left(z^2 - 1\right)^{1/2}\right) \qquad \cosh^{-1}(z) = \ln\left(z + \left(z^2 - 1\right)^{1/2}\right)$$

$$\tan^{-1}(z) = \frac{i}{2}\ln\frac{i+z}{i-z} \qquad\qquad \tanh^{-1}(z) = \frac{1}{2}\ln\frac{1+z}{1-z}$$

For the triangle illustrated in the following figure:

$$\frac{a}{\sin(A)} = \frac{b}{\sin(B)} = \frac{c}{\sin(C)}$$

$$a^2 = b^2 + c^2 - 2bc\cos(A)$$

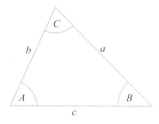

where A is the angle opposite side a, etc.

Appendix C

Expansions, integrals, and mathematical relations

C.1 Expansions and series

The **Taylor expansion** for a smooth function $f(x)$ about x_0 is

$$f(x) = \sum_{n=-\infty}^{n=\infty} \frac{1}{n!} \cdot f^{(n)}(x_0) \cdot (x - x_0)^n$$

Sine and cosine functions expand as

$$\sin(x) = x - \frac{x^3}{3!} + \frac{x^5}{5!} - \cdots \qquad \cos(x) = 1 - \frac{x^2}{2!} + \frac{x^4}{4!} - \cdots$$

and the Taylor expansion for an exponential is

$$e^x = 1 + x + \frac{x^2}{2!} + \frac{x^3}{3!} + \cdots$$

If $|x| < 1$ then

$$\log(1 + x) = x - \frac{x^2}{2} + \frac{x^3}{3} - \frac{x^4}{4} + \cdots$$

A Maclaurin expansion is a special case of the Taylor expansion in which $x_0 = 0$.
Binomial theorem

$$(1 \pm x)^n = 1 \pm nx + \frac{n(n-1)}{2!}x^2 \pm \frac{n(n-1)(n-2)}{3!}x^3 + \cdots$$

Stirling's formula

$$\log(n!) \cong n \log(n) - n \text{ for } n \gg 1$$

C.2 Differentiation

The chain rule for the product of two differentiable functions $f(x)$ and $g(x)$ may be expressed as:

$$\frac{d}{dx}((f(x)g(x))) = (f(x)g(x))' = \left(\frac{d}{dx}(f(x))\right)g(x) + f(x)\left(\frac{d}{dx}(g(x))\right)$$

$$\frac{d}{dx}((f(x)g(x))) = (f(x))'g(x) + f(x)(g(x))'$$

If the ratio of two differentiable functions $f(x)$ and $g(x)$ takes on the indeterminate form $0/0$ at postion $x = x_0$, then one can show, using a Taylor expansion, that

$$\left.\frac{f(x)}{g(x)}\right|_{x \to x_0} = \left.\frac{\frac{d}{dx}f(x)}{\frac{d}{dx}g(x)}\right|_{x \to x_0}$$

which is L'Hospital's rule.

C.3 Integration

If an integrand may be written as the product of two functions, UV', where $'$ indicates a derivative, it is often useful to consider the method of *integration by parts*, for which $\int UV'dx = UV - \int U'Vdx$.

Useful *standard integrals* include:

$$\int \cos(x)dx = \sin(x)$$

$$\int \sin(x)dx = -\cos(x)$$

$$\int \left(\frac{1}{x}\right)dx = \ln(x)$$

$$\int a^x dx = \frac{a^x}{\ln(a)}$$

$$\int \sinh(x)dx = \cosh(x)$$

$$\int \cosh(x)dx = \sinh(x)$$

$$\int_0^\infty e^{-ax^2}dx = \frac{1}{2}\sqrt{\frac{\pi}{a}} \qquad \int_0^\infty xe^{-ax^2}dx = \frac{1}{2a}$$

$$\int_0^\infty x^2 e^{-ax^2} dx = \frac{1}{4}\sqrt{\frac{\pi}{a^3}} \qquad \int_0^\infty x^4 e^{-ax^2} dx = \frac{1}{2a^2}$$

$$\int_0^\infty \frac{x^4 e^x}{(e^x - 1)^2} dx = \frac{4\pi^4}{15} \qquad \int_0^\infty \frac{x^3}{(e^x - 1)^2} dx = \frac{\pi^4}{15}$$

$$\int_0^\infty \frac{x^{p-1}}{e^{rx} - q} dx = \frac{1}{qr^p}\Gamma(p)\sum_{k=1}^\infty \frac{q^k}{k^p} \quad \text{for } p > 0, r > 0, -1 < q < 1$$

C.4 The Dirac delta function

In one dimension:

$$\int_{-\infty}^\infty \delta(x - x')dx = 1 \quad \text{and} \quad \delta(x - x') = \int_{-\infty}^\infty \frac{dk}{2\pi} e^{ik(x-x')}$$

In three dimensions:

$$\int_{-\infty}^\infty d\mathbf{r}^3 \delta(\mathbf{r} - \mathbf{r}') = 1 \quad \text{and} \quad \delta(\mathbf{r} - \mathbf{r}') = \int_{-\infty}^\infty \frac{d^3k}{(2\pi)^3} e^{i\mathbf{k}\cdot(\mathbf{r}-\mathbf{r}')}$$

Other expressions of the delta function in one dimension are:

$$\delta(x - x') = \int_0^\infty \frac{1}{\pi}\cos(k(x - x'))dk$$

$$\delta(x - x') = \frac{1}{\pi}\lim_{\eta\to\infty} \frac{\sin(\eta(x - x'))}{x - x'}$$

$$\delta(x - x') = \frac{1}{\pi}\lim_{\eta\to\infty} \frac{2\sin^2(\eta(x - x')/2)}{\eta(x - x')^2}$$

$$\delta(x - x') = \frac{1}{\pi}\lim_{\varepsilon\to\infty} \frac{\varepsilon}{(x - x')^2 + \varepsilon^2} = \frac{1}{\pi}\lim_{\varepsilon\to\infty} \frac{1}{(x - x') + i\varepsilon}$$

C.5 Root of a quadratic equation

The roots of $ax^2 + bx + c = 0$ are

$$x = \frac{-b \pm \sqrt{b^2 - 4ac}}{2a}$$

C.6 Fourier integral

$$F(k) = \frac{1}{\sqrt{2\pi}} \int\limits_{x=-\infty}^{x=\infty} f(x)e^{-ikx}dx$$

$$f(x) = \frac{1}{\sqrt{2\pi}} \int\limits_{k=-\infty}^{k=\infty} F(k)e^{ikx}dk$$

Examples

$F(k) = \dfrac{1}{\sqrt{2\pi}}$ is a constant that gives $f(x) = \delta(x)$

$F(k) = (2\pi)^{1/2}\delta(k+a)$ gives $f(x) = e^{-iax}$, where a is real.

$F(k) = \dfrac{1}{|k|}$ gives $f(x) = \dfrac{1}{|x|}$.

$F(k) = \dfrac{a(2/\pi)^{1/2}}{a^2 + k^2}$ is a Lorentzian function that gives $f(x) = e^{-a|x|}$ where $a > 0$.

$F(k) = \left(a\sqrt{2}\right)^{-1}e^{-k^2/4a^2}$ is a Gaussian function that gives $f(x) = e^{-a^2x^2}$, where $a > 0$.

C.7 Correlation functions

$$f(t) = \langle E_1^*(t)E_2(t+\tau)\rangle = \int\limits_{\tau=-\infty}^{\tau=\infty} E_1^*(t)E_2(t+\tau)d\tau \quad \text{and} \quad F(\omega) = E_1^*(\omega)E_2(\omega)$$

$$g^{(1)}(\tau) = \frac{\langle E^*(t)E(t+\tau)\rangle}{\langle E^*(t)E(t)\rangle}$$

$$g^{(2)}(\tau) = \frac{\langle E^*(t)E^*(t+\tau)E(t+\tau)E(t)\rangle}{\langle E^*(t)E(t)\rangle^2}$$

Appendix D

Linear algebra

D.1 Matrices

Inverse of matrix $\mathbf{A} = \begin{bmatrix} a_{11} & a_{12} \\ a_{21} & a_{22} \end{bmatrix}$ is $\mathbf{A}^{-1} = \dfrac{1}{|\mathbf{A}|} \begin{bmatrix} a_{22} & -a_{12} \\ -a_{21} & a_{11} \end{bmatrix}$, so that $\mathbf{A}\mathbf{A}^{-1} = \mathbf{I}$, where I is the identity matrix.

The determinant of a 2×2 matrix is $|\mathbf{A}| = a_{11}a_{22} - a_{12}a_{21}$. For a 3×3 matrix, $\mathbf{A} = \begin{bmatrix} a_{11} & a_{12} & a_{13} \\ a_{21} & a_{22} & a_{23} \\ a_{31} & a_{32} & a_{33} \end{bmatrix}$. Expanding along the first column gives

$$|\mathbf{A}| = a_{11} \begin{vmatrix} a_{22} & a_{23} \\ a_{32} & a_{33} \end{vmatrix} - a_{21} \begin{vmatrix} a_{12} & a_{13} \\ a_{32} & a_{33} \end{vmatrix} + a_{31} \begin{vmatrix} a_{12} & a_{13} \\ a_{22} & a_{23} \end{vmatrix} = a_{11}M_{11} - a_{21}M_{21} + a_{31}M_{31}$$

where M_{ik} is the minor of the element a_{ik}.

In general, the determinant of an $n \times n$ matrix $\mathbf{A} = \sum_{i,k}^{n} a_{ik}$ is

$$|\mathbf{A}| = \sum_{k=1}^{n} (-1)^{i+k} a_{ik} M_{ik}$$

where $(i = 1, 2, \ldots, n)$. The inverse is

$$\mathbf{A}^{-1} = \frac{1}{|\mathbf{A}|} \left[\sum_{i,k}^{n} (-1)^{i+k} M_{ki} \right]$$

where $(-1)^{i+k} M_{ki}$ is the cofactor.

Appendix E

Vector calculus and Maxwell's equations

E.1 Vector calculus

Cartesian coordinates (x, y, z)

$$\nabla V = \hat{\mathbf{x}}\frac{\partial V}{\partial x} + \hat{\mathbf{y}}\frac{\partial V}{\partial y} + \hat{\mathbf{z}}\frac{\partial V}{\partial z}$$

$$\nabla \cdot \mathbf{A} = \frac{\partial A_x}{\partial x} + \frac{\partial A_y}{\partial y} + \frac{\partial A_z}{\partial z}$$

$$\nabla \times \mathbf{A} = \begin{bmatrix} \hat{\mathbf{x}} & \hat{\mathbf{y}} & \hat{\mathbf{z}} \\ \dfrac{\partial}{\partial x} & \dfrac{\partial}{\partial y} & \dfrac{\partial}{\partial z} \\ A_x & A_y & A_z \end{bmatrix}$$

$$= \hat{\mathbf{x}}\left(\frac{\partial A_z}{\partial y} - \frac{\partial A_y}{\partial z}\right) + \hat{\mathbf{y}}\left(\frac{\partial A_x}{\partial z} - \frac{\partial A_z}{\partial x}\right) + \hat{\mathbf{z}}\left(\frac{\partial A_y}{\partial x} - \frac{\partial A_x}{\partial y}\right)$$

$$\nabla^2 V = \frac{d^2 V}{dx^2} + \frac{d^2 V}{dy^2} + \frac{d^2 V}{dz^2}$$

Spherical coordinates (r, θ, ϕ)

$$\nabla V = \hat{\mathbf{r}}\frac{\partial V}{\partial x} + \hat{\theta}\frac{1}{r}\frac{\partial}{\partial \theta}(V) + \hat{\phi}\frac{1}{r\sin(\theta)}\frac{\partial}{\partial \phi}(V)$$

$$\nabla \cdot \mathbf{A} = \frac{1}{r^2}\frac{\partial}{\partial r}(r^2 A_r) + \frac{1}{r\sin(\theta)}\frac{\partial}{\partial \theta}(A_\theta \sin(\theta)) + \frac{1}{r\sin(\theta)}\frac{\partial A_\phi}{\partial \phi}$$

$$\nabla \times \mathbf{A} = \frac{1}{r^2\sin(\theta)}\begin{vmatrix} \hat{\mathbf{r}} & r\hat{\theta} & r\sin(\theta)\hat{\phi} \\ \dfrac{\partial}{\partial r} & \dfrac{\partial}{\partial \theta} & \dfrac{\partial}{\partial \phi} \\ A_r & rA_\theta & r\sin(\theta)A_\phi \end{vmatrix}$$

$$\nabla \times \mathbf{A} = \frac{\hat{\mathbf{r}}}{r\sin(\theta)} \left(\frac{\partial}{\partial \theta} A_\phi \sin(\theta) - \frac{\partial A_\theta}{\partial \phi} \right) + \frac{\hat{\theta}}{r} \left(\frac{1}{\sin(\theta)} \frac{\partial A_r}{\partial \phi} - \frac{\partial}{\partial r}(rA_\phi) \right)$$
$$+ \frac{\hat{\phi}}{r} \left(\frac{\partial}{\partial r}(rA_\theta) - \frac{\partial A_r}{\partial \theta} \right)$$

$$\nabla^2 V = \frac{1}{r^2}\frac{\partial}{\partial r}\left(r^2 \frac{\partial V}{\partial r} \right) + \frac{1}{r^2 \sin(\theta)}\frac{\partial}{\partial \theta}\left(\sin(\theta)\frac{\partial V}{\partial \theta} \right) + \frac{1}{r^2 \sin(\theta)}\frac{\partial^2 V}{\partial \phi^2}$$

Useful vector relationships for the vector fields **a**, **b**, and **c** are

$$\nabla \cdot (\nabla \times \mathbf{a}) = 0$$
$$\nabla \times \nabla \times \mathbf{a} = \nabla(\nabla \cdot \mathbf{a}) - \nabla^2 \mathbf{a}$$
$$\nabla \cdot (\mathbf{a} \times \mathbf{b}) = \mathbf{b} \cdot (\nabla \times \mathbf{a}) - \mathbf{a} \cdot (\nabla \times \mathbf{b})$$
$$\mathbf{a} \times (\mathbf{b} \times \mathbf{c}) = (\mathbf{a} \cdot \mathbf{c})\mathbf{b} - (\mathbf{a} \cdot \mathbf{b})\mathbf{c}$$
$$\mathbf{a} \cdot \mathbf{b} \times \mathbf{c} = \mathbf{b} \cdot \mathbf{c} \times \mathbf{a} = \mathbf{c} \cdot \mathbf{a} \times \mathbf{b}$$

Other useful relations in vector calculus are the **divergence theorem** relating volume and surface integrals

$$\int_V \nabla \cdot \mathbf{a}\, d^3r = \int_S \mathbf{a} \cdot \mathbf{n}\, ds$$

where **n** is the unit-normal vector to the surface S and **Stokes's theorem**, which relates surface and line integrals

$$\int_S (\nabla \times \mathbf{a}) \cdot \mathbf{n}\, ds = \oint_C \mathbf{a} \cdot d\mathbf{l}$$

where $d\mathbf{l}$ is the vector line element on the closed loop C.

E.2 Maxwell's equations

In SI-MKS units

$$\nabla \cdot \mathbf{D} = \rho \qquad\qquad \text{Coulomb's law.}$$

$$\nabla \cdot \mathbf{B} = 0 \qquad\qquad \text{No magnetic monopoles.}$$

$$\nabla \times \mathbf{E} = -\frac{\partial \mathbf{B}}{\partial t} \qquad\qquad \text{Faraday's law.}$$

$$\nabla \times \mathbf{H} = \mathbf{J} + \frac{\partial \mathbf{D}}{\partial t} \qquad\qquad \text{Modified Ampere's law.}$$

Current continuity requires that $\nabla \cdot \mathbf{J} + \partial\rho/\partial t = 0$ and in these equations $\mathbf{B} = \mu\mathbf{H}$ and $\mathbf{D} = \varepsilon\mathbf{E} = \varepsilon_0(1 + \chi_e)\mathbf{E} = \varepsilon_0\mathbf{E} + \mathbf{P}$.

In SI units, the permittivity of free space is $\varepsilon_0 = 8.8541878 \times 10^{-12}$ F m^{-1} exactly, and the permeability of free space is $\mu_0 = 4\pi \times 10^{-7}$ H m^{-1}.

In Gaussian or CGS units, Maxwell's equations take on a different form. In this case

$\nabla \cdot \mathbf{D} = 4\pi\rho$ where $\mathbf{D} = \varepsilon\mathbf{E} = (1 + 4\pi\chi_e)\mathbf{E} = \mathbf{E} + 4\pi\mathbf{P}$

$\nabla \cdot \mathbf{B} = 0$

$\nabla \times \mathbf{E} = -\dfrac{1}{c}\dfrac{\partial \mathbf{B}}{\partial t}$

$\nabla \times \mathbf{H} = \dfrac{4\pi}{c}\mathbf{J} + \dfrac{1}{c}\dfrac{\partial \mathbf{D}}{\partial t}$

Appendix F

The Greek alphabet

F.1 The Greek alphabet

A	α	alpha	=	a
B	β	beta	=	b
Γ	γ	gamma	=	g
Δ	δ	delta	=	d
E	ε	epsilon	=	e
Z	ζ	zeta	=	z
H	η	eta	=	e
Θ	θ	theta	=	th (*th*)
I	ι	iota	=	i
K	κ	kappa	=	k
Λ	λ	lambda	=	l
M	μ	mu	=	m
N	ν	nu	=	n
Ξ	ξ	xi	=	x (*ks*)
Π	π	pi	=	p
P	ρ	rho	=	r
Σ	σ	sigma	=	s
T	τ	tau	=	t
Y	υ	upsilon	=	u
Φ	ϕ	phi	=	pf (*f*)
X	χ	chi	=	kh (*hh*)
Ψ	ψ	psi	=	ps
Ω	ω	omega	=	o

Index